Lecture Notes in Mathematics 2229

More information about this series at http://www.springer.com/series/304

Sebastian Klein

A Spectral Theory for Simply Periodic Solutions of the Sinh-Gordon Equation

Springer

Sebastian Klein
School of Business Informatics &
Mathematics
University of Mannheim
Mannheim, Germany

ISSN 0075-8434 ISSN 1617-9692 (electronic)
Lecture Notes in Mathematics
ISBN 978-3-030-01275-5 ISBN 978-3-030-01276-2 (eBook)
https://doi.org/10.1007/978-3-030-01276-2

Library of Congress Control Number: 2018961014

Mathematics Subject Classification (2010): Primary: 37K10, 53A10, 58E12; Secondary: 35P05, 35R30, 37K15, 37K25, 40A05, 40A10, 40A20, 42B05, 46B45, 46C05, 46E15, 46E35

This Springer imprint is published by the registered company Springer Nature Switzerland AG
The registered company address is: Gewerbestrasse 11, 6330 Cham, Switzerland

Preface

This book on a spectral theory for simply periodic solutions of the sinh-Gordon equation is the result of my habilitation project at the University of Mannheim.

I would like to express my sincerest gratitude to Professor Martin U. Schmidt, who has advised me during the creation of this work. His steady support and help has been invaluable to me. I have learned a lot from him. I would also like to thank Professor C. Hertling for the introductory seminary on algebraic geometry and for further advice, Dr. Markus Knopf for discussions on spectral theory and especially on Darboux coordinates, Dr. Alexander Klauer for advice on functional analysis, and Dr. Tobias Simon and Dr. Eva Lübcke for many mathematical discussions.

Mannheim, Germany
August 2018

Sebastian Klein

Contents

Part I Spectral Data

1 **Introduction** 3

2 **Minimal Immersions into the 3-Sphere and the Sinh-Gordon Equation** 21

3 **Spectral Data for Simply Periodic Solutions of the Sinh-Gordon Equation** 25

Part II The Asymptotic Behavior of the Spectral Data

4 **The Vacuum Solution** 41

5 **The Basic Asymptotic of the Monodromy** 47

6 **Basic Behavior of the Spectral Data** 71

7 **The Fourier Asymptotic of the Monodromy** 85

8 **The Consequences of the Fourier Asymptotic for the Spectral Data** 101

Part III The Inverse Problem for the Monodromy

9 **Asymptotic Spaces of Holomorphic Functions** 113

10 **Interpolating Holomorphic Functions** 119

11 **Final Description of the Asymptotic of the Monodromy** 147

12 **Non-special Divisors and the Inverse Problem for the Monodromy** 161

Part IV The Inverse Problem for Periodic Potentials (Cauchy Data)

13 Divisors of Finite Type ... 175

14 Darboux Coordinates for the Space of Potentials ... 189

15 The Inverse Problem for Cauchy Data Along the Real Line ... 209

Part V The Jacobi Variety of the Spectral Curve

16 Estimate of Certain Integrals ... 221

17 Asymptotic Behavior of 1-Forms on the Spectral Curve ... 239

18 Construction of the Jacobi Variety for the Spectral Curve ... 259

19 The Jacobi Variety and Translations of the Potential ... 287

20 Asymptotics of Spectral Data for Potentials on a Horizontal Strip ... 303

21 Perspectives ... 307

A Some Infinite Sums and Products ... 309

B Index of Notations ... 321

References ... 327

Part I
Spectral Data

Chapter 1
Introduction

The Sinh-Gordon Equation and Constant Mean Curvature Surfaces

The objective of the present work is to study periodic, complex-valued solutions $u : X \to \mathbb{C}$ of the 2-dimensional (i.e. $X \subset \mathbb{C}$) sinh-Gordon equation

$$\Delta u + \sinh(u) = 0 \tag{1.1}$$

by means of the theory of integrable systems. In particular, we will develop a spectral theory for periodic solutions of the sinh-Gordon equation. We call such solutions *simply periodic* when we wish to emphasize the difference to *doubly periodic* solutions (which have two linear independent periods).

We first discuss the importance of the sinh-Gordon equation. One of the most significant reasons why the sinh-Gordon equation is interesting is that real-valued solutions of the sinh-Gordon equation arise from minimal surfaces resp. constant mean curvature surfaces (CMC surfaces) without umbilical points (=zeros of the Hopf differential) in the 3-dimensional real space forms. More precisely, the conformal factor of the induced metric yields (in appropriate coordinates and possibly after scaling) a solution of the sinh-Gordon equation for non-minimal constant mean curvature surfaces in the Euclidean space $\mathbb{R}^3$, for (minimal or non-minimal) constant mean curvature surfaces in the 3-sphere S^3, and for constant mean curvature surfaces in the hyperbolic 3-space H^3 where the absolute value of the mean curvature is >1.

For example, as we will describe more explicitly in Chap. 2, for a conformal, minimal immersion $f : X \to S^3$ there exist conformal parameters z so that the Hopf differential is "constant", i.e. equal to $E\,\mathrm{d}z^2$ with some constant $E \in S^1$. With respect to such coordinates, the Gauss equation for the conformal metric $g = \mathrm{e}^{u/2}\,\mathrm{d}z\,\mathrm{d}\bar{z}$ with the conformal factor $u : X \to \mathbb{R}$ of f reduces to the sinh-Gordon

S. Klein, *A Spectral Theory for Simply Periodic Solutions of the Sinh-Gordon Equation*, Lecture Notes in Mathematics 2229,
https://doi.org/10.1007/978-3-030-01276-2_1

equation, thus u is a solution of the sinh-Gordon equation. Moreover the Codazzi equation reduces to $0 = 0$. In this setting umbilical points of the surface (i.e. zeros of its Hopf differential) correspond to coordinate singularities of the solution u. Therefore studying minimal surfaces or CMC surfaces in the 3-dimensional space forms means to study solutions of the sinh-Gordon equation. When we exclude umbilical points of the surface, we get singularity-free solutions of the sinh-Gordon equation.

Examples and Classifications for Constant Mean Curvature Surfaces

Simple examples of CMC surfaces are surfaces with parallel second fundamental form, in $\mathbb{R}^3$ these are the planes, spheres and (circular) cylinders (embedded in the usual way). More interesting examples of CMC surfaces in $\mathbb{R}^3$ are the Delaunay surfaces, a class of immersed CMC surfaces which are topologically cylinders and which were first considered by Delaunay in [Del]. Alexandrov showed in [Al] that the only *embedded* compact CMC surfaces in $\mathbb{R}^3$ are the round 2-spheres, and Hopf showed in [Ho] that the only immersed compact CMC surfaces in $\mathbb{R}^3$ of genus $g = 0$ (i.e. homeomorphic to the 2-sphere) are again the round 2-spheres; he in fact conjectured that there are no immersed compact CMC surfaces in $\mathbb{R}^3$ besides the round 2-spheres at all. Interest in compact immersed CMC surfaces in the 3-dimensional space forms soared, however, after Wente showed in 1986 (see [W]) this conjecture to be false by constructing what is now known as the Wente tori: a family of compact, immersed CMC surfaces in $\mathbb{R}^3$ of genus $g = 1$. Later, Kapouleas found compact immersed CMC surfaces in $\mathbb{R}^3$ at first for every genus $g \geq 3$ [Kapo-1], and then also for $g = 2$ [Kapo-2, Kapo-3].

The theorem of Ruh and Vilms [Ru-V] states that a surface immersion into $\mathbb{R}^3$ is CMC if and only if its Gauss map (i.e. its unit normal field, regarded as a map into the 2-sphere) is harmonic.

A celebrated result was the classification of all immersed CMC tori in $\mathbb{R}^3$ resp. S^3 by Pinkall and Sterling in [Pi-S] and independently by Hitchin in [Hi]. The classification by Pinkall and Sterling is based on proving that CMC tori can be described by polynomial Killing fields; a summary of their proof is found in [Hele, Chapter 9]. Pinkall's and Sterling's work was put into a more general context by Fox and Wang [Fox-W, especially Sections 6 and 13].

While also Pinkall's and Sterling's classification can be interpreted in terms of integrable systems and spectral theory, Hitchin's classification is based on this theory in a more explicit way, and is a significant precursor for the methods employed in the present work. Thus we will discuss the latter classification in some more detail below. Unlike for CMC tori, there still is no classification or parameterization of the moduli space of compact immersed CMC surfaces of genus $g \geq 2$.

In S^3 there exist *embedded* compact CMC surfaces of genus $g \geq 1$. One such surface is the *Clifford torus* $\{ (x_1, x_2, x_3, x_4) \in S^3 \mid x_1^2 + x_2^2 = x_3^2 + x_4^2 = \frac{1}{2} \}$. In 1970, Lawson [Law-2] showed that for any embedded CMC torus in S^3 there exists a diffeomorphism $S^3 \to S^3$ that maps that torus onto the Clifford torus. Moreover he conjectured that any embedded CMC torus in S^3 is in fact congruent to the Clifford torus (i.e. can be mapped onto the Clifford torus by an isometry of S^3). This conjecture is now known as the *Lawson conjecture*. After attracting significant interest over several decades, the conjecture was finally proved by Brendle [Br] in 2013, by means of very clever and intricate, but ultimately surprisingly elementary calculations leading to an application of Bony's strict maximum principle for degenerate elliptic equations [Bon]. An alternative proof of the Lawson conjecture based on integrable systems and deformations methods, and providing more insight into the geometry of the situation, was found by Hauswirth et al. [Hau-K-S-3, Hau-K-S-4].

A family of compact embedded minimal surfaces in S^3 is comprised by the *Lawson surfaces*: they exist for every genus $g \geq 2$, have a large (discrete) symmetry group and were obtained by Lawson [Law-1, Law-3] by applying a reflection principle to a suitably chosen solution of Plateau's Problem of finding a minimal surface with prescribed boundary. The spectral theory for Lawson surfaces and more generally for *Lawson-symmetric surfaces*, i.e. for compact CMC surfaces in S^3 which have the same symmetry group as some Lawson surfaces of the same genus, has been studied extensively by S. Heller, L. Heller, N. Schmitt et al., see for example [Hell-1, Hell-2, Hell-H-S, Hell-S].

Among the most important methods used to study the sinh-Gordon equation is the integrable systems and spectral theory. This is also the most fundamental approach taken in the present book. For this reason we will discuss these methods in detail below, and also give references associated to them.

Another important technique, introduced in 1998 by Dorfmeister, Pedit and Wu [Do-P-W], is called the DPW method (for the initials of the names of its inventors). It uses loop group methods, in particular the Iwasawa and Birkhoff splittings for loop groups, to associate to each solution u of the sinh-Gordon equation a holomorphic 1-form, called the DPW potential of u. The resulting representation of solutions u is to some extent analogous to the classical Weierstrass representation of minimal surfaces in $\mathbb{R}^3$.

Partial Differential Equations and Integrable Systems

It turns out that certain partial differential equations, for example the Korteweg-de Vries (KdV) equation and the non-linear Schrödinger (NLS) equation, but also the sinh-Gordon equation which is at the focus of interest in the present work, are each related to an integrable system on an infinite-dimensional (function) space. This relationship is a very powerful method for the investigation of those partial differential equations to which it applies, and we will apply this method to the sinh-Gordon equation throughout this book. Here an integrable system is a dynamical system which has in some sense "sufficiently many" conserved quantities. For an overview of integrable systems and the mentioned relationship in general, see for example [Hi-S-W] or [Du-K-N].

It is an unfortunate fact that there does not yet exist a general theory describing the integrable systems approach for all partial differential equations to which it applies in a unified manner. Rather, the relationship between the partial differential equation and the integrable system has to be developed separately for each individual equation, by methods which build on similar ideas and in general yield similar results, but differ in their details and the associated calculations. For this reason, we will describe the relationship first for the KdV equation, for which it was first discovered and where it can be described in perhaps its "purest" form, and also say a few words about the NLS equation, before describing the relationship for the sinh-Gordon equation.

The Integrable System of the Korteweg-de Vries (KdV) Equation

The KdV equation is the partial differential equation

$$u_t = -u_{xxx} + 6\,u\,u_x \tag{1.2}$$

in the two real variables t and x. It is named for D. Korteweg and G. de Vries, who encountered this equation in relation to the modeling of water waves in narrow and shallow water channels. The relationship of the KdV equation to integrable systems was discovered by Lax [Lax] when he realized that it can be written as what is now known as a *Lax equation*, i.e. in the form

$$\frac{\mathrm{d}}{\mathrm{d}t} L = [B, L] \tag{1.3}$$

with respect to two other differential operators L and B and where the commutator $[B, L] = B \cdot L - L \cdot B$ is a multiplication operator. In this situation, the pair (L, B) is called a *Lax pair*. Explicitly, for the KdV equation, $L = -\frac{\mathrm{d}^2}{\mathrm{d}x^2} + u$ is the 1-dimensional Schrödinger operator, and $B = -4\frac{\mathrm{d}^3}{\mathrm{d}x^3} + 3u\,\frac{\mathrm{d}}{\mathrm{d}x} + 3\,\frac{\mathrm{d}}{\mathrm{d}x}\,u$ is another differential operator of third order. Because $[B, L]$ is a multiplication operator, the Lax equation (1.3) defines a flow on the space of functions in the one variable x which generates solutions $u = u(x, t)$ of the KdV equation from initial data $u(x, t = 0)$. Note that if the initial data $u(x, t = 0)$ is periodic in x, then the Lax operators L and B are also periodic, and hence the generated solution of the KdV equation will maintain that period.

One can show easily that in this situation, the eigenvalues of L are preserved under the flow of L and B (and also under an infinite family of "higher flows"). If we restrict the dynamical system defined by Eq. (1.3) to a function space that is chosen such that the spectrum of L under this restriction is a pure discrete point spectrum, then the points of the spectrum therefore become conserved quantities of the dynamical system (1.3). We consider one of two typical methods to achieve this, namely to consider solutions u which are periodic, and then to consider the quasi-periodic spectrum of L. (The other method would be to consider solutions u which decrease sufficiently rapidly for $x \to \pm\infty$, leading for example to soliton solutions and the inverse scattering method; see for example [Hi-S-W, Chapter 3] or [No-M-P-Z, Chapter I].) It turns out that at least for the KdV equation, one obtains by this method a discrete point spectrum for L. Thus the spectrum is defined by the eigenvalue equation $L\psi = \lambda\psi$, where $\lambda \in \mathbb{C}$ and ψ is a quasi-periodic function. Here each eigenvalue λ is a conserved quantity of the dynamical system (1.3). The periodic spectrum does not determine u uniquely, however one can define additional data associated to the quasi-periodic eigenfunctions ψ (which move under the flows of L and B), so that the spectrum along with these additional data determine the original solution u of the KdV equation uniquely. Because of this fact, the dynamical system defined by Eq. (1.3) is regarded as an integrable system. The spectrum together with the mentioned additional data are called the *spectral data* of u.

In relation to the correspondence between the solution u and the spectral data there are two fundamental problems: first the *direct problem* to study the spectral data corresponding to a solution u and to derive properties of the spectral data from properties of the solution u; and second, the *inverse problem* to study u by its spectral data, especially to show that u is determined uniquely by the spectral data and to reconstruct u from its spectral data.

Note that this approach to the study of periodic solutions u of the KdV equation depends almost exclusively on the study of the spectrum of the 1-dimensional Schrödinger operator L with potential u along a period (i.e. only with respect to the single variable x). The other Lax operator B comes into play only when one reconstructs the two-dimensional solution u of the KdV equation from its values along the x-line by applying the flow of B.

An important point of view is that the Lax equation can be interpreted as a zero curvature condition in the following way: The second order differential operator L can be rewritten as the 2-dimensional first order differential operator $-\frac{\mathrm{d}}{\mathrm{d}x} + \left(\begin{smallmatrix} 0 & 1 \\ u & 0 \end{smallmatrix}\right)$ and then the eigenvalue equation $L\psi = \lambda\psi$ takes the form

$$\frac{\mathrm{d}}{\mathrm{d}x}\Psi = U_\lambda \Psi \tag{1.4}$$

with the 2-dimensional periodic operator

$$U_\lambda := \begin{pmatrix} 0 & 1 \\ u - \lambda & 0 \end{pmatrix} \quad \text{and} \quad \Psi = \begin{pmatrix} \psi_1 \\ \psi_2 \end{pmatrix} = \begin{pmatrix} \psi \\ \frac{\mathrm{d}}{\mathrm{d}x}\psi \end{pmatrix}.$$

Moreover there exists another 2-dimensional periodic operator V_λ also depending on the parameter λ so that the Lax equation $\frac{\mathrm{d}}{\mathrm{d}t}L = [B, L]$ is equivalent to the condition

$$\frac{\partial U_\lambda}{\partial t} - \frac{\partial V_\lambda}{\partial x} + [U_\lambda, V_\lambda] = 0. \tag{1.5}$$

Equation (1.5) is called a *zero curvature condition* because it is equivalent to the condition that the connection $\mathrm{d}+\alpha_\lambda$ on the trivial bundle $\mathbb{C}^2$ defined by the 1-form $\alpha_\lambda := U_\lambda\,\mathrm{d}x + V_\lambda\,\mathrm{d}t$ is flat. It is also equivalent to the Maurer-Cartan equation for the linear partial differential equation $\mathrm{d}F_\lambda = \alpha_\lambda\,F_\lambda$. Thus for a given solution u of the KdV equation on $\mathbb{R}^2$ there is for every $\lambda \in \mathbb{C}$ a unique solution $F_\lambda : \mathbb{R}^2 \to \mathrm{SL}(2,\mathbb{C})$ of the initial value problem $\mathrm{d}F_\lambda = \alpha_\lambda F_\lambda,\ F(0) = \mathbb{1}$. If we consider a solution $u = u(x,t)$ of the KdV equation (1.2) that is periodic with respect to x, then the corresponding F_λ are in general not periodic; the extent of their non-periodicity is measured by the monodromy $M(\lambda) := F_\lambda(T)$, where $T > 0$ is the minimal period of u. One can show that some $\lambda \in \mathbb{C}$ is a quasi-periodic eigenvalue of L, where $\mu \in \mathbb{C}$ is the quasi-periodic factor of the corresponding eigenvector, if and only if μ is an eigenvalue of $M(\lambda)$. The complex curve

$$\Sigma := \{\, (\lambda,\mu) \in \mathbb{C}^2 \,\big|\, \mu \text{ is an eigenvalue of } M(\lambda) \,\}$$

thereby defined is called the *spectral curve* of u. In the case of the KdV equation, Σ is a hyperelliptic curve above $\mathbb{C}$ that can be extended at $\lambda = \infty$ in a certain sense (see [Sch, Chapter 2]) by adding a single point to Σ above $\lambda = \infty$; this point then becomes a double point singularity of the curve. The eigenvectors of $M(\lambda)$ define a holomorphic line bundle on a suitable partial normalisation of Σ; the corresponding divisor D on Σ is called the *spectral divisor* of u. The pair (Σ, D) is an equivalent description of the spectral data of the solution u. In these names, the adjective "spectral" refers to the fact that the data are derived from a specific eigenvalue problem; the term *spectral theory* in the context of integrable systems refers to the study of the spectral data in this sense.

The KdV equation is also the example of a partial differential equation with an integrable system that has been studied most extensively and for which there are the most diverse publications. Many ideas and techniques that were first developed for the integrable system of the KdV equation have subsequently been successfully applied to the integrable system of other partial differential equations. Also for the present work, results on the KdV integrable system have been inspirational in many ways, and it is also for this reason that we now mention some important publications on KdV.

An overview of the investigation of the integrable systems and spectral theory for the KdV equation can for example be found in [Kapp-P, Chapter III], in [No-M-P-Z, Chapter II] or in [Ma-2, Chapter 4]. A very accessible introduction to the spectral theory of the 1-dimensional Schrödinger operator, i.e. the Lax operator L of the KdV equation, is the book [Pö-T], which however considers only solutions y of the Schrödinger equation with the Dirichlet boundary conditions $y(0) = y(1) = 0$. The book [Pö-T] has been especially influential for the treatment of the sinh-Gordon equation in the present work, and several of the results in the present work correspond to analogous theorems in the theory of the KdV equation resp. the 1-dimensional Schrödinger operator as they are described there; the correspondences will be pointed out.

The existence of periodic solutions for the KdV equation was first shown by Marchenko [Ma-1]. Later the moduli space of periodic solutions of the KdV equation has been characterised by Marchenko and Ostrovskii [Ma-O-1] for real-valued solutions and by Tkachenko [T] for complex-valued solutions. An angular action parameterization on the space of smooth periodic solutions has been described by Kappeler [Kapp]. The best known estimates for the function spaces in which the KdV dynamics work were found by Kappeler and Topalov [Kapp-T].

A periodic solution u is said to be *finite gap* or *finite zone* (for the KdV equation), or *finite type* (in the general situation) if its spectral curve has finite geometric genus and the spectral divisor has finite degree on the partial normalisation of the spectral curve. Such solutions play a special role because in this case the spectral curve can be compactified, and the solution can be described by finitely many parameters and reconstructed from the spectral data by an essentially finite process, providing what is often called an *algebro-geometric correspondence* between the spectral data and the solution. Finite gap solutions have first been considered by Novikov [No], and a good overview source on them is [Bel, Chapter 3]. Finite gap solutions can be re-obtained from the spectral data by the explicit *Its-Matveev formula* which uses the theta function of the spectral curve, see [I-M-2] (in Russian, with the English summary [I-M-1]).

It was shown by Marchenko that the finite gap solutions of the KdV equation are dense in the space of all periodic solutions; a simplified and improved version of the proof was given by Marchenko and Ostrovskii [Ma-O-2] (in Russian, with an English translation [Ma-O-3] and an Erratum [Ma-O-4]). This result is extremely important because it permits to transfer certain results that have been proven for finite gap solutions to general periodic solutions by a limit argument.

On the other hand, when one wishes to prove results about general periodic KdV solutions directly, i.e. without using the finite gap theory, one faces the problem that the spectral curve of such a solution is non-compact and generally has infinite geometric genus, moreover the spectral divisor generally has infinite degree. To handle this situation, one best considers the spectral data as members of suitably chosen Banach spaces. The corresponding estimates are obtained by asymptotic methods, i.e. by estimating the difference between the spectral data for a given solution u and the spectral data of the "vacuum solution" $u = 0$. In this manner, the spectral theory for KdV was investigated by McKean and van Moerbeke in [McK-M]; this work was later extended by McKean and Trubowitz to include also the study of the Jacobi variety of the spectral curve in [McK-T]. Related estimates for the solutions in terms of the gap lengths were given by Korotyaev

[Ko-1]. Kargaev/Korotyaev solved the inverse problem for periodic KdV solutions, i.e. the reconstruction of the solution from its spectral data, in [Kar-Ko-1] (with an Erratum [Kar-Ko-2]) and [Ko-2]. Another solution of the inverse problem was given by Müller et al. in [Mü-S-S] by generalizing the Its-Matveev formula to theta functions for complex curves of infinite geometric genus; using their method they also construct some new infinite gap solutions of the KdV equation.

The spectral theory for an n-dimensional generalisation of the 1-dimensional Schrödinger operator that occurs as the Lax operator for the KdV equation was developed by Schmidt in [Sch].

The Integrable System of the Non-linear Schrödinger (NLS) Equation

Another partial differential equation that has been studied extensively by integrable systems methods is the (self-focusing) non-linear Schrödinger (NLS) equation

$$u_{xx} - 2\mathrm{i} \cdot u_t + 2|u|^2 \cdot u = 0 \tag{1.6}$$

for complex-valued functions $u = u(x,t)$ in two variables. This partial differential equation first arose in physics; for example it describes stationary light beams in the plane under non-linear refraction. An inner-mathematical motivation is that it describes the vortex filament flow of a closed curve in a 3-dimensional space form: Suppose that $\gamma = \gamma(x)$ is a closed curve in one of the 3-dimensional space forms, let κ resp. τ be the (geodesic) curvature resp. the torsion of γ, and let B be the binormal field of γ. We then let γ flow in time t by the vortex filament flow (binormal flow), i.e. according to the differential equation $\frac{\partial \gamma}{\partial t} = \kappa \cdot B$. This flow conserves the closedness and the length of the curve γ. In this setting, Hasimoto [Has] realized that the so-called *Hasimoto curvature* or *complex curvature*

$$u(x,t) = \kappa(x,t) \cdot \exp\left(\mathrm{i} \int_0^x \tau(\xi,t)\,\mathrm{d}\xi\right)$$

is a solution of the NLS equation (1.6). It should be noted that although the solution u of the NLS equation corresponding to the vortex filament flow of a closed curve is always quasi-periodic in x, not every quasi-periodic or even periodic solution u of the NLS equation gives rise to the vortex filament flow of a closed curve in a 3-dimensional space form. This is the case only if u satisfies an additional closing condition which determines two real degrees of freedom.

Like the KdV equation, also the NLS equation (1.6) can be written as a Lax equation (1.3). In this case however, the Lax operators L and B are (2×2)-matrix valued:

$$L = \begin{pmatrix} -\mathrm{i} & 0 \\ 0 & \mathrm{i} \end{pmatrix} \frac{\mathrm{d}}{\mathrm{d}x} + \begin{pmatrix} 0 & u \\ -\bar{u} & 0 \end{pmatrix} \quad \text{and}$$

$$B = \begin{pmatrix} -\mathrm{i} & 0 \\ 0 & \mathrm{i} \end{pmatrix} \frac{\mathrm{d}^2}{\mathrm{d}x^2} + \begin{pmatrix} 0 & u \\ -\bar{u} & 0 \end{pmatrix} \frac{\mathrm{d}}{\mathrm{d}x} + \frac{1}{2} \begin{pmatrix} -\mathrm{i}|u|^2 & u_x \\ -\overline{u_x} & \mathrm{i}|u|^2 \end{pmatrix}.$$

These Lax operators define an integrable system that is associated to the NLS equation in an analogous way as was described for the KdV equation above. Also the representation of the Lax equation as a zero-curvature condition and the construction of the spectral data carries over to the NLS equation. Like for KdV, also the NLS spectral curve is at first a hyperelliptic curve above $\mathbb{C}$ and can (again in the sense of [Sch, Chapter 2]) be extended at $\lambda = \infty$. The most salient difference between the spectral data for KdV and the spectral data for NLS is that whereas a KdV spectral curve has a single point above $\lambda = \infty$ which is a double point singularity, a NLS spectral curve has two points above $\lambda = \infty$ and these are regular points. An overview of the integrable systems theory of NLS can be found in [Bel, Chapter 4].

Grinevich and Schmidt [Gri-S-1, Gri-S-2] discussed deformations of NLS solutions that maintain periodicity. Calini and Ivey [Cal-I-1, Cal-I-2] also used deformation theory to study the topology of solutions of the vortex filament flow equation.

Similarly as for the KdV equation, finite gap periodic solutions of the NLS equation play an important role due to them being determined by finitely many parameters. In particular, they can be reconstructed explicitly from their spectral data in terms of theta functions or Baker-Akhiezer functions, see for example [Cal-I-1, Section 2]. It was shown by Grinevich [Gri] that the finite gap periodic NLS solutions are dense in the space of all periodic NLS solutions. Moreover he also showed that the finite gap periodic NLS solutions which satisfy the closing condition for curves in $\mathbb{R}^3$ are dense in the space of all periodic NLS solutions with this closing condition. Kilian and the author of the present book gave in [Kl-K] another proof of these two statements via integrable systems theory and asymptotic methods, and proved the analogous statement for closed curves in the 3-sphere S^3. In a forthcoming paper they will also show the denseness statement for closed curves in the hyperbolic 3-space H^3.

The Integrable System of the Sinh-Gordon Equation

We now turn our attention to the sinh-Gordon equation (1.1) again. It is the objective of the present work to study simply periodic solutions of the sinh-Gordon equation by means of integrable systems theory. When considering the analogy of the sinh-Gordon equation to the KdV equation or the NLS equation, we note that unlike the latter two equations, which involve two variables x and t with different geometric meaning ("space" resp. "time"), the two variables of the sinh-Gordon equation are essentially equivalent (two "space" coordinates). More importantly, it turns out that the sinh-Gordon equation can *not* be written in the form of a Lax equation (1.3) (even if t is replaced by one of the space coordinates). Therefore the construction of conserved quantities from periodic eigenvalues of a Lax operator that has been described above for the KdV equation does *not* carry over to the sinh-Gordon equation. However, it *is* possible to write the sinh-Gordon equation in the form of a zero-curvature condition; in fact, the sinh-Gordon equation is equivalent to Eq. (1.5), where the families of operators U_λ, V_λ are defined by

$$U_\lambda := \frac{1}{4}\begin{pmatrix} \mathrm{i}\,u_y & -\mathrm{e}^{u/2} - \lambda^{-1}\,\mathrm{e}^{-u/2} \\ \mathrm{e}^{u/2} + \lambda\,\mathrm{e}^{-u/2} & -\mathrm{i}\,u_y \end{pmatrix} \quad \text{and}$$

$$V_\lambda := \frac{\mathrm{i}}{4}\begin{pmatrix} -u_x & \mathrm{e}^{u/2} - \lambda^{-1}\,\mathrm{e}^{-u/2} \\ \mathrm{e}^{u/2} - \lambda\,\mathrm{e}^{-u/2} & u_x \end{pmatrix}. \tag{1.7}$$

Note that U_λ and V_λ are only defined for $\lambda \in \mathbb{C}^*$ rather than for $\lambda \in \mathbb{C}$. By means of these operators it *is* possible to define spectral data, i.e. a spectral curve and a spectral divisor, for simply periodic solutions of the sinh-Gordon equation. Because the dependency of the operators U_λ and V_λ on λ is more complicated in this setting, Eq. (1.4) can no longer be interpreted as the eigenvalue equation for a differential operator L as was the case for the KdV equation, and thus also the parameter λ can no longer be interpreted as an eigenvalue resp. a spectral value. Nevertheless, the analogy to the situation for the KdV equation is very close via the interpretation of the Lax equation as a zero curvature condition, which is the reason why I take the liberty of applying the term *spectral theory* also to the study of periodic solutions of the sinh-Gordon equation by this method; likewise we will use the adjective "spectral" for the parameter λ and similar objects.

Such zero-curvature representations have been first described for harmonic maps into Lie groups or symmetric spaces by Pohlmeyer [Poh] and Uhlenbeck [U]. By the theorem of Ruh/Vilms, solutions of the sinh-Gordon equation correspond to harmonic maps into the 2-sphere; therefore one can obtain the zero-curvature representation for solutions of the sinh-Gordon equation given above from these works. McIntosh [McI] has described spectral data from the point of view of considering harmonic maps from a torus into S^2, and Burstall et al. [Bu-F-P-P] have used such methods to construct harmonic maps from a torus into any symmetric space. It appears that Hitchin [Hi] was the first to apply such a representation and the associated spectral theory to the study of solutions of the sinh-Gordon equation. Bobenko [Bo-1, Bo-2, Bo-3] realized that the operators U_λ and V_λ, and thereby the construction of the spectral data for the sinh-Gordon equation have a geometric interpretation in terms of the relationship between solutions of the sinh-Gordon equation and minimal resp. constant mean curvature surfaces. We now describe the construction of spectral data for simply periodic solutions of the sinh-Gordon equation in terms of this geometric interpretation for minimal immersions into the 3-sphere.

Consider a minimal conformal immersion $f : X \to S^3$ without umbilical points into the 3-sphere, corresponding to a solution $u : X \to \mathbb{R}$ of the sinh-Gordon equation. Then the frame $\underline{F} : X \to \mathrm{SO}(4)$ of the immersion f induces the $\mathfrak{so}(4)$-valued flat connection form $\underline{\alpha} = \underline{F}^{-1} \cdot \mathrm{d}\underline{F}$ on X, which is the Maurer-Cartan form of $\underline{F}$. The Maurer-Cartan equation for $\underline{F}$ is equivalent to the equations of Gauss and Codazzi for the minimal immersion f into S^3, which are in turn equivalent to the sinh-Gordon equation for the *conformal factor* u, i.e. for the map $u : X \to \mathbb{R}$ so that the induced Riemannian metric on X is given by $g = \mathrm{e}^{u/2}\,(\mathrm{d}x^2 + \mathrm{d}y^2)$. In view of the universal covering group $\mathrm{SU}(2) \times \mathrm{SU}(2)$ of $\mathrm{SO}(4)$, the frame $\underline{F}$ locally lifts to a map $F = (F^{(1)}, F^{(2)})$ into $\mathrm{SU}(2) \times \mathrm{SU}(2)$, and it is clear that the components $F^{(k)}$ each again satisfy a partial differential equation, namely their Maurer-Cartan equation. However it was a remarkable observation due to Bobenko, that the two components $F^{(1)}$ and $F^{(2)}$ are governed by essentially the same differential equation. More explicitly, if we define $\alpha_\lambda := U_\lambda\,\mathrm{d}x + V_\lambda\,\mathrm{d}y$ for $\lambda \in \mathbb{C}^*$ with the operators U_λ and V_λ given by Eq. (1.7), and again consider the extended frame F_λ defined by the initial value problem $\mathrm{d}F_\lambda = \alpha_\lambda\, F_\lambda$, $F(0) = \mathbb{1}$, then we have $F = (F^{(1)}, F^{(2)}) = (F_{\lambda=1}, F_{\lambda=-1})$. From F resp. from the extended frame F_λ, the original minimal immersion f can be reconstructed via the *Sym-Bobenko formula*: $f = F_{\lambda=-1} \cdot F_{\lambda=1}^{-1} : N \to \mathrm{SU}(2) \cong S^3$. The two spectral values $\lambda = \pm 1$ which are involved in this reconstruction are called the *Sym points* of the surface.

If we suppose in this setting that the conformal factor u is periodic, we again consider the monodromy $M(\lambda) := F_\lambda(z_1) \cdot F_\lambda(z_0)^{-1}$ along a period. Then the hyperelliptic complex curve (possibly with singularities) defined by the eigenvalues of $M(\lambda)$

$$\Sigma := \{\, (\lambda, \mu) \in \mathbb{C}^* \times \mathbb{C} \,\big|\, \det(M(\lambda) - \mu \cdot \mathbb{1}) = 0 \,\}$$

will be called the *spectral curve* of u in the present work (but is called the multiplier curve by Hitchin, who reserves the term "spectral curve" for a different object, see below), and the divisor D defining the eigenline bundle of $M(\lambda)$ on Σ is the *spectral divisor* of u. (Σ, D) is called the *spectral data* for the solution u of the sinh-Gordon equation. The branch points of Σ are analogous to the periodic spectrum of the 1-dimensional Schrödinger operator L in the treatment of the KdV equation via spectral theory described above, and the spectral data (Σ, D) are analogous to the spectral data for L. It again turns out that the complex curve Σ is hyperelliptic over $\mathbb{C}^*$, and in the analogous sense as in [Sch, Chapter 2], Σ can be compactified by adding one point each above $\lambda = 0$ and above $\lambda = \infty$; these additional points then become ordinary double points of the complex curve.

An overview of the integrable systems theory for the sinh-Gordon equation (and also the closely related sine-Gordon equation) with reality condition is [McK]. Another good overview is contained in [Hau-K-S-1]. An overview for the sine-Gordon equation is also [Bel, Section 4.2].

Like for other integrable systems, the periodic solutions of finite type play a special role for the sinh-Gordon integrable system. One reason is that again, they are described by finitely many parameters, and can be reconstructed from their spectral data by explicit formulas, for example in terms of theta functions. One particularly nice reconstruction formula, based on the vector-valued Baker-Akhiezer function on the spectral curve, was obtained by Knopf in his dissertation [Kn-1] (with summary [Kn-3]). Note that for the sinh-Gordon integrable system, it has not yet been shown that the finite type solutions are dense in the space of all periodic solutions; this is one of the results proven in the present book (Chap. 13).

Another reason why finite type solutions of the sinh-Gordon equation are important is the remarkable fact that all doubly periodic solutions of the sinh-Gordon equation are of finite type. This fact was proven by Hitchin [Hi] as an important step of his classification of the minimal tori in S^3. In more detail, his strategy is as follows: Minimal tori in S^3 correspond to doubly periodic solutions u of the sinh-Gordon equation. The two minimal periods of u induce two different monodromies and therefore Hitchin obtains at first two different sets (Σ, D) of spectral data for u. However, because the homotopy group of the torus is abelian, the two monodromies commute, and therefore the two curves Σ are in fact the same. This curve Σ is the spectral curve of u in our terminology, but is called the multiplier curve by Hitchin. He reserves the term "spectral curve" for the partial normalisation $\widetilde{\Sigma}$ of Σ (called the *middleding* in [Kl-L-S-S]) on which the eigenline bundles of the two monodromies are holomorphic line bundles, corresponding to free divisors. Hitchin proves that the resulting curve $\widetilde{\Sigma}$ has finite geometric genus, and can therefore be compactified as a complex curve by adding a point each above $\lambda = 0$ and $\lambda = \infty$. This implies that all doubly periodic solutions u of the sinh-Gordon equation are of finite type. Using this result, Hitchin can now classify the doubly periodic solutions of the sinh-Gordon equation by applying classical results of the theory of compact Riemann surfaces. Finally he obtains a classification of the minimal tori in S^3 by investigating the "closing condition" that characterises whether a doubly periodic solution u of the sinh-Gordon equation corresponds to a (closed) minimal torus in S^3.

Carberry and Schmidt [Car-S-1, Car-S-2] proved that the spectral curves of CMC tori in $\mathbb{R}^3$ are dense in the space of all smooth finite-gap spectral curves.

Strategy of the Present Work

In the present work, we develop the theory of spectral data for solutions $u : X \to \mathbb{C}$ of the sinh-Gordon equation (with $X \subset \mathbb{C}$) which have only a single period (are *simply periodic*). Note that we include the case of complex-valued u throughout the work. By a linear transformation of X we may suppose without loss of generality that $0 \in X$ holds and that the period of u is 1. Then X is a horizontal strip in $\mathbb{C}$ and

$$u(z+1) = u(z) \quad \text{holds for all } z \in X\,.$$

It is possible to construct the spectral data also for bare Cauchy data (u, u_y) for a solution of the sinh-Gordon equation, here u and u_y are periodic functions defined only on the interval $[0, 1]$, where u_y gives the derivative of u in normal direction. Such Cauchy data correspond to data for the so-called *Björling problem*, which is the problem of finding a minimal surface passing through a prescribed curve and being tangential to a prescribed unit normal field to the curve.

The viewpoint we will take throughout most of the work is that we will study the relationship between Cauchy data (u, u_y) and the corresponding spectral data.

In this context we are interested in requiring only as weak regularity conditions for (u, u_y) as possible. There are two reasons: First, we are interested in characterizing precisely which divisors on a spectral curve are spectral divisors of some Cauchy data (u, u_y); it turns out that every additional differentiability condition imposed on (u, u_y) reduces the space of divisors by an intricate relationship between its divisor points. By not imposing more regularity than necessary, we obtain a description of the space of divisors that is as simple as possible. Second, while any solution u of the sinh-Gordon equation is infinitely differentiable (in fact even real analytic, because the sinh-Gordon equation is elliptic) on the interior of its domain, we are also interested in the behavior of the solution on the boundary of its domain, where its behavior can be worse. For these reasons we only require $(u, u_y) \in W^{1,2}([0, 1]) \times L^2([0, 1])$.

The definition of the spectral data (Σ, D) described above carries over to our situation without difficulty. As spectral curve, we use the complex curve Σ defined by the eigenvalues of the monodromy (called the multiplier curve by Hitchin), not the partial desingularization $\widetilde{\Sigma}$ that Hitchin calls the spectral curve. There are several reasons for this choice:

The most important reason is that I take the point of view of describing variations of u within its spectral class by variation of the spectral divisor, not by change of the curve on which the divisor is defined. Because Hitchin's partial desingularization $\widetilde{\Sigma}$ needs to be chosen dependent on the spectral divisor, spectral divisors on the same complex curve Σ can induce different partial desingularizations $\widetilde{\Sigma}$. Thus using the partial desingularization would complicate the approach of working on a fixed curve.

Another reason is that in our setting (with u only simply periodic) Hitchin's partial desingularization $\widetilde{\Sigma}$ would still have infinite geometric genus in general, so passing from Σ to $\widetilde{\Sigma}$ is not quite as useful in the present situation as it is for Hitchin. Moreover we note that the existence of a partial desingularization of Σ which parameterizes the eigenline bundle of the monodromy is a property of only the sinh-Gordon integrable system (and a few others), as a consequence of the fact that the monodromy takes values in (2×2)-matrices, which causes Σ to be hyperelliptic. In the spectral theory for other integrable systems, the monodromies generally take values in larger endomorphism spaces, hence Σ is no longer hyperelliptic (instead it covers the λ-plane with order ≥ 3), and then there does not generally exist such a partial desingularization, see Remark 3.4.

However, there is one complication that is caused by using the curve Σ as spectral curve, instead of its partial desingularization, and that is that we need to deal with the fact that it is possible for singularities of Σ to occur in the support of the spectral divisor. In this case, the divisor, in the classical sense, does not convey enough information to determine the monodromy $M(\lambda)$ uniquely, which is of course what we expect of the spectral data. The solution to this problem is to generalize the concept of a divisor following a concept of Hartshorne [Har-2]. In this sense, a *generalized divisor* is a subsheaf $\mathscr{D}$ of the sheaf $\mathscr{M}$ of meromorphic functions on Σ which is finitely generated over the sheaf $\mathscr{O}$ of holomorphic functions on Σ. Such a generalized divisor is exactly equivalent to a classical divisor in the regular points of Σ, but it conveys more information in the singular points. We will see that this is the proper concept of a divisor to ensure that the spectral curve Σ along with a generalized divisor $\mathscr{D}$ uniquely determines the monodromy $M(\lambda)$.

In our setting, the geometric genus of the spectral curve Σ is in general infinite—the branch points accumulate near $\lambda = \infty$ and near $\lambda = 0$—so Σ can no longer be compactified. For this reason, the classical results on compact Riemann surfaces which played a significant role for the study of solutions of finite type, are no longer applicable to Σ in the case of infinite type.

We need to replace these standard results with arguments for non-compact surfaces. To make such arguments feasible, one needs to prescribe the asymptotic behavior of the spectral curve Σ and the spectral divisor "at its ends", i.e. near the points $\lambda = \infty$ and $\lambda = 0$. The information on the asymptotic behavior of the spectral data is the replacement for the fact from the finite type theory that (the partial desingularization of) Σ can be compactified at $\lambda = \infty$ and $\lambda = 0$. It is for this reason that asymptotic estimates for the quantities involved in the construction of the spectral data are a very important, perhaps the most important tool for the conduction of the present study.

Unfortunately there are only very few results on open Riemann surfaces with prescribed asymptotics found in the literature. One example would be the book [Fe-K-T] by Feldman et al. From that book, we will indeed use one of the results from Chap. 1 that applies to general surfaces that are parabolic in the sense of Ahlfors [Ah] and Nevanlinna [Ne], i.e. that do not have a Green's function. However the results in the later part of that book, which would be very useful to us, depend on very strict geometric hypotheses for the surface under consideration, see [Fe-K-T, Section 5], and especially the hypothesis GH2(iv), which requires that the "size of the handles" t_j (comparable to the distance between the pairs of branch points) satisfies $\sum t_j^{\beta} < \infty$ for every $\beta > 0$. This condition implies in particular that the t_j are decreasing more rapidly than j^{-n} for every $n \geq 1$. Because of our relatively mild requirements on the differentiability of (u, u_y), the t_j do not decrease sufficiently rapidly by far for our spectral curve Σ, and thus the results from the latter part of [Fe-K-T] are not applicable in our situation. For this reason we develop some results analogous to classical results on compact Riemann surfaces here, as needed.

We mention that Schmidt [Sch] develops the spectral theory for a different integrable system, namely for Lax operators of NLS-like partial differential equations, and in particular studies the asymptotics and the spectral curves of infinite genus associated to this integrable system.

Outline of the Contents of This Book

We now give an overview of the results of the work. In Chap. 2 we explicitly describe the relationship between solutions u of the sinh-Gordon equation, and minimal immersions $f : X \to S^3$ without umbilical points. This relationship is of importance because it provides the $\mathfrak{sl}(2, \mathbb{C})$-valued flat connection form α_λ (depending on the spectral parameter $\lambda \in \mathbb{C}^*$) and thereby the extended frame $F_\lambda : X \to \mathrm{SL}(2, \mathbb{C})$ as the solution of the differential equation $\mathrm{d}F_\lambda = \alpha_\lambda F_\lambda$. In the case where u is simply periodic, the monodromy $M(\lambda) \in \mathrm{SL}(2, \mathbb{C})$ of the extended frame along the period is then used in Chap. 3 to construct the spectral data (Σ, D) for u. Here Σ is the spectral curve, defined by the eigenvalues of $M(\lambda)$, which turns out to be a hyperelliptic complex curve, and D is the divisor on Σ defining the eigenline bundle of $M(\lambda)$. With respect to D, a difficulty arises in view of the fact that the spectral curve Σ can be singular: If the support of D contains singular points, then the classical divisor does not convey enough information at these points to uniquely determine the monodromy $M(\lambda)$ anymore, and thus we replace D with a generalized divisor $\mathscr{D}$ in the sense of Hartshorne as explained above. We will see in Chap. 12 that $(\Sigma, \mathscr{D})$ uniquely determines the monodromy $M(\lambda)$ (up to a choice of sign).

The first main part of the work then is to describe the asymptotic behavior of the monodromy $M(\lambda)$ and thereby of the spectral data $(\Sigma, \mathscr{D})$ near $\lambda = \infty$ and near $\lambda = 0$. As explained above, this is the most important instrument for understanding and controlling the spectral curve in our situation.

We begin in Chap. 4 by calculating the spectral data for the "vacuum solution" $u = 0$ of the sinh-Gordon equation. This example is of importance because we will describe the asymptotic behavior of the various objects near $\lambda \to \infty$ and $\lambda \to 0$ by comparing the objects for any given simply periodic solution u to the corresponding (explicitly calculated) object for the vacuum solution. In this way, one can view the spectral data for the vacuum solution as being "typical" for the spectral data for any simply periodic solution, where the extent of typical-ness is quantified precisely by the asymptotic estimates of the chapters to come.

For the investigation of the asymptotic behavior of $M(\lambda)$ and of the spectral data $(\Sigma, \mathscr{D})$, we do not consider a solution u of the sinh-Gordon equation defined on a horizontal strip in $\mathbb{C}$ of positive height, but rather periodic Cauchy data (u, u_y) defined on $[0, 1]$. For the reasons explained above, we only require $(u, u_y) \in W^{1,2}([0, 1]) \times L^2([0, 1])$; the condition of periodicity then reduces to

$u(0) = u(1)$. Note that for many of the asymptotic estimates we also admit Cauchy data without the condition of periodicity. Such "non-periodic potentials" of course do not correspond to simply-periodic solutions of the sinh-Gordon equation; they are considered for the sole purpose of deriving the asymptotic behavior of the extended frame F_λ in Proposition 11.7.

The strategy for obtaining asymptotic estimates for the spectral data is to find asymptotic estimates for the monodromy $M(\lambda)$ first and then to derive asymptotic estimates for the spectral data from them. Similarly as in the case of the 1-dimensional Schrödinger operator [Pö-T, Chapter 2], one needs to carry out this process in two stages. The final result on the asymptotic of the spectral data will permit to solve the inverse problem for the monodromy $M(\lambda)$, and by this description of $M(\lambda)$ also give a final refinement of the asymptotic description of the monodromy.

This process begins by deriving a relatively mild asymptotic for the monodromy (called the "basic asymptotic" in Chap. 5), which then yields first information on the spectral data (Chap. 6): It turns out that the points in the support of the spectral divisor $\mathscr{D}$ can be enumerated by a sequence $(\lambda_k, \mu_k)_{k\in\mathbb{Z}}$ of points on Σ, where for $|k|$ large, each (λ_k, μ_k) is in a certain neighborhood of a corresponding point $(\lambda_{k,0}, \mu_{k,0})$ in the divisor of the vacuum; we call these neighborhoods the *excluded domains* $\widehat{U}_{k,\delta}$. Note that the divisor points are (asymptotically) free to move in the excluded domains, they are not restricted to intervals or curves (as it is the case, for example, for real-valued solutions of the KdV equation, where these intervals are called "gaps"); the reason is that we are considering complex-valued, not only real-valued, solutions of the sinh-Gordon equation.

Similarly, Σ has two branch points in each $\widehat{U}_{k,\delta}$ for $|k|$ large; if they coincide, then a singularity of Σ occurs. This singularity is an ordinary double point; for $|k|$ large this is the only type of singularity that can occur, but for small $|k|$, the spectral curve Σ can also have singularities of higher order.

These asymptotic assessments can still be refined: By studying the Fourier transform of the given Cauchy data (u, u_y), in Chap. 7 it is shown that the difference between the monodromy $M(\lambda)$ and the monodromy $M_0(\lambda)$ follows for $\lambda = \lambda_{k,0}$ a law of square-summability, which we call the "Fourier asymptotic" for the monodromy. From this fact, it follows that the divisor points and the branch points also follow a square-summability law with respect to the difference to their counterparts for the vacuum. We show this for the divisor points in Chap. 8, whereas the corresponding result for the branch points of the spectral curve is postponed to Chap. 11 to avoid technical difficulties.

Our next objective is to solve the inverse problem for the monodromy: Suppose that spectral data $(\Sigma, \mathscr{D})$ are given, where the support of $\mathscr{D}$ satisfies the square-summability law just mentioned, then one would like to reconstruct the monodromy $M(\lambda)$ from which these spectral data were induced; in particular, we need to show that $M(\lambda)$ is determined uniquely by the spectral data. If we write the $\mathrm{SL}(2, \mathbb{C})$-valued monodromy $M(\lambda)$ in the form $M(\lambda) = \begin{pmatrix} a(\lambda) & b(\lambda) \\ c(\lambda) & d(\lambda) \end{pmatrix}$ with holomorphic functions $a, b, c, d : \mathbb{C}^* \to \mathbb{C}$, the spectral divisor directly conveys information on the zeros of c and on the function values of d (or a) at the zeros of c. The inverse problem for the monodromy is, in essence, to reconstruct the functions a, b, c, d from these information and from their known asymptotic behavior.

To facilitate the discussion of this problem, we introduce in Chap. 9 spaces of holomorphic functions on $\mathbb{C}^*$ or on Σ which have a prescribed asymptotic behavior near $\lambda = \infty$ and/or near $\lambda = 0$. The difference between the functions comprising $M(\lambda)$ and the corresponding functions of the vacuum will turn out to be contained in certain of these spaces, and it is for this reason that these spaces are extremely important in the sequel.

In Chap. 10 we address the mentioned problem of reconstructing a holomorphic function on $\mathbb{C}^*$ with a prescribed asymptotic behavior from the sequence of its zeros, or else from a sequence of its values at certain points. The first of these problems (reconstruction from the zeros) is reminiscent of Hadamard's Factorization Theorem (see for example [Co, Theorem XI.3.4, p. 289]), but here we have a holomorphic function defined on $\mathbb{C}^*$ with two accumulation points of the sequence of zeros ($\lambda = \infty$ and $\lambda = 0$), instead of a function on $\mathbb{C}$ with the zeros accumulating only at $\lambda = \infty$. For both of the reconstruction problems, we obtain explicit descriptions of the solution function as an infinite product resp. sum. For the proof of the corresponding statements, we require several relatively elementary results on the convergence and estimate of infinite sums and products, these results are collected in Appendix A.

By applying the explicit product resp. sum formulas from Chap. 10 to the functions comprising the monodromy $M(\lambda)$, we can improve the asymptotic description for $M(\lambda)$ one final time: in the description of the Fourier asymptotic in Chap. 7, the points $\lambda = \lambda_{k,0}$ played a special role. In Chap. 11 we are now liberated from this condition and obtain a square-summability law for $M - M_0$ with respect to arbitrary points e.g. in excluded domains. We also obtain the asymptotic behavior of the branch points of the spectral curve Σ, and an asymptotic estimate for the extended frame F_λ itself.

Finally in Chap. 12 we characterize those generalized divisors $\mathscr{D}$ on a spectral curve Σ for which $(\Sigma, \mathscr{D})$ is the spectral data of a monodromy $M(\lambda)$. It turns out that besides the asymptotic behavior that was described before, one needs a certain compatibility condition (Definition 12.1(1)) and (as is to be expected) that the divisor $\mathscr{D}$ is non-special. Under these circumstances it turns out that $(\Sigma, \mathscr{D})$ corresponds to a monodromy $M(\lambda)$ and that $M(\lambda)$ is determined uniquely by $(\Sigma, \mathscr{D})$ (up to a change of sign of the off-diagonal entries). This concludes the solution of the inverse problem for the monodromy.

The next question is the inverse problem for the Cauchy data (u, u_y) themselves, i.e. to reconstruct (u, u_y) from the spectral data $(\Sigma, \mathscr{D})$. The corresponding problem has been solved for the potentials of finite type, and it appears natural to try to apply that result for our infinite type potentials.

For this purpose we show in Chap. 13 (by a fixed point argument) that the set of finite type divisors is dense in the space of all asymptotic divisors. Moreover, in Chap. 14 we equip the space of potentials with a symplectic form and construct Darboux coordinates on this symplectic space, building on previous results by

Knopf [Kn-2]. Using these two results, we are then able to show relatively easily in Chap. 15 that for the open and dense subset Pot_{tame} of "tame" potentials, the map $\mathsf{Pot}_{tame} \to \mathsf{Div}$ associating to each tame potential (Cauchy data) the corresponding spectral divisor is a diffeomorphism onto an open and dense subset of the space of all asymptotic divisors. This essentially solves the inverse problem for the Cauchy data (u, u_y).

The final part of the work is concerned with constructing a Jacobi variety and an Abel map for the spectral curve Σ. For the sake of simplicity, we will here restrict ourselves to spectral curves Σ which do not have any singularities other than ordinary double points. Also in this part we admit spectral data corresponding to complex solutions of the sinh-Gordon equation, it is for this reason that the individual Jacobi coordinates take values in $\mathbb{C}$ (and not only in S^1, which would be familiar for example from the Jacobi variety for (real) solutions of the 1-dimensional Schrödinger operator constructed in [McK-T, Sections 11ff.]).

The construction of the Jacobi variety needs to be prepared with more asymptotic estimates: In Chap. 16 we prove estimates for certain contour integrals on the spectral curve Σ, and in Chap. 17 we study holomorphic 1-forms on Σ. In particular we construct holomorphic 1-forms which have zeros in all the excluded domains, with the exception of a single one, and find sufficient conditions for such 1-forms to be square-integrable on Σ.

With help of these results we will construct a canonical basis (ω_n) for the space of holomorphic 1-forms explicitly in Chap. 18. In the case where the spectral curve Σ has no singularities, the existence of such a basis follows from the general theory of parabolic (in the sense of Ahlfors/Nevanlinna) Riemann surfaces described for example in [Fe-K-T, Chapter 1]. However this is not enough for us, not only because we also need to handle the case where Σ has singularities, but also because we derive asymptotic estimates for the ω_n from their explicit representation which are stronger than those one can obtain abstractly; we need these stronger estimates in the construction of the Jacobi variety and the Abel map for Σ.

The work is concluded with two applications of this construction: In Chap. 19 we describe the motions in the Jacobi variety that correspond to translations of the solution u of the sinh-Gordon equation. Similarly as is known for the finite type setting, it turns out that these translations correspond to linear motions in the Jacobi variety. And finally, in Chap. 20 we describe the asymptotic behavior of the spectral data (Σ, D) for an actual simply periodic solution $u : X \to \mathbb{C}$ of the sinh-Gordon equation, given on a horizontal strip of positive height $X \subset \mathbb{C}$. We see that in this case the spectral data satisfy an exponential asymptotic law, much steeper than the asymptotic behavior of the spectral data for Cauchy data (u, u_y), as is to be expected, solutions of the sinh-Gordon equation being real analytic in the interior of their domain.

Chapter 21 outlines a perspective how the present work of research might be extended to study minimal or CMC surfaces with umbilical points. Such points correspond to coordinate singularities of the corresponding solution u of the sinh-Gordon equation. Results on the spectral theory of minimal surfaces with umbilical points might be used to gain insight into the behavior of *compact* minimal surfaces in S^3, especially those of genus $g \geq 2$.

Chapter 2
Minimal Immersions into the 3-Sphere and the Sinh-Gordon Equation

We begin by describing the relationship between minimal immersions without umbilical points into the 3-sphere and solutions of the sinh-Gordon equation explicitly, especially to obtain the $\mathfrak{sl}(2,\mathbb{C})$-valued connection form α_λ corresponding to the zero-curvature representation of the sinh-Gordon equation; from the integration of α_λ we will obtain spectral data for periodic solutions of the sinh-Gordon equation.

Suppose that $f : X \to S^3$ is a conformal immersion of some Riemann surface X into the 3-sphere $S^3(\varkappa) \subset \mathbb{R}^4$ of curvature $\varkappa > 0$. Then there exists a function $u : X \to \mathbb{R}$ such that the pull-back Riemannian metric $g = f^*\langle\,\cdot\,,\cdot\,\rangle$ induced by f on X is given by $g = \mathrm{e}^{u/2}\,\mathrm{d}z\,\mathrm{d}\overline{z} = \mathrm{e}^{u/2}\,(\mathrm{d}x^2 + \mathrm{d}y^2)$. u is called the *conformal factor* of f. If we denote the mean curvature function of f by H, and the Hopf differential of f by $E\,\mathrm{d}z^2$, the equations of Gauss and Codazzi read

$$u_{z\overline{z}} + \frac{1}{2}(H^2+\varkappa)\,\mathrm{e}^u - 2\,E\,\overline{E}\,\mathrm{e}^{-u} = 0$$

$$H_z\,\mathrm{e}^u = 2\,E_{\overline{z}}.$$

We now specialize to the situation where f is minimal, i.e. $H = 0$, and maps into the sphere of radius 2, i.e. $\varkappa = \frac{1}{4}$. Moreover let us suppose that f has no umbilical points. Then we can choose a holomorphic chart z of X such that the Hopf differential's function E is constant and of modulus $\frac{1}{4}$. In this situation the equation of Codazzi reduces to $0 = 0$, whereas the equation of Gauss reduces to the sinh-Gordon equation

$$\Delta u + \sinh(u) = 0. \tag{2.1}$$

This shows that minimal immersions $f : X \to S^3(\varkappa)$ give rise to solutions of the sinh-Gordon equation.

S. Klein, *A Spectral Theory for Simply Periodic Solutions of the Sinh-Gordon Equation*, Lecture Notes in Mathematics 2229,
https://doi.org/10.1007/978-3-030-01276-2_2

We now describe the relationship between minimal immersions into $S^3 := S^3(\varkappa = \frac{1}{4})$ and solutions of the sinh-Gordon equation in more detail. For this purpose, we suppose X to be simply connected.

We denote by f_x and f_y the derivative of f in the direction x resp. y, and put $e_x := \frac{f_x}{\|f_x\|}$, $e_y := \frac{f_y}{\|f_y\|}$. Moreover, we let N be the unit normal field of the immersion f. Then $\left(e_x, e_y, N\right)$ is an orthonormal basis field of TS^3 along f, hence the *frame* $\underline{F} := \left(e_x, e_y, N, \frac{1}{2}f\right) : X \to \mathrm{SO}(4)$ is a positively oriented orthonormal basis field of $\mathbb{R}^4$ along f. X being simply connected, we can lift $\underline{F}$ to the universal covering group $\mathrm{SU}(2) \times \mathrm{SU}(2)$ of $\mathrm{SO}(4)$, obtaining $(F^{[1]}, F^{[2]}) : X \to \mathrm{SU}(2) \times \mathrm{SU}(2)$ with $F^{[1]}, F^{[2]} : X \to \mathrm{SU}(2)$.

It is clear that the "integrability condition" $((F^{[1]}, F^{[2]})_z)_{\bar{z}} = ((F^{[1]}, F^{[2]})_{\bar{z}})_z$ translates into differential equations for $F^{[1]}$ and $F^{[2]}$. But it is an insight due to Bobenko [Bo-2] that the two components $F^{[1]}, F^{[2]}$ of the frame are governed by essentially the *same* differential equation. Indeed, there exists $\lambda \in S^1$ so that we have $(F^{[1]}, F^{[2]}) = (F_\lambda, F_{-\lambda})$, where F_λ satisfies the differential equation $\mathrm{d}F_\lambda = \alpha_\lambda\, F_\lambda$ with

$$\alpha_\lambda := \frac{1}{4}\begin{pmatrix} -u_z & -\lambda^{-1}\,\mathrm{e}^{-u/2} \\ \mathrm{e}^{u/2} & u_z \end{pmatrix} \mathrm{d}z + \frac{1}{4}\begin{pmatrix} u_{\bar{z}} & -\mathrm{e}^{u/2} \\ \lambda\,\mathrm{e}^{-u/2} & -u_{\bar{z}} \end{pmatrix} \mathrm{d}\bar{z}.$$

An explicit calculation shows that

$$\mathrm{d}\alpha_\lambda + [\alpha_\lambda \wedge \alpha_\lambda] = \frac{1}{8}(\Delta u + \sinh(u)) \begin{pmatrix} 1 & 0 \\ 0 & -1 \end{pmatrix} \mathrm{d}z \wedge \mathrm{d}\bar{z},$$

thus the Maurer-Cartan equation $\mathrm{d}\alpha_\lambda + [\alpha_\lambda \wedge \alpha_\lambda] = 0$ for α_λ is equivalent to the sinh-Gordon equation for u. Therefore we have the following equivalence:

Proposition 2.1 *Suppose $u : X \to \mathbb{C}$ is any differentiable function. Then the following three statements are equivalent:*

(a) The function u solves the sinh-Gordon equation $\Delta u + \sinh(u) = 0$.
(b) The metric $\mathrm{e}^{u/2}\,\mathrm{d}z\,\mathrm{d}\bar{z}$ satisfies the equations of Gauss and Codazzi for a minimal immersion into S^3 with constant Hopf differential of modulus $\frac{1}{4}$.
(c) For any, or for all $\lambda \in \mathbb{C}^*$, α_λ (as defined above) satisfies the Maurer-Cartan equation $\mathrm{d}\alpha_\lambda + [\alpha_\lambda \wedge \alpha_\lambda] = 0$.

If any of the statements in the Proposition holds, then (c) shows that there is a unique solution $F_\lambda : X \to \mathrm{SU}(2)$ to the differential equation $\mathrm{d}F_\lambda = \alpha_\lambda\, F_\lambda$ with $F_\lambda(z_0) = \mathbb{1}$ (where $z_0 \in X$ is fixed) for every $\lambda \in \mathbb{C}^*$. It is clear that F_λ depends holomorphically on $\lambda \in \mathbb{C}^*$. In this setting, the family $(F_\lambda)_{\lambda \in \mathbb{C}^*}$ is called the *extended frame* of u or of f; the parameter λ is called the *spectral parameter*.

On the other hand, if $u : X \to \mathbb{C}$ is a solution of the sinh-Gordon equation, then the metric $\mathrm{e}^{u/2}\,\mathrm{d}z\,\mathrm{d}\bar{z}$ together with any constant Hopf differential $E\,\mathrm{d}z^2$ of modulus $\frac{1}{4}$ satisfies the equations of Gauss and Codazzi for a minimal immersion into S^3. Thus, if u is real-valued in this setting, then u corresponds to a minimal conformal immersion $f_E : X \to S^3$ with that metric and that Hopf differential. Thus we see that minimal immersions into S^3 come in families $(f_E)_{E\in(1/4)\,S^1}$; each such family is called an *associated family*.

Therefore solutions $u : X \to \mathbb{R}$ of the sinh-Gordon equation are in one-to-one correspondence with associated families of minimal immersions into S^3.

Note that α_λ has certain symmetries, described in Proposition 2.2 below. In the case where the solution u is simply periodic, the symmetries of α_λ also induce symmetries for the extended frame F_λ and for the *monodromy* $M(\lambda) = F_\lambda(z_1)\cdot F_\lambda(z_0)^{-1}$ of F_λ along the period. For the description of the symmetries, we suppose for the sake of simplicity that $X \subset \mathbb{C}$ holds, that we have $0 \in X$, that the period of u is 1 (so the real interval $[0,1]$ is contained in X), and that we consider the base point $z_0 = 0$. Then the monodromy of the extended frame is given by $M(\lambda) = F_\lambda(z_0+1)\cdot F_\lambda(z_0)^{-1} = F_\lambda(1)$.

One of the applications of these symmetries will be to derive asymptotic estimates for $\lambda \to 0$ from asymptotic estimates for $\lambda \to \infty$.

Proposition 2.2 *Let $\Phi : z \mapsto \overline{z}$. If $u : X \to \mathbb{C}$ is a simply periodic solution of the sinh-Gordon equation as above, then $\widetilde{u} := -u\circ\Phi$ and $\overline{u} : z \mapsto \overline{u(z)}$ also are simply periodic solutions of the sinh-Gordon equation. Moreover we put $g := \left(\begin{smallmatrix}1&0\\0&\lambda\end{smallmatrix}\right)$. Then we have*

(1) $\alpha_{\lambda^{-1},u} = g^{-1}\cdot\Phi^*\alpha_{\lambda,\widetilde{u}}\cdot g,$
$F_{\lambda^{-1},u} = g^{-1}\cdot(F_{\lambda,\widetilde{u}}\circ\Phi)\cdot g,$
$M_u(\lambda^{-1}) = g^{-1}\cdot M_{\widetilde{u}}(\lambda)\cdot g.$

(2) $\alpha_{\overline{\lambda}^{-1},u} = -\overline{\alpha^t_{\lambda,\overline{u}}}$ *(where α^t denotes the transpose of α),*
$F_{\overline{\lambda}^{-1},u} = \overline{(F^{-1}_{\lambda,\overline{u}})^t},$
$M_u(\overline{\lambda}^{-1}) = \overline{(M_{\overline{u}}(\lambda)^{-1})^t}.$

Proof For (1). We have

$$\Phi^*\mathrm{d}z = \mathrm{d}\overline{z} \quad\text{and}\quad \Phi^*\mathrm{d}\overline{z} = \mathrm{d}z\,,$$

and also

$$\widetilde{u}_z\circ\Phi = -u_{\overline{z}} \quad\text{and}\quad \widetilde{u}_{\overline{z}}\circ\Phi = -u_z.$$

Using these equations, we calculate:

$$g^{-1}\cdot(\Phi^*\alpha_{\lambda,\widetilde{u}})(z)\cdot g$$
$$= g^{-1}\cdot\left[\frac{1}{4}\begin{pmatrix}-\widetilde{u}_z\circ\Phi & -\lambda^{-1}\,\mathrm{e}^{-(\widetilde{u}\circ\Phi)/2}\\ \mathrm{e}^{(\widetilde{u}\circ\Phi)/2} & \widetilde{u}_z\circ\Phi\end{pmatrix}\Phi^*\mathrm{d}z\right.$$

$$
\begin{aligned}
&+ \frac{1}{4}\begin{pmatrix} \widetilde{u}_{\bar z}\circ \Phi & -\mathrm{e}^{(\widetilde{u}\circ\Phi)/2} \\ \lambda\,\mathrm{e}^{-(\widetilde{u}\circ\Phi)/2} & -\widetilde{u}_{\bar z}\circ\Phi \end{pmatrix} \Phi^*\mathrm{d}\bar z\Bigg]\cdot g \\
&= g^{-1}\cdot\left[\frac{1}{4}\begin{pmatrix} u_{\bar z} & -\lambda^{-1}\,\mathrm{e}^{u/2} \\ \mathrm{e}^{-u/2} & -u_{\bar z}\end{pmatrix}\mathrm{d}\bar z + \frac{1}{4}\begin{pmatrix} -u_z & -\mathrm{e}^{-u/2} \\ \lambda\,\mathrm{e}^{u/2} & u_z\end{pmatrix}\mathrm{d}z\right]\cdot g \\
&= \frac{1}{4}\begin{pmatrix} u_{\bar z} & -\mathrm{e}^{u/2} \\ \lambda^{-1}\,\mathrm{e}^{-u/2} & -u_{\bar z}\end{pmatrix}\mathrm{d}\bar z + \frac{1}{4}\begin{pmatrix} -u_z & -\lambda\,\mathrm{e}^{-u/2} \\ \mathrm{e}^{u/2} & u_z\end{pmatrix}\mathrm{d}z \\
&= \alpha_{\lambda^{-1},u}(z).
\end{aligned}
$$

Now let $\widetilde{F} := g^{-1}\cdot(F_{\lambda,\widetilde{u}}\circ\Phi)\cdot g$. Then we have

$$
\begin{aligned}
\mathrm{d}\widetilde{F} &= g^{-1}\cdot \mathrm{d}(F_{\lambda,\widetilde{u}}\circ\Phi)\cdot g = g^{-1}\cdot\Phi^*\,\mathrm{d}F_{\lambda,\widetilde{u}}\cdot g = g^{-1}\cdot\Phi^*(\alpha_{\lambda,\widetilde{u}}\,F_{\lambda,\widetilde{u}})\cdot g \\
&= g^{-1}\cdot(\Phi^*\alpha_{\lambda,\widetilde{u}})\cdot g\cdot g^{-1}\cdot(F_{\lambda,\widetilde{u}}\circ\Phi)\cdot g = \alpha_{\lambda^{-1},u}\,\widetilde{F}
\end{aligned}
$$

as well as $\widetilde{F}(0) = \mathbb{1}$. In other words, $\widetilde{F}$ solves the same initial value problem as $F_{\lambda^{-1},u}$, and thus we have $\widetilde{F} = F_{\lambda^{-1},u}$.

Finally, we have $M_u(\lambda^{-1}) = F_{\lambda^{-1},u}(1) = g^{-1}\cdot F_{\lambda,\widetilde{u}}(\Phi(1))\cdot g = g^{-1}\cdot M_{\widetilde{u}}(\lambda)\cdot g$.

For (2). We have

$$
\begin{aligned}
\overline{\alpha_{\bar\lambda^{-1},\bar u}} &= \overline{\frac{1}{4}\begin{pmatrix} -\bar u_z & -\bar\lambda\,\mathrm{e}^{-\bar u/2} \\ \mathrm{e}^{\bar u/2} & \bar u_z\end{pmatrix}\mathrm{d}z + \frac{1}{4}\begin{pmatrix} \bar u_{\bar z} & -\mathrm{e}^{\bar u/2} \\ \bar\lambda^{-1}\,\mathrm{e}^{-\bar u/2} & -\bar u_{\bar z}\end{pmatrix}\mathrm{d}\bar z} \\
&= \frac{1}{4}\begin{pmatrix} -u_{\bar z} & -\lambda\,\mathrm{e}^{-u/2} \\ \mathrm{e}^{u/2} & u_{\bar z}\end{pmatrix}\mathrm{d}\bar z + \frac{1}{4}\begin{pmatrix} u_z & -\mathrm{e}^{u/2} \\ \lambda^{-1}\,\mathrm{e}^{-u/2} & -u_z\end{pmatrix}\mathrm{d}z \\
&= -\alpha^t_{\lambda,u}.
\end{aligned}
$$

Now let $\widetilde{F} := \overline{(F^{-1}_{\lambda,u})^t}$. We then have

$$
\begin{aligned}
\mathrm{d}\widetilde{F} &= \overline{(\mathrm{d}F^{-1}_{\lambda,u})^t} = -\overline{(F^{-1}_{\lambda,u}\,\mathrm{d}F_{\lambda,u}\,F^{-1}_{\lambda,u})^t} = -\overline{(F^{-1}_{\lambda,u}\,\alpha_{\lambda,u}\,F_{\lambda,u}\,F^{-1}_{\lambda,u})^t} \\
&= -\overline{(F^{-1}_{\lambda,u}\,\alpha_{\lambda,u})^t} = -\overline{\alpha^t_{\lambda,u}}\,\widetilde{F} = \alpha_{\bar\lambda^{-1},\bar u}\,\widetilde{F}
\end{aligned}
$$

as well as $\widetilde{F}(0) = \mathbb{1}$. It follows that $\widetilde{F}$ solves the initial value problem for $F_{\bar\lambda^{-1},\bar u}$, and hence $F_{\bar\lambda^{-1},\bar u} = \widetilde{F}$ holds.

Finally, we have $(M_u(\lambda)^{-1})^t = (F_{\lambda,u}(1)^{-1})^t = \overline{F_{\bar\lambda^{-1},\bar u}(1)} = \overline{M_{\bar u}(\bar\lambda^{-1})}$. □

Chapter 3
Spectral Data for Simply Periodic Solutions of the Sinh-Gordon Equation

We now suppose that a *simply periodic* solution $u : X \to \mathbb{C}$ of the sinh-Gordon equation $\Delta u + \sinh(u) = 0$ on an (open or closed) domain $X \subset \mathbb{C}$ is given. Without loss of generality, we suppose that $0 \in X$ and that $1 \in \mathbb{C}$ is the period of u. Then X is a horizontal strip in $\mathbb{C}$ that contains the real axis $\mathbb{R}$ and we have

$$u(z+1) = u(z) \quad \text{for all } z \in X.$$

Figure 3.1 illustrates the situation.

We let F_λ be the extended frame corresponding to u, i.e. for each $\lambda \in \mathbb{C}^*$, $F_\lambda : X \to \mathrm{SU}(2)$ satisfies

$$\mathrm{d}F_\lambda = \alpha_\lambda \, F_\lambda \, , \quad F_\lambda(0) = \mathbb{1}$$

with

$$\alpha_\lambda := \frac{1}{4}\begin{pmatrix} -u_z & -\lambda^{-1}\,\mathrm{e}^{-u/2} \\ \mathrm{e}^{u/2} & u_z \end{pmatrix} \mathrm{d}z + \frac{1}{4}\begin{pmatrix} u_{\overline{z}} & -\mathrm{e}^{u/2} \\ \lambda\,\mathrm{e}^{-u/2} & -u_{\overline{z}} \end{pmatrix} \mathrm{d}\overline{z} \tag{3.1}$$

$$= \frac{1}{4}\begin{pmatrix} \mathrm{i}\,u_y & -\mathrm{e}^{u/2} - \lambda^{-1}\,\mathrm{e}^{-u/2} \\ \mathrm{e}^{u/2} + \lambda\,\mathrm{e}^{-u/2} & -\mathrm{i}\,u_y \end{pmatrix} \mathrm{d}x$$

$$+ \frac{\mathrm{i}}{4}\begin{pmatrix} -u_x & \mathrm{e}^{u/2} - \lambda^{-1}\,\mathrm{e}^{-u/2} \\ \mathrm{e}^{u/2} - \lambda\,\mathrm{e}^{-u/2} & u_x \end{pmatrix} \mathrm{d}y. \tag{3.2}$$

Although u is periodic, the extended frame F_λ generally is not. To measure the deviance of F_λ from being periodic, we are lead to consider the *monodromy* $M_z(\lambda) := F_\lambda(z+1) \cdot F_\lambda(z)^{-1}$. It should be noted that the monodromy M itself satisfies a differential equation with respect to differentiation in the base point z.

S. Klein, *A Spectral Theory for Simply Periodic Solutions of the Sinh-Gordon Equation*, Lecture Notes in Mathematics 2229,
https://doi.org/10.1007/978-3-030-01276-2_3

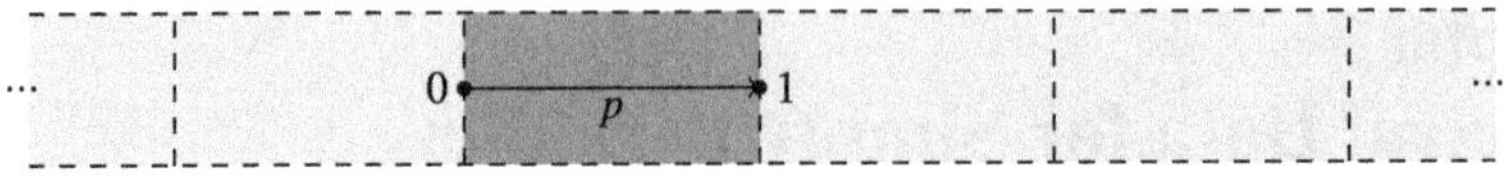

Fig. 3.1 The domain X and the minimal period $p = 1$ of the solution u. The more darkly shaded rectangle is a fundamental domain, and the values of u repeat in the other rectangles drawn

Indeed, we have

$$\begin{aligned} d_z M_z(\lambda) &= d\big(F_\lambda(z+1) \cdot F_\lambda(z)^{-1}\big) \\ &= dF_\lambda(z+1) \cdot F_\lambda(z)^{-1} - F_\lambda(z+1) \cdot F_\lambda(z)^{-1} \cdot dF_\lambda(z) \cdot F_\lambda(z)^{-1} \\ &= \alpha_\lambda(z+1) \cdot F_\lambda(z+1) \cdot F_\lambda(z)^{-1} - F_\lambda(z+1) \cdot F_\lambda(z)^{-1} \cdot \alpha_\lambda(z) \\ &= [\alpha_\lambda(z), M_z(\lambda)] \,, \end{aligned} \tag{3.3}$$

because $\alpha_\lambda(z+1) = \alpha_\lambda(z)$ holds due to the periodicity of u. Here $[A, B] := A \cdot B - B \cdot A$ is the commutator of two quadratic matrices. Consequently, we have

$$M_z(\lambda) = F_\lambda(z) \cdot M_{z=0}(\lambda) \cdot F_\lambda(z)^{-1}.$$

We use the monodromy $M(\lambda) := M_{z=0}(\lambda)$ at $z = 0$ to define *spectral data* for u. We first define the *spectral curve* by the eigenvalues of $M(\lambda)$:

$$\Sigma := \{ (\lambda, \mu) \in \mathbb{C}^* \times \mathbb{C} \,\big|\, \det(M(\lambda) - \mu \cdot \mathbb{1}) = 0 \}. \tag{3.4}$$

Σ is a non-compact complex curve (i.e. a 1-dimensional complex analytic space), possibly with singularities. The natural functions λ and μ on Σ are holomorphic, and they generate the sheaf of holomorphic functions on Σ, i.e. a function $f : \Sigma \to \mathbb{C}$ is holomorphic if and only if it can locally be extended to a holomorphic function in λ and μ on a neighborhood in $\mathbb{C}^* \times \mathbb{C}$.

To phrase the definition (3.4) of Σ more explicitly, we write the monodromy $M(\lambda)$ in the form

$$M(\lambda) = \begin{pmatrix} a(\lambda) & b(\lambda) \\ c(\lambda) & d(\lambda) \end{pmatrix} \tag{3.5}$$

with the holomorphic functions $a, b, c, d : \mathbb{C}^* \to \mathbb{C}$. Then we have

$$\Delta(\lambda) := \operatorname{tr}(M(\lambda)) = a(\lambda) + d(\lambda). \tag{3.6}$$

Moreover, because α_λ is trace-free, we have

$$a(\lambda) \cdot d(\lambda) - b(\lambda) \cdot c(\lambda) = \det(M(\lambda)) = 1. \tag{3.7}$$

Equations (3.6) and (3.7) show that the characteristic polynomial of $M(\lambda)$ is $\mu^2 - \Delta(\lambda) \cdot \mu + 1$, and thus we obtain

$$\Sigma = \{ (\lambda, \mu) \in \mathbb{C}^* \times \mathbb{C} \,|\, \mu^2 - \Delta(\lambda) \cdot \mu + 1 = 0 \}. \tag{3.8}$$

It follows from this presentation of Σ that for any $(\lambda, \mu) \in \Sigma$, $\mu \neq 0$ holds, that there are at least one and at most two points $(\lambda, \mu) \in \Sigma$ that are above some given $\lambda \in \mathbb{C}^*$, and that the holomorphic involution

$$\sigma : \Sigma \to \Sigma, \ (\lambda, \mu) \mapsto (\lambda, \mu^{-1}) \tag{3.9}$$

maps Σ onto itself. Therefore the complex curve Σ is hyperelliptic[1] above $\mathbb{C}^*$ with the hyperelliptic involution σ.

More generally, for a given $\lambda \in \mathbb{C}^*$ we have $(\lambda, \mu) \in \Sigma$ if and only if

$$\mu = \frac{1}{2} \left(\Delta(\lambda) \pm \sqrt{\Delta(\lambda)^2 - 4} \right) \tag{3.10}$$

holds; equivalently $(\lambda, \mu) \in \Sigma$ is characterized by the eigenvalue equation

$$(a(\lambda) - \mu) \cdot (d(\lambda) - \mu) = b(\lambda) \cdot c(\lambda). \tag{3.11}$$

Note that $\sqrt{\Delta^2 - 4} = \mu - \mu^{-1}$ is a holomorphic function on Σ, which is anti-symmetric with respect to σ, i.e. we have $\sqrt{\Delta^2 - 4} \circ \sigma = -\sqrt{\Delta^2 - 4}$.

It also follows from Eq. (3.10) that $(\lambda_*, \mu_*) \in \Sigma$ is a fixed point of the hyperelliptic involution σ if and only if $\Delta(\lambda_*) = \pm 2$ (or $\Delta(\lambda_*)^2 - 4 = 0$) holds; this is the case if and only if $\mu_* \in \{\pm 1\}$ holds, and in this case, there is only one point in Σ above λ_*. Such a point is a branch point of Σ if the zero of $\Delta \mp 2$ (or $\Delta^2 - 4$) is of odd order, and it is a singularity of Σ if the zero of $\Delta \mp 2$ (or $\Delta^2 - 4$) is of order ≥ 2 (an ordinary double point if the zero is of order 2); in the latter case the order of zero of $\Delta \mp 2$ (or $\Delta^2 - 4$) is the order of the singularity. There are no other branch points or singularities of Σ.

In this situation, the zeros of $\Delta^2 - 4$ (corresponding to the branch points and singularities of Σ) are analogous to the Dirichlet spectrum in the spectral theory of the 1-dimensional Schrödinger equation (see for example [Pö-T, Chapter 2, p. 25f.]).

We will now describe the local complex space structure of Σ, in particular near its singularities. We let $\mathcal{O}$ resp. $\mathcal{M}$ be the sheaf of holomorphic resp. of meromorphic functions on Σ. The sheaf of unitary rings $\mathcal{O}$ is generated by the holomorphic functions λ and μ (or by λ and $\mu - \mu^{-1}$) in the sense that a function

[1] Generally one calls a complex curve hyperelliptic if it is a (branched) two-fold covering above $S^2 = \mathbb{C} \cup \{\infty\}$. In extension of this concept, we will take the liberty of calling a complex curve hyperelliptic above $\mathbb{C}$ or $\mathbb{C}^*$ if it is a branched two-fold covering above $\mathbb{C}$ resp. $\mathbb{C}^*$.

f on Σ is holomorphic if and only if can locally be written in the form

$$f = f_+(\lambda) + f_-(\lambda) \cdot (\mu - \mu^{-1})$$

with holomorphic functions f_+ and f_- in λ; note that $(\mu - \mu^{-1})^2 = \Delta^2 - 4$ is a function in λ. Equivalently, f is holomorphic at some $(\lambda_*, \mu_*) \in \Sigma$ if and only if it can be extended to a holomorphic function in λ and μ on a neighborhood of (λ_*, μ_*) in $\mathbb{C}^* \times \mathbb{C}^*$.

Near any point (λ_*, μ_*) of Σ that is not a fixed point of σ (i.e. that is neither a branch point nor a singularity), the holomorphic function $\lambda - \lambda_*$ is a local coordinate on Σ near (λ_*, μ_*), and near any regular branch point (λ_*, μ_*) of Σ, the function $\mu - \mu^{-1}$ is a local coordinate.

To study the structure of Σ at singular points, we consider the normalization $\widehat{\Sigma}$ of Σ together with the one-sheeted covering $\pi : \widehat{\Sigma} \to \Sigma$ (see for example [deJ-P, Section 4.4, p. 161ff.]). We let $\mathcal{O}_{\widehat{\Sigma}}$ resp. $\mathcal{M}_{\widehat{\Sigma}}$ be the sheaf of holomorphic functions resp. of meromorphic functions on $\widehat{\Sigma}$. Then the abelian group $H^0(\Sigma, \mathcal{M})$ of meromorphic functions on Σ is isomorphic to the abelian group $H^0(\Sigma, \pi_* \mathcal{M}_{\widehat{\Sigma}})$ of sections in the direct image sheaf $\pi_* \mathcal{M}_{\widehat{\Sigma}}$ via $f \mapsto f \circ \pi$, and we identify $\pi_* \mathcal{M}_{\widehat{\Sigma}}$ with $\mathcal{M}$ in this way. Under this identification, $\widehat{\mathcal{O}} := \pi_* \mathcal{O}_{\widehat{X}}$ corresponds to the germs of meromorphic functions on Σ which are locally bounded [deJ-P, Theorem 4.4.15, p. 167].

If Σ has at (λ_*, μ_*) a singularity of odd order $2n+1$ (i.e. $\Delta^2 - 4$ has at λ_* a zero of order $2n+1$), then there is exactly one point in $\widehat{\Sigma}$ above (λ_*, μ_*), this point is a branch point of the normalization, $\sqrt{\lambda - \lambda_*}$ is a coordinate of $\widehat{\Sigma}$ near that branch point, $d\lambda$ vanishes at (λ_*, μ_*) of first order, and we have (on $\widehat{\Sigma}$)

$$\mu - \mu^{-1} = \left(\sqrt{\lambda - \lambda_*}\right)^{2n+1} \cdot \varphi \tag{3.12}$$

with a non-zero holomorphic function germ $\varphi \in \widehat{\mathcal{O}}$. It follows that the δ-invariant $\dim(\widehat{\mathcal{O}}_{(\lambda_*,\mu_*)}/\mathcal{O}_{(\lambda_*,\mu_*)})$ of Σ at (λ_*, μ_*) (see [deJ-P, Definition 5.2.1(1), p. 186]) equals n. A meromorphic function germ $f \in \mathcal{M}$ on Σ at (λ_*, μ_*) can be described as a meromorphic function germ $\widehat{f}$ (i.e. a Laurent series) in $\sqrt{\lambda - \lambda_*}$; in this setting we define the order of a zero (or the order of a pole) of f at (λ_*, μ_*), denoted by $\mathrm{ord}^{\Sigma}(f)$ (or by $\mathrm{polord}^{\Sigma}(f) := -\mathrm{ord}^{\Sigma}(f)$), as the order of the zero (or pole) of $\widehat{f}$ as a function in $\sqrt{\lambda - \lambda_*}$. In particular, for a function g that is meromorphic in λ, we have $\mathrm{ord}^{\Sigma}(g) = 2 \cdot \mathrm{ord}^{\mathbb{C}}(g)$.

On the other hand, if Σ has at (λ_*, μ_*) a singularity of even order $2n$ (i.e. $\Delta^2 - 4$ has at λ_* a zero of order $2n$), then there are two points in $\widehat{\Sigma}$ above (λ_*, μ_*), these points are not branch points of $\widehat{\Sigma}$, $\lambda - \lambda_*$ is a coordinate near each of these two points, so $d\lambda$ does not vanish at either of these points. A meromorphic function germ $f \in \mathcal{M}$ on Σ at (λ_*, μ_*) can be described as a pair (f_1, f_2) of germs of functions meromorphic in $\lambda - \lambda_*$, and in this sense we have

$$\mu - \mu^{-1} = \big((\lambda - \lambda_*)^n \cdot \varphi \, , \; -(\lambda - \lambda_*)^n \cdot \varphi\big) \tag{3.13}$$

with a non-zero holomorphic function germ φ in λ. It follows that the δ-invariant $\dim(\widehat{\mathcal{O}}_{(\lambda_*,\mu_*)}/\mathcal{O}_{(\lambda_*,\mu_*)})$ of Σ at (λ_*, μ_*) again equals n. We define the order of a zero (or the order of a pole) of a meromorphic function germ f at (λ_*, μ_*), denoted by $\mathrm{ord}^{\Sigma}(f)$ (or by $\mathrm{polord}^{\Sigma}(f) := -\mathrm{ord}^{\Sigma}(f)$), as the sum of the two zero (or pole) orders

$$\mathrm{ord}^{\Sigma}(f) := \mathrm{ord}^{\mathbb{C}}(f_1) + \mathrm{ord}^{\mathbb{C}}(f_2) \quad \text{or}$$

$$\mathrm{polord}^{\Sigma}(f) := \mathrm{polord}^{\mathbb{C}}(f_1) + \mathrm{polord}^{\mathbb{C}}(f_2),$$

where f is represented by (f_1, f_2) as above. In particular, for a function g that is meromorphic in λ, we again have $\mathrm{ord}^{\Sigma}(g) = 2 \cdot \mathrm{ord}^{\mathbb{C}}(g)$.

The spectral curve Σ alone does not fully characterize the monodromy $M(\lambda)$. It describes the eigenvalues of $M(\lambda)$, but not the corresponding eigenvectors. Therefore we define a second spectral object, namely the divisor associated to the bundle Λ of eigenvectors of $M(\lambda)$, which is a holomorphic line bundle at least on Σ', the Riemann surface of regular points of Σ.

To obtain an explicit description of Λ, we note that $(v_1, v_2) \in \mathbb{C}^2$ is an eigenvector of $M(\lambda)$ corresponding to the eigenvalue $\mu \in \mathbb{C}^*$ if and only if either of the two equivalent (by the eigenvalue equation (3.11)) equations

$$(a(\lambda) - \mu) \cdot v_1 + b(\lambda) \cdot v_2 = 0 \quad \text{or} \quad c(\lambda) \cdot v_1 + (d(\lambda) - \mu) \cdot v_2 = 0$$

holds. Therefore $(\frac{\mu - d(\lambda)}{c(\lambda)}, 1)$ is a global meromorphic section of Λ, and thus the divisor on Σ characterizing the line bundle Λ is the pole divisor D of the meromorphic function $\frac{\mu-d}{c}$ on Σ.

The divisor D is analogous in the spectral theory for the 1-dimensional Schrödinger operator L to the constants κ_n associated to the periodic eigenfunctions of L, that uniquely determine the potential of L together with the periodic spectrum of L (see for example, [Pö-T, p. 59 and Theorem 3.5, p. 62]). We preliminarily call D the *spectral divisor* of the monodromy $M(\lambda)$; this definition will be modified below in the case that the support of D contains singular points of Σ.

Proposition 3.1 *If (λ_*, μ_*) is a regular point of Σ that is a member of the support of D, say with multiplicity $m \geq 1$, then $\mathrm{ord}^{\mathbb{C}}_{\lambda_*}(c) = m$ and $\mathrm{ord}^{\Sigma}_{(\lambda_*,\mu_*)}(\mu - a) \geq m$ holds. In particular $c(\lambda_*) = 0$ and $a(\lambda_*) = \mu_*$ holds. Moreover, if (λ_*, μ_*) is a regular branch point of Σ, then $m = 1$ holds.*

Proof Let $(\lambda_*, \mu_*) \in \Sigma$ be a regular point of Σ that is in the support of D with multiplicity m, and therefore a pole of $\frac{\mu-d}{c}$ of order m.

Let us first suppose that (λ_*, μ_*) is not a branch point of Σ, i.e. that $\Delta^2 - 4$ is non-zero at λ_*. Then both λ and μ are local coordinates of Σ near (λ_*, μ_*), and therefore the order of zeros or poles of holomorphic functions at (λ_*, μ_*) do not depend on whether we view the functions on Σ or on $\mathbb{C}^*$. Moreover, it is not possible for both of the functions $\mu - a$ and $\mu - d$ to be zero at (λ_*, μ_*); it follows by Eq. (3.11) that either $\mu - a$ or $\mu - d$ has a zero of order at least m. If $\mu - d$

had a zero of order at least m, then $\frac{\mu-d}{c}$ would not have a pole at (λ_*, μ_*), and therefore (λ_*, μ_*) would not be in the support of D. Thus we see that $\mu - d$ is not zero at (λ_*, μ_*), and therefore the pole order m of $\frac{\mu-d}{c}$ equals the exact order of the zero of c. Moreover Eq. (3.11) shows that $\mu - a$ has a zero of order at least m.

Now consider the case that (λ_*, μ_*) is a regular branch point of Σ, i.e. that $\Delta^2 - 4$ has a simple zero at λ_*. Then $\mu_* \in \{\pm 1\}$ and therefore $a(\lambda_*) = d(\lambda_*) = \mu_*$ holds. Thus we see that as function on $\mathbb{C}^*$, $(a-d)^2$ has a zero of order at least 2; the equation (which follows from Eqs. (3.6) and (3.7))

$$\Delta^2 - 4 = (a+d)^2 - 4(ad - bc) = (a-d)^2 + 4bc$$

therefore shows that bc can have a zero of order at most 1. Thus $\operatorname{ord}^{\mathbb{C}}_{\lambda_*}(c) = 1$ and $b(\lambda_*) \neq 0$ holds. Because of $\operatorname{ord}^{\Sigma}_{(\lambda_*,\mu_*)}(\lambda) = 2$, we thus obtain $\operatorname{ord}^{\Sigma}_{(\lambda_*,\mu_*)}(c) = 2$ and $\operatorname{ord}^{\Sigma}_{(\lambda_*,\mu_*)}(\mu - d) \geq 1$. Because $\frac{\mu-d}{c}$ has a pole at (λ_*, μ_*), the only possibility is in fact $\operatorname{ord}^{\Sigma}_{(\lambda_*,\mu_*)}(\mu - d) = 1$. Thus the pole order m of $\frac{\mu-d}{c}$ equals $2 - 1 = 1$, and hence $\operatorname{ord}^{\mathbb{C}}_{\lambda_*}(c) = m$ holds. □

The preceding proposition shows in particular that if the support of D is contained in the set of regular points of Σ, then we have

$$D = \{\, (\lambda, \mu)^m \in \mathbb{C}^* \times \mathbb{C} \mid c(\lambda) = 0, \mu = a(\lambda), m = \operatorname{ord}^{\mathbb{C}}_{\lambda}(c) \,\}. \tag{3.14}$$

However, in general it is possible for poles of $\frac{\mu-d}{c}$ to lie in singularities of Σ, and in these points, the spectral divisor D as defined above does not contain enough information to completely characterize the behavior of the monodromy $M(\lambda)$. To handle this case, we need to generalize the concept of a divisor in an appropriate way such that the necessary additional information at the singularities of Σ at which $\frac{\mu-d}{c}$ is not holomorphic is included. It turns out that the most suitable generalization of the concept of a divisor for the present problem has been introduced by Hartshorne in [Har-2, §1]. We now describe Hartshorne's concept of a generalized divisor (which he introduced for general Gorenstein curves in [Har-2], and later in even more generality) in our setting, i.e. on a hyperelliptic complex plane curve Σ. Compare also [Kl-L-S-S, Section 3].

As before, we denote by $\mathscr{O}$ resp. by $\mathscr{M}$ the sheaf of holomorphic functions resp. of meromorphic functions on Σ. A *generalized divisor* is a subsheaf $\mathscr{D}$ of $\mathscr{M}$ that is finitely generated over $\mathscr{O}$, and we say that $\mathscr{D}$ is *positive* if $\mathscr{O} \subset \mathscr{D}$ holds. For a positive generalized divisor $\mathscr{D}$, the *support* of $\mathscr{D}$ is the set of points $(\lambda, \mu) \in \Sigma$ for which $\mathscr{D}_{(\lambda,\mu)} \neq \mathscr{O}_{(\lambda,\mu)}$ holds. For a positive generalized divisor $\mathscr{D}$, the map $\Sigma \to \mathbb{N}_0$, $(\lambda, \mu) \mapsto \dim\left(\mathscr{D}_{(\lambda,\mu)}/\mathscr{O}_{(\lambda,\mu)}\right)$ defines a divisor on Σ in the usual sense, which we call the *underlying classical divisor* of $\mathscr{D}$.

Definition 3.2 The *(generalized) spectral divisor* of the monodromy

$$M(\lambda) = \begin{pmatrix} a(\lambda) & b(\lambda) \\ c(\lambda) & d(\lambda) \end{pmatrix}$$

is the subsheaf $\mathscr{D}$ of $\mathscr{M}$ on Σ generated by the meromorphic functions 1 and $\frac{\mu-d}{c}$ over $\mathscr{O}$.

The generalized spectral divisor $\mathscr{D}$ is the proper replacement for the pole divisor of $\frac{\mu-d}{c}$ in the classical sense, which we preliminarily regarded as spectral divisor above. It is obviously positive, and at regular points of Σ, $\mathscr{D}$ is equivalent to that pole divisor. It should be noted that it is possible for a singular point (λ_*, μ_*) to be in the support of $\mathscr{D}$ even if $\frac{\mu-d}{c}$ does not have a pole at this point; this happens if $\frac{\mu-d}{c}$ is locally bounded (and therefore holomorphic on the normalization $\widehat{\Sigma}$) at (λ_*, μ_*), but not holomorphic on Σ at this point. Therefore $\frac{\mu-d}{c}$ is not necessarily a function of maximal pole order in $\mathscr{D}_{(\lambda_*,\mu_*)}$. From now on, the *classical spectral divisor* D is always the underlying classical divisor of $\mathscr{D}$ (rather than the pole divisor of $\frac{\mu-d}{c}$ in the classical sense). We will see below that D is given by Eq. (3.14), now without the restriction to regular points.

We introduced the generalized spectral divisor because it conveys more information than a classical divisor at the singular points of Σ. We will now study in more detail exactly which additional information is conveyed. The following proposition gives the key for understanding the structure of generalized divisors on the hyperelliptic complex curve Σ in general. The sheaf of rings $\mathscr{R}$ that is constructed there for a generalized divisor $\mathscr{D}$ is the sheaf of holomorphic functions (structure sheaf) of a partial normalisation of Σ. This situation has been investigated in [Kl-L-S-S, Section 4], where this partial normalisation was called the *middleding* of $(\Sigma, \mathscr{D})$.

Proposition 3.3 *Any generalized divisor $\mathscr{D}$ on Σ is locally free over a unique subsheaf of rings $\mathscr{R}$ of $\widehat{\mathscr{O}}$.*

More specifically, if (λ_*, μ_*) is a singular point of Σ, and $f \in \mathscr{D}_{(\lambda_*,\mu_*)}$ is of maximal pole order, let $j_0 \geq 0$ be the largest integer so that $\frac{\mu-\mu^{-1}}{(\lambda-\lambda_*)^{j_0}} \cdot f \in \mathscr{D}_{(\lambda_*,\mu_*)}$ holds, and let $\mathscr{R}_{(\lambda_*,\mu_*)}$ be the subring of $\widehat{\mathscr{O}}_{(\lambda_*,\mu_*)}$ generated by $\frac{\mu-\mu^{-1}}{(\lambda-\lambda_*)^{j_0}}$ and $\mathscr{O}_{(\lambda_*,\mu_*)}$. Then $\mathscr{D}_{(\lambda_*,\mu_*)}$ is generated by f over $\mathscr{R}_{(\lambda_*,\mu_*)}$, or by f and $\frac{\mu-\mu^{-1}}{(\lambda-\lambda_*)^{j_0}} \cdot f$ over $\mathscr{O}_{(\lambda_*,\mu_*)}$. Here we have $j_0 \in \{0, \dots, n\}$, where the order of the singularity (λ_*, μ_*) is either $2n$ or $2n+1$.

Note that in this proposition, the case $j_0 = 0$ is equivalent to $\mathscr{R}_{(\lambda_*,\mu_*)} = \mathscr{O}_{(\lambda_*,\mu_*)}$; in this case $\mathscr{D}$ is at (λ_*, μ_*) a locally free divisor on Σ. The case $j_0 = n$ is equivalent to $\mathscr{R}_{(\lambda_*,\mu_*)} = \widehat{\mathscr{O}}_{(\lambda_*,\mu_*)}$; in this case $\mathscr{D}$ corresponds at (λ_*, μ_*) to a locally free divisor on the normalization $\widehat{\Sigma}$ of Σ. Moreover, if the maximal pole order of function germs in $\mathscr{D}_{(\lambda_*,\mu_*)}$ is 0 (then the germs in $\mathscr{D}_{(\lambda_*,\mu_*)}$ are locally bounded), then 1 is a function of maximal pole order in $\mathscr{D}_{(\lambda_*,\mu_*)}$ and therefore $\mathscr{R}_{(\lambda_*,\mu_*)} = \mathscr{D}_{(\lambda_*,\mu_*)}$ holds.

Remark 3.4 As noted in the Introduction, Hitchin uses in his classification of the minimal tori in S^3 in [Hi] a certain partial desingularization of Σ as spectral curve; his spectral curve is precisely the complex curve $\widetilde{\Sigma}$ "between" Σ and its normalization $\widehat{\Sigma}$ whose sheaf of holomorphic functions is the sheaf of rings $\mathscr{R}$ from Proposition 3.3.

We also note that for the validity of Proposition 3.3, the fact that Σ is hyperelliptic is of crucial importance. For every $m \geq 3$, there are examples of complex curves which are m-fold (branched) coverings above $\mathbb{P}^1$ and on which there exist generalized divisors that are not locally free. An example for $m = 3$ is described in [Sch, Example 9.3, p. 93f.] and in [Kl-L-S-S, Example 4.7].

Proof (of Proposition 3.3) We need to consider the situation only at singularities of Σ. Let $(\lambda_*, \mu_*) \in \Sigma$ be a singular point of Σ, say of order $2n+1$ or $2n$ with $n \geq 1$. In what follows, we consider the stalks at (λ_*, μ_*) of the sheaves under consideration, and omit the subscript $_{(\lambda_*,\mu_*)}$ for them. We suppose that (λ_*, μ_*) is in the support of $\mathscr{D}$, i.e. that $\mathscr{D}/\mathscr{O} \neq \{0\}$ holds.

We now choose $f \in \mathscr{D}$ such that its pole order (as defined above) is maximal among the elements of $\mathscr{D}$, and let $\widehat{\mathscr{D}}$ resp. $\widetilde{\mathscr{D}}$ be the generalized divisor generated by f over $\widehat{\mathscr{O}}$ resp. over $\mathscr{O}$. Because $\mathscr{D}$ is an $\mathscr{O}$-module, we have $\widetilde{\mathscr{D}} \subset \mathscr{D}$. We also have $\mathscr{D} \subset \widehat{\mathscr{D}}$ (in the case where (λ_*, μ_*) is a singularity of even order, both components f_ν of the representant (f_1, f_2) of f have maximal pole order); from this inclusion it also follows that $\widehat{\mathscr{D}}$ is the generalized divisor generated by $\mathscr{D}$ over $\widehat{\mathscr{O}}$.

Because of $\dim(\widehat{\mathscr{O}}/\mathscr{O}) = n$, we have $\dim(\widehat{\mathscr{D}}/\widetilde{\mathscr{D}}) = n$. Moreover, $(\varphi_j)_{j=1,\dots,n}$ with $\varphi_j := \frac{\mu-\mu^{-1}}{(\lambda-\lambda_*)^j}$ is a basis of $\widehat{\mathscr{O}}/\mathscr{O}$, and therefore $(f_j)_{j=1,\dots,n}$ with $f_j := \varphi_j \cdot f$ is a basis of $\widehat{\mathscr{D}}/\widetilde{\mathscr{D}}$.

For $g \in \widehat{\mathscr{D}}/\widetilde{\mathscr{D}}$ written as $g = \sum_{j=1}^n t_j f_j$ with $t_j \in \mathbb{C}$, we let $j_0(g) \in \{0, \dots, n\}$ be the largest $j_0(g) \geq 1$ so that $t_{j_0(g)} \neq 0$, or $j_0(g) = 0$ for $g = 0$. Further we let $j_0 \in \{0, \dots, n\}$ be the largest value of $j_0(g)$ that occurs for any $g \in \mathscr{D}/\widetilde{\mathscr{D}}$. Then we claim that $(f_j)_{j=1,\dots,j_0}$ is a basis of $\mathscr{D}/\widetilde{\mathscr{D}}$ (with $\mathscr{D}/\widetilde{\mathscr{D}} = \{0\}$ if $j_0 = 0$). It only needs to be shown that $f_j \in \mathscr{D}/\widetilde{\mathscr{D}}$ holds for $j \in \{1, \dots, j_0\}$, and to show this, we let $g = \sum_{j=1}^n t_j f_j \in \mathscr{D}/\widetilde{\mathscr{D}}$ be with $t_{j_0} \neq 0$ and $t_j = 0$ for all $j > j_0$. Then $g \cdot (\lambda - \lambda_0)^{j_0 - 1} = t_{j_0} f_1 \in \mathscr{D}/\widetilde{\mathscr{D}}$ and therefore $f_1 \in \mathscr{D}/\widetilde{\mathscr{D}}$ holds. Now suppose that we have already shown that $f_1, \dots, f_{j_1-1} \in \mathscr{D}/\widetilde{\mathscr{D}}$ holds for some $j_1 \in \{2, \dots, j_0\}$. Then we also have $(g - \sum_{j=1}^{j_1-1} t_j f_j) \cdot (\lambda - \lambda_*)^{j_0 - j_1} = t_{j_0} f_{j_1} \in \mathscr{D}/\widetilde{\mathscr{D}}$ and thus $f_{j_1} \in \mathscr{D}/\widetilde{\mathscr{D}}$.

Therefore $\mathscr{D}$ is locally free at (λ_*, μ_*); it is generated by f over the subring $\mathscr{R}$ of $\widehat{\mathscr{O}}$ generated by $\frac{\mu-\mu^{-1}}{(\lambda-\lambda_k)^{j_0}} \in \widehat{\mathscr{O}}$ and $\mathscr{O}$. □

Proposition 3.5 *Let $\mathscr{D}$ be a positive generalized divisor on Σ and (λ_*, μ_*) be a singular point of Σ that is in the support of $\mathscr{D}$. We let $m := \dim(\mathscr{D}_{(\lambda_*,\mu_*)}/\mathscr{O}_{(\lambda_*,\mu_*)})$ be the degree of $\mathscr{D}$ at (λ_*, μ_*), $s \geq 0$ be the maximal pole order that occurs for a germ in $\mathscr{D}_{(\lambda_*,\mu_*)}$, and j_0 be the number from Proposition 3.3. Then we have $m = s + j_0$.*

Proof As before, we work in the stalks of sheaves at (λ_*, μ_*) and omit the subscript $_{(\lambda_*,\mu_*)}$.

Like in the proof of Proposition 3.3 we let $\widehat{\mathscr{D}}$ be the generalized divisor generated by $\mathscr{D}$ over $\widehat{\mathscr{O}}$. Because $\mathscr{D}$ and $\widehat{\mathscr{D}}$ are positive, we then have the inclusions $\mathscr{O} \subset \mathscr{D} \subset \widehat{\mathscr{D}}$ and $\mathscr{O} \subset \widehat{\mathscr{O}} \subset \widehat{\mathscr{D}}$, which imply the short exact sequences

$$\begin{aligned} & 0 \longrightarrow \mathscr{D}/\mathscr{O} \longrightarrow \widehat{\mathscr{D}}/\mathscr{O} \longrightarrow \widehat{\mathscr{D}}/\mathscr{D} \longrightarrow 0 \\ \text{and} \quad & 0 \longrightarrow \widehat{\mathscr{O}}/\mathscr{O} \longrightarrow \widehat{\mathscr{D}}/\mathscr{O} \longrightarrow \widehat{\mathscr{D}}/\widehat{\mathscr{O}} \longrightarrow 0 , \end{aligned}$$

and thereby the equations

$$\begin{aligned} & \dim(\mathscr{D}/\mathscr{O}) - \dim(\widehat{\mathscr{D}}/\mathscr{O}) + \dim(\widehat{\mathscr{D}}/\mathscr{D}) = 0 \\ \text{and} \quad & \dim(\widehat{\mathscr{O}}/\mathscr{O}) - \dim(\widehat{\mathscr{D}}/\mathscr{O}) + \dim(\widehat{\mathscr{D}}/\widehat{\mathscr{O}}) = 0 , \end{aligned}$$

whence

$$m = \dim(\mathscr{D}/\mathscr{O}) = \dim(\widehat{\mathscr{O}}/\mathscr{O}) + \dim(\widehat{\mathscr{D}}/\widehat{\mathscr{O}}) - \dim(\widehat{\mathscr{D}}/\mathscr{D})$$

follows. We have $\dim(\widehat{\mathscr{O}}/\mathscr{O}) = n$ (this is the δ-invariant of the singularity (λ_*, μ_*) of Σ), $\dim(\widehat{\mathscr{D}}/\widehat{\mathscr{O}}) = s$ and $\dim(\widehat{\mathscr{D}}/\mathscr{D}) = n - j_0$ (compare the proof of Proposition 3.3). Thus we obtain

$$m = n + s - (n - j_0) = s + j_0.$$

□

The following two propositions give a "geometric" interpretation of the data $\mathscr{R}$ and j_0 introduced by Proposition 3.3 in the case where $\mathscr{D}$ is the spectral divisor of a monodromy $M(\lambda)$. In the terminology of [Kl-L-S-S, Section 4] we will show that the middleding of $(\Sigma, \mathscr{D})$, i.e. the partial normalisation of Σ with sheaf of holomorphic functions $\mathscr{R}$, is equal to the middleding of the holomorphic matrix $M(\lambda)$.

We let $\mathscr{O}_M$ be the sheaf of (2×2)-matrices of holomorphic functions in λ which commute with the monodromy $M(\lambda)$ for every $\lambda \in \mathbb{C}^*$. We say that a meromorphic function φ on Σ is an *eigenvalue* of $\mathscr{O}_M$, if there exists locally a section $N(\lambda)$ of $\mathscr{O}_M$ so that φ is an eigenvalue of the matrix $N(\lambda)$, and we denote by $\mathscr{R}_M$ the sheaf of eigenvalues of $\mathscr{O}_M$. It is clear that $\lambda \cdot \mathbb{1}$ and $M(\lambda)$ are global sections in $\mathscr{O}_M$, and therefore the functions λ and μ are eigenvalues of $\mathscr{O}_M$. Hence we have $\mathscr{O} \subset \mathscr{R}_M$.

Conversely, if $N(\lambda)$ is a (2×2)-matrix of holomorphic functions in λ that commutes with $M(\lambda)$, then $N(\lambda)$ acts on the eigenvector $\binom{\mu-d}{c}$ of $M(\lambda)$ by

$$N \cdot \begin{pmatrix} \mu - d \\ c \end{pmatrix} = \varphi \cdot \begin{pmatrix} \mu - d \\ c \end{pmatrix}$$

with a meromorphic function φ in λ, and in this situation φ is a section in $\mathscr{R}_M$.

Proposition 3.6 *Let $\mathscr{D}$ be the generalized spectral divisor of the monodromy $M(\lambda)$ and $(\lambda_*, \mu_*) \in \Sigma$, and let $\mathscr{R}_{(\lambda_*,\mu_*)}$ and j_0 be the data associated to $\mathscr{D}_{(\lambda_*,\mu_*)}$ in Proposition 3.3. Then we have $\mathscr{R}_{(\lambda_*,\mu_*)} = \mathscr{R}_{M,(\lambda_*,\mu_*)}$, and j_0 is the order of the zero of $M(\lambda) - \frac{1}{2}\Delta(\lambda)\,\mathbb{1}$ at λ_* (defined as the minimum of the order of the zeros of the entries of this (2×2)-matrix).*

Proof As before, we omit the subscript $_{(\lambda_*,\mu_*)}$.

By Proposition 3.3, $\mathscr{R}$ is uniquely characterized as the subring of $\widehat{\mathscr{O}}$ over which $\mathscr{D}$ is locally free. Because of this uniqueness property, $\mathscr{R}$ is the largest subring of $\widehat{\mathscr{O}}$ that acts on $\mathscr{D}$. To show that $\mathscr{R} = \mathscr{R}_M$ holds, it therefore suffices to show that $\mathscr{R}_M$ acts on $\mathscr{D}$, and that for any meromorphic, locally bounded function φ with $\varphi \cdot \mathscr{D} \subset \mathscr{D}$, we have $\varphi \in \mathscr{R}_M$.

We first show that $\mathscr{R}_M$ acts on $\mathscr{D}$. Let $\varphi \in \mathscr{R}_M$ be given. By definition, there exists $N \in \mathscr{O}_M$ so that $\varphi \begin{pmatrix} f_1 \\ f_2 \end{pmatrix} = N \cdot \begin{pmatrix} f_1 \\ f_2 \end{pmatrix}$ holds, where $f_1 := \frac{\mu - d}{c}$ and $f_2 := 1$ are the two generators of $\mathscr{D}$ over $\mathscr{O}$. Because $\mathscr{D}$ is a $\mathscr{O}$-module, and the entries of N are in $\mathscr{O}$, it follows that $\varphi\, f_1, \varphi\, f_2 \in \mathscr{D}$ holds. Because f_1, f_2 generate $\mathscr{D}$, we in fact have $\varphi \cdot \mathscr{D} \subset \mathscr{D}$.

We now show that $\mathscr{R}_M$ is the largest subring of $\widehat{\mathscr{O}}$ that acts on $\mathscr{D}$. For this purpose, let φ be a meromorphic, locally bounded function with $\varphi \cdot \mathscr{D} \subset \mathscr{D}$. We will show that $\varphi \in \mathscr{R}_M$ holds. We decompose φ into its symmetric and its anti-symmetric part, i.e. we write

$$\varphi(\lambda, \mu) = \varphi_+(\lambda) + \varphi_-(\lambda) \cdot (\mu - \mu^{-1})$$

with meromorphic functions φ_+ and φ_- in λ. The symmetric part $\varphi_+(\lambda)$ is meromorphic in λ and locally bounded, and therefore holomorphic in λ. Because of $\mathscr{O} \subset \mathscr{R}_M$, we thus have $\varphi_+ \in \mathscr{R}_M$. It therefore remains to show that $\varphi_-(\lambda) \cdot (\mu - \mu^{-1}) \in \mathscr{R}_M$ holds, and for this, it suffices to show that the entries of the (2×2)-matrix $N(\lambda) := \varphi_-(\lambda) \cdot (M(\lambda) - M(\lambda)^{-1})$ are holomorphic in λ, i.e. that $N(\lambda)$ does not have a pole.

Because the vectors $\begin{pmatrix} f_1(\lambda,\mu) \\ f_2(\lambda,\mu) \end{pmatrix} = \begin{pmatrix} \frac{\mu - d(\lambda)}{c(\lambda)} \\ 1 \end{pmatrix}$ and $\begin{pmatrix} f_1(\lambda,\mu^{-1}) \\ f_2(\lambda,\mu^{-1}) \end{pmatrix} = \begin{pmatrix} \frac{\mu^{-1} - d(\lambda)}{c(\lambda)} \\ 1 \end{pmatrix}$ are linear independent in $\mathbb{C}^2$ for any $(\lambda, \mu) \in \Sigma$ with $\mu \neq \mu^{-1}$ and $c(\lambda) \neq 0$, any meromorphic function h can uniquely be represented in the form

$$h(\lambda, \mu) = h_1(\lambda) \cdot f_1(\lambda, \mu) + h_2(\lambda) \cdot f_2(\lambda, \mu) \tag{3.15}$$

with meromorphic functions h_1, h_2. In this situation we have $h \in \mathscr{D}$ if and only if h_1 and h_2 are holomorphic. (Indeed, if h_1, h_2 are holomorphic, then we have $h \in \mathscr{D}$ simply because $\mathscr{D}$ is an $\mathscr{O}$-module generated by f_1 and f_2. Conversely, if $h \in \mathscr{D}$ holds, we have a representation $h = \widetilde{h}_1\, f_1 + \widetilde{h}_2\, f_2$ with holomorphic $\widetilde{h}_1, \widetilde{h}_2$ because $\mathscr{D}$ is generated by f_1 and f_2; and because of the uniqueness of the representation (3.15), we see that $h_1 = \widetilde{h}_1$, $h_2 = \widetilde{h}_2$ are holomorphic.)

Now we have

$$N(\lambda) \cdot \begin{pmatrix} f_1 \\ f_2 \end{pmatrix} = \varphi_-(\lambda) \cdot (\mu - \mu^{-1}) \cdot \begin{pmatrix} f_1 \\ f_2 \end{pmatrix} , \tag{3.16}$$

where $(\mu - \mu^{-1}) \begin{pmatrix} f_1 \\ f_2 \end{pmatrix} \in \mathscr{D}$ holds. Because of the hypothesis $\varphi_- \cdot \mathscr{D} \subset \mathscr{D}$, it follows that the components of the vector in (3.16) are in $\mathscr{D}$, whence it follows by the preceding statement that the entries of $N(\lambda)$ are holomorphic.

This completes the proof that $\mathscr{R} = \mathscr{R}_M$ holds.

$j_0 \geq 0$ is by definition the largest number so that $\frac{\mu-\mu^{-1}}{(\lambda-\lambda_*)^{j_0}} \in \mathscr{R}_M$ holds. For any $j \geq 0$, we have $\frac{\mu-\mu^{-1}}{(\lambda-\lambda_*)^{j}} \in \mathscr{R}_M$ if and only if $\frac{1}{(\lambda-\lambda_*)^{j}} \cdot (M(\lambda) - M(\lambda)^{-1})$ is holomorphic and therefore a section in $\mathscr{O}_M$. This is the case if and only if the entries of $M(\lambda) - M(\lambda)^{-1}$ all have zeros of multiplicity at least j at λ_*. Thus j_0 is the multiplicity of the zero of $M(\lambda) - M(\lambda)^{-1}$ at λ_*.

Because of $\det(M(\lambda)) = 1$, we have

$$M(\lambda) - M(\lambda)^{-1} = \begin{pmatrix} a(\lambda) - d(\lambda) & 2b(\lambda) \\ 2c(\lambda) & d(\lambda) - a(\lambda) \end{pmatrix} = 2 \cdot \left(M(\lambda) - \tfrac{1}{2}\, \Delta(\lambda)\, \mathbb{1} \right) ,$$

and therefore j_0 is the multiplicity of the zero of $M(\lambda) - \frac{1}{2}\,\Delta(\lambda)\,\mathbb{1}$ at λ_*. □

The following proposition describes further properties of the spectral divisors belonging to monodromies of the kind studied here. Proposition 3.7(1) shows that the underlying classical divisor D of the generalized spectral divisor $\mathscr{D}$ of a monodromy $M(\lambda)$ is given by Eq. (3.14), now without the restriction that the support of D be contained in the regular points of Σ. We will use the property of Proposition 3.7(2) in Chap. 12 as one of several properties that characterize the spectral divisors among all the generalized divisors on Σ.

Proposition 3.7 *Let $\mathscr{D}$ be the generalized spectral divisor of the monodromy $M(\lambda) = \begin{pmatrix} a(\lambda) & b(\lambda) \\ c(\lambda) & d(\lambda) \end{pmatrix}$ and $(\lambda_*, \mu_*) \in \Sigma$ a singular point in the support of $\mathscr{D}$. Let $m := \dim(\mathscr{D}_{(\lambda_*,\mu_*)}/\mathscr{O}_{(\lambda_*,\mu_*)})$ be the degree of (λ_*, μ_*) in $\mathscr{D}$.*

(1) $m = \mathrm{ord}^{\mathbb{C}}(c)$.

(2) There exists $g \in \mathscr{D}_{(\lambda_*,\mu_*)}$ so that $\eta := g - \frac{\mu-\mu^{-1}}{(\lambda-\lambda_*)^m}$ is a meromorphic function germ solely in λ, with $\mathrm{polord}^{\Sigma}(\eta) \leq \mathrm{polord}^{\Sigma}(g)$.

(3) If $m = 1$, then we have $j_0 = 1$ for the number j_0 from Proposition 3.3, and every germ in $\mathscr{D}_{(\lambda_*,\mu_*)}$ is locally bounded (i.e. we have $s = 0$ for the maximal pole order occurring for germs in $\mathscr{D}_{(\lambda_*,\mu_*)}$).

Proof As before, we omit the subscript $_{(\lambda_*,\mu_*)}$.

For (1). Let us abbreviate $\ell := \operatorname{ord}^{\mathbb{C}}(c)$, then we have $c = \gamma\cdot(\lambda-\lambda_*)^\ell$ with some invertible $\gamma \in \mathscr{O}$. Because $\mathscr{D}$ is generated by 1 and $\frac{\mu-d}{c}$ over $\mathscr{O}$, it is therefore also generated by 1 and $\frac{\mu-d}{(\lambda-\lambda_*)^\ell}$ over $\mathscr{O}$. Thus $\mathscr{D}/\mathscr{O}$ is spanned as a linear space by $\frac{\mu-d}{(\lambda-\lambda_*)^j}$ with $j = 1, \dots, \ell$. In fact, $\left(\frac{\mu-d}{(\lambda-\lambda_*)^j}\right)_{j=1,\dots,\ell}$ is a basis of $\mathscr{D}/\mathscr{O}$: Otherwise there would exist a non-trivial linear combination of $\left(\frac{\mu-d}{(\lambda-\lambda_*)^j}\right)_{j=1,\dots,\ell}$ that is a member of $\mathscr{O}$; by multiplying this linear combination with an appropriate power of $(\lambda-\lambda_*)$, we would obtain $\frac{\mu-d}{\lambda-\lambda_*} \in \mathscr{O}$. Hence the anti-symmetric part of $\frac{\mu-d}{c}$, which equals $\frac{\mu-\mu^{-1}}{2\,(\lambda-\lambda_*)}$, would also be a member of $\mathscr{O}$. But this is a contradiction to the hypothesis that (λ_*,μ_*) is a singular point of Σ. Therefore we have $m = \dim(\mathscr{D}/\mathscr{O}) = \ell$.

For (2). We again write $c = \gamma\cdot(\lambda-\lambda_*)^m$, then we have $\frac{1}{\gamma}\cdot\frac{\mu-d}{(\lambda-\lambda_*)^m} = \frac{\mu-d}{c} \in \mathscr{D}$ and therefore also $g := 2\cdot\frac{\mu-d}{(\lambda-\lambda_*)^m} \in \mathscr{D}$. With this choice of g,

$$\eta := g - \frac{\mu-\mu^{-1}}{(\lambda-\lambda_*)^m} = \frac{2\cdot(\mu-d)-(\mu-\mu^{-1})}{(\lambda-\lambda_*)^m} = \frac{\mu+\mu^{-1}-2d}{(\lambda-\lambda_*)^m} = \frac{\Delta-2d}{(\lambda-\lambda_*)^m}$$

is a meromorphic function in λ.

The decomposition of g into its even and odd parts is

$$g = \frac{\Delta-2d}{(\lambda-\lambda_*)^m} + \frac{\mu-\mu^{-1}}{(\lambda-\lambda_*)^m},$$

and therefore we have

$$\begin{aligned}\operatorname{polord}^\Sigma(g) &= \max\left\{\operatorname{polord}^\Sigma\left(\frac{\Delta-2d}{(\lambda-\lambda_*)^m}\right), \operatorname{polord}^\Sigma\left(\frac{\mu-\mu^{-1}}{(\lambda-\lambda_*)^m}\right)\right\}\\ &= \max\left\{\operatorname{polord}^\Sigma(\eta), 2m-\widehat{n}\right\} \geq \operatorname{polord}^\Sigma(\eta),\end{aligned}$$

where $\widehat{n}$ denotes the order of the singularity at (λ_*,μ_*), i.e. the multiplicity of the zero of Δ^2-4 at λ_*.

For (3). By Proposition 3.5 we have $s + j_0 = m = 1$ and therefore either $s = 0, j_0 = 1$ or $s = 1, j_0 = 0$. Assume $s = 1, j_0 = 0$. Because of $s > 0$, the function $\frac{\mu-d}{c}$ is then of maximal pole order in $\mathscr{D}$ and thus has a pole of order $s = 1$ at (λ_*,μ_*). On the other hand we have

$$\frac{\mu-d}{c} = \frac{\Delta-2d}{2c} + \frac{\mu-\mu^{-1}}{2c}.$$

Here it follows from (1) that $\operatorname{ord}^{\mathbb{C}}(c) = m = 1$ and therefore $\operatorname{ord}^\Sigma(c) \leq 2$ holds. On the other hand, $\operatorname{ord}^\Sigma(\mu-\mu^{-1}) \geq 2$ holds because (λ_*,μ_*) is a singular point of Σ, and therefore $\frac{\mu-\mu^{-1}}{2c}$ is locally bounded near (λ_*, μ_*). Moreover we have $\Delta(\lambda_*) = \pm 2 = 2d(\lambda_*)$, hence $\Delta - 2d$ has a zero at λ_*, and therefore $\frac{\Delta-2d}{2c}$ is also locally bounded near (λ_*, μ_*). It follows that $\frac{\mu-d}{c}$ is locally bounded near (λ_*, μ_*), which is a contradiction. Thus we have $s = 0$, $j_0 = 1$. □

Part II
The Asymptotic Behavior of the Spectral Data

Chapter 4
The Vacuum Solution

The most obvious solution of the sinh-Gordon equation $\Delta u + \sinh(u) = 0$ is the "vacuum" $u = 0$, corresponding to a minimal surface of zero sectional curvature, i.e. to a minimal cylinder. It is remarkable that the monodromy and the spectral data can be calculated explicitly for this solution, which we will do in the present chapter. The vacuum solution is of particular importance to us because in the coming chapters, we will describe the asymptotic behavior of the spectral data in the general situation by comparing these data to the corresponding spectral data of the vacuum; it will turn out that the spectral data of any given solution is closely approximated by the spectral data of the vacuum for $\lambda \to \infty$ and for $\lambda \to 0$.

By plugging $u = 0$ into Eq. (3.2) we see that the connection form $\alpha_0 := \alpha_{u=0}$ corresponding to the vacuum is given by

$$\alpha_0 = \frac{1}{4}\begin{pmatrix} 0 & -(1+\lambda^{-1}) \\ 1+\lambda & 0 \end{pmatrix} \mathrm{d}x + \frac{\mathrm{i}}{4}\begin{pmatrix} 0 & 1-\lambda^{-1} \\ 1-\lambda & 0 \end{pmatrix} \mathrm{d}y. \tag{4.1}$$

Because α_0 does not depend on the point $z = x + \mathrm{i}y$, and its $\mathrm{d}x$-component and its $\mathrm{d}y$-component

$$A := \frac{1}{4}\begin{pmatrix} 0 & -(1+\lambda^{-1}) \\ 1+\lambda & 0 \end{pmatrix} \quad \text{resp.} \quad \widetilde{A} := \frac{\mathrm{i}}{4}\begin{pmatrix} 0 & 1-\lambda^{-1} \\ 1-\lambda & 0 \end{pmatrix}$$

commute, we can compute the extended frame $F_0 = F_0(z, \lambda)$ corresponding to the vacuum simply by

$$F_0(x + \mathrm{i}y, \lambda) = \exp(xA) \cdot \exp(y\widetilde{A}).$$

S. Klein, *A Spectral Theory for Simply Periodic Solutions of the Sinh-Gordon Equation*, Lecture Notes in Mathematics 2229,
https://doi.org/10.1007/978-3-030-01276-2_4

We carry out this computation explicitly, using

$$A^2 = -\frac{1}{16}(1+\lambda)(1+\lambda^{-1})\,\mathbb{1} = -\zeta(\lambda)^2\cdot\mathbb{1} \quad \text{with} \quad \zeta(\lambda) := \frac{1}{4}\left(\lambda^{1/2}+\lambda^{-1/2}\right) \tag{4.2}$$

and

$$\widetilde{A}^2 = -\frac{1}{16}(1-\lambda)(1-\lambda^{-1})\,\mathbb{1} = \widetilde{\zeta}(\lambda)^2\cdot\mathbb{1} \quad \text{with} \quad \widetilde{\zeta}(\lambda) := \frac{1}{4}\left(\lambda^{1/2}-\lambda^{-1/2}\right). \tag{4.3}$$

The function $\zeta(\lambda)$ that occurs here for the first time is a local coordinate for $\mathbb{C}^*$ provided that the branch of the square root function is chosen suitably. It will be extremely useful for this reason later on. The preceding formulas imply that we have for every $n \in \mathbb{N}_0$

$$A^{2n} = (-1)^n\,\zeta(\lambda)^{2n}\cdot\mathbb{1} \quad \text{and} \quad \widetilde{A}^{2n} = \widetilde{\zeta}(\lambda)^{2n}\cdot\mathbb{1}$$

and

$$A^{2n+1} = (-1)^n\,\zeta(\lambda)^{2n}\,A = (-1)^n\,\zeta(\lambda)^{2n+1}\left(\begin{smallmatrix}0 & -\lambda^{-1/2}\\ \lambda^{1/2} & 0\end{smallmatrix}\right)$$
$$\text{and} \quad \widetilde{A}^{2n+1} = \widetilde{\zeta}(\lambda)^{2n}\,\widetilde{A} = -\mathrm{i}\,\widetilde{\zeta}(\lambda)^{2n+1}\left(\begin{smallmatrix}0 & \lambda^{-1/2}\\ \lambda^{1/2} & 0\end{smallmatrix}\right),$$

and therefore

$$\begin{aligned}\exp(xA) &= \sum_{n=0}^{\infty}\frac{x^n}{n!}A^n\\ &= \sum_{n=0}^{\infty}\frac{x^{2n}}{(2n)!}(-1)^n\zeta(\lambda)^{2n}\mathbb{1}\\ &\quad + \sum_{n=0}^{\infty}\frac{x^{2n+1}}{(2n+1)!}(-1)^n\,\zeta(\lambda)^{2n+1}\left(\begin{smallmatrix}0 & -\lambda^{-1/2}\\ \lambda^{1/2} & 0\end{smallmatrix}\right)\\ &= \cos(x\,\zeta(\lambda))\cdot\mathbb{1} + \sin(x\,\zeta(\lambda))\cdot\left(\begin{smallmatrix}0 & -\lambda^{-1/2}\\ \lambda^{1/2} & 0\end{smallmatrix}\right)\end{aligned}$$

and

$$\exp(y\widetilde{A}) = \sum_{n=0}^{\infty} \frac{y^n}{n!}\widetilde{A}^n = \sum_{n=0}^{\infty} \frac{y^{2n}}{(2n)!}\widetilde{\zeta}(\lambda)^{2n}\mathbb{1}$$
$$- \mathrm{i}\sum_{n=0}^{\infty} \frac{y^{2n+1}}{(2n+1)!}\widetilde{\zeta}(\lambda)^{2n+1}\begin{pmatrix} 0 & \lambda^{-1/2} \\ \lambda^{1/2} & 0 \end{pmatrix}$$
$$= \cosh(y\,\widetilde{\zeta}(\lambda))\cdot\mathbb{1} - \mathrm{i}\,\sinh(y\,\widetilde{\zeta}(\lambda))\cdot\begin{pmatrix} 0 & \lambda^{-1/2} \\ \lambda^{1/2} & 0 \end{pmatrix}.$$

Thus we obtain

$$\begin{aligned} &F_0(x+\mathrm{i}y,\lambda) \\ &= \exp(xA)\cdot\exp(y\widetilde{A}) \\ &= \left(\cos(x\,\zeta(\lambda))\cdot\mathbb{1} + \sin(x\,\zeta(\lambda))\begin{pmatrix} 0 & -\lambda^{-1/2} \\ \lambda^{1/2} & 0 \end{pmatrix}\right) \\ &\quad\cdot\left(\cosh(y\,\widetilde{\zeta}(\lambda))\cdot\mathbb{1} - \mathrm{i}\,\sinh(y\,\widetilde{\zeta}(\lambda))\begin{pmatrix} 0 & \lambda^{-1/2} \\ \lambda^{1/2} & 0 \end{pmatrix}\right) \\ &= \begin{pmatrix} \cos\left(x\zeta(\lambda) - \mathrm{i}y\widetilde{\zeta}(\lambda)\right) & -\lambda^{-1/2}\sin\left(x\zeta(\lambda) + \mathrm{i}y\widetilde{\zeta}(\lambda)\right) \\ \lambda^{1/2}\cdot\sin\left(x\zeta(\lambda) - \mathrm{i}y\widetilde{\zeta}(\lambda)\right) & \cos\left(x\zeta(\lambda) + \mathrm{i}y\widetilde{\zeta}(\lambda)\right) \end{pmatrix}. \end{aligned} \tag{4.4}$$

Note that the entries of F_0 are even in $\lambda^{1/2}$, and therefore indeed define holomorphic functions in $\lambda \in \mathbb{C}^*$.

In particular, we have for the monodromy of the vacuum with respect to the base point $z_0 = 0$

$$M_0(\lambda) = F_0(1,\lambda) = \begin{pmatrix} \cos(\zeta(\lambda)) & -\lambda^{-1/2}\sin(\zeta(\lambda)) \\ \lambda^{1/2}\sin(\zeta(\lambda)) & \cos(\zeta(\lambda)) \end{pmatrix} =: \begin{pmatrix} a_0(\lambda) & b_0(\lambda) \\ c_0(\lambda) & d_0(\lambda) \end{pmatrix}. \tag{4.5}$$

We will use the names $a_0, \ldots, d_0$ for the component functions of the monodromy of the vacuum throughout the entire book without any further reference, and likewise we will use the definition

$$\varDelta_0(\lambda) := \mathrm{tr}(M_0(\lambda)) = 2\cos(\zeta(\lambda)). \tag{4.6}$$

As consequence of Eq. (3.8), the spectral curve of the vacuum is given by

$$\begin{aligned} \Sigma_0 &= \{\,(\lambda,\mu)\in\mathbb{C}^*\times\mathbb{C}\,\big|\,\mu = \tfrac{1}{2}\left(\varDelta_0(\lambda)\pm\sqrt{\varDelta_0(\lambda)^2-4}\right)\} \\ &= \{\,(\lambda,\mu)\in\mathbb{C}^*\times\mathbb{C}\,\big|\,\mu = \cos(\zeta(\lambda))\pm\sqrt{\cos(\zeta(\lambda))^2-1}\,\} \\ &= \{\,(\lambda,\mu)\in\mathbb{C}^*\times\mathbb{C}\,\big|\,\mu = \mathrm{e}^{\pm\mathrm{i}\zeta(\lambda)}\,\}. \end{aligned} \tag{4.7}$$

This curve has no branch points above $\mathbb{C}^*$. It has double points at all those $\lambda \in \mathbb{C}^*$ for which $\zeta(\lambda)$ is an integer multiple of π; these values of λ are exactly the following:

$$\lambda_{k,0} := 8\pi^2 k^2 + 4\pi k\sqrt{4\pi^2 k^2 - 1} - 1 \quad \text{with } k \in \mathbb{Z}. \tag{4.8}$$

Proof Let us choose for $\sqrt{\cdots}$ the standard branch of the square root function (so that we have $\mathrm{Re}(\sqrt{z}) \geq 0$ for all $z \in \mathbb{C}$). Then the equation $\zeta(\lambda) = k\pi$ with $k \in \mathbb{Z}$ has a solution only if $k \geq 0$. In this case we have $\zeta(\lambda) = \frac{1}{4}(\lambda^{1/2} + \lambda^{-1/2}) = k\pi$ if and only if $\lambda - 4k\pi\sqrt{\lambda} + 1 = 0$ holds. The two possible values of $\sqrt{\lambda}$ that correspond to this equation are $\sqrt{\lambda} = 2k\pi \pm \sqrt{4k^2\pi^2 - 1}$, and this yields

$$\lambda = \left(2k\pi \pm \sqrt{4k^2\pi^2 - 1}\right)^2 = 8\pi^2 k^2 \pm 4\pi k\sqrt{4\pi^2 k^2 - 1} - 1.$$

Therefore the entirety of solutions of $\zeta(\lambda) \in \mathbb{Z}\pi$ is given by (4.8), where now k again runs through all of $\mathbb{Z}$. □

We have $\lambda_{k,0} \in \mathbb{R}$ for all $k \in \mathbb{Z}$, and $|\lambda_{k,0}| > 1$ resp. $|\lambda_{k,0}| < 1$ if $k > 0$ resp. $k < 0$. Moreover, $\lambda_{k,0}$ tends to ∞ resp. to 0 for $k \to \infty$ resp. $k \to -\infty$. By using the Taylor expansion $\sqrt{1+x} = 1 + \frac{1}{2}x - \frac{1}{8}x^2 + \frac{1}{16}x^3 + O(x^4)$, we obtain the more specific asymptotic assessments:

$$\lambda_{k,0} = 16\pi^2 k^2 - 2 + O(k^{-2}) \text{ for } k \to \infty$$

$$\text{and} \quad \lambda_{k,0} = \frac{1}{16\pi^2}k^{-2} + \frac{1}{128\pi^4}k^{-4} + O(k^{-6}) \text{ for } k \to -\infty. \tag{4.9}$$

Figure 4.1 depicts the position of the zeros $\lambda_{k,0}$ in λ-coordinates and in ζ-coordinates. Note that for the depiction of $\zeta(\lambda_{k,0})$ we have applied to the calculation of the square root the following convention: We use the standard branch of the square root function ($\mathrm{Re}(\sqrt{z}) \geq 0$) when k is positive, but its negative ($\mathrm{Re}(\sqrt{z}) \leq 0$) when k is negative.

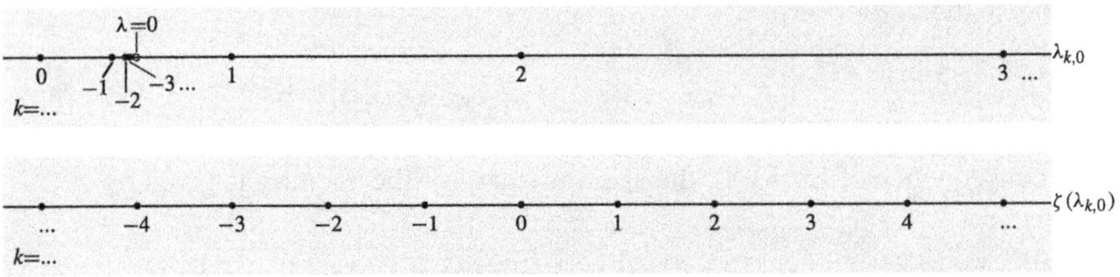

Fig. 4.1 Depiction of $\lambda_{k,0}$ and $\zeta(\lambda_{k,0})$. The numbers below the points indicate the corresponding value of k

We finally calculate the spectral divisor of the vacuum. The points $(\lambda, \mu) \in \Sigma_0$ of the classical spectral divisor are determined by the equations

$$c_0(\lambda) := \sqrt{\lambda} \cdot \sin(\zeta(\lambda)) = 0 \quad \text{and} \quad \mu = a_0(\lambda) := \cos(\zeta(\lambda)).$$

The equation $c_0(\lambda) = 0$ holds if and only if $\lambda = \lambda_{k,0}$ for some $k \in \mathbb{Z}$, all these zeros of c_0 are of simple multiplicity, and we have

$$\mu_{k,0} := a_0(\lambda_{k,0}) = \cos(k\pi) = (-1)^k. \tag{4.10}$$

Thus, the classical spectral divisor of the vacuum is given by the divisor on Σ_0

$$D_0 := \{ (\lambda_{k,0}, \mu_{k,0}) \mid k \in \mathbb{Z} \} = \{ (\lambda_{k,0}, (-1)^k) \mid k \in \mathbb{Z} \}. \tag{4.11}$$

Notice that for the vacuum, the support of the spectral divisor coincides with the set of double points of the spectral curve.

To determine the generalized spectral divisor $\mathscr{D}_0$ of the vacuum, we look at the associated data $\mathscr{R}$ and j_0 from Proposition 3.3. Because every point $(\lambda_{k,0}, \mu_{k,0})$ in the support of $\mathscr{D}_0$ is of degree $m = 1$, Proposition 3.7(3) shows that we have $j_0 = 1$, and thus j_0 equals the δ-invariant of that point in Σ. Hence we have $\mathscr{R}_{(\lambda_{k,0},\mu_{k,0})} = \widehat{\mathscr{O}}_{(\lambda_{k,0},\mu_{k,0})}$. By Proposition 3.7(3) the maximal pole order s occurring in $(\mathscr{D}_0)_{(\lambda_{k,0},\mu_{k,0})}$ is $s = 0$, and therefore $(\mathscr{D}_0)_{(\lambda_{k,0},\mu_{k,0})}$ is generated by 1 over $\mathscr{R}_{(\lambda_{k,0},\mu_{k,0})} = \widehat{\mathscr{O}}_{(\lambda_{k,0},\mu_{k,0})}$. Thus $(\mathscr{D}_0)_{(\lambda_{k,0},\mu_{k,0})} = (\widehat{\mathscr{O}}_0)_{(\lambda_{k,0},\mu_{k,0})}$ holds, where $\widehat{\mathscr{O}}_0$ is the direct image in Σ_0 of the sheaf of holomorphic functions on the normalization $\widehat{\Sigma}_0$ of Σ_0. Therefore we have $\mathscr{D}_0 = \widehat{\mathscr{O}}_0$.

Chapter 5
The Basic Asymptotic of the Monodromy

In the present chapter, we will prove basic asymptotic estimates for the monodromy. We will refine our asymptotic assessment in Chap. 7 and again in Chap. 11. The asymptotic estimate of Theorem 5.4 (resp. Proposition 5.9, where we require one more degree of differentiability) is analogous to the basic estimates of [Pö-T, Theorem 1.3, p. 13] in the treatment of the 1-dimensional Schrödinger equation.

Previously we considered simply periodic solutions u of the sinh-Gordon equation on horizontal strips X, compare Fig. 3.1. Beginning with this chapter, we suppose that for the potential (simply periodic solution of the sinh-Gordon equation) u, only periodic *Cauchy data* (u, u_y) on the segment $[0, 1]$ of the real line are given. Note that this segment could be in the interior, but just as well on the boundary of the original domain X, see Fig. 5.1. Such periodic Cauchy data correspond to the data for a Björling problem for minimal surfaces with respect to a closed curve. For the reasons explained in the Introduction, we want the requirements on the differentiability of u and u_y to be as relaxed as possible. Specifically, we only require that u is in the Sobolev-space of weakly once-differentiable functions with square-integrable derivative, i.e. $u \in W^{1,2}([0, 1])$ and that u_y is square-integrable, i.e. $u_y \in L^2([0, 1])$. Note that u is in particular continuous, so individual function values $u(x)$ of u are well-defined.

We regard u and u_y as being extended periodically to the real line, and we define "mixed derivatives" of u in the natural way by using both u and u_y, e.g. $u_z := \frac{1}{2}(u_x - \mathrm{i}\, u_y)$, where u_x is the Sobolev-derivative of u, and u_y is the given function.

We denote the space of such *non-periodic potentials* by $\mathsf{Pot}_{np} := \{ (u, u_y) \mid u \in W^{1,2}([0, 1]), u_y \in L^2([0, 1]) \}$. Via the norm

$$\|(u, u_y)\|_{\mathsf{Pot}} := \sqrt{\|u\|^2_{W^{1,2}} + \|u_y\|^2_{L^2}} \,,$$

Pot_{np} becomes a Banach space.

S. Klein, *A Spectral Theory for Simply Periodic Solutions of the Sinh-Gordon Equation*, Lecture Notes in Mathematics 2229,
https://doi.org/10.1007/978-3-030-01276-2_5

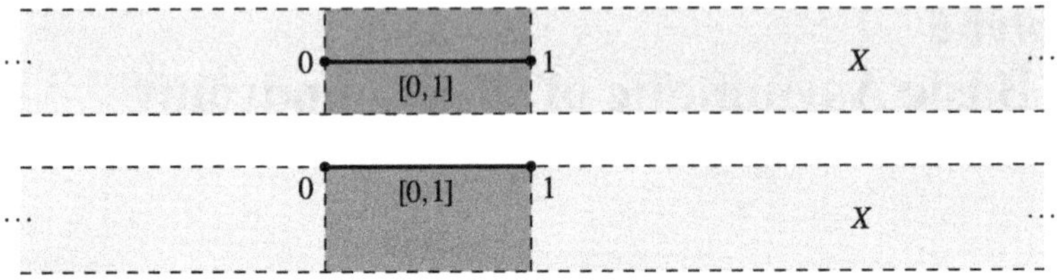

Fig. 5.1 Two possible positions for the domain of definition X of the original simply periodic solution u with respect to the interval $[0, 1]$, which is the domain of definition of our Cauchy data (u, u_y). The darkly shaded rectangle is a fundamental domain of u

Because our ultimate interest is with periodic solutions of the sinh-Gordon equation, we are of course most interested in *periodic potentials*. u and u_y are at first only defined on $[0, 1]$, so the condition of periodicity for u is simply $u(0) = u(1)$; this condition is well-defined because u is continuous. u_y is only square-integrable, so we cannot access individual function values of u_y, and for this reason we do not impose a similar condition of periodicity on u_y. We then regard u and u_y also as extended periodically to the real line. The space of *periodic potentials* is thus given by

$$\mathsf{Pot} := \{ (u, u_y) \in \mathsf{Pot}_{np} \,\big|\, u(0) = u(1) \}, \tag{5.1}$$

it is a complex hyperplane in Pot_{np}.

Also for such Cauchy data $(u, u_y) \in \mathsf{Pot}_{np}$ (instead of a solution u of the sinh-Gordon equation defined on a horizontal strip in $\mathbb{C}$), we can define the $\mathrm{d}x$-part of the flat connection 1-form α

$$\alpha = \frac{1}{4} \begin{pmatrix} \mathrm{i}\, u_y & -\mathrm{e}^{u/2} - \lambda^{-1}\,\mathrm{e}^{-u/2} \\ \mathrm{e}^{u/2} + \lambda\,\mathrm{e}^{-u/2} & -\mathrm{i}\, u_y \end{pmatrix} \mathrm{d}x. \tag{5.2}$$

Although α is in general only square-integrable and not locally Lipschitz continuous (as a consequence of our conditions of differentiability on (u, u_y)), the following Lemma 5.2 shows that we still have an extended frame $F_\lambda : \mathbb{R} \to \mathrm{SL}(2, \mathbb{C})$ along the real line associated to (u, u_y), i.e. a solution to the initial value problem

$$F_\lambda'(x) = \alpha_\lambda(x) \cdot F_\lambda(x) \quad \text{and} \quad F_\lambda(0) = \mathbb{1}. \tag{5.3}$$

From F_λ we obtain the monodromy at $x = 0$ as $M(\lambda) := F_\lambda(1)$. If (u, u_y) is periodic ($(u, u_y) \in \mathsf{Pot}$), we define spectral data (Σ, D) resp. $(\Sigma, \mathscr{D})$ for (u, u_y) exactly as in Chap. 3. If (u, u_y) is not periodic, we do not define a spectral curve Σ for (u, u_y) (if one were to define Σ by Eq. (3.4), it would turn out that the branch points of Σ do not satisfy any reasonable asymptotic law, so such a definition would be of no interest), but we still consider the classical spectral divisor D of (u, u_y) as a point multi-set in $\mathbb{C}^* \times \mathbb{C}^*$ defined by Eq. (3.14).

Remark 5.1 Lemma 5.2 shows that to ensure the existence of the extended frame F_λ as a solution of Eq. (5.3), it in fact suffices for α to be Lebesgue-integrable (instead of square-integrable). Most of the asymptotic estimates of the present chapter, and of Chap. 7, could also be carried out in an analogous way, if the space of potentials were extended from $W^{1,2}([0,1]) \times L^2([0,1])$ to $W^{1,p}([0,1]) \times L^p([0,1])$ with a fixed $p > 1$. We will phrase several of the following lemmas (but not the full proof of the asymptotic estimates) in such a way that this possibility becomes apparent.

In what follows, we denote by $|\ldots|$ also an arbitrary sub-multiplicative matrix norm on the space $\mathbb{C}^{2\times 2}$ of complex (2×2)-matrices.

Lemma 5.2 *Let $\alpha \in L^p([0,1],\mathbb{C}^{2\times 2})$ with $p \geq 1$. Then the sum*

$$F : [0,1] \to \mathbb{C}^{2\times 2},\ x \mapsto \mathbb{1} + \sum_{n=1}^{\infty} \int_{t_1=0}^{x} \int_{t_2=0}^{t_1} \cdots \int_{t_n=0}^{t_{n-1}} \alpha(t_1)\,\alpha(t_2)\,\ldots\alpha(t_n)\,\mathrm{d}^n t$$

converges in $W^{1,p}([0,1],\mathbb{C}^{2\times 2})$ to the solution of the initial value problem

$$F' = \alpha\, F \quad \textit{and} \quad F(0) = \mathbb{1},$$

and we have $|F(x)| \leq \exp(\|\alpha\|_{L^1})$ for every $x \in [0,1]$.

Proof For every $n \geq 1$, we put

$$F_n(x) := \int_{t_1=0}^{x} \int_{t_2=0}^{t_1} \cdots \int_{t_n=0}^{t_{n-1}} \alpha(t_1)\,\alpha(t_2)\,\ldots\alpha(t_n)\,\mathrm{d}^n t.$$

We then have $F_n \in W^{1,p}([0,1],\mathbb{C}^{2\times 2})$ with

$$F_n'(x) = \alpha(x)\cdot F_{n-1}(x)$$

(where we put $F_0(x) := \mathbb{1}$). Moreover, we have

$$\begin{aligned}
|F_n(x)| &= \left| \int_{t_1=0}^{x} \int_{t_2=0}^{t_1} \cdots \int_{t_n=0}^{t_{n-1}} \alpha(t_1)\,\alpha(t_2)\,\ldots\alpha(t_n)\,\mathrm{d}^n t \right| \\
&\leq \int_{t_1=0}^{x} \int_{t_2=0}^{t_1} \cdots \int_{t_n=0}^{t_{n-1}} |\alpha(t_1)|\cdot|\alpha(t_2)|\cdot\ldots\cdot|\alpha(t_n)|\mathrm{d}^n t \\
&= \frac{1}{n!} \int_{t_1=0}^{x} \int_{t_2=0}^{x} \cdots \int_{t_n=0}^{x} |\alpha(t_1)|\cdot|\alpha(t_2)|\cdot\ldots\cdot|\alpha(t_n)|\mathrm{d}^n t \\
&= \frac{1}{n!} \left(\int_0^x |\alpha(t)|\,\mathrm{d}t \right)^n \leq \frac{1}{n!}\,\|\alpha\|_{L^1}^n .
\end{aligned}$$

It follows that the series $\sum_{n\geq 1} F_n(x)$ converges absolutely and uniformly for $x \in [0,1]$, whereas the series $\sum_{n\geq 1} F_n' = \alpha \cdot \sum_{n\geq 0} F_n$ converges in $L^p([0,1],\mathbb{C}^{2\times 2})$. Therefore the series defining F converges in $W^{1,p}([0,1])$, and we have $F' = \alpha \cdot F$ and $F(0) = \mathbb{1}$. Moreover, it follows from the preceding estimate that $|F(x)| \leq \exp(\|\alpha\|_{L^1})$ holds for every $x \in [0,1]$. □

We now turn to the description of the asymptotic behavior of the monodromy $M(\lambda)$. For the estimate of the error of the asymptotic approximation of $M(\lambda)$, the function

$$w(\lambda) := |\cos(\zeta(\lambda))| + |\sin(\zeta(\lambda))| \tag{5.4}$$

will be important. We jot down a few facts about this function:

Proposition 5.3

(1) We have $\frac{1}{2}\,\mathrm{e}^{|\operatorname{Im}(\zeta(\lambda))|} \leq w(\lambda) \leq 2\,\mathrm{e}^{|\operatorname{Im}(\zeta(\lambda))|}$ *for all* $\lambda \in \mathbb{C}^*$.
(2) w *is bounded on every horizontal strip in the* ζ*-plane.*

Proof For (1). For any $\lambda \in \mathbb{C}^*$, we have

$$\begin{aligned}|\cos(\zeta(\lambda))| &= \left|\frac{1}{2}\left(\mathrm{e}^{\mathrm{i}\zeta(\lambda)} + \mathrm{e}^{-\mathrm{i}\zeta(\lambda)}\right)\right| \leq \max\left\{\left|\mathrm{e}^{\mathrm{i}\zeta(\lambda)}\right|, \left|\mathrm{e}^{-\mathrm{i}\zeta(\lambda)}\right|\right\} \\ &= \max\left\{\mathrm{e}^{\operatorname{Im}\zeta(\lambda)}, \mathrm{e}^{-\operatorname{Im}\zeta(\lambda)}\right\} = \mathrm{e}^{|\operatorname{Im}\zeta(\lambda)|}\end{aligned}$$

and likewise

$$|\sin(\zeta(\lambda))| \leq \mathrm{e}^{|\operatorname{Im}\zeta(\lambda)|}\,,$$

hence

$$w(\lambda) = |\cos(\zeta(\lambda))| + |\sin(\zeta(\lambda))| \leq 2\,\mathrm{e}^{|\operatorname{Im}(\zeta(\lambda))|}.$$

For the lower bound for $w(\lambda)$, we have

$$\begin{aligned}w(\lambda) &= |\cos(\zeta(\lambda))| + |\sin(\zeta(\lambda))| \\ &= \left|\frac{1}{2}\left(\mathrm{e}^{\mathrm{i}\zeta(\lambda)} + \mathrm{e}^{-\mathrm{i}\zeta(\lambda)}\right)\right| + \left|\frac{1}{2\mathrm{i}}\left(\mathrm{e}^{\mathrm{i}\zeta(\lambda)} - \mathrm{e}^{-\mathrm{i}\zeta(\lambda)}\right)\right| \\ &= \frac{1}{2}\left|\mathrm{e}^{\mathrm{i}\zeta(\lambda)}\right| \cdot \left(\left|1 + \mathrm{e}^{-2\mathrm{i}\zeta(\lambda)}\right| + \left|1 - \mathrm{e}^{-2\mathrm{i}\zeta(\lambda)}\right|\right) \\ &= \frac{1}{2}\,\mathrm{e}^{|\operatorname{Im}(\zeta(\lambda))|} \cdot \left(\left|1 + \mathrm{e}^{x+\mathrm{i}y}\right| + \left|1 - \mathrm{e}^{x+\mathrm{i}y}\right|\right)\end{aligned} \tag{5.5}$$

with $x := \mathrm{Re}(-2\mathrm{i}\zeta(\lambda))$, $y := \mathrm{Im}(-2\mathrm{i}\zeta(\lambda))$. We further have

$$\begin{aligned} \left|1 \pm \mathrm{e}^{x+\mathrm{i}y}\right|^2 &= \left|(1 \pm \mathrm{e}^x \cos(y)) \pm \mathrm{i} \cdot \mathrm{e}^x \sin(y)\right|^2 \\ &= (1 \pm \mathrm{e}^x \cos(y))^2 + (\mathrm{e}^x \sin(y))^2 \\ &= 1 \pm 2\,\mathrm{e}^x \cos(y) + \mathrm{e}^{2x} \geq 1 \pm 2\,\mathrm{e}^x \cos(y). \end{aligned}$$

Depending on the sign of $\cos(y)$, for at least one choice of the sign $\pm$, we have $1 \pm 2\,\mathrm{e}^x \cos(y) \geq 1$, and therefore it follows from (5.5) that $w(\lambda) \geq \frac{1}{2}\,\mathrm{e}^{|\,\mathrm{Im}(\zeta(\lambda))|}$ holds.

For (2). This is an immediate consequence of (1). □

The next theorem gives the basic asymptotic estimate for the monodromy:

Theorem 5.4 *Let* $(u, u_y) \in \mathsf{Pot}_{np}$ *be given. We put* $\tau := \mathrm{e}^{-(u(0)+u(1))/4}$ *and* $\upsilon := \mathrm{e}^{(u(1)-u(0))/4}$. *(Note that we have* $\upsilon = 1$ *for* $(u, u_y) \in \mathsf{Pot}$.*)*

We compare the monodromy $M(\lambda) = \begin{pmatrix} a(\lambda) & b(\lambda) \\ c(\lambda) & d(\lambda) \end{pmatrix}$ *of* (u, u_y) *to the monodromy* $M_0(\lambda) = \begin{pmatrix} a_0(\lambda) & b_0(\lambda) \\ c_0(\lambda) & d_0(\lambda) \end{pmatrix}$ *of the vacuum, see Eq. (4.5). For every* $\varepsilon > 0$ *there exists* $R > 0$, *such that:*

(1) For all $\lambda \in \mathbb{C}$ *with* $|\lambda| \geq R$, *we have*

$$\begin{aligned} |a(\lambda) - \upsilon\, a_0(\lambda)| &\leq \varepsilon\, w(\lambda) \\ |b(\lambda) - \tau^{-1}\, b_0(\lambda)| &\leq \varepsilon\, |\lambda|^{-1/2}\, w(\lambda) \\ |c(\lambda) - \tau\, c_0(\lambda)| &\leq \varepsilon\, |\lambda|^{1/2}\, w(\lambda) \\ |d(\lambda) - \upsilon^{-1}\, d_0(\lambda)| &\leq \varepsilon\, w(\lambda). \end{aligned}$$

(2) For all $\lambda \in \mathbb{C}^*$ *with* $|\lambda| \leq \frac{1}{R}$, *we have*

$$\begin{aligned} |a(\lambda) - \upsilon^{-1}\, a_0(\lambda)| &\leq \varepsilon\, w(\lambda) \\ |b(\lambda) - \tau\, b_0(\lambda)| &\leq \varepsilon\, |\lambda|^{-1/2}\, w(\lambda) \\ |c(\lambda) - \tau^{-1}\, c_0(\lambda)| &\leq \varepsilon\, |\lambda|^{1/2}\, w(\lambda) \\ |d(\lambda) - \upsilon\, d_0(\lambda)| &\leq \varepsilon\, w(\lambda). \end{aligned}$$

Moreover if P *is a relatively compact subset of* Pot_{np}, *then* R *can be chosen uniformly (depending on* ε*) for* $(u, u_y) \in P$ *in (1) and (2).*

Most of the remainder of the chapter is concerned with the proof of this theorem. One important instrument for the proof will be the following lemma about the comparison of solutions of the differential equations $\mathrm{d}F = \alpha F$ and $\mathrm{d}F = (\alpha+\beta)F$ with suitable α, β.

Lemma 5.5 *Suppose that $p \geq 1$ is fixed, and $\alpha, \beta \in L^p([0,1], \mathbb{C}^{2\times 2})$ are given. We put $\widetilde{\alpha} := \alpha+\beta$, and let $F, \widetilde{F} \in W^{1,p}([0,1], \mathbb{C}^{2\times 2})$ be the solutions of $F' = \alpha F$ and $\widetilde{F}' = \widetilde{\alpha}\widetilde{F}$ with $F(0) = \widetilde{F}(0) = \mathbb{1}$, see Lemma 5.2.*

(1) We have

$$\widetilde{F}(x) = F(x) + \sum_{n=1}^{\infty} \widetilde{F}_n(x)\,, \tag{5.6}$$

where we put for $n \geq 1$

$$\widetilde{F}_n(x) = F(x) \cdot \int_{t_1=0}^{x} \int_{t_2=0}^{t_1} \cdots \int_{t_n=0}^{t_{n-1}} \prod_{j=1}^{n} F(t_j)^{-1}\,\beta(t_j)\,F(t_j)\,\mathrm{d}^n t\,,$$

and the infinite sum in Eq. (5.6) converges in $W^{1,p}([0,1], \mathbb{C}^{2\times 2})$. Moreover, there exists a constant $C_{\|\alpha\|} > 0$ depending only on $\|\alpha\|_{L^1}$, such that for all $x \in [0,1]$ we have

$$|\widetilde{F}(x)| \leq \exp(C_{\|\alpha\|} \cdot \|\beta\|_{L^1}) \cdot |F(x)|. \tag{5.7}$$

If we fix $R > 0$, then the convergence of the sum in (5.6) is uniform for all $\alpha, \beta \in L^p([0,1], \mathbb{C}^{2\times 2})$ with $\|\alpha\|_{L^1}, \|\beta\|_{L^1} \leq R$.

(2) In the previous setting, we now suppose that α is constant with respect to x, and that $\alpha^2 = a \cdot \mathbb{1}$ holds with some $a \in \mathbb{C}^*$. Then we have:

(a) $F(x) = \exp(x\,\alpha)$ for all $x \in [0,1]$. Therefore F extends to the group homomorphism $\mathbb{R} \to \mathrm{GL}(2,\mathbb{C})$, $x \mapsto F(x) := \exp(x\,\alpha)$, i.e. we have besides $F(0) = \mathbb{1}$ for all $x, y \in \mathbb{R}$

$$F(x+y) = F(x) \cdot F(y) \quad \text{and} \quad F(x)^{-1} = F(-x). \tag{5.8}$$

Moreover, $|F(x)| \leq \exp(x\,|\alpha|)$ holds.

(b) There exist unique $\beta_+, \beta_- \in L^p([0,1], \mathbb{C}^{2\times 2})$ with $\beta = \beta_+ + \beta_-$ such that β_+ commutes with α, and β_- anti-commutes with α. We also have the following rule of commutation:

$$\beta_\pm(x) \cdot F(y) = F(\pm y) \cdot \beta_\pm(x). \tag{5.9}$$

Then we have for every $n \geq 1$

$$\widetilde{F}_n(x) = \sum_{\varepsilon \in \{\pm 1\}^n} \int_{t_1=0}^{x} \int_{t_2=0}^{t_1} \cdots \int_{t_n=0}^{t_{n-1}} F(\xi_\varepsilon(t))\,\beta_{\varepsilon_1}(t_1) \cdots \beta_{\varepsilon_n}(t_n)\,\mathrm{d}^n t.$$

Here we define for every $\varepsilon := (\varepsilon_1, \ldots, \varepsilon_n) \in \{\pm 1\}^n$ *and* $t = (t_1, \ldots, t_n) \in [0, x]^n$ *with* $0 \leq t_n \leq t_{n-1} \leq \ldots \leq t_1 \leq x$

$$\xi_\varepsilon(t) := x - 2t_{\nu_1} + 2t_{\nu_2} - + \ldots + 2(-1)^j t_{\nu_j} ,$$

where $1 \leq \nu_1 < \ldots < \nu_j \leq n$ *are those indices* $\nu \in \{1, \ldots, n\}$ *for which we have* $\varepsilon_\nu = -1$*. We have* $\xi_\varepsilon(t) \in [-x, x]$*.*

(c) Now suppose that $\beta^{[1]}, \beta^{[2]} \in L^p([0,1], \mathbb{C}^{2\times 2})$ *are given. We apply the situation of (b) to* $\beta^{[\nu]}$ *(for* $\nu \in \{1, 2\}$*), denoting the quantities corresponding to* $\beta^{[\nu]}$ *by the superscript* ${}^{[\nu]}$*. Then there exists a constant* $C > 0$*, depending only on an upper bound for* $\| \max\{|\beta_\pm^{[\nu]}|\}\|_{L^1}$*, such that*

$$|\widetilde{F}^{[1]}(x) - \widetilde{F}^{[2]}(x)| \leq C \cdot |F(x)| \cdot \Big\| \max\{|\beta_+^{[1]} - \beta_+^{[2]}|, |\beta_-^{[1]} - \beta_-^{[2]}|\}\Big\|_{L^1}$$

holds.

Proof For (1). We begin by showing the claims on the convergence of the infinite sum in (5.6). We note that we have for any $n \geq 1$: $\widetilde{F}_n \in W^{1,p}([0,1], \mathbb{C}^{2\times 2})$ and

$$\begin{aligned}
\widetilde{F}_n'(x) &= F'(x) \cdot \int_{t_1=0}^{x} \int_{t_2=0}^{t_1} \cdots \int_{t_n=0}^{t_{n-1}} \prod_{j=1}^{n} F(t_j)^{-1}\, \beta(t_j)\, F(t_j)\, \mathrm{d}^n t \\
&\quad + F(x) \cdot F(x)^{-1}\, \beta(x)\, F(x) \int_{t_2=0}^{x} \cdots \int_{t_n=0}^{t_{n-1}} \prod_{j=2}^{n} F(t_j)^{-1}\, \beta(t_j)\, F(t_j)\, \mathrm{d}^{n-1} t \\
&= \alpha(x)\, \widetilde{F}_n(x) + \beta(x)\, \widetilde{F}_{n-1}(x)
\end{aligned} \tag{5.10}$$

(with $\widetilde{F}_0 := F$). Moreover, for $x \in [0,1]$, we have

$$\begin{aligned}
|\widetilde{F}_n(x)| &\leq |F(x)| \cdot \int_{t_1=0}^{x} \int_{t_2=0}^{t_1} \cdots \int_{t_n=0}^{t_{n-1}} \prod_{j=1}^{n} |F(t_j)^{-1}| \cdot |\beta(t_j)| \cdot |F(t_j)|\, \mathrm{d}^n t \\
&\leq |F(x)| \cdot \frac{1}{n!} \int_{t_1=0}^{x} \int_{t_2=0}^{x} \cdots \int_{t_n=0}^{x} \prod_{j=1}^{n} |F(t_j)^{-1}| \cdot |\beta(t_j)| \cdot |F(t_j)|\, \mathrm{d}^n t \\
&= |F(x)| \cdot \frac{1}{n!} \left(\int_0^x |F(t)^{-1}| \cdot |\beta(t_j)| \cdot |F(t)|\, \mathrm{d}t \right)^n \\
&\leq |F(x)| \cdot \frac{1}{n!} \left(\exp(\|\alpha\|_{L^1})^2\, \|\beta\|_{L^1} \right)^n ,
\end{aligned}$$

where the last inequality follows from Lemma 5.2. Therefore the series $\sum_{n=1}^{\infty} \widetilde{F}_n(x)$ converges absolutely and uniformly in x, whereas the series

$$\sum_{n=1}^{\infty} \widetilde{F}_n'(x) = \alpha(x) \cdot \sum_{n=1}^{\infty} \widetilde{F}_n(x) + \beta(x) \cdot \sum_{n=0}^{\infty} \widetilde{F}_n(x)$$

converges in $L^p([0,1], \mathbb{C}^{2\times 2})$. Therefore $\sum_{n=1}^{\infty} \widetilde{F}_n$ converges in the Sobolev space $W^{1,p}([0,1], \mathbb{C}^{2\times 2})$. Moreover, these convergences are also uniform when α and β vary with $\|\alpha\|_{L^1}, \|\beta\|_{L^1} \leq R$ (for a fixed $R > 0$). Moreover, we have $1 + \sum_{n=1}^{\infty} |\widetilde{F}_n(x)| \leq |F(x)| \cdot \exp(C_{\|\alpha\|} \cdot \|\beta\|_{L^1})$ with $C_{\|\alpha\|} := \exp(\|\alpha\|_{L^1})^2$.

Therefore Eq. (5.6) defines an element $\widetilde{F} \in W^{1,p}([0,1], \mathbb{C}^{2\times 2})$. We have $\widetilde{F}_n(0) = 0$ for all $n \geq 1$, and therefore $\widetilde{F}(0) = F(0) = \mathbb{1}$. It remains to show that $\widetilde{F}$ solves the differential equation $\widetilde{F}' = \widetilde{\alpha}\widetilde{F}$. Indeed, because the series $\sum_{n=1}^{\infty} \widetilde{F}_n$ converges in $W^{1,p}([0,1], \mathbb{C}^{2\times 2})$, we have

$$\begin{aligned}\widetilde{F}'(x) &= F'(x) + \sum_{n=1}^{\infty} \widetilde{F}_n'(x) \overset{(5.10)}{=} \alpha(x)\, F(x) + \sum_{n=1}^{\infty} \big(\alpha(x)\, \widetilde{F}_n(x) + \beta(x)\, \widetilde{F}_{n-1}(x)\big) \\ &= (\alpha(x) + \beta(x)) \cdot \left(F(x) + \sum_{n=1}^{\infty} \widetilde{F}_n(x) \right) = \widetilde{\alpha}(x) \cdot \widetilde{F}(x).\end{aligned}$$

For (2). (a) is obvious. For (b), we put

$$\beta_{\pm} := \frac{1}{2}\,(\beta \pm \alpha\,\beta\,\alpha^{-1}).$$

Then we have $\beta_+ + \beta_- = \beta$. Because of the hypothesis $\alpha^2 = a \cdot \mathbb{1}$, we also have $\alpha^{-1} = \frac{1}{a} \cdot \alpha$, and therefore

$$\alpha \cdot \beta_{\pm} = \frac{1}{2}\,(\alpha\,\beta \pm \alpha^2\,\beta\,\alpha^{-1}) = \frac{1}{2}\,(\alpha\,\beta \pm \beta\,\alpha) = \pm\frac{1}{2}\,(\beta \pm \alpha\,\beta\,\alpha^{-1})\,\alpha = \pm\beta_{\pm} \cdot \alpha\,,$$

i.e. α commutes with β_+ and anti-commutes with β_-; herefrom also Eq. (5.9) follows. It is clear that this decomposition of β is unique.

We now calculate $\widetilde{F}_n$ in the present situation, using the properties of F given by Eqs. (5.8) and (5.9):

$$\begin{aligned}\widetilde{F}_n(x) &= F(x) \cdot \int_{t_1=0}^{x} \int_{t_2=0}^{t_1} \cdots \int_{t_n=0}^{t_{n-1}} \prod_{j=1}^{n} F(t_j)^{-1}\,\beta(t_j)\,F(t_j)\,\mathrm{d}^n t \\ &= \int_0^x \int_0^{t_1} \cdots \int_0^{t_{n-1}} F(x - t_1)\,\beta(t_1)\,F(t_1 - t_2)\,\beta(t_2) \ldots \beta(t_n)\,F(t_n)\mathrm{d}^n t\end{aligned}$$

$$= \sum_{\varepsilon\in\{\pm 1\}^n} \int_0^x \int_0^{t_1} \cdots \int_0^{t_{n-1}} F(x-t_1)\,\beta_{\varepsilon_1}(t_1)\,F(t_1-t_2)\,\beta_{\varepsilon_2}(t_2)\dots\beta_{\varepsilon_n}(t_n)\,F(t_n)\mathrm{d}^n t$$

$$= \sum_{\varepsilon\in\{\pm 1\}^n} \int_0^x \int_0^{t_1} \cdots \int_0^{t_{n-1}} F(\xi_\varepsilon(t))\,\beta_{\varepsilon_1}(t_1)\,\beta_{\varepsilon_2}(t_2)\dots\beta_{\varepsilon_n}(t_n)\,\mathrm{d}^n t\,, \tag{5.11}$$

where we define for any $\varepsilon = (\varepsilon_1,\dots,\varepsilon_n) \in \{\pm 1\}^n$ and $t \in \mathbb{R}^n$

$$\begin{aligned}\xi_\varepsilon(t) &= (x-t_1) + \varepsilon_1(t_1-t_2) \\ &\quad + \varepsilon_1\varepsilon_2(t_2-t_3) + \dots + \varepsilon_1\cdots\varepsilon_{n-1}(t_{n-1}-t_n) + \varepsilon_1\cdots\varepsilon_n t_n \\ &= x - 2t_{\nu_1} + 2t_{\nu_2} - + \dots + 2(-1)^j t_{\nu_j}\end{aligned}$$

with the ν_j defined as in the statement (2)(b).

For (c), we abbreviate $\beta^* := \max\{|\beta_\pm^{[1]}|, |\beta_\pm^{[2]}|\} \in L^p([0,1])$ and

$$\delta := \max\{|\beta_+^{[1]} - \beta_+^{[2]}|, |\beta_-^{[1]} - \beta_-^{[2]}|\} \in L^p([0,1]).$$

Then we have for any $n \geq 1$, any $\varepsilon = (\varepsilon_1,\dots,\varepsilon_n) \in \{\pm 1\}^n$, and any $0 \leq t_n \leq \dots \leq t_1 \leq x$

$$\begin{aligned}&\beta_{\varepsilon_1}^{[1]}(t_1)\dots\beta_{\varepsilon_n}^{[1]}(t_n) - \beta_{\varepsilon_1}^{[2]}(t_1)\dots\beta_{\varepsilon_n}^{[2]}(t_n) \\ &\quad= \sum_{j=1}^n \beta_{\varepsilon_1}^{[2]}(t_1)\cdots\beta_{\varepsilon_{j-1}}^{[2]}(t_{j-1})\cdot(\beta_{\varepsilon_j}^{[1]}(t_j) - \beta_{\varepsilon_j}^{[2]}(t_j))\cdot\beta_{\varepsilon_{j+1}}^{[1]}(t_{j+1})\cdots\beta_{\varepsilon_n}^{[1]}(t_n)\end{aligned}$$

and therefore by (b)

$$\begin{aligned}&|\widetilde{F}_n^{[1]}(x) - \widetilde{F}_n^{[2]}(x)| \\ &\leq \sum_{\varepsilon\in\{\pm1\}^n} \int_{t_1=0}^x \int_{t_2=0}^{t_1}\cdots \int_{t_n=0}^{t_{n-1}} \underbrace{|F(\xi_\varepsilon(t))|}_{\leq |F(x)|}\cdot|\beta_{\varepsilon_1}^{[1]}(t_1)\cdots\beta_{\varepsilon_n}^{[1]}(t_n) - \beta_{\varepsilon_1}^{[2]}(t_1)\cdots\beta_{\varepsilon_n}^{[2]}(t_n)|\,\mathrm{d}^n t \\ &\leq |F(x)|\cdot\sum_{\varepsilon\in\{\pm1\}^n}\sum_{j=1}^n \int_{t_1=0}^x \int_{t_2=0}^{t_1}\cdots \int_{t_n=0}^{t_{n-1}} |\beta_{\varepsilon_1}^{[2]}(t_1)|\cdots|\beta_{\varepsilon_{j-1}}^{[2]}(t_{j-1})|\cdot|\beta_{\varepsilon_j}^{[1]}(t_j) - \beta_{\varepsilon_j}^{[2]}(t_j)| \\ &\quad\cdot|\beta_{\varepsilon_{j+1}}^{[1]}(t_{j+1})|\cdots|\beta_{\varepsilon_n}^{[1]}(t_n)|\,\mathrm{d}^n t \\ &\leq |F(x)|\cdot\sum_{\varepsilon\in\{\pm1\}^n}\sum_{j=1}^n \int_{t_1=0}^x \int_{t_2=0}^{t_1}\cdots \int_{t_n=0}^{t_{n-1}} \beta^*(t_1)\cdots\beta^*(t_{j-1})\delta(t_j)\beta^*(t_{j+1})\cdots\beta^*(t_n)\,\mathrm{d}^n t \\ &\leq |F(x)|\cdot 2^n\cdot n\cdot\frac{1}{(n-1)!}\cdot\|\beta^*\|_{L^1}^{n-1}\cdot\|\delta\|_{L^1}\,,\end{aligned}$$

whence

$$
\begin{aligned}
&|\widetilde{F}^{[1]}(x) - \widetilde{F}^{[2]}(x)| \\
&\quad \leq \sum_{n=1}^{\infty} |\widetilde{F}_n^{[1]}(x) - \widetilde{F}_n^{[2]}(x)| \leq |F(x)| \cdot \|\delta\|_{L^1} \sum_{n=1}^{\infty} \frac{2^n \cdot n}{(n-1)!} \cdot \|\beta^*\|_{L^1}^{n-1} \\
&\quad \leq 8\,|F(x)| \cdot \|\delta\|_{L^1} \cdot \|\beta^*\|_{L^1} \cdot (1 + \exp(2\,\|\beta^*\|_{L^1})) \\
&\quad = C \cdot |F(x)| \cdot \|\delta\|_{L^1}
\end{aligned}
$$

follows with $C := 8\|\beta^*\|_{L^1} \cdot (1 + \exp(2\|\beta^*\|_{L^1}))$. □

Proof (of Theorem 5.4) For the purposes of the proof, we fix an arbitrary branch of the square root function on some slitted complex plane, denoted by $\lambda^{1/2}$ (or $\sqrt{\lambda}$). Below, we will show the claimed asymptotic assessment for λ in this slitted plane; it then extends to all of $\mathbb{C}^*$ by an argument of continuity.

We will primarily prove the asymptotic behavior for $\lambda \to \infty$; only in the end will we use Proposition 2.2 to transfer that result to the case $\lambda \to 0$.

The first important step in the proof is to *regauge* α, that is to say, we choose a function $g = g_\lambda : \mathbb{R} \to \mathrm{GL}(2, \mathbb{C})$ depending holomorphically on the spectral parameter $\lambda \in \mathbb{C}^*$ such that $g_\lambda \in W^{1,2}([0,1], \mathbb{C}^{2\times 2})$ (g should be periodic if u is periodic), and then pass from α and F to

$$\widetilde{\alpha} = g^{-1}\,\alpha\,g - g^{-1}\,g' \quad \text{and} \quad \widetilde{F}(x) = g(x)^{-1} \cdot F(x) \cdot g(0). \tag{5.12}$$

We then again have $\widetilde{\alpha} \in L^2([0,1], \mathbb{C}^{2\times 2})$ and $\widetilde{F} \in W^{1,2}([0,1], \mathbb{C}^{2\times 2})$, and $\widetilde{F}$ satisfies the following ordinary initial value problem, which is analogous to Eq. (5.3)

$$\widetilde{F}'(x) = \widetilde{\alpha}(x) \cdot \widetilde{F}(x) \quad \text{and} \quad \widetilde{F}(0) = \mathbb{1}. \tag{5.13}$$

We also put $\widetilde{M}(\lambda) = \widetilde{F}_\lambda(1)$.

The g we are going to use for the regauging of α is

$$g := \begin{pmatrix} 1 & 0 \\ 0 & \lambda^{1/2} \end{pmatrix} \cdot \begin{pmatrix} e^{u/4} & 0 \\ 0 & e^{-u/4} \end{pmatrix} = \begin{pmatrix} e^{u/4} & 0 \\ 0 & \lambda^{1/2}\,e^{-u/4} \end{pmatrix}. \tag{5.14}$$

In the definition of g, the first factor represents the transition from an "untwisted" to a "twisted" representation of the minimal surface; it is also needed to ensure that the leading term (with respect to λ) of $\widetilde{\alpha}$ is no longer nilpotent (which would imply that its adjunct operator is not invertible, and would cause problems later on). The second factor serves to make the leading term of $\widetilde{\alpha}$ independent of u, thereby making asymptotic assessments feasible.

Via Eq. (3.2) and the equations

$$g^{-1} = \begin{pmatrix} e^{-u/4} & 0 \\ 0 & \lambda^{-1/2}\, e^{u/4} \end{pmatrix} \quad \text{and} \quad g' = \frac{1}{4} \begin{pmatrix} u_x\, e^{u/4} & 0 \\ 0 & -\lambda^{1/2}\, u_x\, e^{-u/4} \end{pmatrix}, \tag{5.15}$$

we obtain

$$\begin{aligned} \widetilde{\alpha} = g^{-1}\, \alpha\, g - g^{-1}\, g' &= \begin{pmatrix} -\frac{1}{2}\, u_z & -\frac{1}{4}\, \lambda^{1/2} - \frac{1}{4}\, \lambda^{-1/2}\, e^{u} \\ \frac{1}{4}\, \lambda^{1/2} + \frac{1}{4}\, \lambda^{-1/2}\, e^{-u} & \frac{1}{2}\, u_z \end{pmatrix} \\ &= \widetilde{\alpha}_0 + \beta + \gamma \end{aligned} \tag{5.16}$$

with

$$\widetilde{\alpha}_0 := \zeta(\lambda) \begin{pmatrix} 0 & -1 \\ 1 & 0 \end{pmatrix}, \quad \beta := -\frac{1}{2} u_z \begin{pmatrix} 1 & 0 \\ 0 & -1 \end{pmatrix}, \quad \gamma := \frac{1}{4}\, \lambda^{-1/2} \begin{pmatrix} 0 & -e^{u} + 1 \\ e^{-u} - 1 & 0 \end{pmatrix}.$$

We will now asymptotically compare the monodromy $\widetilde{M}(\lambda)$ with the monodromy $\widetilde{M}_0(\lambda) := \widetilde{F}_{0,\lambda}(1)$ corresponding to $\widetilde{F}_0' = \widetilde{\alpha}_0\, \widetilde{F}_0$, $\widetilde{F}_0(0) = \mathbb{1}$. More specifically, if we denote by $|\ldots|$ also the maximum absolute row sum norm[1] for (2×2)-matrices, we will show below that for given $\varepsilon > 0$ there exists $R > 0$ so that for $\lambda \in \mathbb{C}$ with $|\lambda| \geq R$, we have

$$|\widetilde{M}(\lambda) - \widetilde{M}_0(\lambda)| \leq \varepsilon \cdot w(\lambda). \tag{5.17}$$

Note that

$$\widetilde{M}_0(\lambda) = \begin{pmatrix} \cos(\zeta(\lambda)) & -\sin(\zeta(\lambda)) \\ \sin(\zeta(\lambda)) & \cos(\zeta(\lambda)) \end{pmatrix} \tag{5.18}$$

and therefore $w(\lambda) = |\widetilde{M}_0(\lambda)|$ holds.

To see that the claimed estimate in Theorem 5.4(1) follows from (5.17), we write the non-regauged monodromy as $M(\lambda) = \left(\begin{smallmatrix} a(\lambda) & b(\lambda) \\ c(\lambda) & d(\lambda) \end{smallmatrix}\right)$ as in the statement of the theorem. Then we have

$$\begin{aligned} \widetilde{M}(\lambda) &= g(1)^{-1}\, M(\lambda)\, g(0) \\ &= \begin{pmatrix} e^{-(u(1)-u(0))/4}\, a(\lambda) & \lambda^{1/2}\, e^{-(u(0)+u(1))/4}\, b(\lambda) \\ \lambda^{-1/2}\, e^{(u(0)+u(1))/4}\, c(\lambda) & e^{(u(1)-u(0))/4}\, d(\lambda) \end{pmatrix}. \end{aligned}$$

[1]Note that the maximum absolute row sum norm is the operator norm associated to the maximum norm on $\mathbb{C}^2$. Therefore it is sub-multiplicative.

By plugging this equation, as well as Eq. (5.18) into the estimate (5.17), we obtain Theorem 5.4(1).

It thus suffices to prove the estimate (5.17).

The expression γ can be neglected in our asymptotic consideration. The reason for this is, roughly speaking, that γ is of order $|\lambda|^{-1/2}$ for $|\lambda| \to \infty$. To prove more precisely that γ can be neglected, we apply Lemma 5.5(2)(c) with $\alpha = \alpha_0$, $\beta^{[1]} = \beta$ and $\beta^{[2]} = \beta + \gamma$ to compare the resulting solutions $\widetilde{F}^{[1]}$ and $\widetilde{F}^{[2]}$. Notice that β anti-commutes with α_0, thus we have $\beta_+^{[1]} = 0$ and $\beta_-^{[1]} = \beta$. Moreover, we have

$$\gamma_+ = \frac{1}{4}\,\lambda^{-1/2}\,(\cosh(u) - 1)\begin{pmatrix} 0 & -1 \\ 1 & 0 \end{pmatrix} \quad \text{and} \quad \gamma_- = -\frac{1}{4}\,\lambda^{-1/2}\,\sinh(u)\begin{pmatrix} 0 & 1 \\ 1 & 0 \end{pmatrix}.$$

It follows that we have $\beta_\pm^{[1]} - \beta_\pm^{[2]} = \gamma_\pm = O(|\lambda|^{-1/2})$, and that $\|\max\{|\beta_\pm^{[\nu]}|\}\|_{L^1}$ is bounded for $\lambda \to \infty$. It now follows from Lemma 5.5(2)(c) that there exists $C > 0$ such that $|\widetilde{F}^{[1]} - \widetilde{F}^{[2]}| \leq C \cdot |\lambda|^{-1/2} \cdot w(\lambda)$ holds. By choosing $|\lambda|$ large, we can therefore ensure $|\widetilde{F}^{[1]} - \widetilde{F}^{[2]}| \leq \frac{1}{2}\,\varepsilon\, w(\lambda)$.

In other words, it follows from this argument that it suffices to show that for given $\varepsilon > 0$ there exists $R > 0$ so that for $\lambda \in \mathbb{C}$ with $|\lambda| > R$ and $x \in [0, 1]$ we have

$$|E_\lambda(1) - E_{0,\lambda}(1)| \leq \varepsilon \cdot w(\lambda) \tag{5.19}$$

for the solution $E = E_\lambda$ of the initial value problem

$$E'(x) = (\widetilde{\alpha}_0 + \beta(x))\,E(x) \quad \text{with} \quad E(0) = \mathbb{1} \tag{5.20}$$

and for the solution $E_0(x) = E_{0,\lambda}(x) = \exp(x\,\widetilde{\alpha}_0)$ of

$$E_0'(x) = \widetilde{\alpha}_0\,E_0(x) \quad \text{with} \quad E_0(0) = \mathbb{1}\ ;$$

we have

$$E_0(x) = \widetilde{F}_0(x) = \begin{pmatrix} \cos(\zeta(\lambda)\,x) & -\sin(\zeta(\lambda)\,x) \\ \sin(\zeta(\lambda)\,x) & \cos(\zeta(\lambda)\,x) \end{pmatrix}. \tag{5.21}$$

Because β anti-commutes with $\widetilde{\alpha}_0$, we have by Lemma 5.5(2)(b)

$$E = E_0 + \sum_{n=1}^{\infty} E_n \tag{5.22}$$

with

$$\begin{aligned} E_n(x) &= \int_0^x \int_0^{t_1} \cdots \int_0^{t_{n-1}} E_0(x - t_1)\, \beta(t_1)\, E_0(t_1 - t_2) \\ &\quad \beta(t_2) E_0(t_2 - t_3) \cdots \beta(t_n)\, E_0(t_n)\, \mathrm{d}^n t \\ &= \int_0^x \int_0^{t_1} \cdots \int_0^{t_{n-1}} E_0(\xi(t))\, \beta(t_1)\, \beta(t_2) \cdots \beta(t_n)\, \mathrm{d}^n t\,, \end{aligned} \tag{5.23}$$

where

$$\xi(t) := x - 2t_1 + 2t_2 - 2t_3 + - \ldots + 2(-1)^n\, t_n.$$

In the integral of Eq. (5.23), we now carry out the substitution $(t_1, \ldots, t_n) \mapsto (s_1, \ldots, s_n)$ with $s_j = t_j - s_{j+1}$ for $j \le n-1$ and $s_n = t_n$, i.e.

$$\begin{aligned} s_n &= t_n \\ s_{n-1} &= t_{n-1} - s_n = t_{n-1} - t_n \\ s_{n-2} &= t_{n-2} - s_{n-1} = t_{n-2} - t_{n-1} + t_n \\ s_{n-3} &= t_{n-3} - s_{n-2} = t_{n-3} - t_{n-2} + t_{n-1} - t_n \\ &\ldots \\ s_2 &= t_2 - s_3 = t_2 - t_3 + t_4 - + \ldots + (-1)^n\, t_n \\ s_1 &= t_1 - s_2 = t_1 - t_2 + t_3 - + \ldots + (-1)^{n+1}\, t_n. \end{aligned}$$

Then we have $t_j = s_j + s_{j+1}$ for $j \le n-1$ and $t_n = s_n$. Thus, the corresponding mapping $\Phi : (s_j) \mapsto (t_j)$ is a diffeomorphism with $\det \Phi' = 1$ from

$$\begin{aligned} U := \{\, (s_1, \ldots, s_n) \in \mathbb{R}^n \mid \forall j = 1, \ldots, n : s_j \ge 0, \\ s_1 + s_2 \le x,\ \forall j = 3, \ldots, n : s_j \le s_{j-2} \,\} \end{aligned}$$

onto the simplex

$$O := \{\, (t_1, \ldots, t_n) \in \mathbb{R}^n \mid 0 \le t_n \le t_{n-1} \le \cdots \le t_1 \le x \,\}$$

which is the domain of integration in (5.23). We also see

$$\xi(t) = x - 2s_1.$$

By carrying out the substitution in (5.23), we therefore obtain

$$\begin{aligned} E_n(x) &= \int_U E_0(x-2s_1)\,\beta(s_1+s_2)\,\beta(s_2+s_3)\,\ldots\,\beta(s_{n-1}+s_n)\,\beta(s_n)\,\mathrm{d}^n s \\ &= \int_{s_1=0}^{x} E_0(x-2s_1) \int_{s_2=0}^{x-s_1} \beta(s_1+s_2) \int_{s_3=0}^{s_1} \beta(s_2+s_3) \int_{s_4=0}^{s_2} \beta(s_3+s_4) \\ &\qquad \cdots \int_{s_n=0}^{s_{n-2}} \beta(s_{n-1}+s_n)\,\beta(s_n)\,\mathrm{d}^n s \\ &= \int_{s_1=0}^{x} E_0(x-2s_1)\,G_n(s_1)\,\mathrm{d}s_1 \end{aligned} \tag{5.24}$$

with

$$G_n(s_1) := H_n(x-s_1, s_1)\;, \tag{5.25}$$

where H_n is defined by

$$H_1(s_0, s_1) = \beta(s_1) \tag{5.26}$$

and

$$H_n(s_0, s_1) = \int_{s_2=0}^{s_0} \beta(s_1+s_2)\,H_{n-1}(s_1, s_2)\,\mathrm{d}s_2 \quad \text{for } n \geq 2.$$

Therefore we have for $n \geq 2$

$$\begin{aligned} H_n(s_0, s_1) := &\int_{s_2=0}^{s_0} \beta(s_1+s_2) \int_{s_3=0}^{s_1} \beta(s_2+s_3) \cdots \\ &\int_{s_n=0}^{s_{n-2}} \beta(s_{n-1}+s_n)\,\beta(s_n)\,\mathrm{d}^{n-1}s. \end{aligned} \tag{5.27}$$

We will now show[2] for all $n \geq 1$

$$G_n \in L^2([0,1], \mathbb{C}^{2\times 2}) \quad \text{and} \quad \|G_n\|_2 \leq \frac{1}{(\lfloor n/2 \rfloor - 1)!} \cdot \|\beta\|_2^n. \tag{5.28}$$

Indeed, for $n = 1$ we have $G_1 = \beta$, and thus (5.28) holds in this case. For $n = 2$, we have

$$G_2(s_1) = H_2(x-s_1, s_1) = \int_0^{x-s_1} \beta(s_1+s_2)\,\beta(s_2)\,\mathrm{d}s_2$$

[2] The proof given below shows that we in fact have $G_n \in L^\infty([0,1], \mathbb{C}^{2\times 2})$ for $n \geq 2$.

and therefore $G_2 \in W^{1,1}([0,1], \mathbb{C}^{2\times 2}) \subset L^2([0,1], \mathbb{C}^{2\times 2})$ and $\|G_2\|_2 \leq \|\beta\|_2^2$ by the Cauchy-Schwarz inequality. For even $n \geq 4$, we fix $s_0, s_1 \in [0,1]$ and define the simplices

$$S_+ := \{ (s_2, s_4, \dots, s_n) \in \mathbb{R}^{n/2} \,|\, 0 \leq s_n \leq \dots \leq s_4 \leq s_2 \leq s_0 \},$$
$$S_- := \{ (s_3, s_5, \dots, s_{n-1}) \in \mathbb{R}^{n/2-1} \,|\, 0 \leq s_{n-1} \leq \dots \leq s_5 \leq s_3 \leq s_1 \}.$$

Then we have

$$H_n(s_0, s_1) = \int_{S_-}\int_{S_+} \beta(s_2+s_1)\,\beta(s_2+s_3)\,\beta(s_4+s_3)\,\beta(s_4+s_5)\,\dots \\ \beta(s_n+s_{n-1})\,\beta(s_n)\,\mathrm{d}^{n/2}s\,\mathrm{d}^{n/2-1}s.$$

For the inner integral, we have by the Cauchy-Schwarz inequality

$$\left| \int_{S_+} \beta(s_2+s_1)\,\beta(s_2+s_3)\,\beta(s_4+s_3)\,\beta(s_4+s_5)\,\dots\,\beta(s_n+s_{n-1})\,\beta(s_n)\,\mathrm{d}^{n/2}s \right| \leq \|\beta\|_2^n,$$

and thus we have

$$|H_n(s_0,s_1)| \leq \mathrm{vol}(S_-)\cdot\|\beta\|_2^n = \frac{s_1^{n/2-1}}{(n/2-1)!}\cdot\|\beta\|_2^n \leq \frac{1}{(n/2-1)!}\cdot\|\beta\|_2^n.$$

Thus we see $G_n(s) = H_n(x-s,s) \in L^\infty([0,1], \mathbb{C}^{2\times 2}) \subset L^2([0,1], \mathbb{C}^{2\times 2})$ and $\|G_n\|_2 \leq \|H_n\|_\infty \leq \frac{1}{(n/2-1)!}\cdot\|\beta\|_2^n$. Therefore we obtain (5.28) for even $n \geq 4$. For odd $n \geq 3$, we can argue similarly, obtaining

$$|H_n(s_0,s_1)| \leq \frac{1}{((n-1)/2)!}\cdot\|\beta\|_2^n,$$

and therefore also in this case (5.28), completing its proof.

As a consequence of (5.28), we have

$$\sum_{n=1}^{\infty} \|G_n\|_2 \leq \sum_{n=1}^{\infty} \frac{1}{(\lfloor n/2\rfloor - 1)!}\cdot\|\beta\|_2^n < \infty$$

(because the power series $r \mapsto \sum_{n=1}^{\infty} \frac{1}{(\lfloor n/2\rfloor-1)!}\cdot r^n$ has convergence radius $\rho = \infty$), and thus we have

$$G := \sum_{n=1}^{\infty} G_n \in L^2([0,1], \mathbb{C}^{2\times 2}).$$

It follows via Eqs. (5.22) and (5.24) that the series $\sum_{n=0}^{\infty} E_n$ converges in $W^{1,2}([0,1],\mathbb{C}^{2\times 2})$ and that we have

$$E(x) - E_0(x) = \int_0^x E_0(x-2s)\,G(s)\,\mathrm{d}s.$$

We now use this representation of $E - E_0$ to show the estimate (5.17). For this purpose, we set $x = 1$, and fix $\delta > 0$ at first arbitrarily. Then we have for all $s \in [0,1]$: $|E_0(1-2s)| \leq |E_0(1)| = w(\lambda)$, and for all $s \in [\delta, 1-\delta]$: $|E_0(1-2s)| \leq |E_0(1-2\delta)| \leq \mathrm{e}^{-2\delta\,|\operatorname{Im}(\zeta(\lambda))|} \cdot w(\lambda)$. Thus we obtain

$$\begin{aligned}
&|E(1) - E_0(1)| \\
&\quad \leq \int_0^1 |E_0(1-2s)| \cdot |G(s)|\,\mathrm{d}s \\
&\quad = \int_0^{\delta} |E_0(1-2s)| \cdot |G(s)|\,\mathrm{d}s + \int_{\delta}^{1-\delta} |E_0(1-2s)| \cdot |G(s)|\,\mathrm{d}s \\
&\qquad + \int_{1-\delta}^{1} |E_0(1-2s)| \cdot |G(s)|\,\mathrm{d}s \\
&\quad \leq w(\lambda) \cdot \|G|[0,\delta]\|_1 + \mathrm{e}^{-2\delta\,|\operatorname{Im}(\zeta(\lambda))|} \cdot w(\lambda) \cdot \|G\|_1 + w(\lambda) \cdot \|G|[1-\delta,1]\|_1 .
\end{aligned}$$

Because of $G \in L^1([0,1],\mathbb{C}^{2\times 2})$, we can now choose $\delta > 0$ such that we have $\|G|[0,\delta]\|_1 \leq \frac{\varepsilon}{4}$ and $\|G|[1-\delta,1]\|_1 \leq \frac{\varepsilon}{4}$. Dependent on this δ, there further exists $C > 0$ such that $\mathrm{e}^{-2\delta C} \cdot \|G\|_1 \leq \frac{\varepsilon}{2}$. For all $\lambda \in \mathbb{C}^*$ with $|\operatorname{Im}(\zeta(\lambda))| \geq C$ we then obtain

$$|E(1) - E_0(1)| \leq \varepsilon \cdot w(\lambda) \tag{5.29}$$

and therefore (5.19).

It remains to show that (5.19) also holds within $\{\, \lambda \in \mathbb{C}^* \,\big|\, |\operatorname{Im}(\zeta(\lambda))| \leq C \,\}$ for λ of sufficiently large absolute value. We have

$$E(1) - E_0(1) = \int_0^1 E_0(1-2s)\,G(s)\,\mathrm{d}s = \frac{1}{2}\int_{-1}^{1} E_0(s)\,G\left(\tfrac{1-s}{2}\right)\,\mathrm{d}s\,,$$

where the entries of $E_0(s)$ are

$$\begin{aligned}
\cos(\zeta(\lambda)s) &= \frac{1}{2}(\mathrm{e}^{\mathrm{i}s\zeta(\lambda)} + \mathrm{e}^{-\mathrm{i}s\zeta(\lambda)}) \quad \text{and} \\
\pm\sin(\zeta(\lambda)s) &= \pm\frac{1}{2i}(\mathrm{e}^{\mathrm{i}s\zeta(\lambda)} - \mathrm{e}^{-\mathrm{i}s\zeta(\lambda)}),
\end{aligned} \tag{5.30}$$

therefore it follows from the variant of Riemann-Lebesgue's Lemma given in the following lemma (Lemma 5.6), applied with the compact set $N \subset L^1([-1,1])$

comprised of the four component functions of $G(\frac{1-s}{2})$, that there exists $R > 0$ so that (5.19) holds for $\lambda \in \mathbb{C}^*$ with $|\operatorname{Re}(\zeta(\lambda))| \geq R$ and $|\operatorname{Im}(\zeta(\lambda))| \leq C$. This completes the proof of (1).

To prove that R can be chosen uniformly depending on ε for relatively compact subsets P of Pot_{np}, we note that for $(u, u_y) \in P$, $\|(u, u_y)\|_{\mathsf{Pot}}$ is bounded, and therefore $\|\beta\|_{L^1}$ and $\|\gamma\|_{L^1}$ are also bounded for $(u, u_y) \in P$. For this reason, the estimates leading up to (5.29) are uniform for $(u, u_y) \in P$, and thus we obtain (5.19) for $\lambda \in \mathbb{C}^*$ with $|\operatorname{Im}(\zeta(\lambda))| \geq C$ for all $(u, u_y) \in P$, where $C > 0$ is a constant depending on ε. Moreover, we let N be the topological closure of the set of the component functions of $G(\frac{1-s}{2})$ where (u, u_y) now runs through all of P; N is compact, because P is relatively compact. By applying Lemma 5.6 with this N, we obtain (5.19) for $|\operatorname{Re}(\zeta(\lambda))| \geq R$ and $|\operatorname{Im}(\zeta(\lambda))| \leq C$ for all $(u, u_y) \in P$.

To prove part (2) of the theorem (i.e. the case $\lambda \to 0$), we let $\widetilde{u}(x) := -u(x)$ and $\widetilde{u}_y(x) := u_y(x)$; then we have $(\widetilde{u}, \widetilde{u}_y) \in \mathsf{Pot}$. If we denote the monodromy corresponding to u resp. to $\widetilde{u}$ by

$$M_u(\lambda) = \begin{pmatrix} a(\lambda) & b(\lambda) \\ c(\lambda) & d(\lambda) \end{pmatrix} \quad \text{resp. by} \quad M_{\widetilde{u}}(\lambda) = \begin{pmatrix} \widetilde{a}(\lambda) & \widetilde{b}(\lambda) \\ \widetilde{c}(\lambda) & \widetilde{d}(\lambda) \end{pmatrix},$$

then we have by Proposition 2.2(1)

$$M_u(\lambda) = \begin{pmatrix} 1 & 0 \\ 0 & \lambda \end{pmatrix} \cdot M_{\widetilde{u}}(\lambda^{-1}) \cdot \begin{pmatrix} 1 & 0 \\ 0 & \lambda^{-1} \end{pmatrix},$$

i.e.

$$\begin{pmatrix} a(\lambda) & b(\lambda) \\ c(\lambda) & d(\lambda) \end{pmatrix} = \begin{pmatrix} \widetilde{a}(\lambda^{-1}) & \lambda^{-1} \cdot \widetilde{b}(\lambda^{-1}) \\ \lambda \cdot \widetilde{c}(\lambda^{-1}) & \widetilde{d}(\lambda^{-1}) \end{pmatrix}. \tag{5.31}$$

By the result for $\lambda \to \infty$, applied to $\widetilde{u}$, we have for $|\lambda^{-1}| \geq R$

$$\begin{aligned}
|\widetilde{a}(\lambda^{-1}) - e^{(\widetilde{u}(1)-\widetilde{u}(0))/4} \cos(\zeta(\lambda^{-1}))| &\leq \varepsilon\, w(\lambda^{-1}) \\
|\widetilde{b}(\lambda^{-1}) - (-\lambda^{1/2}\, e^{(\widetilde{u}(0)+\widetilde{u}(1))/4} \sin(\zeta(\lambda^{-1}))| &\leq \varepsilon\, |\lambda|^{1/2}\, w(\lambda^{-1}) \\
|\widetilde{c}(\lambda^{-1}) - \lambda^{-1/2}\, e^{-(\widetilde{u}(0)+\widetilde{u}(1))/4} \sin(\zeta(\lambda^{-1}))| &\leq \varepsilon\, |\lambda|^{-1/2}\, w(\lambda^{-1}) \\
|\widetilde{d}(\lambda^{-1}) - e^{-(\widetilde{u}(1)-\widetilde{u}(0))/4} \cos(\zeta(\lambda^{-1}))| &\leq \varepsilon\, w(\lambda^{-1}).
\end{aligned}$$

By Eq. (5.31), and $\zeta(\lambda^{-1}) = \zeta(\lambda)$, $w(\lambda^{-1}) = w(\lambda)$, $\widetilde{u}(0) = -u(0)$, we now obtain for $|\lambda| \leq 1/R$

$$\begin{aligned}
|a(\lambda) - e^{-(u(1)-u(0))/4} \cos(\zeta(\lambda))| &\leq \varepsilon\, w(\lambda) \\
|\lambda \cdot b(\lambda) - (-\lambda^{1/2})\, e^{-(u(0)+u(1))/4} \sin(\zeta(\lambda))| &\leq \varepsilon\, |\lambda|^{1/2}\, w(\lambda)
\end{aligned}$$

$$|\lambda^{-1} \cdot c(\lambda) - \lambda^{-1/2} \, e^{(u(0)+u(1))/4} \, \sin(\zeta(\lambda))| \le \varepsilon \, |\lambda|^{-1/2} \, w(\lambda)$$
$$|d(\lambda) - e^{(u(1)-u(0))/4} \, \cos(\zeta(\lambda))| \le \varepsilon \, w(\lambda)$$

and therefore the statement claimed for $\lambda \to 0$. It is clear that also the statement on the uniformness of the estimate for $(u, u_y) \in P$ transfers to the case $\lambda \to 0$. □

The following Lemma, which is a variant of the Riemann-Lebesgue Lemma, has been used in the preceding proof:

Lemma 5.6 *Let N be a compact subset of $L^1([a, b])$. For any given $\varepsilon, C > 0$ there then exists $R > 0$ such that for every $\zeta \in \mathbb{C}$ with $|\operatorname{Im}(\zeta)| \le C$, $|\operatorname{Re}(\zeta)| \ge R$ and every $g \in N$ we have*

$$\left| \int_a^b e^{-2\pi i \zeta t} \, g(t) \, dt \right| \le \varepsilon.$$

Proof We extend the functions in $L^1([a, b])$ to $\mathbb{R}$ by zero, then we have $N \subset L^1(\mathbb{R})$. The map

$$\Phi : L^1(\mathbb{R}) \times [-C, C] \to L^1(\mathbb{R}), \ (g, y) \mapsto (t \mapsto e^{2\pi y t} \cdot g(t))$$

is continuous, hence the image $\widetilde{N} := \Phi(N \times [-C, C])$ is a compact set in $L^1(\mathbb{R})$.

Therefore there exist finitely many $f_1, \dots, f_n \in L^1(\mathbb{R})$ with $\widetilde{N} \subset \bigcup_{k=1}^n B(f_k, \frac{\varepsilon}{2})_{L^1}$. By the classical Riemann-Lebesgue Lemma (see e.g. [Gra, Proposition 2.2.17, p. 105]), there exists $R > 0$, so that we have $|\widehat{f_k}(x)| \le \frac{\varepsilon}{2}$ for every $x \in \mathbb{R}$ with $|x| \ge R$ and every $k \in \{1, \dots, n\}$; here we denote for any $f \in L^1(\mathbb{R})$ by

$$\widehat{f}(x) := \int_{\mathbb{R}} e^{-2\pi i x t} \, f(t) \, dt$$

the Fourier transform of f.

Now let $g \in N$ and $\zeta = x + iy \in \mathbb{C}$ be given with $|x| \ge R$, $|y| \le C$. By construction, there exists some $k \in \{1, \dots, n\}$ with $\|\Phi(g, y) - f_k\|_1 \le \frac{\varepsilon}{2}$, and with this k we have

$$\|\widehat{\Phi(g, y)} - \widehat{f_k}\|_\infty \le \|\Phi(g, y) - f_k\|_1 \le \frac{\varepsilon}{2}$$

by the Hausdorff-Young inequality for the case $p' = \infty$, $p = 1$ [Gra, Proposition 2.2.16, p. 104]. We now have

$$\left| \int_a^b e^{-2\pi i \zeta t} \, g(t) \, dt \right| = \left| \int_a^b e^{-2\pi i x t} \, e^{2\pi y t} \, g(t) \, dt \right| = \left| \int_a^b e^{-2\pi i x t} \, \Phi(g, y)(t) \, dt \right|$$
$$= \left| \widehat{\Phi(g, y)}(x) \right| \le \left| \widehat{\Phi(g, y)}(x) - \widehat{f_k}(x) \right| + \left| \widehat{f_k}(x) \right| \le \frac{\varepsilon}{2} + \frac{\varepsilon}{2} = \varepsilon.$$

□

Corollary 5.7 *In the setting of Theorem 5.4 we have the following additional facts:*

(1) There exists $C > 0$ *(dependent on* (u, u_y)*) so that we have*

$$|a(\lambda)| \leq C\, w(\lambda)$$
$$|b(\lambda)| \leq C\, |\lambda|^{-1/2}\, w(\lambda)$$
$$|c(\lambda)| \leq C\, |\lambda|^{1/2}\, w(\lambda)$$
$$|d(\lambda)| \leq C\, w(\lambda).$$

(2) There exists for every $\varepsilon > 0$ *some* $R > 0$*, so that for every* $\lambda \in \mathbb{C}^*$ *with* $|\lambda| \geq R$ *or* $|\lambda| \leq \frac{1}{R}$ *we have*

(a) $|b(\lambda)\, c(\lambda) - b_0(\lambda)\, c_0(\lambda)| \leq \varepsilon\, w(\lambda)^2.$

We now suppose $(u, u_y) \in \mathsf{Pot}$ *and define* $\Delta(\lambda) := \mathrm{tr}(M(\lambda)) = a(\lambda) + d(\lambda)$*. For every* $\varepsilon > 0$ *there exists some* $R > 0$*, so that for every* $\lambda \in \mathbb{C}^*$ *with* $|\lambda| \geq R$ *or* $|\lambda| \leq \frac{1}{R}$ *we have*

(b) $|\Delta(\lambda) - \Delta_0(\lambda)| \leq \varepsilon\, w(\lambda).$
(c) $|(\Delta(\lambda)^2 - 4) - (\Delta_0(\lambda)^2 - 4)| \leq \varepsilon\, w(\lambda)^2.$

The constants C *resp.* R *can be chosen uniformly for* $(u, u_y) \in P$*, where* P *is a relatively compact subset of* Pot_{np}*.*

Proof For (1). By Theorem 5.4 (applied with $\varepsilon = 1$) there exists $R > 0$ so that we have for $\lambda \in \mathbb{C}^*$ with $|\lambda| \geq R$ or $|\lambda| \leq \frac{1}{R}$

$$|a(\lambda) - \upsilon\, a_0(\lambda)| \leq w(\lambda)$$

and therefore

$$|a(\lambda)| \leq |\upsilon| \cdot |a_0(\lambda)| + |a(\lambda) - \upsilon\, a_0(\lambda)| \leq |\upsilon| \cdot w(\lambda) + w(\lambda) = (|\upsilon| + 1) \cdot w(\lambda).$$

Because $\{\frac{1}{R} \leq |\lambda| \leq R\}$ is compact, a is bounded on this set by some constant $C_1 > 0$. Because $w(\lambda) \geq 1$ holds for all $\lambda \in \mathbb{C}^*$, we then have $|a(\lambda)| \leq C\, w(\lambda)$ for all $\lambda \in \mathbb{C}^*$ with $C := \max\{|\upsilon| + 1, C_1\}$. The claims on b, c and d are shown similarly.

For (2). For any $\widetilde{\varepsilon} > 0$ we have by Theorem 5.4 some $R > 0$ so that the estimates from that theorem hold for $|\lambda| \geq R$ resp. for $|\lambda| \leq \frac{1}{R}$.

We then have for $|\lambda| \geq R$

$$\begin{aligned}
&|b(\lambda)\, c(\lambda) - b_0(\lambda)\, c_0(\lambda)| \\
&\quad = |b(\lambda)\, c(\lambda) - \tau^{-1}\, b_0(\lambda)\, \tau\, c_0(\lambda)| \\
&\quad \leq |b(\lambda) - \tau^{-1}\, b_0(\lambda)| \cdot |c(\lambda)| + |\tau|^{-1}\, |b_0(\lambda)| \cdot |c(\lambda) - \tau\, c_0(\lambda)|
\end{aligned}$$

$$\leq \widetilde{\varepsilon}\,|\lambda|^{-1/2}\,w(\lambda)\cdot C\,|\lambda|^{1/2}\,w(\lambda) + |\tau|^{-1}\,|\lambda|^{-1/2}\,w(\lambda)\cdot\widetilde{\varepsilon}\,|\lambda|^{1/2}\,w(\lambda)$$
$$= (C + |\tau|^{-1})\,\widetilde{\varepsilon}\,w(\lambda)^2.$$

A similar calculation gives for $|\lambda| \leq \frac{1}{R}$

$$|b(\lambda)\,c(\lambda) - b_0(\lambda)\,c_0(\lambda)| \leq (C + |\tau|)\,\widetilde{\varepsilon}\,w(\lambda)^2.$$

By choosing $\widetilde{\varepsilon} := (C + \max\{|\tau|, |\tau|^{-1}\})^{-1}\cdot\varepsilon$, we obtain (2)(a).

We now suppose that we have $(u, u_y) \in \mathsf{Pot}$ and therefore $\upsilon = 1$. We then have

$$|\Delta(\lambda) - \Delta_0(\lambda)| \leq |a(\lambda) - a_0(\lambda)| + |d(\lambda) - d_0(\lambda)| \leq 2\widetilde{\varepsilon}\,w(\lambda)\,,$$

therefore we obtain (2)(b) by choosing $\widetilde{\varepsilon} := \varepsilon/2$. Finally, we have

$$\begin{aligned}|(\Delta(\lambda)^2 - 4) - (\Delta_0(\lambda)^2 - 4)| &= |\Delta(\lambda) + \Delta_0(\lambda)|\cdot|\Delta(\lambda) - \Delta_0(\lambda)|\\ &\leq (2|\Delta_0(\lambda)| + |\Delta(\lambda) - \Delta_0(\lambda)|)\cdot|\Delta(\lambda) - \Delta_0(\lambda)|\\ &\overset{(b)}{\leq} (4\,w(\lambda) + \widetilde{\varepsilon}\,w(\lambda))\cdot\widetilde{\varepsilon}\,w(\lambda) = (4 + \widetilde{\varepsilon})\,\widetilde{\varepsilon}\,w(\lambda)^2.\end{aligned}$$

By choosing $\widetilde{\varepsilon} > 0$ such that $(4 + \widetilde{\varepsilon})\,\widetilde{\varepsilon} = \varepsilon$, we obtain (2)(c). □

The following corollary shows that the derivatives (with respect to λ) of the functions comprising the monodromy can also be estimated via Theorem 5.4. This is basically a consequence of Cauchy's inequality, which is applicable because the monodromy depends holomorphically on λ. For reference, we note

$$a_0'(\lambda) = d_0'(\lambda) = \frac{1-\lambda}{8\,\lambda^2}\,c_0(\lambda) \tag{5.32}$$

$$b_0'(\lambda) = \frac{1-\lambda}{8\,\lambda}\,a_0(\lambda) - \frac{1}{2\,\lambda}\,b_0(\lambda) \tag{5.33}$$

$$c_0'(\lambda) = \frac{\lambda-1}{8\,\lambda}\,a_0(\lambda) + \frac{1}{2\,\lambda}\,c_0(\lambda). \tag{5.34}$$

Corollary 5.8 *In the situation of Theorem 5.4, for every $\varepsilon > 0$ there exists $R' > 0$ such that:*

(1) For all $\lambda \in \mathbb{C}$ with $|\lambda| \geq R'$, we have

$$|a'(\lambda) - \upsilon\,a_0'(\lambda)| \leq \varepsilon\,|\lambda|^{-1/2}\,w(\lambda)$$
$$|b'(\lambda) - \tau^{-1}\,b_0'(\lambda)| \leq \varepsilon\,|\lambda|^{-1}\,w(\lambda)$$
$$|c'(\lambda) - \tau\,c_0'(\lambda)| \leq \varepsilon\,w(\lambda)$$
$$|d'(\lambda) - \upsilon^{-1}\,d_0'(\lambda)| \leq \varepsilon\,|\lambda|^{-1/2}\,w(\lambda).$$

(2) For all $\lambda \in \mathbb{C}^*$ *with* $|\lambda| \leq \frac{1}{R'}$, *we have*

$$\begin{aligned}
|a'(\lambda) - \upsilon^{-1} a_0'(\lambda)| &\leq \varepsilon\, |\lambda|^{-3/2}\, w(\lambda) \\
|b'(\lambda) - \tau\, b_0'(\lambda)| &\leq \varepsilon\, |\lambda|^{-2}\, w(\lambda) \\
|c'(\lambda) - \tau^{-1} c_0'(\lambda)| &\leq \varepsilon\, |\lambda|^{-1}\, w(\lambda) \\
|d'(\lambda) - \upsilon\, d_0'(\lambda)| &\leq \varepsilon\, |\lambda|^{-3/2}\, w(\lambda).
\end{aligned}$$

The constant R' *can be chosen uniformly for* $(u, u_y) \in P$, *where* P *is a relatively compact subset of* Pot_{np}.

Proof We show the estimates for the function a', the other functions are handled analogously. Let $\varepsilon > 0$ be given, and fix $\delta > 0$. By Theorem 5.4 there exists $R > 0$ such that for $\lambda \in \mathbb{C}$ with $|\lambda| \geq R$ we have

$$|a(\lambda) - \upsilon\, a_0(\lambda)| \leq \frac{\varepsilon\, \delta}{2\, e^{\delta}}\, w(\lambda).$$

There exists $R' > R$ such that for all $\lambda \in \mathbb{C}$ with $|\lambda| \geq R'$, the closure $\overline{U}$ of $U := \{\lambda' \in \mathbb{C} \,\big|\, |\lambda' - \lambda| < \delta\, |\lambda|^{1/2}\}$ is entirely contained in $\{|\lambda'| \geq R\}$. For $\lambda' \in \overline{U}$, we have $w(\lambda') \leq 2\, e^{\delta}\, w(\lambda)$ by Proposition 5.3(1), and therefore by Cauchy's inequality, applied to the holomorphic function $a - a_0$:

$$\begin{aligned}
|a'(\lambda) - \upsilon\, a_0'(\lambda)| &\leq \frac{1}{\delta\, |\lambda|^{1/2}} \max_{\lambda' \in \partial U} |a(\lambda') - \upsilon\, a_0(\lambda')| \\
&\leq \frac{1}{\delta\, |\lambda|^{1/2}} \cdot \frac{\varepsilon\, \delta}{2\, e^{\delta}} \cdot 2\, e^{\delta}\, w(\lambda) = \varepsilon\, |\lambda|^{-1/2}\, w(\lambda).
\end{aligned}$$

For the case $\lambda \to 0$, we take $U := \{\lambda' \in \mathbb{C}^* \,\big|\, |\lambda' - \lambda| < \delta\, |\lambda|^{3/2}\}$, and again choose $R' > R$, such that for $|\lambda| \leq \frac{1}{R'}$, we have $\overline{U} \subset \{|\lambda'| \leq \frac{1}{R}\}$. We again have $w(\lambda') \leq 2\, e^{\delta}\, w(\lambda)$ for $\lambda' \in \overline{U}$, and therefore by Cauchy's inequality

$$\begin{aligned}
|a'(\lambda) - \upsilon^{-1} a_0'(\lambda)| &\leq \frac{1}{\delta\, |\lambda|^{3/2}} \max_{\lambda' \in \partial U} |a(\lambda') - \upsilon^{-1} a_0(\lambda')| \\
&\leq \frac{1}{\delta\, |\lambda|^{3/2}} \cdot \frac{\varepsilon\, \delta}{2\, e^{\delta}} \cdot 2\, e^{\delta}\, w(\lambda) = \varepsilon\, |\lambda|^{-3/2}\, w(\lambda).
\end{aligned}$$

□

For Cauchy data (u, u_y) that are once more differentiable, we obtain an even better asymptotic estimate. For stating this asymptotic estimate, we introduce the space of (non-periodic) once more differentiable potentials

$$\mathsf{Pot}_{np}^{1} := \{\, (u, u_y) \,|\, u \in W^{2,2}([0, 1]),\, u_y \in W^{1,2}([0, 1]) \,\}\,, \tag{5.35}$$

and phrase the statement in terms of the regauged monodromy E from the proof of Theorem 5.4:

Proposition 5.9 *Suppose that we have $(u, u_y) \in \mathsf{Pot}^1_{np}$. Then there exist constants $R, C > 0$, such that for every $\lambda \in \mathbb{C}$ with $|\lambda| \geq R$ and $x \in [0, 1]$, we have*

$$|E_\lambda(x) - E_{0,\lambda}(x)| \leq \frac{C}{\sqrt{|\lambda|}}\, w(\lambda).$$

If P is a relatively compact subset of Pot^1_{np}, then the constants R, C can be chosen uniformly for $(u, u_y) \in P$.

Proof We use the objects and notations from the proof of Theorem 5.4. In particular we use the regauging described at the beginning at that proof. But now we can use the higher degree of differentiability of (u, u_y) to eliminate the λ^0-component from $\widetilde{\alpha}$ by regauging $\widetilde{\alpha}$ resp. $\widetilde{F}$ once more, with

$$h := \begin{pmatrix} \lambda^{-1/2}\, u_z & 1 \\ -1 & -\lambda^{-1/2}\, u_z \end{pmatrix}. \tag{5.36}$$

We have

$$h^{-1} = \frac{1}{1 - \lambda^{-1}\, u_z^2} \begin{pmatrix} -\lambda^{-1/2}\, u_z & -1 \\ 1 & \lambda^{-1/2}\, u_z \end{pmatrix} \quad \text{and}$$

$$\mathrm{d}h = \begin{pmatrix} \lambda^{-1/2}\, u_{zx} & 0 \\ 0 & -\lambda^{-1/2}\, u_{zx} \end{pmatrix}, \tag{5.37}$$

and from this, we obtain for the again regauged $\widetilde{\alpha}$ by an explicit calculation

$$\overline{\alpha} := h^{-1}\, \widetilde{\alpha}\, h - h^{-1}\, \mathrm{d}h = \widetilde{\alpha}_0 + \lambda^{-1/2} \cdot (\overline{\beta}_+ + \overline{\beta}_-) + O(|\lambda|^{-1}) \text{ with}$$

$$\overline{\beta}_+ := \begin{pmatrix} 0 & \frac{1}{2}\, u_z^2 - \frac{1}{4}\, \cosh(u) + 1 \\ -\frac{1}{2}\, u_z^2 + \frac{1}{4}\, \cosh(u) - 1 & 0 \end{pmatrix} \text{ and}$$

$$\overline{\beta}_- := \begin{pmatrix} 0 & -u_{zx} - \frac{1}{4}\, \sinh(u) \\ -u_{zx} - \frac{1}{4}\, \sinh(u) & 0 \end{pmatrix}.$$

We also consider $\overline{E}(x) := h^{-1}(x)\, E(x)\, h(0)$. From Eqs. (5.21), (5.36) and (5.37), we obtain by an explicit calculation

$$h^{-1}\, E_0\, h = E_0 + O(|\lambda|^{-1/2}). \tag{5.38}$$

By Lemma 5.5(2)(b) we have $\overline{E} = E_0 + \sum_{n=1}^{\infty} \overline{E}_n$ with

$$\overline{E}_n(x) = \lambda^{-n/2} \sum_{\varepsilon \in \{\pm 1\}^n} \int_0^x \int_0^{t_1} \cdots \int_0^{t_{n-1}} E_0(\xi_\varepsilon(t))\, \overline{\beta}_{\varepsilon_1}(t_1)\, \overline{\beta}_{\varepsilon_2}(t_2) \ldots \overline{\beta}_{\varepsilon_n}(t_n)\, \mathrm{d}^n t + O(|\lambda|^{-1}),$$

where ξ_ε is defined as in Lemma 5.5(2)(b). We have $\xi_\varepsilon(t) \in [-x, x]$ and therefore

$$|E_0(\xi_\varepsilon(t))| \leq |E_0(x)| \leq |E_0(1)| = w(\lambda).$$

We therefore obtain

$$|\overline{E}_n(x)| \leq |\lambda|^{-n/2} \cdot 2^n \cdot \frac{1}{n!} \cdot w(\lambda) \cdot (\|\overline{\beta}_+\|_\infty + \|\overline{\beta}_-\|_\infty)^n.$$

It follows that the sum $\sum_{n=0}^{\infty} \overline{E}_n$ converges uniformly to some $\overline{E}$, which is a solution of $\overline{E}'(x) = \overline{\alpha}(x)\,\overline{E}(x)$, $\overline{E}(0) = \mathbb{1}$, and we have

$$\begin{aligned} |\overline{E}(x) - \overline{E}_0(x)| &\leq \left(\exp(2\,|\lambda|^{-1/2}\,(\|\overline{\beta}_+\|_\infty + \|\overline{\beta}_-\|_\infty)) - 1\right) \\ &\quad \cdot w(\lambda) \leq C_1\,|\lambda|^{-1/2}\,w(\lambda) \end{aligned} \tag{5.39}$$

with a suitable constant $C_1 > 0$.

Because we have

$$E(x) - E_0(x) = h(x)\left(\overline{E}(x) - \overline{E}_0(x) + E_0(x) - h(x)^{-1}\,E_0(x)\,h(0)\right) h(0)^{-1}\,,$$

it follows from the estimates (5.39) and (5.38), along with the fact that h and h^{-1} are bounded with respect to $\lambda \in \mathbb{C}^*$, that the estimate given in the Proposition holds. □

Chapter 6
Basic Behavior of the Spectral Data

We would like to show that the spectral data $(\Sigma, \mathscr{D})$ corresponding to a potential behave like the spectral data $(\Sigma_0, \mathscr{D}_0)$ of the vacuum (as described in Chap. 4) "asymptotically" for λ near ∞, and for λ near 0. In particular, we would like to show that the classical spectral divisor D corresponding to $\mathscr{D}$, like the classical spectral divisor D_0 of the vacuum, is composed of a $\mathbb{Z}$-sequence of points $(\lambda_k, \mu_k)_{k\in\mathbb{Z}}$, and that for $|k|$ large, $(\lambda_k, \mu_k) \in D$ is near $(\lambda_{k,0}, \mu_{k,0}) \in D_0$. Similarly, we will show that the set of zeros of $\Delta^2 - 4$ with multiplicities (corresponding to the branch points resp. singularities of the spectral curve Σ) is enumerated by two $\mathbb{Z}$-sequences $(\varkappa_{k,1})$ and $(\varkappa_{k,2})$ such that for $|k|$ large, $\varkappa_{k,1}$ and $\varkappa_{k,2}$ are near $\lambda_{k,0}$.

For a first quantified description of these asymptotic claims, we introduce the concept of so-called *excluded domains* around the double points of Σ_0.

For this, let $\delta > 0$ be given. If δ is sufficiently small, certainly for $\delta < \pi - \frac{1}{2}$, the open set $\{\, \lambda \in \mathbb{C}^* \mid |\zeta(\lambda) - \zeta(\lambda_{k,0})| < \delta \,\}$ (where $\zeta(\lambda)$ is here and henceforth again defined as in Eq. (4.2), and used as a local coordinate on $\mathbb{C}^*$) has two connected components for every $k \in \mathbb{Z} \setminus \{0\}$, one of them contained in $\{|\lambda| > 1\}$, the other in $\{|\lambda| < 1\}$; whereas it is connected for $k = 0$.

Proof $O := \{\, \lambda \in \mathbb{C}^* \mid |\zeta(\lambda) - \zeta(\lambda_{k,0})| < \delta \,\}$ is invariant under $\lambda \mapsto \lambda^{-1}$, thus it can be connected only if O contains at least one $\lambda \in \mathbb{C}^*$ with $|\lambda| = 1$. Because of $\zeta(e^{it}) = \frac{1}{4}(e^{it/2} + e^{-it/2}) = \frac{1}{2}\cos(t/2)$, we see that $|\lambda| = 1$ implies $|\zeta(\lambda)| \in [0, \frac{1}{2}]$. Therefore O cannot contain any λ with $|\lambda| = 1$ if $\delta < \pi - \frac{1}{2}$. □

Definition 6.1 Let $0 < \delta < \pi - \frac{1}{2}$ be given. Then we put for $k \in \mathbb{Z}$

$$U_{k,\delta} := \begin{cases} \{\, \lambda \in \mathbb{C}^* \mid |\zeta(\lambda) - \zeta(\lambda_{k,0})| < \delta,\ |\lambda| > 1 \,\} & \text{for } k > 0 \\ \{\, \lambda \in \mathbb{C}^* \mid |\zeta(\lambda)| < \delta \,\} & \text{for } k = 0 \\ \{\, \lambda \in \mathbb{C}^* \mid |\zeta(\lambda) - \zeta(\lambda_{k,0})| < \delta,\ |\lambda| < 1 \,\} & \text{for } k < 0 \end{cases} .$$

S. Klein, *A Spectral Theory for Simply Periodic Solutions of the Sinh-Gordon Equation*, Lecture Notes in Mathematics 2229,
https://doi.org/10.1007/978-3-030-01276-2_6

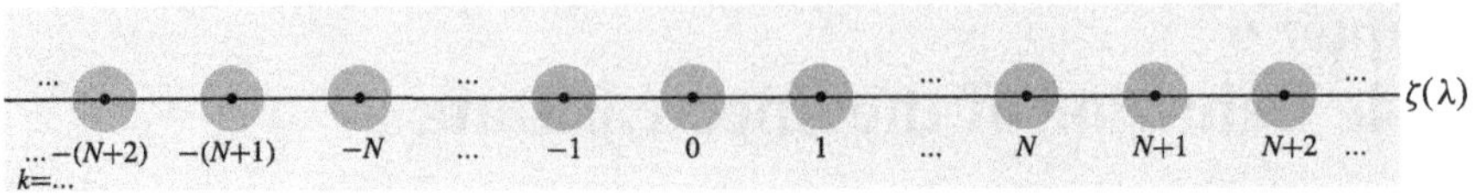

Fig. 6.1 The excluded domains $U_{k,\delta}$ in ζ-coordinates. The black points are the $\zeta(\lambda_{k,0})$. The numbers below the neighborhoods indicate the corresponding value of the index k

We call $U_{k,\delta}$ the *excluded domain* near $\lambda_{k,0}$ of radius δ. We also consider the *union of all excluded domains* $U_\delta := \bigcup_{k\in\mathbb{Z}} U_{k,\delta}$ and the *area outside the excluded domains* $V_\delta := \mathbb{C}^* \setminus U_\delta$.

We will see that the excluded domains $U_{k,\delta}$ are the appropriate neighborhoods of the $\lambda_{k,0}$ to use for the purpose of describing the basic asymptotic behavior of the spectral data. Note that in λ-coordinates, the diameter of $U_{k,\delta}$ increases on the order of k when k is positive, and decreases on the order of k^{-1} when k is negative. In ζ-coordinates however, the diameter of $U_{k,\delta}$ does not depend on k. Figure 6.1 depicts the situation in the ζ-coordinate, where we have again adopted the convention of using the standard branch of the square root function where k is positive, and its negative when k is negative.

In the sequel, we will use the domains $U_{k,\delta}$, U_δ and V_δ permanently without further explicit reference.

Proposition 6.2 *There exist constants* $C_1, C_2 > 0$, *such that for every* $0 < \delta < \pi - \frac{1}{2}$, *every* $k \in \mathbb{Z}$ *and every* $\lambda_1, \lambda_2 \in U_{k,\delta}$ *we have*

(1) For $k > 0$: $C_1 \cdot |k|^{-1} \cdot |\lambda_1 - \lambda_2| \leq |\zeta(\lambda_1) - \zeta(\lambda_2)| \leq C_2 \cdot |k|^{-1} \cdot |\lambda_1 - \lambda_2|$
For $k < 0$: $C_1 \cdot |k|^3 \cdot |\lambda_1 - \lambda_2| \leq |\zeta(\lambda_1) - \zeta(\lambda_2)| \leq C_2 \cdot |k|^3 \cdot |\lambda_1 - \lambda_2|$

(2) For $k > 0$: $B(\lambda_{k,0}, \delta C_1 |k|) \subset U_{k,\delta} \subset B(\lambda_{k,0}, \delta C_2 |k|)$
For $k < 0$: $B(\lambda_{k,0}, \delta C_1 |k|^{-3}) \subset U_{k,\delta} \subset B(\lambda_{k,0}, \delta C_2 |k|^{-3})$
Here $B(z_0, r) := \{ z \in \mathbb{C} \mid |z - z_0| < r \}$ *is the open Euclidean ball of radius* $r > 0$ *around* $z_0 \in \mathbb{C}$ *in* $\mathbb{C}$.

Proof For (1). For $k > 0$ we have

$$\zeta(\lambda_1) - \zeta(\lambda_2) = \frac{1}{4} \cdot \left(\sqrt{\lambda_1} - \sqrt{\lambda_2} + \frac{1}{\sqrt{\lambda_1}} - \frac{1}{\sqrt{\lambda_2}} \right)$$
$$= \frac{1}{4} \cdot \left(1 - \frac{1}{\sqrt{\lambda_1 \lambda_2}} \right) \cdot \frac{1}{\sqrt{\lambda_1} + \sqrt{\lambda_2}} \cdot (\lambda_1 - \lambda_2).$$

Because of $\lambda_\nu \in U_{k,\delta}$ for $\nu \in \{1,2\}$ and $k > 0$, we have $|\lambda_\nu| > 2$, and therefore $\frac{1}{2} \leq \left| 1 - \frac{1}{\sqrt{\lambda_1 \lambda_2}} \right| \leq \frac{3}{2}$, and moreover, $\left| \frac{1}{\sqrt{\lambda_1}+\sqrt{\lambda_2}} \right|$ is of the order of k. Herefrom, the claimed statement follows.

For $k < 0$, we use the fact that $\zeta(\lambda_\nu) = \zeta(\lambda_\nu^{-1})$ holds, and that we have $\lambda_\nu^{-1} \in U_{-k,\delta}$ where $-k > 0$. By application of the first part of the proof, we thus obtain

$$C_1 \cdot |k|^{-1} \cdot |\lambda_1^{-1} - \lambda_2^{-1}| \leq |\zeta(\lambda_1) - \zeta(\lambda_2)| \leq C_2 \cdot |k|^{-1} \cdot |\lambda_1^{-1} - \lambda_2^{-1}|.$$

Because of $|\lambda_1^{-1} - \lambda_2^{-1}| = \frac{|\lambda_1 - \lambda_2|}{|\lambda_1 \lambda_2|}$ and because $\frac{1}{|\lambda_\nu|}$ is of order k^2, we obtain the claimed statement also in this case.

For (2). This is an immediate consequence of (1). □

Proposition 6.3 *Let $0 < \delta < \pi - \frac{1}{2}$ be given.*

(1) $w(\lambda)$ is bounded on U_δ, more precisely we have for every $\lambda \in U_\delta$:

$$\tfrac{1}{2} \leq w(\lambda) \leq 2\,e^\delta.$$

(2) $\cot \circ \zeta$ is bounded on V_δ.
(3) There exists a constant $C > 1$ (dependent on δ) so that $|\sin(\zeta(\lambda))| \leq w(\lambda) \leq C\,|\sin(\zeta(\lambda))|$ holds for all $\lambda \in V_\delta$.

Proof For (1). Let $\lambda \in U_\delta$ be given. Then there exists $k \in \mathbb{Z}$ with $\lambda \in U_{k,\delta}$ and therefore $|\zeta(\lambda) - \zeta(\lambda_{k,0})| < \delta$. We have $\zeta(\lambda_{k,0}) = |k|\pi \in \mathbb{R}$, and therefore it follows that $|\operatorname{Im}(\zeta(\lambda))| < \delta$. Thus we obtain from Proposition 5.3(1):

$$\tfrac{1}{2} \leq \tfrac{1}{2}\,e^{|\operatorname{Im}(\zeta(\lambda))|} \leq w(\lambda) \leq 2\,e^{|\operatorname{Im}(\zeta(\lambda))|} \leq 2\,e^\delta.$$

For (2). It suffices to show that $\cot$ is bounded on $\mathbb{C} \setminus \bigcup_{k\in\mathbb{Z}} B(k\pi, \delta)$, and for this, it in turn suffices to show that $\cot$ is bounded on the sets

$$M_1 := \{\, x + iy \in \mathbb{C} \,\big|\, |x - k\pi| \geq \tfrac{\delta}{2} \text{ for all } k \in \mathbb{Z} \text{ and } |y| \leq \tfrac{\delta}{2} \,\}$$

and

$$M_2 := \{\, x + iy \in \mathbb{C} \,\big|\, |y| \geq \tfrac{\delta}{2} \,\}.$$

M_1 is the union of translations by integer multiples of π of the rectangle $Q := [\frac{\delta}{2}, \pi - \frac{\delta}{2}] \times [-\frac{\delta}{2}, \frac{\delta}{2}]$. $\cot$ is bounded on the compact set Q; because it is π-periodic, it follows that $\cot$ is also bounded on all of M_1.

To show that $\cot$ is bounded on M_2, we use the formula

$$\cot(x + iy) = \frac{\sin(2x)}{\cosh(2y) - \cos(2x)} - i\,\frac{\sinh(2y)}{\cosh(2y) - \cos(2x)}.$$

If $x + iy \in M_1$ is given, we have

$$\left|\frac{\sin(2x)}{\cosh(2y) - \cos(2x)}\right| \leq \frac{1}{\cosh(\delta) - 1}$$

and

$$\left| \frac{\sinh(2y)}{\cosh(2y)-\cos(2x)} \right| \leq \frac{|\sinh(2y)|}{\cosh(2y)-1};$$

note that $y \mapsto \frac{|\sinh(2y)|}{\cosh(2y)-1}$ is real for $y \neq 0$, and tends to 1 for $y \to \pm\infty$, hence is bounded on $\{|y| \geq \frac{\delta}{2}\}$. Thus it follows that cot is bounded on M_2.

For (3). Because of

$$w(\lambda) = |\cos(\zeta(\lambda))| + |\sin(\zeta(\lambda))| , \tag{6.1}$$

the inequality $|\sin(\zeta(\lambda))| \leq w(\lambda)$ is obvious. On the other hand, $\cot \circ \zeta$ is bounded on V_δ by (2), thus there exists $C > 1$ with $|\cot(\zeta(\lambda))| \leq C - 1$ for $\lambda \in V_\delta$. It follows from Eq. (6.1) that we have

$$w(\lambda) = |\sin(\zeta(\lambda))| \cdot (|\cot(\zeta(\lambda))| + 1) \leq |\sin(\zeta(\lambda))| \cdot C.$$

□

We now suppose that "a monodromy" that has the asymptotic behavior of Theorem 5.4 is given, that is to say: Suppose that a (2×2)-matrix

$$M(\lambda) = \begin{pmatrix} a(\lambda) & b(\lambda) \\ c(\lambda) & d(\lambda) \end{pmatrix}$$

of holomorphic functions $a, b, c, d : \mathbb{C}^* \to \mathbb{C}$ is given, and denote by

$$M_0(\lambda) = \begin{pmatrix} a_0(\lambda) & b_0(\lambda) \\ c_0(\lambda) & d_0(\lambda) \end{pmatrix} = \begin{pmatrix} \cos(\zeta(\lambda)) & -\lambda^{-1/2} \sin(\zeta(\lambda)) \\ \lambda^{1/2} \sin(\zeta(\lambda)) & \cos(\zeta(\lambda)) \end{pmatrix},$$

the monodromy of the vacuum, compare Eq. (4.5). Then we require of $M(\lambda)$ that $\det(M(\lambda)) = 1$ holds for all $\lambda \in \mathbb{C}^*$ and that there exist constants $\tau, \upsilon \in \mathbb{C}^*$, such that for every $\varepsilon > 0$ there exists $R > 0$ so that for every $\lambda \in \mathbb{C}^*$ with $|\lambda| \geq R$ resp. with $|\lambda| \leq \frac{1}{R}$ the estimates of Theorem 5.4(1) resp. (2) hold. We note that in this setting, the Corollaries 5.7 and 5.8 also hold.

We also use the trace functions $\Delta(\lambda) := a(\lambda) + d(\lambda)$ and $\Delta_0(\lambda) = a_0(\lambda) + d_0(\lambda) = 2\cos(\zeta(\lambda))$ again. In the case $\upsilon = 1$ we define the spectral curve Σ, the generalized spectral divisor $\mathscr{D}$ associated to $M(\lambda)$ and the underlying classical divisor D in the way described in Chap. 3. For $\upsilon \neq 1$ we avoid defining the spectral curve (as it would not be of any interest, because it would not be asymptotically close to any "reasonable" curve), but we still define the classical spectral divisor D as a multi-set of points in $\mathbb{C}^* \times \mathbb{C}^*$ by Eq. (3.14).

Theorem 5.4 shows that this situation is at hand in particular when a potential $(u, u_y) \in \mathsf{Pot}_{np}$ is given and $M(\lambda)$ is the monodromy associated to it; for $(u, u_y) \in \mathsf{Pot}$ we have $\upsilon = 1$.

Whenever we have a spectral curve Σ, we consider also the pre-images $\widehat{U}_{k,\delta}$, $\widehat{U}_\delta$ and $\widehat{V}_\delta$ in Σ of the excluded domains $U_{k,\delta}$, their union $U_\delta = \bigcup_{k\in\mathbb{Z}} U_{k,\delta}$ and the complement $V_\delta = \mathbb{C}^* \setminus U_\delta$. More explicitly, we define for $\delta > 0$ and $k \in \mathbb{Z}$

$$\begin{aligned}
\widehat{U}_{k,\delta} &:= \{ (\lambda, \mu) \in \Sigma \mid \lambda \in U_{k,\delta} \} , \\
\widehat{U}_\delta &:= \{ (\lambda, \mu) \in \Sigma \mid \lambda \in U_\delta \} = \bigcup_{k\in\mathbb{Z}} \widehat{U}_{k,\delta} , \\
\widehat{V}_\delta &:= \{ (\lambda, \mu) \in \Sigma \mid \lambda \in V_\delta \} = \Sigma \setminus \widehat{U}_\delta .
\end{aligned}$$

We call also the $\widehat{U}_{k,\delta}$ *excluded domains* (in Σ).

Definition 6.4 Let $X \subset \mathbb{C}^*$ resp. $\widehat{X} \subset \Sigma$ be any subset, $\delta > 0$ and $n \in \mathbb{N}_0$.

(1) We say that the excluded domains $U_{k,\delta}$ resp. $\widehat{U}_{k,\delta}$ *asymptotically* contain (exactly) n elements of X resp. of $\widehat{X}$, if there exists $k_0 \in \mathbb{N}$ such that for all $k \in \mathbb{Z}$ with $|k| \geq k_0$ we have $\#(X \cap U_{k,\delta}) = n$ resp. $\#(\widehat{X} \cap \widehat{U}_{k,\delta}) = n$.

(2) We say that the excluded domains $U_{k,\delta}$ resp. $\widehat{U}_{k,\delta}$ *asymptotically and totally* contain (exactly) n elements of X resp. of $\widehat{X}$, if there exists $k_0 \in \mathbb{N}$ such that for all $k \in \mathbb{Z}$ with $|k| \geq k_0$ we have

$$\#(X \cap U_{k,\delta}) = n \quad \text{and} \quad \#(X \setminus \bigcup_{|k|\geq k_0} U_{k,\delta}) = n \cdot (2k_0 - 1)$$

resp.

$$\#(\widehat{X} \cap \widehat{U}_{k,\delta}) = n \quad \text{and} \quad \#(\widehat{X} \setminus \bigcup_{|k|\geq k_0} \widehat{U}_{k,\delta}) = n \cdot (2k_0 - 1).$$

The following statement, which establishes the fundamental structure of the spectral data, is analogous to the "Counting Lemma" in the treatment of the 1-dimensional Schrödinger equation in [Pö-T, Lemma 2.2, p. 27].

Proposition 6.5

(1) In the case $\upsilon = 1$, the excluded domains $U_{k,\delta}$ asymptotically and totally contain exactly two zeros (with multiplicity) of $\Delta^2 - 4$.

Therefore the zeros of Δ^2-4 can be enumerated by two sequences $(\varkappa_{k,1})_{k\in\mathbb{Z}}$ and $(\varkappa_{k,2})_{k\in\mathbb{Z}}$ in $\mathbb{C}^*$ in such a way that for every $0 < \delta < \pi - \frac{1}{2}$ there exists $N \in \mathbb{N}$ so that $\varkappa_{k,1}, \varkappa_{k,2} \in U_{k,\delta}$ for all $k \in \mathbb{Z}$ with $|k| \geq N$.

(2) The excluded domains $U_{k,\delta}$ asymptotically and totally contain exactly one zero (with multiplicity) of c.

Therefore the zeros of c can be enumerated by a sequence $(\lambda_k)_{k\in\mathbb{Z}}$ in $\mathbb{C}^*$ in such a way that for every $0 < \delta < \pi - \frac{1}{2}$ there exists $N \in \mathbb{N}$ so that $\lambda_k \in U_{k,\delta}$ holds for all $k \in \mathbb{Z}$ with $|k| \geq N$.

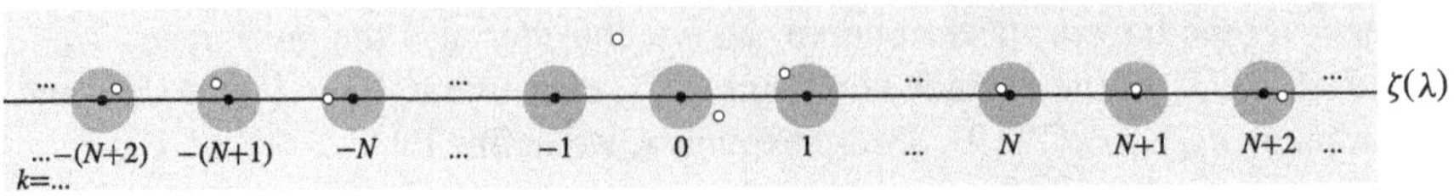

Fig. 6.2 One possible position for the λ_k in ζ-coordinates. The gray disks are the excluded domains $U_{k,\delta}$, the black dots are the $\lambda_{k,0}$, the white dots are the λ_k, and the numbers below the excluded domains indicate the corresponding value of the index k

(3) We have $D = \{ (\lambda_k, \mu_k) \mid k \in \mathbb{Z} \}$, where we put $\mu_k := a(\lambda_k)$. Then we have

$$\lim_{k\to\infty} (\mu_k - \upsilon \cdot \mu_{k,0}) = \lim_{k\to-\infty} (\mu_k - \upsilon^{-1} \cdot \mu_{k,0}) = 0.$$

These estimates hold uniformly for the set of monodromies $M(\lambda)$ corresponding to a relatively compact set P of potentials in Pot_{np}.

Proposition 6.5(2) shows, for example, that for each zero $\lambda_{k,0}$ of c_0 there is exactly one associated zero λ_k of c, and that there exists some $N \in \mathbb{N}$ (depending on δ) so that λ_k lies in the excluded domain $U_{k,\delta}$ around $\lambda_{k,0}$ for all k with $|k| \geq N$. Note that for $|k| < N$, the numbering of the zeros of c is essentially arbitrary, and there does not exist a "natural" bijection from the $\lambda_{k,0}$ to the λ_k in this part of the sequence. Figure 6.2 illustrates the situation in the ζ-coordinates.

Proof (of Proposition 6.5) For (1). Let $0 < \delta < \pi - \frac{1}{2}$ be given, and let $C_1 > 1$ be the constant from Proposition 6.3(3) so that

$$\forall \lambda \in V_\delta \;:\; |\sin(\zeta(\lambda))| \leq w(\lambda) \leq C_1 \cdot |\sin(\zeta(\lambda))| \tag{6.2}$$

holds. Because of $\upsilon = 1$, by Corollary 5.7(2)(c) there then exists $R > 0$, such that for every $\lambda \in \mathbb{C}^*$ with $|\lambda| \geq R$ or $|\lambda| \leq \frac{1}{R}$, we have

$$|(\Delta(\lambda)^2 - 4) - (\Delta_0(\lambda)^2 - 4)| \leq \frac{2}{C_1^2}\, w(\lambda)^2.$$

If we additionally have $\lambda \in V_\delta$, then it follows by the estimate (6.2) that

$$\begin{aligned} |(\Delta(\lambda)^2 - 4) - (\Delta_0(\lambda)^2 - 4)| &\leq \frac{2}{C_1^2} \cdot (C_1\, |\sin(\zeta(\lambda))|)^2 \\ &= \frac{1}{2} |\Delta_0(\lambda)^2 - 4| < |\Delta_0(\lambda)^2 - 4| \end{aligned} \tag{6.3}$$

holds.

There exists $N \in \mathbb{N}$ such that the boundaries of the excluded domains $U_{k,\delta}$ with $|k| > N$ and moreover the boundaries of the topological annuli

$$T_k := \{\, \lambda \in \mathbb{C}^* \mid |\zeta(\lambda)| \leq \pi \left(k + \tfrac{1}{2}\right) \}$$

with $k \geq N$ are contained in $V_\delta \cap \{|\lambda| \geq R \text{ or } |\lambda| \leq \frac{1}{R}\}$. It then follows from (6.3) by Rouché's Theorem that $\Delta^2 - 4$ and $\Delta_0^2 - 4$ have the same number of zeros in each $U_{k,\delta}$ and in each T_k with $|k| \geq N$. Therefore $\Delta^2 - 4$ has two zeros on each $U_{k,\delta}$ with $|k| \geq N$, and $2 \cdot (2k+1)$ zeros on each T_k with $|k| \geq N$.

Herefrom (1) follows: For $|k| \geq N$, we let $\varkappa_{k,1}$ and $\varkappa_{k,2}$ be the two zeros of $\Delta^2 - 4$ on $U_{k,\delta}$, and then we name the remaining $2 \cdot (2N+1)$ zeros as $\varkappa_{k,1}, \varkappa_{k,2}$, where $k \in \{-N, \dots, N\}$.

For (2). We would like to apply a similar argument to c as we used for $\Delta^2 - 4$ in the proof of (1). However, in the basic asymptotics from Theorem 5.4, c in general approximates different functions for $\lambda \to \infty$ and for $\lambda \to 0$, namely $\tau\, c_0$ resp. $\tau^{-1}\, c_0$. But in order to be able to apply Rouché's Theorem for the topological annuli T_k, we need a comparison function that approximates c both near $\lambda \to \infty$ and $\lambda \to 0$.

To obtain such a comparison function, consider the constant potential $(\widetilde{u}, \widetilde{u}_y) \in \mathsf{Pot}$ with $\widetilde{u} = -2\ln(\tau)$, $\widetilde{u}_y = 0$, together with the associated monodromy $\widetilde{M}(\lambda) = \begin{pmatrix} \widetilde{a}(\lambda) & \widetilde{b}(\lambda) \\ \widetilde{c}(\lambda) & \widetilde{d}(\lambda) \end{pmatrix}$. Then we also have $\mathrm{e}^{-(\widetilde{u}(0)+\widetilde{u}(1))/4} = \tau$, and therefore for given $\varepsilon > 0$ there exists $R > 0$ such that for $|\lambda| \geq R$ we have

$$\text{both} \quad |c(\lambda) - \tau\, c_0(\lambda)| \leq \varepsilon\, |\lambda|^{1/2}\, w(\lambda) \quad \text{and} \quad |\widetilde{c}(\lambda) - \tau\, c_0(\lambda)| \leq \varepsilon\, |\lambda|^{1/2}\, w(\lambda)$$

(the first inequality by our hypothesis on $M(\lambda)$, and the second inequality from Theorem 5.4), wherefrom we obtain

$$|c(\lambda) - \widetilde{c}(\lambda)| \leq 2\varepsilon\, |\lambda|^{1/2}\, w(\lambda). \tag{6.4}$$

If we additionally have $\lambda \in V_\delta$, then we have $|\tau\, c_0(\lambda)| \leq |\widetilde{c}(\lambda)| + |\widetilde{c}(\lambda) - \tau\, c_0(\lambda)| \leq |\widetilde{c}(\lambda)| + \varepsilon\, C_1\, |c_0(\lambda)|$ and therefore

$$|\lambda|^{1/2}\, w(\lambda) \leq C_1\, |c_0(\lambda)| \leq \frac{C_1}{|\tau| - \varepsilon\, C_1}\, |\widetilde{c}(\lambda)|.$$

By combining this estimate with (6.4), we obtain

$$|c(\lambda) - \widetilde{c}(\lambda)| \leq \frac{2\, C_1\, \varepsilon}{|\tau| - \varepsilon\, C_1}\, |\widetilde{c}(\lambda)|.$$

By choosing $\varepsilon > 0$ sufficiently small, we can thus achieve for all $\lambda \in V_\delta$, $|\lambda| \geq R$

$$|c(\lambda) - \widetilde{c}(\lambda)| \leq \frac{1}{2}\, |\widetilde{c}(\lambda)| < |\widetilde{c}(\lambda)|.$$

By a similar argument, we obtain the same estimate also for $\lambda \in V_\delta$ with $|\lambda| \leq \frac{1}{R}$. By an analogous application of Rouché's Theorem as in the proof of (1), it follows that there exists $N \in \mathbb{N}$, so that c and $\widetilde{c}$ have the same number of zeros on the excluded domains $U_{k,\delta}$ with $|k| \geq N$ and on the topological annuli T_k with $k \geq N$. To complete the proof of (2), it therefore suffices to show that the statement of (2) holds for $\widetilde{c}$ in the place of c.

For this purpose, we calculate $\widetilde{c}$ and its zeros explicitly. Proceeding similarly as in the investigation of the vacuum in Chap. 4, we note that the 1-form $\widetilde{\alpha}_\lambda$ associated to $(\widetilde{u}, \widetilde{u}_y)$ (Eq. (5.2)) is independent of x, and therefore $\widetilde{M}(\lambda) = \exp(\widetilde{\alpha}_\lambda)$ holds. By an explicit calculation similar to the one in Chap. 4, we obtain

$$\widetilde{c}(\lambda) = \sqrt{\frac{\lambda\,\tau + \tau^{-1}}{\lambda^{-1}\,\tau + \tau^{-1}}} \cdot \sin(\xi(\lambda)) \quad \text{with} \quad \xi(\lambda) = \frac{1}{4}\sqrt{(\lambda\,\tau + \tau^{-1}) \cdot (\lambda^{-1}\,\tau + \tau^{-1})}.$$

$\lambda \in \mathbb{C}^*$ is a zero of $\widetilde{c}$ if and only if

$$\text{either} \quad \xi(\lambda)^2 = (k\pi)^2 \text{ with } k \in \mathbb{N} \quad \text{or} \quad \sqrt{\frac{\lambda\,\tau + \tau^{-1}}{\lambda^{-1}\,\tau + \tau^{-1}}} \cdot \xi(\lambda) = 0$$

holds. For $k \in \mathbb{N}$, the equation $\xi(\lambda)^2 = (k\pi)^2$ yields the quadratic equation

$$\lambda^2 - (16\pi^2 k^2 - \tau^2 - \tau^{-2}) \cdot \lambda + 1 = 0$$

and thereby the two zeros of $\widetilde{c}$

$$\widetilde{\lambda}_{\pm k} := \frac{1}{2} \cdot \left(16\pi^2 k^2 - \tau^2 - \tau^{-2} \pm \sqrt{(16\pi^2 k^2 - \tau^2 - \tau^{-2})^2 - 4} \right)$$

By comparing $\widetilde{\lambda}_{\pm k}$ with the asymptotic assessments of $\lambda_{k,0}$ in (4.9), we see that for $k \to \infty$ $\widetilde{\lambda}_k - \lambda_{k,0} = O(1)$ and $\widetilde{\lambda}_{-k} - \lambda_{-k,0} = O(k^{-4})$ holds. Therefore we have $\widetilde{\lambda}_k \in U_{k,\delta}$ for $|k|$ sufficiently large.

Finally, the equation $\sqrt{\frac{\lambda\,\tau+\tau^{-1}}{\lambda^{-1}\,\tau+\tau^{-1}}} \cdot \xi(\lambda) = 0$ is equivalent to $\lambda\,\tau + \tau^{-1} = 0$, which yields one further zero: $\widetilde{\lambda}_0 = -\tau^{-2}$. This shows that $\widetilde{c}$ satisfies the property of (2), and therefore the proof of (2) is completed.

For (3). Let $\varepsilon > 0$ be given. For $k > 0$, we have

$$\begin{aligned} \mu_k - \upsilon\,\mu_{k,0} &= a(\lambda_k) - \upsilon\,a_0(\lambda_{k,0}) = a(\lambda_k) - \upsilon\,a_0(\lambda_k) + \upsilon \cdot \big(a_0(\lambda_k) - a_0(\lambda_{k,0})\big) \\ &= a(\lambda_k) - \upsilon\,a_0(\lambda_k) + \upsilon \cdot \int_{\lambda_{k,0}}^{\lambda_k} a_0'(\lambda)\,\mathrm{d}\lambda. \end{aligned}$$

We now fix $0 < \delta < \pi - \frac{1}{2}$ (to be chosen later). By the asymptotic behavior of a and the fact that $w(\lambda)$ is bounded on U_δ (Proposition 6.3(1)), there exists $N \in \mathbb{N}$

such that for all $k \geq N$ we have

$$|a(\lambda_k) - \upsilon\, a_0(\lambda_k)| \leq \frac{\varepsilon}{2}.$$

We now estimate $\int_{\lambda_{k,0}}^{\lambda_k} a_0'(\lambda)\,\mathrm{d}\lambda$:

$$\left| \int_{\lambda_{k,0}}^{\lambda_k} a_0'(\lambda)\,\mathrm{d}\lambda \right| \leq |\lambda_k - \lambda_{k,0}| \cdot \max_{\lambda \in U_{k,\delta}} |a_0'(\lambda)|.$$

We have $a_0'(\lambda) = -\frac{1}{8}(\lambda^{-1/2} - \lambda^{-3/2})\sin(\zeta(\lambda))$ by Eq. (5.32); because sin is bounded on any horizontal strip in the ζ-plane, there exists $C > 0$ (independent of the choice of $\delta < \pi - \frac{1}{2}$), such that $\max_{\lambda \in U_{k,\delta}} |a_0'(\lambda)| \leq \frac{C}{k}$ holds for all $k \geq N$. We now set $\delta := \frac{\varepsilon}{2C}$, then it follows from (2) and Proposition 6.2(2) that by increasing N if needed, we can obtain $|\lambda_k - \lambda_{k,0}| \leq \frac{\varepsilon}{2C\,|\upsilon|}\,k$ for $k \geq N$. For such k we then have $\left|\int_{\lambda_{k,0}}^{\lambda_k} a_0'(\lambda)\,\mathrm{d}\lambda\right| \leq \frac{\varepsilon}{2\,|\upsilon|}$ and thus

$$|\mu_k - \upsilon \cdot \mu_{k,0}| \leq |a(\lambda_k) - \upsilon\, a_0(\lambda_k)| + |\upsilon| \cdot \left| \int_{\lambda_{k,0}}^{\lambda_k} a_0'(\lambda)\,\mathrm{d}\lambda \right| \leq \frac{\varepsilon}{2} + |\upsilon| \cdot \frac{\varepsilon}{2\,|\upsilon|} = \varepsilon.$$

This shows that $\lim_{k\to\infty}(\mu_k - \upsilon \cdot \mu_{k,0}) = 0$ holds. The limit $\lim_{k\to-\infty}(\mu_k - \upsilon^{-1} \cdot \mu_{k,0}) = 0$ is shown similarly.

Finally, we note that the fact that the estimates (1)–(3) hold uniformly for a relatively compact set P of potentials follows from the corresponding statements in the propositions in Chap. 5. □

Remark 6.6 The trace function $\widetilde{\Delta}$ corresponding to the constant potential $(\widetilde{u}, \widetilde{u}_y)$ studied in the proof of Proposition 6.5(2) equals $\widetilde{\Delta}(\lambda) = 2\cos(\xi(\lambda))$, and thus we have $\widetilde{\Delta}^2(\lambda) - 4 = -4\sin(\xi(\lambda))^2$. It follows that for the zeros $\widetilde{\varkappa}_{k,\nu}$ of $\widetilde{\Delta}^2 - 4$ we have for $k \neq 0$: $\widetilde{\varkappa}_{k,1} = \widetilde{\varkappa}_{k,2} = \widetilde{\lambda}_k$. However, for $k = 0$, we have $\widetilde{\varkappa}_{k,1} = -\tau^{-2} \neq -\tau^2 = \widetilde{\varkappa}_{k,2}$ (for $\tau \notin \{\pm 1, \pm \mathrm{i}\}$).

This shows that the spectral curve $\widetilde{\Sigma}$ corresponding to $\widetilde{u}$ has geometric genus 1: Compared to the spectral curve of the vacuum, which has only double points, no branch points, and therefore geometric genus 0, one double point (namely the one for $k = 0$) has been "opened" into a pair of branch points.

The minimal surfaces in S^3 of spectral genus 1, and how they arise from surfaces of spectral genus 0 by means of deformation, have been studied extensively by Kilian et al. [Ki-S-S]. Such minimal surfaces are "bubbletons" (i.e. solitons of the sinh-Gordon equation); they are constant along all lines parallel to the x-axis. Moreover, in [Ki-S-S, Section 3.3], an explicit presentation of the solution of the sinh-Gordon equation corresponding to our potential $(\widetilde{u}, \widetilde{u}_y)$ in terms of elliptic functions is given.

Proposition 6.7 (Uniqueness Statements)

(1) Suppose $c, \widetilde{c} : \mathbb{C}^* \to \mathbb{C}$ *are two holomorphic functions with the following asymptotic behavior: There are constants* $\tau, \widetilde{\tau} \in \mathbb{C}^*$ *such that for every* $\varepsilon > 0$ *there exists* $R > 0$*, such that we have for all* $\lambda \in \mathbb{C}$ *with* $|\lambda| \geq R$

$$|c(\lambda) - \tau\, c_0(\lambda)| \leq \varepsilon\, |\lambda|^{1/2}\, w(\lambda) \quad \textit{resp.} \quad |\widetilde{c}(\lambda) - \widetilde{\tau}\, c_0(\lambda)| \leq \varepsilon\, |\lambda|^{1/2}\, w(\lambda)$$

and for all $\lambda \in \mathbb{C}^*$ *with* $|\lambda| \leq \frac{1}{R}$

$$|c(\lambda) - \tau^{-1}\, c_0(\lambda)| \leq \varepsilon\, |\lambda|^{1/2}\, w(\lambda) \quad \textit{resp.} \quad |\widetilde{c}(\lambda) - \widetilde{\tau}^{-1}\, c_0(\lambda)| \leq \varepsilon\, |\lambda|^{1/2}\, w(\lambda).$$

If c *and* $\widetilde{c}$ *have the same zeros (counted with multiplicity), then either* $c = \widetilde{c}$, $\tau = \widetilde{\tau}$ *or* $c = -\widetilde{c}$, $\tau = -\widetilde{\tau}$ *holds.*

(2) Suppose further that $a, \widetilde{a} : \mathbb{C}^* \to \mathbb{C}$ *are two holomorphic functions with the following asymptotic behavior: There are constants* $\upsilon, \widetilde{\upsilon} \in \mathbb{C}^*$ *such that for every* $\varepsilon > 0$ *there exists* $R > 0$*, such that we have for all* $\lambda \in \mathbb{C}$ *with* $|\lambda| \geq R$

$$|a(\lambda) - \upsilon\, a_0(\lambda)| \leq \varepsilon\, w(\lambda) \quad \textit{resp.} \quad |\widetilde{a}(\lambda) - \widetilde{\upsilon}\, a_0(\lambda)| \leq \varepsilon\, w(\lambda)$$

and for all $\lambda \in \mathbb{C}^*$ *with* $|\lambda| \leq \frac{1}{R}$

$$|a(\lambda) - \upsilon^{-1}\, a_0(\lambda)| \leq \varepsilon\, w(\lambda) \quad \textit{resp.} \quad |\widetilde{a}(\lambda) - \widetilde{\upsilon}^{-1}\, a_0(\lambda)| \leq \varepsilon\, w(\lambda).$$

If we have $a(\lambda) = \widetilde{a}(\lambda)$ *and* $\mathrm{ord}_\lambda(a - \widetilde{a}) \geq \mathrm{ord}_\lambda(c)$ *for every* $\lambda \in \mathbb{C}^*$ *with* $c(\lambda) = 0$*, then* $a = \widetilde{a}$ *and* $\upsilon = \widetilde{\upsilon}$ *holds.*

(3) Suppose that $f, \widetilde{f} : \mathbb{C}^* \to \mathbb{C}$ *are two holomorphic functions with the following asymptotic behavior: There are constants* $\tau_\pm, \widetilde{\tau}_\pm \in \mathbb{C}^*$ *such that for every* $\varepsilon > 0$ *there exists* $R > 0$ *such that we have for all* $\lambda \in \mathbb{C}$ *with* $|\lambda| \geq R$

$$|f(\lambda) - \tau_+\, (\Delta_0(\lambda)^2 - 4)| \leq \varepsilon\, w(\lambda)^2 \quad \textit{resp.} \quad |\widetilde{f}(\lambda) - \widetilde{\tau}_+\, (\Delta_0(\lambda)^2 - 4)| \leq \varepsilon\, w(\lambda)^2$$

and for all $\lambda \in \mathbb{C}^*$ *with* $|\lambda| \leq \frac{1}{R}$

$$|f(\lambda) - \tau_-\, (\Delta_0(\lambda)^2 - 4)| \leq \varepsilon\, w(\lambda)^2 \quad \textit{resp.} \quad |\widetilde{f}(\lambda) - \widetilde{\tau}_-\, (\Delta_0(\lambda)^2 - 4)| \leq \varepsilon\, w(\lambda)^2.$$

If f *and* $\widetilde{f}$ *have the same zeros (counted with multiplicity), then there is a constant* $A \in \mathbb{C}^*$ *such that* $f = A \cdot \widetilde{f}$ *and* $\tau_\pm = A \cdot \widetilde{\tau}_\pm$ *holds.*

Proof For (1). Because c and $\widetilde{c}$ have the same zeros, $\varphi := c/\widetilde{c}$ is holomorphic and non-zero on all of $\mathbb{C}^*$. For fixed $\varepsilon, \delta > 0$, there exists $R > 0$ so that for $\lambda \in V_\delta$ with $|\lambda| \geq R$ we have

$$|c(\lambda) - \tau\, c_0(\lambda)| \leq \varepsilon\, |\lambda|^{1/2}\, w(\lambda) \leq \varepsilon\, C_1\, |c_0(\lambda)|\,,$$

where $C_1 > 1$ is the constant from Proposition 6.3(3), and hence

$$|c(\lambda)| \leq (|\tau| + \varepsilon\, C_1) \cdot |c_0(\lambda)|. \tag{6.5}$$

Likewise, from

$$|\widetilde{c}(\lambda) - \widetilde{\tau}\, c_0(\lambda)| \leq \varepsilon\, |\lambda|^{1/2}\, w(\lambda) \leq \varepsilon\, C_1\, |c_0(\lambda)|$$

we obtain

$$|\widetilde{c}(\lambda)| \geq (|\widetilde{\tau}| - \varepsilon\, C_1) \cdot |c_0(\lambda)|. \tag{6.6}$$

From the estimates (6.5) and (6.6) we see that for $\lambda \in V_\delta$ with $|\lambda| \geq R$, we have

$$|\varphi(\lambda)| = \frac{|c(\lambda)|}{|\widetilde{c}(\lambda)|} \leq \frac{|\tau| + \varepsilon\, C_1}{|\widetilde{\tau}| - \varepsilon\, C_1} =: C_2.$$

If we now choose $\varepsilon > 0$ with $|\widetilde{\tau}| - \varepsilon\, C_1 > 0$, it follows that φ is bounded on V_δ near $\lambda \to \infty$. If $k \in \mathbb{N}$ is so large that $|\lambda| \geq R$ holds for all $\lambda \in U_{k,\delta}$, then we have by the maximum principle for holomorphic functions for every $\lambda \in U_{k,\delta}$

$$|\varphi(\lambda)| \leq \max_{\lambda' \in \partial U_{k,\delta}} |\varphi(\lambda')| \leq C_2.$$

Hence φ is bounded on a neighborhood of $\lambda = \infty$, and thus can be extended holomorphically at $\lambda = \infty$.

An analogous argument shows that φ can be extended holomorphically also at $\lambda = 0$. Thus φ is a holomorphic function on the Riemann sphere $\mathbb{P}^1(\mathbb{C})$, and hence constant. In other words, there exists $A \in \mathbb{C}^*$ with $\widetilde{c} = A \cdot c$.

Let us now consider the sequence of real numbers $(\xi_k)_{k\in\mathbb{N}}$ with $|\xi_k| > 1$, $\zeta(\xi_k) = (2k + \frac{1}{2})\pi$. Then we have $\lim \xi_k = \infty$, $\sin(\zeta(\xi_k)) = 1$, $\cos(\zeta(\xi_k)) = 0$, and thus $c_0(\xi_k) = \xi_k^{1/2}$ and $w(\xi_k) = 1$. Therefore there exists for any $\varepsilon > 0$ an $R > 0$ such that

$$|c(\xi_k) - \tau\, \xi_k^{1/2}| \leq \varepsilon\, |\xi_k|^{1/2} \quad \text{and} \quad |A\, c(\xi_k) - \widetilde{\tau}\, \xi_k^{1/2}| \leq \varepsilon\, |\xi_k|^{1/2}$$

holds. From the second inequality, we get $|c(\xi_k) - \frac{\widetilde{\tau}}{A}\, \xi_k^{1/2}| \leq \frac{\varepsilon}{|A|}\, |\xi_k|^{1/2}$, hence we obtain from the first inequality

$$\left| \left(\tau - \frac{\widetilde{\tau}}{A} \right) \xi_k^{1/2} \right| \leq \varepsilon \left(1 + \frac{1}{|A|} \right) |\xi_k|^{1/2},$$

and therefore $\left|\tau - \frac{\widetilde{\tau}}{A}\right| \leq \varepsilon\,(1 + \frac{1}{|A|})$ follows. Because the latter inequality holds true for every $\varepsilon > 0$, we have $\left|\tau - \frac{\widetilde{\tau}}{A}\right| = 0$ and thus $\widetilde{\tau} = A \cdot \tau$. The analogous argument for a sequence (ξ_k) with $\xi_k \to 0$ gives $\widetilde{\tau}^{-1} = A \cdot \tau^{-1}$. Therefrom we conclude $A \in \{\pm 1\}$, and the claim follows.

For (2). We first note that if we denote the sequence of the zeros of c by $(\lambda_k)_{k\in\mathbb{Z}}$ as in Proposition 6.5(2) and put $\mu_k := a(\lambda_k) = \widetilde{a}(\lambda_k)$, then we have by Proposition 6.5(3) (applied to both the function a and the function $\widetilde{a}$)

$$\lim_{k\to\infty} (\mu_k - \upsilon\, \mu_{k,0}) = \lim_{k\to\infty} (\mu_k - \widetilde{\upsilon}\, \mu_{k,0}) = 0$$

and therefore

$$\lim_{k\to\infty} \big((\widetilde{\upsilon} - \upsilon) \cdot \mu_{k,0}\big) = 0.$$

Because of $\mu_{k,0} = (-1)^k$, $\upsilon = \widetilde{\upsilon}$ follows.

Now consider $\psi(\lambda) := \lambda \cdot \frac{a(\lambda)-\widetilde{a}(\lambda)}{c(\lambda)}$. Because of the hypothesis $\mathrm{ord}_\lambda(a - \widetilde{a}) \geq \mathrm{ord}_\lambda(c)$ for all $\lambda \in \mathbb{C}^*$ with $c(\lambda) = 0$, ψ is holomorphic on all of $\mathbb{C}^*$, even at the zeros of c.

We again fix $\varepsilon, \delta > 0$. Then there exists $R > 0$ so that for every $\lambda \in \mathbb{C}^*$ with $|\lambda| \geq R$ or $|\lambda| \leq \frac{1}{R}$ we have

$$|a(\lambda) - \upsilon\, a_0(\lambda)| \leq \varepsilon\, w(\lambda) \quad \text{and} \quad |\widetilde{a}(\lambda) - \upsilon\, a_0(\lambda)| \leq \varepsilon\, w(\lambda)$$

and therefore

$$|a(\lambda) - \widetilde{a}(\lambda)| \leq 2\varepsilon\, w(\lambda).$$

If we additionally have $\lambda \in V_\delta$ and let $C_1 > 0$ be the constant from Proposition 6.3(3), then we deduce

$$|a(\lambda) - \widetilde{a}(\lambda)| \leq 2\varepsilon\, C_1\, |\lambda|^{-1/2}\, |c_0(\lambda)|.$$

We then also have by the asymptotic for c (after suitably enlarging R if necessary)

$$|c(\lambda) - \tau\, c_0(\lambda)| \leq \varepsilon\, |\lambda|^{1/2}\, w(\lambda) \leq \varepsilon\, C_1\, |c_0(\lambda)|$$

and therefore

$$|c(\lambda)| \geq (|\tau| - \varepsilon\, C_1)\, |c_0(\lambda)|.$$

Thus we obtain on $V_\delta \cap \{|\lambda| \geq R \text{ or } |\lambda| \leq \frac{1}{R}\}$:

$$|\psi(\lambda)| = \left|\lambda \cdot \frac{a(\lambda) - \widetilde{a}(\lambda)}{c(\lambda)}\right| \leq |\lambda| \cdot \frac{2\varepsilon\, C_1\, |\lambda|^{-1/2}\, |c_0(\lambda)|}{(|\tau| - \varepsilon\, C_1)\, |c_0(\lambda)|} = \frac{2\varepsilon\, C_1}{|\tau| - \varepsilon\, C_1}\, |\lambda|^{1/2}.$$

By applying the maximum principle to ψ on the excluded domains contained in $\{|\lambda| \geq R \text{ or } |\lambda| \leq \frac{1}{R}\}$ and then on $\{\frac{1}{R} \leq |\lambda| \leq R\}$, it follows that there exists a constant $C_2 > 0$ such that

$$|\psi(\lambda)| \leq C_2 \cdot |\lambda|^{1/2} \quad \text{holds for all } \lambda \in \mathbb{C}^*.$$

From this estimate, it follows first that ψ can be extended holomorphically at $\lambda = 0$ by setting $\psi(0) = 0$, and then that the holomorphic function ψ on the complex plane $\mathbb{C}$ is constant. We therefore have $\psi = 0$, whence $a = \widetilde{a}$ follows.

For (3). The proof is analogous to that of (1). We abbreviate $f_0(\lambda) := \Delta_0(\lambda)^2 - 4 = -4\,\sin(\zeta(\lambda))^2$.

Because f and $\widetilde{f}$ have the same zeros, $\varphi := f/\widetilde{f}$ is holomorphic and non-zero on all of $\mathbb{C}^*$. For fixed $\varepsilon, \delta > 0$, there exists $R > 0$ so that for $\lambda \in V_\delta$ with $|\lambda| \geq R$ we have

$$|f(\lambda) - \tau_+\, f_0(\lambda)| \leq \varepsilon\, w(\lambda)^2 \leq \varepsilon\, C_1^2\, |\sin(\zeta(\lambda))|^2 = \tfrac{1}{4}\,\varepsilon\, C_1^2\, |f_0(\lambda)|\,,$$

where $C_1 > 1$ again is the constant from Proposition 6.3(3), and hence

$$|f(\lambda)| \leq \left(|\tau_+| + \tfrac{1}{4}\,\varepsilon\, C_1^2\right) \cdot |f_0(\lambda)|. \tag{6.7}$$

From the analogous inequality for $\widetilde{f}$,

$$|\widetilde{f}(\lambda) - \widetilde{\tau}_+\, f_0(\lambda)| \leq \tfrac{1}{4}\,\varepsilon\, C_1^2\, |f_0(\lambda)|,$$

we obtain

$$|\widetilde{f}(\lambda)| \geq \left(|\widetilde{\tau}_+| - \frac{1}{4}\,\varepsilon\, C_1^2\right) \cdot |f_0(\lambda)|. \tag{6.8}$$

From the estimates (6.7) and (6.8) we see that for $\lambda \in V_\delta$ with $|\lambda| \geq R$, we have

$$|\varphi(\lambda)| = \frac{|f(\lambda)|}{|\widetilde{f}(\lambda)|} \leq \frac{|\tau_+| + \frac{1}{4}\,\varepsilon\, C_1^2}{|\widetilde{\tau}_+| - \frac{1}{4}\,\varepsilon\, C_1^2} =: C_3.$$

If we now choose $\varepsilon > 0$ with $|\widetilde{\tau}_+| - \frac{1}{4}\,\varepsilon\, C_1^2 > 0$, it follows that φ is bounded on V_δ near $\lambda \to \infty$. If $k \in \mathbb{N}$ is so large that $|\lambda| \geq R$ holds for all $\lambda \in U_{k,\delta}$, then we have by the maximum principle for holomorphic functions for every $\lambda \in U_{k,\delta}$

$$|\varphi(\lambda)| \leq \max_{\lambda' \in \partial U_{k,\delta}} |\varphi(\lambda')| \leq C_3.$$

Hence φ is bounded on $\mathbb{C}^*$ near $\lambda = \infty$, and thus can be extended holomorphically in $\lambda = \infty$.

An analogous argument shows that φ can be extended holomorphically also in $\lambda = 0$. Thus φ is a holomorphic function on the Riemann sphere $\mathbb{P}^1(\mathbb{C})$, and hence constant. That is, there exists $A \in \mathbb{C}^*$ with $f = A \cdot \widetilde{f}$.

Let us now consider the sequence of real numbers $(\xi_k)_{k\in\mathbb{N}}$ with $|\xi_k| > 1$, $\zeta(\xi_k) = (2k + \frac{1}{2})\pi$. Then we have $\lim \xi_k = \infty$, $\sin(\zeta(\xi_k)) = 1$, $\cos(\zeta(\xi_k)) = 0$ and $w(\xi_k) = 1$. For any $\varepsilon > 0$ there thus exists $R > 0$ so that

$$|A \cdot \widetilde{f}(\xi_k) + 4\,\tau_+| \leq \varepsilon \quad \text{and} \quad |\widetilde{f}(\xi_k) + 4\,\widetilde{\tau}_+| \leq \varepsilon$$

holds. From the second inequality, we get $|A \cdot \widetilde{f}(\xi_k) + 4\,A\,\widetilde{\tau}_+| \leq \varepsilon \cdot |A|$, hence

$$|\tau_+ - A \cdot \widetilde{\tau}_+| \leq (1 + |A|) \cdot \varepsilon \;,$$

whence $\tau_+ = A \cdot \widetilde{\tau}_+$ follows. The analogous argument also gives $\tau_- = A \cdot \widetilde{\tau}_-$. □

Chapter 7
The Fourier Asymptotic of the Monodromy

Beyond the "basic" asymptotic of the monodromy described in the Chap. 5, which applies to all λ for which $|\lambda|$ is sufficiently large resp. small, we will also need another type of asymptotic estimate that specifically relates $(M(\lambda_{k,0}))_{k\in\mathbb{Z}}$ to certain Fourier coefficients. In particular, that series is square-summable. Because of the relation to Fourier coefficients, we will call this type of asymptotic "Fourier asymptotic". Via the Fourier asymptotic, we will prove a refinement of the basic asymptotic description of the spectral data from Chap. 6. Later, in Chap. 11, we will liberate the Fourier asymptotics for the monodromy described here from the special choice $\lambda = \lambda_{k,0}$.

In the treatment of the 1-dimensional Schrödinger equation, our Fourier asymptotic as described in the following Theorem is analogous to the refined estimate of [Pö-T, Theorem 1.4, p. 16].

Theorem 7.1 *Let $(u, u_y) \in \mathsf{Pot}_{np}$ be given and put $\tau := \mathrm{e}^{-(u(0)+u(1))/4}$ and $\upsilon := \mathrm{e}^{(u(1)-u(0))/4}$. We denote by a_k resp. b_k the cosine resp. the sine Fourier coefficients of u_z, and by $\widetilde{a}_k, \widetilde{b}_k$ the Fourier coefficients of $-u_{\bar{z}}$:*

$$a_k := \int_0^1 u_z(t)\,\cos(2\pi kt)\,\mathrm{d}t \qquad \widetilde{a}_k := -\int_0^1 u_{\bar{z}}(t)\,\cos(2\pi kt)\,\mathrm{d}t$$

$$b_k := \int_0^1 u_z(t)\,\sin(2\pi kt)\,\mathrm{d}t \qquad \widetilde{b}_k := -\int_0^1 u_{\bar{z}}(t)\,\sin(2\pi kt)\,\mathrm{d}t. \tag{7.1}$$

Let $(r_{ij,k})_{k\in\mathbb{Z}}$ for $i, j \in \{1, 2\}$ be defined by the following equations, where $k \geq 0$:

$$(-1)^k\,M(\lambda_{k,0}) = \begin{pmatrix} \upsilon & 0 \\ 0 & \upsilon^{-1} \end{pmatrix} + \frac{1}{2}\begin{pmatrix} -\upsilon\,a_k & -\lambda_{k,0}^{-1/2}\,\tau^{-1}\,b_k \\ \lambda_{k,0}^{1/2}\,\tau\,b_k & \upsilon^{-1}\,a_k \end{pmatrix} + \begin{pmatrix} r_{11,k} & \lambda_{k,0}^{-1/2}\,r_{12,k} \\ \lambda_{k,0}^{1/2}\,r_{21,k} & r_{22,k} \end{pmatrix}$$

$$(-1)^k\,M(\lambda_{-k,0}) = \begin{pmatrix} \upsilon^{-1} & 0 \\ 0 & \upsilon \end{pmatrix} + \frac{1}{2}\begin{pmatrix} -\upsilon^{-1}\,\widetilde{a}_k & -\lambda_{-k,0}^{-1/2}\,\tau\,\widetilde{b}_k \\ \lambda_{-k,0}^{1/2}\,\tau^{-1}\,\widetilde{b}_k & \upsilon\,\widetilde{a}_k \end{pmatrix} + \begin{pmatrix} r_{11,-k} & \lambda_{-k,0}^{-1/2}\,r_{12,-k} \\ \lambda_{-k,0}^{1/2}\,r_{21,-k} & r_{22,-k} \end{pmatrix}.$$

S. Klein, *A Spectral Theory for Simply Periodic Solutions of the Sinh-Gordon Equation*, Lecture Notes in Mathematics 2229,
https://doi.org/10.1007/978-3-030-01276-2_7

For every $p > 1$ we then have $(r_{ij,k})_{k\in\mathbb{Z}} \in \ell^p(\mathbb{Z})$, and there exists a constant $C > 0$ (dependent on p) with $\|r_{ij}\|_{\ell^p} \leq C \cdot \|(u, u_y)\|^2_{\mathsf{Pot}}$.

The remainder of this chapter is concerned with the proof of the above theorem. The strategy for the proof is to decompose the given potential (u, u_y) into two summands, one with a higher order of regularity and one that is small with respect to $\|\cdot\|_{\mathsf{Pot}}$. We will derive asymptotics for these two kinds of potentials separately, in Lemmas 7.4 and 7.3, and finally combine the two to obtain the proof of Theorem 7.1.

In the proof, we will make use of the Banach spaces $\ell^p(\mathbb{Z})$ and $\ell^p(\mathbb{Z}^2)$. We note that for $1 \leq p < q \leq \infty$, we have $\ell^p \subset \ell^q$ and for any $(a_k) \in \ell^p$:

$$\|a_k\|_{\ell^q} \leq \|a_k\|_{\ell^p}. \tag{7.2}$$

Proof (of Eq. (7.2)) Let $(a_k) \in \ell^p$ be given. Without loss of generality we may suppose $\|a_k\|_{\ell^p} = 1$, then we have in particular $|a_k| \leq 1$ for all k. Therefore we have $|a_k|^q \leq |a_k|^p$ because of $p < q$, and hence

$$\|a_k\|^q_{\ell^q} = \sum_k |a_k|^q \leq \sum_k |a_k|^p = \|a_k\|^p_{\ell^p} = 1 ,$$

whence $\|a_k\|_{\ell^q} \leq 1 = \|a_k\|_{\ell^p}$ follows. □

There are two important operations for ℓ^p-sequences:

First, the component-wise multiplication $(a_k \cdot b_k)$ of two sequences a_k and b_k. The Hölder inequality states that whenever $a_k \in \ell^p$ and $b_k \in \ell^{p'}$, where $p, p' \geq 1$ and $\frac{1}{p} + \frac{1}{p'} = 1$, we have $(a_k \cdot b_k) \in \ell^1$ and $\|a_k \cdot b_k\|_{\ell^1} \leq \|a_k\|_{\ell^p} \cdot \|b_k\|_{\ell^{p'}}$.

The second important operation is the convolution of two sequences (a_k) and (b_k). The convolution $a * b$ is defined by

$$(a * b)_k := \sum_{j\in\mathbb{Z}} a_{k-j}\, b_j = (b * a)_k \tag{7.3}$$

for any sequences (a_k) and (b_k) such that the above infinite sum converges for all k. Young's inequality states that for $p, q, r \geq 1$ with $\frac{1}{r} + 1 = \frac{1}{p} + \frac{1}{q}$, and any sequences $(a_k) \in \ell^p$ and $(b_k) \in \ell^q$, the convolution $a * b$ is well-defined, we have $a * b \in \ell^r$ and $\|a * b\|_{\ell^r} \leq \|a\|_{\ell^p} \cdot \|b\|_{\ell^q}$ [Ben-S, Theorem 4.2.4, p. 199]. Moreover, the convolution with the sequences $(\frac{1}{k})_{k\in\mathbb{Z}}$ and $(\frac{1}{|k|})_{k\in\mathbb{Z}}$ (which are only weakly ℓ^1), where we assign $\frac{1}{0}$ an arbitrary value, will be important for us. In relation to these convolutions, we have for any $1 < p \leq 2$ (not for $p = 1$, however!) and any $a_k \in \ell^p(\mathbb{Z})$ also $(\frac{1}{k} * a), (\frac{1}{|k|} * a) \in \ell^p(\mathbb{Z})$ and

$$\left\|\tfrac{1}{k} * a\right\|_{\ell^p}, \left\|\tfrac{1}{|k|} * a\right\|_{\ell^p} \leq C_{Y,p} \cdot \|a\|_{\ell^p} \tag{7.4}$$

with a constant $C_{Y,p} > 0$ depending only on p.

Proof (of the Preceding Statement) Let T be either of the linear operators $(a_k) \mapsto (\frac{1}{k} * a_k)$ or $(a_k) \mapsto (\frac{1}{|k|} * a_k)$. Then T maps $\ell^1(\mathbb{Z})$ into the Lorentz space $\ell^{1,\infty}(\mathbb{Z})$, and it maps $\ell^2(\mathbb{Z})$ into $\ell^2(\mathbb{Z})$. The reason for the latter inclusion is that up to a factor, $\frac{1}{k}$ is the Fourier transform of $g(x) := x \in L^\infty([0,1])$. For a given $(a_k) \in \ell^2(\mathbb{Z})$ there exists some $f \in L^2([0,1])$ whose Fourier transform is (a_k). Then the L^2-function $f \cdot g$ has the Fourier transform $(\frac{1}{k} * a_k) = T((a_k))$, and therefore we have $T((a_k)) \in \ell^2(\mathbb{Z})$. The claim now follows by application of the Marcinkiewicz interpolation theorem (see [Ben-S, Corollary 4.4.14, p. 226]) to the operator T. □

In the sequel, we will consider the Fourier transform for integrable functions $f \in L^1([0,1])$ on the circle resp. functions $g \in L^1([0,1]^2)$ on the torus. For $k \in \mathbb{Z}$ resp. for $k_1, k_2 \in \mathbb{Z}$ we define the Fourier transform

$$\widehat{f}(k) := \int_0^1 f(t)\,\mathrm{e}^{-2\pi i k t}\,\mathrm{d}t \quad \text{resp.}$$

$$\widehat{g}(k_1,k_2) := \int_{[0,1]^2} g(t_1,t_2)\,\mathrm{e}^{-2\pi \mathrm{i}(k_1 t_1 + k_2 t_2)}\,\mathrm{d}^2 t. \tag{7.5}$$

It is important to note that for $f \in L^2([0,1])$, we have $\widehat{f} \in \ell^2(\mathbb{Z})$, and the Fourier map $L^2([0,1]) \to \ell^2(\mathbb{Z})$ is an isomorphism of Banach (actually Hilbert) spaces, i.e. it is bijective and preserves the norm by Plancherel's identity (see [Gra, Proposition 3.1.16(1), p. 170]): $\|\widehat{f}\|_{\ell^2(\mathbb{Z})} = \|f\|_{L^2([0,1])}$ for any $f \in L^2([0,1])$. Moreover, the Fourier series $\sum_{k\in\mathbb{Z}} \widehat{f}(k) \cdot \mathrm{e}^{2\pi i m x}$ converges to f with respect to the L^2-norm [Gra, Proposition 3.1.16(2), p. 170]. Analogous statements hold for functions on the torus $[0,1]^2$.

We also note that the multiplication of functions corresponds under Fourier transformation to the convolution of the corresponding Fourier series: For any $f_1, f_2 \in L^2([0,1])$, we have $\widehat{f_1 \cdot f_2} = \widehat{f_1} * \widehat{f_2}$. If $f \in W^{1,2}([0,1])$ is periodic, we also have $\widehat{f'}(k) = 2\pi i k\,\widehat{f}(k)$ for $k \in \mathbb{Z}$. The following Lemma concerns the inverse of the latter operation, namely the Fourier coefficients of the anti-derivative of a function f.

For this, and in the sequel, we will have use besides the sequence spaces $\ell^p(N)$ (with $N \subset \mathbb{Z}$, $1 \leq p \leq \infty$) also for the spaces $\ell^p_n(N)$ ($n \in \mathbb{Z}$). We say that a sequence $(a_k)_{k\in\mathbb{N}}$ is in $\ell^p_n(N)$ if and only if the sequence $(k^n\,a_k)_{k\in\mathbb{N}}$ is in $\ell^p(N)$; here (and also in associated formulas in the sequel) we put $0^n := 1$ for all $n \in \mathbb{Z}$ to simplify notation. Of course, we equip $\ell^p_n(N)$ with the norm

$$\|a_k\|_{\ell^p_n(N)} := \|k^n\,a_k\|_{\ell^p(N)}.$$

In this way, $\ell^p_n(N)$ becomes a Banach space (a Hilbert space for $p = 2$).

Lemma 7.2 *Let $f \in L^1([0,1])$ be such that $\widehat{f} \in \ell^1_{-1}(\mathbb{Z})$. We put $F(x) := \int_0^x f(x)\,\mathrm{d}x$. Then we have*

$$\widehat{F}(k) = \begin{cases} \frac{\widehat{f}(k)-\widehat{f}(0)}{2\pi \mathrm{i}k} & \text{for } k \neq 0 \\ \frac{1}{2}\widehat{f}(0) - \sum_{j\in\mathbb{Z}\setminus\{0\}} \frac{\widehat{f}(j)}{2\pi\mathrm{i}j} & \text{for } k = 0 \end{cases}. \tag{7.6}$$

Proof Note that $F(0) = 0$ and $F(1) = \widehat{f}(0)$ holds. For $k \in \mathbb{Z}\setminus\{0\}$ we thus have

$$\begin{aligned}
\widehat{F}(k) &= \int_0^1 F(t)\,\mathrm{e}^{-2\pi \mathrm{i}kt}\,\mathrm{d}t \\
&= -\frac{1}{2\pi\mathrm{i}k} F(t)\,\mathrm{e}^{-2\pi\mathrm{i}kt}\Big|_{t=0}^{t=1} - \int_0^1 f(t)\left(-\frac{1}{2\pi\mathrm{i}k}\right)\mathrm{e}^{-2\pi\mathrm{i}kt}\,\mathrm{d}t \\
&= -\frac{1}{2\pi\mathrm{i}k}\widehat{f}(0) + \frac{1}{2\pi\mathrm{i}k}\int_0^1 f(t)\,\mathrm{e}^{-2\pi\mathrm{i}kt}\,\mathrm{d}t \\
&= \frac{\widehat{f}(k)-\widehat{f}(0)}{2\pi\mathrm{i}k}.
\end{aligned}$$

To calculate $\widehat{F}(0)$, we consider for $j \in \mathbb{Z}$ the functions $f_j(x) := \mathrm{e}^{2\pi\mathrm{i}jx}$ and

$$F_j(x) := \int_0^x f_j(t)\,\mathrm{d}t = \begin{cases} \frac{\mathrm{e}^{2\pi\mathrm{i}jx}-1}{2\pi\mathrm{i}j} & \text{for } j \neq 0 \\ x & \text{for } j = 0 \end{cases}.$$

Then we have for $j \neq 0$

$$\widehat{F}_j(0) = \int_0^1 F_j(x)\,\mathrm{d}x = \frac{\frac{1}{2\pi\mathrm{i}j}\mathrm{e}^{2\pi\mathrm{i}jx} - x}{2\pi\mathrm{i}j}\Bigg|_{x=0}^{x=1} = -\frac{1}{2\pi\mathrm{i}j},$$

and for $j = 0$

$$\widehat{F}_0(0) = \int_0^1 F_0(x)\,\mathrm{d}x = \frac{1}{2}x^2\Big|_{x=0}^{x=1} = \frac{1}{2}.$$

It follows that we have

$$\widehat{F}(0) = \frac{1}{2}\widehat{f}(0) - \sum_{j\in\mathbb{Z}\setminus\{0\}} \frac{\widehat{f}(j)}{2\pi\mathrm{i}j}.$$

□

The following two lemmas each prove a sort of a Fourier asymptotic in a special case with respect to the potential (u, u_y): Lemma 7.3 deals with the case where $\|(u, u_y)\|_{\mathsf{Pot}}$ is small, and Lemma 7.4 deals with the case where $(u, u_y) \in \mathsf{Pot}^1_{np}$ holds, i.e. (u, u_y) is once more differentiable. Because every $(u, u_y) \in \mathsf{Pot}_{np}$ can be decomposed as a sum of one potential each of these two kinds, we will be able to combine these two lemmas into a full proof of Theorem 7.1.

Both lemmas operate in the setting of the proof of Theorem 5.4. In particular they concern the regauged and modified monodromy E introduced in that proof and its series representation $E(x) = \sum_{n=0}^{\infty} E_n(x)$, see Eq. (5.22). In the following Lemma 7.3, note that the constant term $\mathbb{1}$ equals $(-1)^k E_{0,\lambda_{k,0}}(1)$ and the Fourier term $\frac{1}{2}\begin{pmatrix} -a_k & b_k \\ b_k & a_k \end{pmatrix}$ equals $(-1)^k E_{1,\lambda_{k,0}}(1)$. Therefore the main point of Lemma 7.3 is to estimate the sum $\sum_{n\geq 2} E_{n,\lambda_{k,0}}(1)$ of the remaining terms in $E_{\lambda_{k,0}}(1)$. It is relatively easy to see that each individual of these summands is in $\ell^p(k, \mathbb{C}^{2\times 2})$, but it is a remarkable fact that the behavior is not worsened by taking the infinite sum over all $n \geq 2$.

Lemma 7.3 *Suppose* $1 < p \leq 2$. *There exists a constant* $R_0 > 0$ *such that for all* $(u, u_y) \in \mathsf{Pot}_{np}$ *with* $\|(u, u_y)\|_{\mathsf{Pot}} \leq R_0$, *we have*

$$(-1)^k E_{\lambda_{k,0}}(1) = \mathbb{1} + \frac{1}{2}\begin{pmatrix} -a_k & b_k \\ b_k & a_k \end{pmatrix} + \ell^p(k, \mathbb{C}^{2\times 2}),$$

and the ℓ^p*-norm of the* ℓ^p*-sequence is* $\leq C \cdot \|(u, u_y)\|^2_{\mathsf{Pot}}$ *with some constant* $C > 0$ *(that depends on* p*). Here* a_k *and* b_k *are the Fourier coefficients of* u_z *as in Eq.* (7.1).

Proof We continue to work in the setting of the proof of Theorem 5.4. At first we do not restrict the norm of (u, u_y). We set $\lambda = \lambda_{k,0}$ with $k \geq 1$, and omit the subscript $\lambda_{k,0}$ from E in the sequel.

The strategy we follow is to prepare the situation for a proof by induction on n that $E_n(1) \in \ell^p(k, \mathbb{C}^{2\times 2})$ holds with a specific estimate for $\|E_n(1)\|_{\ell^p(k,\mathbb{C}^{2\times 2})}$. Via this estimate we will show that $\sum_{n\geq 2} E_n(1)$ is bounded in $\ell^p(k, \mathbb{C}^{2\times 2})$ by a geometric sum. This geometric sum converges (only) if $\|(u, u_y)\|_{\mathsf{Pot}}$ is small, yielding the result of the lemma.

For $n \geq 1$ we define functions $h_n(s_0, s_1)$ and $g_n(s)$ by

$$h_1(s_0, s_1) := -\frac{1}{2}\, u_z(s_1)\ ,$$

for $n \geq 2$

$$h_n(s_0, s_1) := \left(-\frac{1}{2}\right)^n \int_{s_2=0}^{s_0} u_z(s_1+s_2) \int_{s_3=0}^{s_1} u_z(s_2+s_3) \cdots \int_{s_n=0}^{s_{n-2}} u_z(s_{n-1}+s_n)\, u_z(s_n)\, \mathrm{d}^{n-1}s\ ,$$

and for $n \geq 1$

$$g_n(s) := h_n(1-s, s).$$

Then we have by Eqs. (5.25), (5.26) and (5.27)

$$G_n(s) = g_n(s)\, L^n \quad \text{and} \quad H_n(s_0, s_1) = h_n(s_0, s_1)\, L^n,$$

where we put

$$L := \begin{pmatrix} 1 & 0 \\ 0 & -1 \end{pmatrix} \quad \text{and therefore have} \quad L^n = \begin{cases} L & \text{for } n \text{ odd} \\ \mathbb{1} & \text{for } n \text{ even} \end{cases}.$$

For $n = 0$ we have by Eq. (5.21)

$$E_0(1) = \begin{pmatrix} \cos(k\pi \cdot 1) & -\sin(k\pi \cdot 1) \\ \sin(k\pi \cdot 1) & \cos(k\pi \cdot 1) \end{pmatrix} = (-1)^k \cdot \mathbb{1}.$$

Moreover we have $\zeta(\lambda) = k\pi$, $w(\lambda) = 1$, and therefore for $n \geq 1$ by Eqs. (5.24) and (5.21)

$$\begin{aligned}
E_n(1) &= \int_0^1 E_0(1-2s)\, G_n(s)\, \mathrm{d}s \\
&= \int_0^1 \begin{pmatrix} \cos((1-2s)k\pi) & -\sin((1-2s)k\pi) \\ \sin((1-2s)k\pi) & \cos((1-2s)k\pi) \end{pmatrix} G_n(s)\, \mathrm{d}s \\
&= (-1)^k \cdot \int_0^1 \begin{pmatrix} \cos(2k\pi s) & \sin(2k\pi s) \\ -\sin(2k\pi s) & \cos(2k\pi s) \end{pmatrix} G_n(s)\, \mathrm{d}s \\
&= (-1)^k \cdot \int_0^1 \begin{pmatrix} \cos(2k\pi s) & \sin(2k\pi s) \\ -\sin(2k\pi s) & \cos(2k\pi s) \end{pmatrix} g_n(s)\, \mathrm{d}s \cdot L^n. \qquad (7.7)
\end{aligned}$$

Specifically for $n = 1$ we have

$$g_1(s) = h_1(1-s, s) = -\frac{1}{2} u_z(s)\,,$$

hence

$$\begin{aligned}
E_1(1) &= -\frac{(-1)^k}{2} \int_0^1 \begin{pmatrix} \cos(2k\pi s) & \sin(2k\pi s) \\ -\sin(2k\pi s) & \cos(2k\pi s) \end{pmatrix} u_z(s)\, \mathrm{d}s \cdot L \\
&= -\frac{(-1)^k}{2} \begin{pmatrix} a_k & b_k \\ -b_k & a_k \end{pmatrix} \cdot L = \frac{(-1)^k}{2} \begin{pmatrix} -a_k & b_k \\ b_k & a_k \end{pmatrix}.
\end{aligned}$$

Thus it remains to show that for $(u, u_y) \in \mathsf{Pot}_{np}$ with $\|(u, u_y)\|_{\mathsf{Pot}} \leq R_0$ for some constant $R_0 > 0$,

$$\sum_{n=2}^{\infty} E_n(1) \in \ell^p(k)$$

holds, where $\| \sum_{n=2}^{\infty} E_n(1)\|_{\ell^p(k)} \leq C \cdot \|(u, u_y)\|_{\mathsf{Pot}}^2$ holds with a constant $C > 0$; and because of Eqs. (7.7) and (5.30), for this it is in turn sufficient to show that for the discrete Fourier coefficients $\widehat{g}_n$ of g_n we have

$$\sum_{n=2}^{\infty} \widehat{g}_n(k) \in \ell^p(k) \quad \text{with} \quad \left\| \sum_{n=2}^{\infty} \widehat{g}_n(k) \right\|_{\ell^p(k)} \leq C \cdot \|(u, u_y)\|_{\mathsf{Pot}}^2 \tag{7.8}$$

with some constant $C > 0$.

Because of $h_n \in L^2([0, 1]^2)$, we can express the bivariate function h_n on the torus by its Fourier series:

$$h_n(s_0, s_1) = \sum_{k_0, k_1 \in \mathbb{Z}} \widehat{h}_n(k_0, k_1)\, \mathrm{e}^{2\pi \mathrm{i}(k_0 s_0 + k_1 s_1)}.$$

Thereby we obtain

$$\begin{aligned} \widehat{g}_n(k) &= \int_0^1 g_n(s)\, \mathrm{e}^{-2\pi \mathrm{i} k s}\, \mathrm{d}s = \int_0^1 h_n(1-s, s)\, \mathrm{e}^{-2\pi \mathrm{i} k s}\, \mathrm{d}s \\ &= \sum_{k_0, k_1 \in \mathbb{Z}} \widehat{h}_n(k_0, k_1) \int_0^1 \mathrm{e}^{2\pi \mathrm{i}(k_0(1-s) + k_1 s)}\, \mathrm{e}^{-2\pi \mathrm{i} k s}\, \mathrm{d}s \\ &= \sum_{k_0, k_1 \in \mathbb{Z}} \widehat{h}_n(k_0, k_1) \int_0^1 \mathrm{e}^{2\pi \mathrm{i}(k_1 - k_0 - k) s}\, \mathrm{d}s \\ &= \sum_{k_0, k_1 \in \mathbb{Z}} \widehat{h}_n(k_0, k_1) \delta_{k_1, k+k_0} = \sum_{\ell \in \mathbb{Z}} \widehat{h}_n(\ell - k, \ell). \end{aligned} \tag{7.9}$$

For the sake of brevity, we put $c_k := -\frac{1}{2}\widehat{u_z}(k)$ for $k \in \mathbb{Z}$; then we have $(c_k) \in \ell^2(\mathbb{Z})$ and $\|c\|_{\ell^2} \leq \|(u, u_y)\|_{\mathsf{Pot}}$.

We now make the following claim: There exists a constant $C_1 > 0$ such that for every $n \geq 2$, we have

$$\forall k \in \mathbb{Z} : \widehat{h}_n(\ell - k, \ell) \in \ell^1(\ell) \tag{7.10}$$

$$\|\widehat{h}_n(\ell - k, \ell)\|_{\ell^1(\ell)} \in \ell^p(k) \tag{7.11}$$

$$\left\| \, \|\widehat{h}_n(\ell - k, \ell)\|_{\ell^1(\ell)} \, \right\|_{\ell^p(k)} \leq (C_1 \cdot \|c\|_{\ell^2})^n. \tag{7.12}$$

The claims (7.10), (7.11) imply that $\widehat{h}_n$ lies in a Bochner space $\ell^1 \otimes \ell^p$.

Before we prove the claim (7.10)–(7.12), we would like to note that the statement (7.8) which we need to show follows from (7.10) to (7.12). Indeed, we then have for $n \geq 2$

$$|g_n(k)| \overset{(7.9)}{\leq} \|\widehat{h}_n(\ell - k, \ell)\|_{\ell^1(\ell)}$$

and therefore by (7.12)

$$\|g_n(k)\|_{\ell^p(k)} \leq (C_1 \cdot \|c\|_{\ell^2})^n .$$

If we now put $R_0 := \frac{1}{2C_1}$ and $C := 2\,C_1^2$, and consider $(u, u_y) \in \mathsf{Pot}$ with $\|c\|_{\ell^2} \leq \|(u, u_y)\|_{\mathsf{Pot}} \leq R_0$, then we thus see

$$\sum_{n=2}^{\infty} \|g_n(k)\|_{\ell^p(k)} \leq \frac{(C_1 \cdot \|c\|_{\ell^2})^2}{1 - C_1 \cdot \|c\|_{\ell^2}} \leq 2\,(C_1 \cdot \|c\|_{\ell^2})^2 \leq C \cdot \|(u, u_y)\|_{\mathsf{Pot}}^2 .$$

Therefore (7.8) follows from the claim (7.10)–(7.12).

We will now prove the claim (7.10)–(7.12) by induction on $n \geq 2$. First, let us look at the case $n = 2$. Then we have

$$h_2(s_0, s_1) = \int_{s_2=0}^{s_0} \varphi_2(s_2, s_1)\,\mathrm{d}s_2 \quad \text{with} \quad \varphi_2(s_0, s_1) = (-\tfrac{1}{2} u_z(s_0)) \cdot (-\tfrac{1}{2} u_z(s_0 + s_1)).$$

For $(k_0, k_1) \in \mathbb{Z}^2$ we thus have

$$\begin{aligned} \widehat{\varphi}_2(k_0, k_1) &= (\delta_{j_1,0}\, c_{j_0} * \delta_{j_0, j_1}\, c_{j_0})_{(k_0,k_1)} \\ &= \sum_{j_0, j_1 \in \mathbb{Z}} \delta_{j_1,0}\, c_{j_0}\, \delta_{(k_0 - j_0),(k_1 - j_1)}\, c_{k_0 - j_0} = c_{k_0 - k_1}\, c_{k_1} , \end{aligned} \tag{7.13}$$

and therefore by Lemma 7.2

$$\begin{aligned} \widehat{h}_2(k_0, k_1) &= \begin{cases} \frac{1}{2\pi i k_0}\, \widehat{\varphi}_2(k_0, k_1) - \frac{1}{2\pi i k_0}\, \widehat{\varphi}_2(0, k_1) & \text{for } k_0 \neq 0 \\ \frac{1}{2} \widehat{\varphi}_2(0, k_1) - \sum_{\ell \neq 0} \frac{\widehat{\varphi}_2(\ell, k_1)}{2\pi i \ell} & \text{for } k_0 = 0 \end{cases} \\ &\overset{(7.13)}{=} \begin{cases} \frac{1}{2\pi i k_0}\, c_{k_0 - k_1}\, c_{k_1} - \frac{1}{2\pi i k_0}\, c_{-k_1}\, c_{k_1} & \text{for } k_0 \neq 0 \\ \frac{1}{2}\, c_{-k_1}\, c_{k_1} - c_{k_1} \sum_{\ell \neq 0} \frac{c_{\ell - k_1}}{2\pi i \ell} & \text{for } k_0 = 0 \end{cases} . \end{aligned} \tag{7.14}$$

Therefore we obtain for $k \in \mathbb{Z}$

$$\begin{aligned}\|\widehat{h}_2(\ell-k,\ell)\|_{\ell^1(\ell)} &= \sum_{\ell\neq k} |\widehat{h}_2(\ell-k,\ell)| + |\widehat{h}_2(0,k)| \\ &\overset{(7.14)}{\leq} \frac{1}{2\pi}|c_{-k}| \sum_{\ell\neq k} \frac{|c_\ell|}{|\ell-k|} + \frac{1}{2\pi}\sum_{\ell\neq k}\frac{|c_{-\ell}|\cdot|c_\ell|}{|\ell-k|} \\ &\quad + \frac{1}{2}|c_{-k}|\cdot|c_k| + \frac{1}{2\pi}|c_k|\sum_{\ell\neq 0}\frac{|c_{\ell-k}|}{|\ell|} \\ &= \tfrac{1}{2\pi}|c_{-k}|\cdot(\tfrac{1}{|\ell|} * |c_\ell|)_k + \tfrac{1}{2\pi}(\tfrac{1}{|\ell|} * |c_{-\ell}\,c_\ell|)_k \\ &\quad + \tfrac{1}{2}|c_{-k}|\cdot|c_k| + \tfrac{1}{2\pi}|c_k|\cdot(\tfrac{1}{|\ell|} * |c_{-\ell}|)_k < \infty.\end{aligned}$$

Thus we have $\widehat{h}_2(\ell-k,\ell) \in \ell^1(\ell)$ and moreover, by use of the Hölder inequality and Young's inequality in the variant for weakly ℓ^1-sequences of (7.4):

$$\begin{aligned}&\Big\| \, \|\widehat{h}_2(\ell-k,\ell)\|_{\ell^1(\ell)} \, \Big\|_{\ell^p(k)} \\ &\leq \tfrac{1}{2\pi}\|c\|_{\ell^2}\, C_{Y,2}\, \|c\|_{\ell^2} + \tfrac{1}{2\pi} C_{Y,p}\, C_{1,p}\, \|c\|_{\ell^2}\, \|c\|_{\ell^2} \\ &\quad + \tfrac{1}{2}\|c\|_{\ell^2}\,\|c\|_{\ell^2} + \tfrac{1}{2\pi}\|_{\ell^2}\, C_{Y,2}\, \|c\|_{\ell^2} \leq (C_1\cdot\|c\|_{\ell^2})^2\end{aligned}$$

with a suitable $C_1 > 0$. This shows that (7.10)–(7.12) hold for $n = 2$.

We now tackle the induction step. We suppose that $n \geq 3$ is given such that (7.10)–(7.12) hold for $n-1$. We then have

$$h_n(s_0,s_1) = \int_{s_2=0}^{s_0} \varphi_n(s_2,s_1)\,\mathrm{d}s_2 \quad \text{with} \quad \varphi_n(s_0,s_1) := -\frac{1}{2}\,u_z(s_0+s_1)\,h_{n-1}(s_1,s_0).$$

Therefore we have

$$\begin{aligned}\widehat{\varphi}_n(k_0,k_1) &= (\delta_{j_0,j_1}\, c_{j_0} * \widehat{h}_{n-1}(j_1,j_0))_{(k_0,k_1)} \\ &= \sum_{j_0,j_1\in\mathbb{Z}} \delta_{j_0,j_1}\, c_{j_0}\, \widehat{h}_{n-1}(k_1-j_1,k_0-j_0) \\ &= \sum_{j\in\mathbb{Z}} c_j\, \widehat{h}_{n-1}(k_1-j,k_0-j).\end{aligned}$$

and hence by Lemma 7.2

$$\widehat{h}_n(k_0,k_1)=\begin{cases}\frac{1}{2\pi \mathrm{i}k_0}\,\widehat{\varphi}_n(k_0,k_1)-\frac{1}{2\pi \mathrm{i}k_0}\,\widehat{\varphi}_n(0,k_1) & \text{for } k_0\neq 0\\ \frac{1}{2}\widehat{\varphi}_n(0,k_1)-\sum_{\ell\neq 0}\frac{\widehat{\varphi}_n(\ell,k_1)}{2\pi \mathrm{i}\ell} & \text{for } k_0=0\end{cases}$$

$$=\begin{cases}\frac{1}{2\pi \mathrm{i}k_0}\sum_j c_j\,\widehat{h}_{n-1}(k_1-j,k_0-j)-\frac{1}{2\pi \mathrm{i}k_0}\sum_j c_j\,\widehat{h}_{n-1}(k_1-j,-j) & \text{for } k_0\neq 0\\ \frac{1}{2}\sum_j c_j\widehat{h}_{n-1}(k_1-j,-j)-\frac{1}{2\pi \mathrm{i}}\sum_{\ell\neq 0,j\in\mathbb{Z}}\frac{1}{\ell}\,c_j\,\widehat{h}_{n-1}(k_1-j,\ell-j) & \text{for } k_0=0\end{cases}. \tag{7.15}$$

We therefore obtain

$$\begin{aligned}\|\widehat{h}_n(\ell-k,\ell)\|_{\ell^1(\ell)}&=\sum_{\ell\neq k}|\widehat{h}_n(\ell-k,\ell)|+|\widehat{h}_n(0,k)|\\ &\leq\underbrace{\frac{1}{2\pi}\sum_{\ell\neq k,j\in\mathbb{Z}}\frac{|c_j|\,|\widehat{h}_{n-1}(\ell-j,\ell-k-j)|}{|\ell-k|}}_{(A)}+\underbrace{\frac{1}{2\pi}\sum_{\ell\neq k,j\in\mathbb{Z}}\frac{|c_j|\,|\widehat{h}_{n-1}(\ell-j,-j)|}{|\ell-k|}}_{(B)}\\ &\quad+\underbrace{\frac{1}{2}\sum_{j\in\mathbb{Z}}|c_j|\,|\widehat{h}_{n-1}(k-j,-j)|}_{(C)}+\underbrace{\frac{1}{2\pi}\sum_{\ell\neq 0,j\in\mathbb{Z}}\frac{|c_j|\,|\widehat{h}_{n-1}(k-j,\ell-j)|}{|\ell|}}_{(D)}\end{aligned} \tag{7.16}$$

We estimate the four sums labeled (A), (B), (C) and (D) separately. First, we have by Cauchy-Schwarz's inequality and the variant (7.4) of Young's inequality

$$\begin{aligned}(A)&=\sum_{\ell\neq k,j\in\mathbb{Z}}\frac{|c_j|\,|\widehat{h}_{n-1}(\ell-j,\ell-k-j)|}{|\ell-k|}=\sum_{\ell\neq -j,j\in\mathbb{Z}}\frac{|c_j|\,|\widehat{h}_{n-1}(\ell+k,\ell)|}{|\ell+j|}\\ &=\sum_{j\in\mathbb{Z}}|c_j|\left(\tfrac{1}{|\ell|}*|\widehat{h}_{n-1}(-\ell+k,-\ell)|\right)_j\leq\|c\|_{\ell^2}\cdot\left\|\tfrac{1}{|\ell|}*|\widehat{h}_{n-1}(\ell+k,\ell)|\right\|_{\ell^2(\ell)}\\ &\leq\|c\|_{\ell^2}\cdot C_{Y,2}\cdot\|\widehat{h}_{n-1}(\ell+k,\ell)\|_{\ell^2(\ell)}\overset{(7.2)}{\leq}\|c\|_{\ell^2}\cdot C_{Y,2}\cdot\|\widehat{h}_{n-1}(\ell+k,\ell)\|_{\ell^1(\ell)}.\end{aligned}$$

By the induction hypothesis, it thus follows that $(A)<\infty$, $(A)\in\ell^p(k)$ and

$$\|(A)\|_{\ell^p(k)}\leq\|c\|_{\ell^2}\cdot C_{Y,2}\cdot\left\|\,\|\widehat{h}_{n-1}(\ell+k,\ell)\|_{\ell^1(\ell)}\right\|_{\ell^p(k)}\leq(C_1\cdot\|c\|_{\ell^2})^n\,,$$

when $C_1>0$ is chosen with $C_1\geq C_{Y,2}$.

Next, we have by Cauchy-Schwarz's inequality and (7.2)

$$
\begin{aligned}
(B) &= \sum_{\ell\neq k, j\in\mathbb{Z}} \frac{|c_j|\,|\widehat{h}_{n-1}(\ell-j,-j)}{|\ell-k|} \leq \sum_{\ell\neq k} \frac{1}{|\ell-k|}\|c\|_{\ell^2}\,\|\widehat{h}_{n-1}(\ell-j,-j)\|_{\ell^2(j)} \\
&\leq \sum_{\ell\neq k} \frac{1}{|\ell-k|}\|c\|_{\ell^2}\,\|\widehat{h}_{n-1}(\ell-j,-j)\|_{\ell^1(j)} \\
&= \|c\|_{\ell^2}\cdot\left(\tfrac{1}{|\ell|} * \|\widehat{h}_{n-1}(\ell-j,-j)\|_{\ell^1(j)}\right)_k.
\end{aligned}
$$

By the induction hypothesis, it thus follows that $(B) < \infty$. Because of the induction hypothesis $\|\widehat{h}_{n-1}(\ell-j,-j)\|_{\ell^1(j)} \in \ell^p(\ell)$, we also have $\frac{1}{|\ell|} * \|\widehat{h}_{n-1}(\ell-j,-j)\|_{\ell^1(j)} \in \ell^p$ by the variant (7.4) of Young's inequality, and therefore $(B) \in \ell^p(k)$ and

$$
\|(B)\|_{\ell^p(k)} \leq \|c\|_{\ell^2}\cdot C_{Y,p}\cdot\left\|\,\|\widehat{h}_{n-1}(\ell-j,-j)\|_{\ell^1(j)}\right\|_{\ell^p(\ell)} \leq (C_1\cdot\|c\|_{\ell^2})^n\,,
$$

when $C_1 > 0$ is chosen with $C_1 \geq C_{Y,p}$.

Third, we estimate

$$
\begin{aligned}
(C) = \sum_{j\in\mathbb{Z}} |c_j|\,|\widehat{h}_{n-1}(k-j,-j)| &\leq \|c\|_{\ell^2}\cdot\|\widehat{h}_{n-1}(k-j,-j)\|_{\ell^2(j)} \\
&\overset{(7.2)}{\leq} \|c\|_{\ell^2}\cdot\|\widehat{h}_{n-1}(k-j,-j)\|_{\ell^1(j)}.
\end{aligned}
$$

By the induction hypothesis, it thus follows that $(C) < \infty$, $(C) \in \ell^p(k)$ and

$$
\|(C)\|_{\ell^p(k)} \leq \|c\|_{\ell^2}\cdot\left\|\,\|\widehat{h}_{n-1}(k-j,-j)\|_{\ell^1(j)}\right\|_{\ell^p(k)} \leq (C_1\cdot\|c\|_{\ell^2})^n\,,
$$

when $C_1 > 0$ is chosen with $C_1 \geq 1$.

Finally, we obtain (using Eq. (7.2))

$$
\begin{aligned}
(D) &= \sum_{\ell\neq 0, j\in\mathbb{Z}} \frac{|c_j|\,|\widehat{h}_{n-1}(k-j,\ell-j)|}{|\ell|} = \sum_{\ell\neq 0, j\in\mathbb{Z}} \frac{|c_{\ell-j}|\,|\widehat{h}_{n-1}(j+(k-\ell),j)|}{|\ell|} \\
&\leq \sum_{\ell\neq 0} \frac{1}{|\ell|}\,\|c\|_{\ell^2}\,\|\widehat{h}_{n-1}(j+(k-\ell),j)\|_{\ell^2(j)} \\
&\leq \|c\|_{\ell^2}\cdot\sum_{\ell\neq 0}\frac{1}{|\ell|}\,\|\widehat{h}_{n-1}(j+(k-\ell),j)\|_{\ell^1(j)} \\
&= \|c\|_{\ell^2}\cdot\left(\tfrac{1}{|\ell|} *_\ell \|\widehat{h}_{n-1}(j+\ell,j)\|_{\ell^1(j)}\right)_k.
\end{aligned}
$$

By the induction hypothesis, it thus follows that $(D) < \infty$. Because of the induction hypothesis $\|\widehat{h}_{n-1}(j+\ell, j)\|_{\ell^1(j)} \in \ell^p(\ell)$, we also have $\frac{1}{|\ell|} *_\ell \|\widehat{h}_{n-1}(j+\ell, j)\|_{\ell^1(j)} \in \ell^p$ by the variant (7.4) of Young's inequality, and therefore $(D) \in \ell^p(k)$ and

$$\|(D)\|_{\ell^p(k)} \leq \|c\|_{\ell^2} \cdot C_{Y,p} \cdot \left\| \|\widehat{h}_{n-1}(j+\ell, j)\|_{\ell^1(j)} \right\|_{\ell^p(\ell)} \leq (C_1 \cdot \|c\|_{\ell^2})^n ,$$

when $C_1 > 0$ is chosen with $C_1 \geq C_{Y,p}$.

By applying these estimates for (A), (B), (C) and (D) to (7.16), we see that the claim (7.10)–(7.12) indeed holds for n, with $C_1 := \max\{1, C_{Y,2}, C_{Y,p}\} > 0$ (independent of n). □

Lemma 7.4 *For every $(u, u_y) \in \mathsf{Pot}^1_{np}$ there exist constants $\sigma_u \in \mathbb{C}$ and $\varrho_u > 0$ so that for $k \geq 1$, the matrices $A_k \in \mathbb{C}^{2\times 2}$ defined by the equation*

$$(-1)^k\, E_{\lambda_{k,0}}(1) = \mathbb{1} + \frac{1}{2}\begin{pmatrix} -a_k & b_k \\ b_k & a_k \end{pmatrix} + \frac{\sigma_u}{k}\begin{pmatrix} 0 & -1 \\ 1 & 0 \end{pmatrix} + A_k$$

satisfy $|A_k| \leq \frac{\varrho_u}{k^2}$. Here both σ_u and ϱ_u are bounded on any closed ball in Pot^1_{np}.

Proof We work in the situation of the proof of Proposition 5.9. When we fix $\lambda = \lambda_{k,0}$ and $x = 1$, we have

$$\overline{E}_0(1) = (-1)^k \cdot \mathbb{1}$$

and

$$\begin{aligned}
\overline{E}_1(1) &= \overline{E}_0(1) \cdot \int_0^1 \overline{E}_0(-t) \cdot \overline{\beta}(t) \cdot \overline{E}_0(t)\, \mathrm{d}t \\
&= (-1)^k\, \lambda_{k,0}^{-1/2} \left(\int_0^1 \overline{\beta}_+(t)\, \mathrm{d}t + \int_0^1 \overline{E}(2t)\, \overline{\beta}_-(t)\, \mathrm{d}t \right) \\
&= (-1)^k\, \lambda_{k,0}^{-1/2} \left(\int_0^1 (-\tfrac{1}{2}\, u_z^2 + \tfrac{1}{4}\, \cosh(u) - 1)\, \mathrm{d}t \begin{pmatrix} 0 & 1 \\ -1 & 0 \end{pmatrix} \right. \\
&\quad \left. + \int_0^1 \begin{pmatrix} \cos(2\pi kt) & -\sin(2\pi kt) \\ \sin(2\pi kt) & \cos(2\pi kt) \end{pmatrix} \begin{pmatrix} 0 & 1 \\ 1 & 0 \end{pmatrix} (-u_{zx} - \tfrac{1}{4}\sinh(u))\, \mathrm{d}t \right) \\
&= \frac{(-1)^k}{4\pi k} \left(\overline{\sigma}_u \begin{pmatrix} 0 & 1 \\ -1 & 0 \end{pmatrix} - \int_0^1 \begin{pmatrix} -\sin(2\pi kt) & \cos(2\pi kt) \\ \cos(2\pi kt) & \sin(2\pi kt) \end{pmatrix} u_{zx}(t)\, \mathrm{d}t \right) + O(k^{-3})
\end{aligned}$$

with $\overline{\sigma}_u := \int_0^1 (-\frac{1}{2}\, u_z^2 + \frac{1}{4}\, \cosh(u) - 1)\, \mathrm{d}t$; note that $\sinh(u) \in W^{2,2}([0,1])$ holds, and therefore the Fourier coefficients of this function are $O(k^{-2})$; see also Eq. (4.9). □

Proof (of Theorem 7.1) We first consider the case $k > 0$, and we regauge α as in the proof of Theorem 5.4. By the analogous argument as in the proof of Theorem 5.4 we see that the contribution of γ to $\widetilde{F}$ is of order $O(|\lambda_{k,0}|^{-1/2}) = O(\frac{1}{k}) \in \ell^p(k)$, and therefore can again be neglected. We have $E_0(1) = (-1)^k \mathbb{1}$ and $E_1(1) = \frac{(-1)^k}{2}\begin{pmatrix} -a_k & b_k \\ b_k & a_k \end{pmatrix}$. In view of the regauging function g, it therefore suffices to show that $\sum_{n=2}^{\infty} E_n(1) \in \ell^p(k, \mathbb{C}^{2\times 2})$.

Because Pot_{np}^1 is dense in Pot_{np}, there exist $(u^{[1]}, u_y^{[1]}) \in \mathsf{Pot}_{np}^1$ and $(u^{[2]}, u_y^{[2]}) \in \mathsf{Pot}_{np}$ with $\|(u^{[2]}, u_y^{[2]})\|_{\mathsf{Pot}} \leq R_0$ (where R_0 is the constant from Lemma 7.3), such that $(u, u_y) = (u^{[1]}, u_y^{[1]}) + (u^{[2]}, u_y^{[2]})$ holds. For $\nu \in \{1, 2\}$, we denote the quantities associated to $(u^{[\nu]}, u_y^{[\nu]})$ by the superscript ${}^{[\nu]}$. Then by Lemma 7.4 we have $\sum_{n=2}^{\infty} E_n^{[1]}(1) \in \ell^p(k, \mathbb{C}^{2\times 2})$. We also note that by Proposition 5.9, there exists a constant $C_1 > 0$ such that we have for all $x \in [0, 1]$

$$|E^{[1]}(x) - E_0(x)| \leq \frac{C_1}{\sqrt{|\lambda|}};$$

note that $w(\lambda_{k,0}) = 1$ holds. We also have $|E_0(x)|, |E_0(x)^{-1}| \leq 2$, and therefore $|E^{[1]}(x)| \leq C_1 + 2 =: C_2$. We moreover obtain

$$\begin{aligned} |(E^{[1]})^{-1}| &\leq |E_0^{-1}| + |(E^{[1]})^{-1} - E_0^{-1}| \leq 2 + |E_0^{-1}|\,|(E^{[1]})^{-1}|\,|E_0 - E^{[1]}| \\ &\leq 2 + 2\,C_1\,|\lambda|^{-1/2}\,|(E^{[1]})^{-1}| \end{aligned}$$

and therefore

$$|(E^{[1]})^{-1}| \leq 2\,(1 - 2\,C_1\,|\lambda|^{-1/2})^{-1}.$$

It follows that for $|\lambda|$ sufficiently large, we have $|(E^{[1]})^{-1}| \leq C_3$ for some $C_3 > 0$. Then we also have

$$|(E^{[1]})^{-1} - E_0^{-1}| \leq |E_0^{-1}|\,|(E^{[1]})^{-1}|\,|E_0 - E^{[1]}| \leq 2 \cdot C_3 \cdot C_1\,|\lambda|^{-1/2} = C_4\,|\lambda|^{-1/2}$$

with $C_4 := 2\,C_1\,C_3$. We put $A := \max\{2, C_1, C_2, C_3, C_4\}$.

We also note that we have $\sum_{n=2}^{\infty} E_n^{[2]}(1) \in \ell^p(k, \mathbb{C}^{2\times 2})$ and $\|\sum_{n=2}^{\infty} E_n^{[2]}(1)\|_{\ell^p} \leq C \cdot \|(u^{[2]}, u_y^{[2]})\|_{\mathsf{Pot}}$ by Lemma 7.3.

To compare E_n with $E_n^{[1]}$, we now apply Lemma 5.5(1) with $\alpha = \widetilde{\alpha}_0 + \widetilde{\beta}^{[1]}$ and $\beta = \widetilde{\beta}^{[2]}$. Therefrom we obtain

$$\begin{aligned} E_n(1) &= E^{[1]}(1) \cdot \int_{t_1=0}^{x} \int_{t_2=0}^{t_1} \cdots \int_{t_n=0}^{t_{n-1}} \prod_{j=1}^{n} E^{[1]}(t_j)^{-1}\,\widetilde{\beta}^{[2]}(t_j)\,E^{[1]}(t_j)\,\mathrm{d}^n t \\ &= E_0(1) \cdot \int_{t_1=0}^{x} \int_{t_2=0}^{t_1} \cdots \int_{t_n=0}^{t_{n-1}} \prod_{j=1}^{n} E_0(t_j)^{-1}\,\widetilde{\beta}^{[2]}(t_j)\,E_0(t_j)\,\mathrm{d}^n t + D_n(1) \\ &= E^{[2]}(1) + D_n(1) \end{aligned}$$

with

$$D_n(x) = (E^{[1]}(x) - E_0(x)) \int_{t_1=0}^{x} \cdots \int_{t_n=0}^{t_{n-1}} \prod_{j=1}^{n} E_0(t_j)^{-1}\, \widetilde{\beta}^{[2]}(t_j)\, E_0(t_j)\, \mathrm{d}^n t$$
$$+ E^{[1]}(x) \sum_{\ell=1}^{n} \int_{t_1=0}^{x} \cdots \int_{t_n=0}^{t_{n-1}} \left(\prod_{j=1}^{\ell-1} E^{[1]}(t_j)^{-1}\, \widetilde{\beta}^{[2]}(t_j)\, E^{[1]}(t_j) \right)$$
$$\cdot \Big((E^{[1]}(t_\ell)^{-1} - E_0(t_\ell)^{-1})\, \widetilde{\beta}^{[2]}(t_\ell)\, E_0(t_\ell)$$
$$+ E^{[1]}(t_\ell)^{-1}\, \widetilde{\beta}^{[2]}(t_\ell)\, (E^{[1]}(t_\ell) - E_0(t_\ell)) \Big)$$
$$\cdot \left(\prod_{j=\ell+1}^{n} E_0(t_j)^{-1}\, \widetilde{\beta}^{[2]}(t_j)\, E_0(t_j) \right) \mathrm{d}^n t,$$

whence it follows

$$|D_n(x)| \leq (2n+1) \cdot \frac{x^n}{n!} \cdot \|\widetilde{\beta}^{[2]}\|^n \cdot A^{2n} \cdot A\, |\lambda_{k,0}|^{-1/2} \leq \frac{(\widetilde{A}\, \|(u^{[2]}, u_y^{[2]})\|_{\mathsf{Pot}}\, x)^n}{n!} \cdot \frac{1}{k}$$

with some $\widetilde{A} > 0$. If $\|(u^{[2]}, u_y^{[2]})\|_{\mathsf{Pot}}$ is sufficiently small, we therefore see that

$$\sum_{n=2}^{\infty} |D_n(x)| \leq C \cdot \|(u^{[2]}, u_y^{[2]})\|_{\mathsf{Pot}}^2 \cdot \frac{1}{k} \in \ell^p(k)$$

holds with some $C > 0$.

We now simply combine

$$\sum_{n=2}^{\infty} |E_n(1)| \leq \sum_{n=2}^{\infty} |E_n^{[1]}(1)| + \sum_{n=2}^{\infty} |D_n(1)| \in \ell^p(k),$$

completing the proof of Theorem 7.1 for the case $k > 0$.

Finally, we use Proposition 2.2(1) to derive the case $k < 0$ from the case $k > 0$. We let $\widetilde{u}(z) := -u(\overline{z})$ and $g(\lambda) := \left(\begin{smallmatrix} 1 & 0 \\ 0 & \lambda \end{smallmatrix}\right)$. Then we have $\widetilde{u}_z(z) = -u_{\overline{z}}(\overline{z})$, and therefore $\widetilde{a}_k$ resp. $\widetilde{b}_k$ (defined in Eq. (7.1)) are the cosine resp. the sine Fourier coefficients of $\widetilde{u}_z$. Also, we have $\mathrm{e}^{-\widetilde{u}(0)/2} = \mathrm{e}^{u(0)/2} = \tau^{-1}$. Thus we obtain from Proposition 2.2(1) and the case $k > 0$ of the present theorem, remembering $\lambda_{-k,0} = \lambda_{k,0}^{-1}$:

$$\begin{aligned}
&(-1)^k\, M_u(\lambda_{-k,0}) \\
&\quad = g(\lambda_{-k,0}^{-1})^{-1} \cdot (-1)^k\, M_{\widetilde{u}}(\lambda_{-k,0}^{-1}) \cdot g(\lambda_{-k,0}^{-1}) \\
&\quad = g(\lambda_{k,0})^{-1} \cdot (-1)^k\, M_{\widetilde{u}}(\lambda_{k,0}) \cdot g(\lambda_{k,0}) \\
&\quad = g(\lambda_{k,0})^{-1} \cdot \left(\mathbb{1} + \tfrac{1}{2} \begin{pmatrix} -\widetilde{a}_k & -\lambda_{k,0}^{-1/2}\,\tau\,\widetilde{b}_k \\ \lambda_{k,0}^{1/2}\,\tau^{-1}\,\widetilde{b}_k & \widetilde{a}_k \end{pmatrix} + \begin{pmatrix} \widetilde{r}_{11,k} & \lambda_{k,0}^{-1/2}\,\widetilde{r}_{12,k} \\ \lambda_{k,0}^{1/2}\,\widetilde{r}_{21,k} & \widetilde{r}_{22,k} \end{pmatrix} \right) \cdot g(\lambda_{k,0}) \\
&\quad = \mathbb{1} + \tfrac{1}{2} \begin{pmatrix} -\widetilde{a}_k & -\lambda_{k,0}^{1/2}\,\tau\,\widetilde{b}_k \\ \lambda_{k,0}^{-1/2}\,\tau^{-1}\,\widetilde{b}_k & \widetilde{a}_k \end{pmatrix} + \begin{pmatrix} \widetilde{r}_{11,k} & \lambda_{k,0}^{1/2}\,\widetilde{r}_{12,k} \\ \lambda_{k,0}^{-1/2}\,\widetilde{r}_{21,k} & \widetilde{r}_{22,k} \end{pmatrix} \\
&\quad = \mathbb{1} + \tfrac{1}{2} \begin{pmatrix} -\widetilde{a}_k & -\lambda_{-k,0}^{-1/2}\,\tau\,\widetilde{b}_k \\ \lambda_{-k,0}^{1/2}\,\tau^{-1}\,\widetilde{b}_k & \widetilde{a}_k \end{pmatrix} + \begin{pmatrix} \widetilde{r}_{11,k} & \lambda_{-k,0}^{-1/2}\,\widetilde{r}_{12,k} \\ \lambda_{-k,0}^{1/2}\,\widetilde{r}_{21,k} & \widetilde{r}_{22,k} \end{pmatrix} .
\end{aligned}$$

By putting $r_{ij,-k} := \widetilde{r}_{ij,k}$, the proof of the case $k < 0$ is completed. □

Chapter 8
The Consequences of the Fourier Asymptotic for the Spectral Data

We can use the Fourier asymptotics of the previous chapter to improve our description of the asymptotic behavior of the (classical) spectral divisor $D = \{(\lambda_k, \mu_k)\}$ of a potential (u, u_y). This is similar to the analogous refinement in the treatment of the 1-dimensional Schrödinger equation found in [Pö-T, Theorem 2.4, p. 35].

We would expect a similar refinement for the asymptotic behavior of the branch points $\varkappa_{k,\nu}$ of the spectral curve Σ. However, we postpone the investigation of this refinement until Chap. 11, to avoid technical difficulties.

The following Proposition 8.1 describes a relationship between the distance between the corresponding points of two spectral divisors, and the function values at $\lambda_{k,0}$ of the functions c and a in the corresponding monodromies. We will then use this relationship in Corollary 8.2 to derive the (improved) asymptotic behavior of the spectral data from the result of Theorem 7.1.

Note that Proposition 8.1 gives more information than what would be needed for Corollary 8.2: First, we compare two arbitrary spectral divisors with each other (instead of one spectral divisor to the divisor of the vacuum), and second, we give in Proposition 8.1(1) two different variants of the estimate, where only the second one (involving the sequence $\widetilde{\varepsilon}_k$) would be needed for Corollary 8.2. We do this to facilitate a second application of Proposition 8.1 in the proof of Lemma 13.3 (where the zeros of the function $\Delta'(\lambda)$ are studied).

We use the notations of the latter part of Chap. 6.

Proposition 8.1

(1) Let $c^{[1]}, c^{[2]} : \mathbb{C}^* \to \mathbb{C}$ be holomorphic functions which satisfy the basic asymptotics of Theorem 5.4, i.e. there exist numbers $\tau^{[1]}, \tau^{[2]} \in \mathbb{C}^*$ so that for

S. Klein, *A Spectral Theory for Simply Periodic Solutions of the Sinh-Gordon Equation*, Lecture Notes in Mathematics 2229,
https://doi.org/10.1007/978-3-030-01276-2_8

every $\varepsilon > 0$ *there exists* $R > 0$ *so that we have*

$$|c^{[\nu]}(\lambda) - \tau^{[\nu]}\, c_0(\lambda)| \leq \varepsilon\, |\lambda|^{1/2}\, w(\lambda) \text{ for } |\lambda| \geq R$$
$$\text{and} \quad |c^{[\nu]}(\lambda) - (\tau^{[\nu]})^{-1}\, c_0(\lambda)| \leq \varepsilon\, |\lambda|^{1/2}\, w(\lambda) \text{ for } |\lambda| \leq \tfrac{1}{R},$$

where $\nu \in \{1,2\}$. *Then let* $(\lambda_k^{[\nu]})_{k\in\mathbb{Z}}$ *be the sequence of zeros of* $c^{[\nu]}$ *as in Proposition 6.5(2).*

In this setting we have for $k > 0$

$$\begin{aligned}&\left|\left(\lambda_k^{[1]} - \lambda_k^{[2]}\right) - 8\,(-1)^k \left((\tau^{[1]})^{-1}\, c^{[1]}(\lambda_{k,0}) - (\tau^{[2]})^{-1}\, c^{[2]}(\lambda_{k,0})\right)\right| \\ &\leq \varepsilon_k \cdot |\lambda_k^{[1]} - \lambda_k^{[2]}| + |\lambda_k^{[1]} - \lambda_{k,0}| \cdot \frac{C}{k} \\ &\quad \cdot \max_{\lambda\in\overline{U_{k,\delta}}} \left|(\tau^{[2]})^{-1}\, c^{[2]}(\lambda) - (\tau^{[1]})^{-1}\, c^{[1]}(\lambda)\right| \\ &\leq \widetilde{\varepsilon}_k \cdot \max_{\lambda\in\overline{U_{k,\delta}}} \left|(\tau^{[2]})^{-1}\, c^{[2]}(\lambda) - (\tau^{[1]})^{-1}\, c^{[1]}(\lambda)\right|\end{aligned}$$

and

$$\begin{aligned}&\left|\left(\lambda_{-k}^{[1]} - \lambda_{-k}^{[2]}\right) - 8\,(-1)^{k+1}\, \lambda_{-k,0} \left(\tau^{[1]}\, c^{[1]}(\lambda_{-k,0}) - \tau^{[2]}\, c^{[2]}(\lambda_{-k,0})\right)\right| \\ &\leq \varepsilon_{-k} \cdot |\lambda_{-k}^{[1]} - \lambda_{-k}^{[2]}| + |\lambda_{-k}^{[1]} - \lambda_{-k,0}| \cdot C \cdot k \\ &\quad \cdot \max_{\lambda\in\overline{U_{-k,\delta}}} \left|\tau^{[2]}\, c^{[2]}(\lambda) - \tau^{[1]}\, c^{[1]}(\lambda)\right| \\ &\leq \widetilde{\varepsilon}_{-k} \cdot \frac{1}{k^2} \cdot \max_{\lambda\in\overline{U_{-k,\delta}}} \left|\tau^{[2]}\, c^{[2]}(\lambda) - \tau^{[1]}\, c^{[1]}(\lambda))\right|,\end{aligned}$$

where $(\varepsilon_k)_{k\in\mathbb{Z}}$, $(\widetilde{\varepsilon}_k)_{k\in\mathbb{Z}}$ *are sequences converging towards zero for* $k \to \pm\infty$, *and* $C > 0$ *is a constant.*

(2) *Suppose that additionally holomorphic functions* $a^{[1]}, a^{[2]} : \mathbb{C}^* \to \mathbb{C}$ *are given that satisfy the corresponding basic asymptotics of Theorem 5.4, i.e. there exist numbers* $\upsilon^{[1]}, \upsilon^{[2]} \in \mathbb{C}^*$ *so that for every* $\varepsilon > 0$ *there exists* $R > 0$ *so that we have*

$$|a^{[\nu]}(\lambda) - \upsilon^{[\nu]}\, a_0(\lambda)| \leq \varepsilon\, w(\lambda) \text{ for } |\lambda| \geq R$$
$$\text{and} \quad |a^{[\nu]}(\lambda) - (\upsilon^{[\nu]})^{-1}\, a_0(\lambda)| \leq \varepsilon\, w(\lambda) \text{ for } |\lambda| \leq \tfrac{1}{R}.$$

Then put $\mu_k^{[\nu]} := a^{[\nu]}(\lambda_k^{[\nu]})$ *for* $k \in \mathbb{Z}$, *see Proposition 6.5(3).*

In this setting there exist $C_1, C_2 > 0$ *so that we have for* $k > 0$

$$\left|\left((\upsilon^{[2]})^{-1}\,\mu_k^{[2]} - (\upsilon^{[1]})^{-1}\,\mu_k^{[1]}\right) - \left((\upsilon^{[2]})^{-1}\,a^{[2]}(\lambda_{k,0}) - (\upsilon^{[1]})^{-1}\,a^{[1]}(\lambda_{k,0})\right)\right|$$

$$\leq C_1 \cdot \frac{|\lambda_k^{[1]} - \lambda_{k,0}|}{k} \cdot \max_{\lambda\in\overline{U_{k,\delta}}} \left|(\upsilon^{[2]})^{-1}\,a^{[2]}(\lambda) - (\upsilon^{[1]})^{-1}\,a^{[1]}(\lambda)\right|$$

$$+ C_2 \cdot \frac{|\lambda_k^{[2]} - \lambda_k^{[1]}|}{k} \cdot \left(1 + \max_{\lambda\in\overline{U_{k,\delta}}} |(\upsilon^{[2]})^{-1}\,a^{[2]}(\lambda) - a_0(\lambda)|\right)$$

and

$$\left|\left(\upsilon^{[2]}\,\mu_{-k}^{[2]} - \upsilon^{[1]}\,\mu_{-k}^{[1]}\right) - \left(\upsilon^{[2]}\,a^{[2]}(\lambda_{-k,0}) - \upsilon^{[1]}\,a^{[1]}(\lambda_{-k,0})\right)\right|$$

$$\leq C_1 \cdot |\lambda_{-k}^{[1]} - \lambda_{-k,0}| \cdot k^3 \cdot \max_{\lambda\in\overline{U_{-k,\delta}}} \left|\upsilon^{[2]}\,a^{[2]}(\lambda) - \upsilon^{[1]}\,a^{[1]}(\lambda)\right|$$

$$+ C_2 \cdot |\lambda_{-k}^{[2]} - \lambda_{-k}^{[1]}| \cdot k^3 \cdot \left(1 + \max_{\lambda\in\overline{U_{-k,\delta}}} |\upsilon^{[2]}\,a^{[2]}(\lambda) - a_0(\lambda)|\right).$$

Proof We first show in both (1) and (2) the estimates for λ_k and μ_k with k positive.

For (1). The strategy of the proof is to express the quantity $(\tau^{[1]})^{-1}\,c^{[1]}(\lambda_{k,0}) - (\tau^{[2]})^{-1}\,c^{[2]}(\lambda_{k,0})$ in terms of integrals, and then use the given asymptotic behavior of the $c^{[\nu]}$ (also in the form of Corollary 5.8, which follows by the application of Cauchy's inequality) to relate these integrals to $\lambda_k^{[1]} - \lambda_k^{[2]}$.

By definition, we have $c^{[\nu]}(\lambda_k^{[\nu]}) = 0$ and therefore

$$\begin{aligned}
&(\tau^{[1]})^{-1}\,c^{[1]}(\lambda_{k,0}) - (\tau^{[2]})^{-1}\,c^{[2]}(\lambda_{k,0})\\
&= -(\tau^{[1]})^{-1}\left(c^{[1]}(\lambda_k^{[1]}) - c^{[1]}(\lambda_{k,0})\right) + (\tau^{[2]})^{-1}\left(c^{[2]}(\lambda_k^{[2]}) - c^{[2]}(\lambda_{k,0})\right)\\
&= (\tau^{[2]})^{-1}\int_{\lambda_{k,0}}^{\lambda_k^{[2]}} c^{[2]\prime}(\lambda)\,d\lambda - (\tau^{[1]})^{-1}\int_{\lambda_{k,0}}^{\lambda_k^{[1]}} c^{[1]\prime}(\lambda)\,d\lambda\\
&= \int_{\lambda_k^{[1]}}^{\lambda_k^{[2]}} c_0'(\lambda)\,d\lambda + \int_{\lambda_{k,0}}^{\lambda_k^{[2]}} \left((\tau^{[2]})^{-1}\,c^{[2]\prime}(\lambda) - c_0'(\lambda)\right)d\lambda\\
&\quad + \int_{\lambda_k^{[1]}}^{\lambda_k^{[2]}} \left((\tau^{[2]})^{-1}\,c^{[2]\prime}(\lambda) - c_0'(\lambda)\right)d\lambda.
\end{aligned} \tag{8.1}$$

We now handle the three resulting integrals separately.

By Eq. (5.34) we have

$$\begin{aligned} c_0'(\lambda) &= \frac{\lambda-1}{8\lambda}\cos(\zeta(\lambda)) + \frac{1}{2\sqrt{\lambda}}\sin(\zeta(\lambda)) \\ &= \frac{(-1)^k}{8} + \frac{1}{8}\left(\cos(\zeta(\lambda)) - (-1)^k\right) - \frac{1}{8\lambda}\cos(\zeta(\lambda)) + \frac{1}{2\sqrt{\lambda}}\sin(\zeta(\lambda)). \end{aligned}$$

It follows that there exists a constant $C_1 > 0$ so that for all $k \geq 1$ and all $\lambda \in U_{k,\delta}$ we have

$$\left| c_0'(\lambda) - \frac{(-1)^k}{8} \right| \leq \frac{C_1}{k},$$

and therefore

$$\left| \int_{\lambda_k^{[1]}}^{\lambda_k^{[2]}} c_0'(\lambda)\,\mathrm{d}\lambda - \frac{(-1)^k}{8}(\lambda_k^{[1]} - \lambda_k^{[2]}) \right| \leq \frac{C_1}{k} \cdot |\lambda_k^{[1]} - \lambda_k^{[2]}|. \tag{8.2}$$

Next, there exists a constant $C_2 > 0$ so that we have

$$\begin{aligned} &\left| \int_{\lambda_{k,0}}^{\lambda_k^{[1]}} \big((\tau^{[2]})^{-1} c^{[2]\prime}(\lambda) - (\tau^{[1]})^{-1} c^{[1]\prime}(\lambda)\big)\mathrm{d}\lambda \right| \\ &\leq |\lambda_k^{[1]} - \lambda_{k,0}| \cdot \max_{\lambda \in [\lambda_{k,0}, \lambda_k^{[1]}]} \left| (\tau^{[2]})^{-1} c^{[2]\prime}(\lambda) - (\tau^{[1]})^{-1} c^{[1]\prime}(\lambda) \right| \\ &\leq |\lambda_k^{[1]} - \lambda_{k,0}| \cdot \frac{C_2}{k} \cdot \max_{\lambda \in \overline{U_{k,\delta}}} \left| (\tau^{[2]})^{-1} c^{[2]}(\lambda) - (\tau^{[1]})^{-1} c^{[1]}(\lambda) \right|, \end{aligned} \tag{8.3}$$

where we denote for $\lambda_1, \lambda_2 \in \mathbb{C}^*$ by $[\lambda_1, \lambda_2]$ the straight line from λ_1 to λ_2 in the complex plane, and where the second $\leq$-sign follows from an application of Cauchy's inequality similar to the one in the proof of Corollary 5.8.

Moreover by Corollary 5.8(1), the sequence

$$\varepsilon_k^{[1]} := \max_{\lambda \in [\lambda_k^{[1]}, \lambda_k^{[2]}]} \left| (\tau^{[2]})^{-1} c^{[2]\prime}(\lambda) - c_0'(\lambda) \right|$$

converges to zero for $k \to \infty$ (note that $w(\lambda)$ is uniformly bounded on $[\lambda_k^{[1]}, \lambda_k^{[2]}] \subset U_{k,\delta}$), and therefore we have

$$\begin{aligned}
&\left| \int_{\lambda_k^{[1]}}^{\lambda_k^{[2]}} \left((\tau^{[2]})^{-1} c^{[2]\prime}(\lambda) - c_0'(\lambda) \right) \mathrm{d}\lambda \right| \\
&\leq |\lambda_k^{[2]} - \lambda_k^{[1]}| \cdot \max_{\lambda \in [\lambda_k^{[1]}, \lambda_k^{[2]}]} \left| (\tau^{[2]})^{-1} c^{[2]\prime}(\lambda) - c_0'(\lambda) \right| \\
&\leq \varepsilon_k^{[1]} \cdot |\lambda_k^{[2]} - \lambda_k^{[1]}|. \qquad (8.4)
\end{aligned}$$

By applying the estimates (8.2), (8.3) and (8.4) to Eq. (8.1), we obtain

$$\begin{aligned}
&\left| \left((\tau^{[1]})^{-1} c^{[1]}(\lambda_{k,0}) - (\tau^{[2]})^{-1} c^{[2]}(\lambda_{k,0}) \right) - \frac{(-1)^k}{8} (\lambda_k^{[1]} - \lambda_k^{[2]}) \right| \\
&\leq \varepsilon_k^{[2]} \cdot |\lambda_k^{[1]} - \lambda_k^{[2]}| + |\lambda_k^{[1]} - \lambda_{k,0}| \cdot \frac{C_2}{k} \\
&\quad \cdot \max_{\lambda \in \overline{U_{k,\delta}}} \left| (\tau^{[2]})^{-1} c^{[2]}(\lambda) - (\tau^{[1]})^{-1} c^{[1]}(\lambda) \right| \qquad (8.5)
\end{aligned}$$

with the sequence $\varepsilon_k^{[2]} := \frac{C_1}{k} + \varepsilon_k^{[1]}$, which converges to zero for $k \to \infty$, and therefore the first claimed estimate for $|\lambda_k^{[1]} - \lambda_k^{[2]}|$.

For the second claimed estimate we note that by Proposition 6.5(2) there exists a sequence $(\varepsilon_k^{[3]})$ which converges to zero for $k \to \infty$ so that $|\lambda_k^{[1]} - \lambda_{k,0}| \leq k \cdot \varepsilon_k^{[3]}$ holds, and therefore it follows from (8.5) that

$$\begin{aligned}
&\left| \left((\tau^{[1]})^{-1} c^{[1]}(\lambda_{k,0}) - (\tau^{[2]})^{-1} c^{[2]}(\lambda_{k,0}) \right) - \frac{(-1)^k}{8} (\lambda_k^{[1]} - \lambda_k^{[2]}) \right| \\
&\leq \varepsilon_k^{[2]} \cdot |\lambda_k^{[1]} - \lambda_k^{[2]}| + C_2 \cdot \varepsilon_k^{[3]} \cdot \max_{\lambda \in \overline{U_{k,\delta}}} \left| (\tau^{[2]})^{-1} c^{[2]}(\lambda) - (\tau^{[1]})^{-1} c^{[1]}(\lambda) \right| \qquad (8.6)
\end{aligned}$$

holds.

It follows from (8.6) that there exists $C_3 > 0$ so that

$$|\lambda_k^{[1]} - \lambda_k^{[2]}| \leq C_3 \cdot \max_{\lambda \in \overline{U_{k,\delta}}} \left| (\tau^{[2]})^{-1} c^{[2]}(\lambda) - (\tau^{[1]})^{-1} c^{[1]}(\lambda) \right|$$

holds, and that we therefore have

$$\begin{aligned}
&\left| \left((\tau^{[1]})^{-1} c^{[1]}(\lambda_{k,0}) - (\tau^{[2]})^{-1} c^{[2]}(\lambda_{k,0}) \right) - \frac{(-1)^k}{8} (\lambda_k^{[1]} - \lambda_k^{[2]}) \right| \\
&\leq \varepsilon_k^{[4]} \cdot \max_{\lambda \in \overline{U_{k,\delta}}} \left| (\tau^{[2]})^{-1} c^{[2]}(\lambda) - (\tau^{[1]})^{-1} c^{[1]}(\lambda) \right|
\end{aligned}$$

with the sequence $\varepsilon_k^{[4]} := \varepsilon_k^{[2]} \cdot C_3 + C_2 \cdot \varepsilon_k^{[3]}$, which converges to zero for $k \to \infty$. Hence the second claimed estimate for $|\lambda_k^{[1]} - \lambda_k^{[2]}|$ follows.

For (2). We have $\mu_k^{[\nu]} = a^{[\nu]}(\lambda_k^{[\nu]})$ by definition and therefore

$$\begin{aligned}
&\left((\upsilon^{[2]})^{-1}\,\mu_k^{[2]} - (\upsilon^{[1]})^{-1}\,\mu_k^{[1]}\right) \;-\; \left((\upsilon^{[2]})^{-1}\,a^{[2]}(\lambda_{k,0}) - (\upsilon^{[1]})^{-1}\,a^{[1]}(\lambda_{k,0})\right)\\
&= (\upsilon^{[2]})^{-1}\cdot\left(a^{[2]}(\lambda_k^{[2]}) - a^{[2]}(\lambda_{k,0})\right) - (\upsilon^{[1]})^{-1}\cdot\left(a^{[1]}(\lambda_k^{[1]}) - a^{[1]}(\lambda_{k,0})\right)\\
&= (\upsilon^{[2]})^{-1}\cdot\int_{\lambda_{k,0}}^{\lambda_k^{[2]}} a^{[2]\prime}(\lambda)\,\mathrm{d}\lambda - (\upsilon^{[1]})^{-1}\cdot\int_{\lambda_{k,0}}^{\lambda_k^{[1]}} a^{[1]\prime}(\lambda)\,\mathrm{d}\lambda\\
&= \int_{\lambda_{k,0}}^{\lambda_k^{[1]}} \left((\upsilon^{[2]})^{-1}\,a^{[2]\prime}(\lambda) - (\upsilon^{[1]})^{-1}\,a^{[1]\prime}(\lambda)\right)\mathrm{d}\lambda + (\upsilon^{[2]})^{-1}\\
&\quad\cdot\int_{\lambda_k^{[1]}}^{\lambda_k^{[2]}} a^{[2]\prime}(\lambda)\,\mathrm{d}\lambda.
\end{aligned}\tag{8.7}$$

There exist constants $C_4, C_5, C_6 > 0$ so that we have

$$\begin{aligned}
&\left|\int_{\lambda_{k,0}}^{\lambda_k^{[1]}} \left((\upsilon^{[2]})^{-1}\,a^{[2]\prime}(\lambda) - (\upsilon^{[1]})^{-1}\,a^{[1]\prime}(\lambda)\right)\mathrm{d}\lambda\right|\\
&\leq |\lambda_k^{[1]} - \lambda_{k,0}|\cdot \max_{\lambda\in[\lambda_{k,0},\lambda_k^{[1]}]}\left|(\upsilon^{[2]})^{-1}\,a^{[2]\prime}(\lambda) - (\upsilon^{[1]})^{-1}\,a^{[1]\prime}(\lambda)\right|\\
&\leq |\lambda_k^{[1]} - \lambda_{k,0}|\cdot \frac{C_4}{k}\cdot \max_{\lambda\in\overline{U_{k,\delta}}}\left|(\upsilon^{[2]})^{-1}\,a^{[2]}(\lambda) - (\upsilon^{[1]})^{-1}\,a^{[1]}(\lambda)\right|
\end{aligned}$$

(where the second inequality follows from Cauchy's inequality), and also

$$\begin{aligned}
&\left|(\upsilon^{[2]})^{-1}\cdot\int_{\lambda_k^{[1]}}^{\lambda_k^{[2]}} a^{[2]\prime}(\lambda)\,\mathrm{d}\lambda\right|\\
&\leq |\lambda_k^{[2]} - \lambda_k^{[1]}|\cdot|\upsilon^{[2]}|^{-1}\cdot \max_{\lambda\in[\lambda_k^{[1]},\lambda_k^{[2]}]}|a^{[2]\prime}(\lambda)|\\
&\leq |\lambda_k^{[2]} - \lambda_k^{[1]}|\cdot\left(\frac{C_5}{k} + \max_{\lambda\in[\lambda_k^{[1]},\lambda_k^{[2]}]}|(\upsilon^{[2]})^{-1}\,a^{[2]\prime}(\lambda) - a_0'(\lambda)|\right)\\
&\leq \frac{|\lambda_k^{[2]} - \lambda_k^{[1]}|}{k}\cdot\left(C_5 + C_6\cdot \max_{\lambda\in\overline{U_{k,\delta}}}|(\upsilon^{[2]})^{-1}\,a^{[2]}(\lambda) - a_0(\lambda)|\right)
\end{aligned}$$

(the last estimate again by Cauchy's inequality). By taking the absolute value of Eq. (8.7) and then applying the preceding estimates, we obtain the result claimed for $\mu_k^{[1]} - \mu_k^{[2]}$ in the Proposition.

The estimates for $\lambda_{-k}^{[\nu]}$ and $\mu_{-k}^{[\nu]}$ in (1) and (2) can now be derived by applying the previously shown results to the holomorphic functions $\widetilde{c}^{[\nu]}, \widetilde{a}^{[\nu]} : \mathbb{C}^* \to \mathbb{C}$ for $\nu \in \{1, 2\}$ given by

$$\widetilde{c}^{[\nu]}(\lambda) := \lambda \cdot c^{[\nu]}(\lambda^{-1}) \quad \text{and} \quad \widetilde{a}^{[\nu]}(\lambda) := a^{[\nu]}(\lambda^{-1}).$$

We denote the quantities associated to $\widetilde{c}^{[\nu]}$ resp. to $\widetilde{a}^{[\nu]}$ by attaching a tilde to the associated symbol. $\widetilde{c}^{[\nu]}$ and $\widetilde{a}^{[\nu]}$ satisfy the asymptotic hypotheses required in (1) resp. (2) of the Proposition, with the constants

$$\widetilde{\tau}^{[\nu]} = (\tau^{[\nu]})^{-1} \quad \text{and} \quad \widetilde{\upsilon}^{[\nu]} = (\upsilon^{[\nu]})^{-1}.$$

Moreover, we have

$$\widetilde{\lambda}_k^{[\nu]} = (\lambda_{-k}^{[\nu]})^{-1} \quad \text{and} \quad \widetilde{\mu}_k^{[\nu]} = \mu_{-k}^{[\nu]}.$$

The estimates for $\lambda_{-k}^{[\nu]}$ and $\mu_{-k}^{[\nu]}$ now follow by applying the previous results to $\widetilde{c}^{[\nu]}$ and $\widetilde{a}^{[\nu]}$. □

To simplify notations in the sequel, we consider besides the space ℓ_n^p also $\ell_{n,m}^p := \ell_{n,m}^p(\mathbb{Z})$ where $m, n \in \mathbb{Z}$. We define the corresponding norm for sequences $(a_k)_{k\in\mathbb{Z}}$ by

$$\|a_k\|_{\ell_{n,m}^p} := \|a_k\|_{\ell_n^p(k>0)} + |a_0| + \|a_{-k}\|_{\ell_m^p(k>0)}.$$

Of course, we put $\ell_{n,m}^p := \{ (a_k)_{k\in\mathbb{Z}} \mid \|a_k\|_{\ell_{n,m}^p} < \infty \}$; in this way, $\ell_{n,m}^p$ becomes a Banach space.

Corollary 8.2 *Let $(u, u_y) \in \mathsf{Pot}_{np}$ (or $M(\lambda)$ a monodromy matrix satisfying the asymptotic properties of Theorems 5.4 and 7.1), and D the classical spectral divisor of (u, u_y) (or of $M(\lambda)$). We enumerate $D = \{(\lambda_k, \mu_k)\}_{k\in\mathbb{Z}}$ as in Proposition 6.5(2),(3), then we have*

$$\lambda_k - \lambda_{k,0} \in \ell_{-1,3}^2(k) \quad \textit{and} \quad \begin{Bmatrix} \mu_k - \upsilon\, \mu_{k,0} & \textit{if } k \geq 0 \\ \mu_k - \upsilon^{-1}\, \mu_{k,0} & \textit{if } k < 0 \end{Bmatrix} \in \ell_{0,0}^2(k),$$

where $\upsilon := \mathrm{e}^{(u(1)-u(0))/4} \in \mathbb{C}^*$.

Proof We write the monodromy $M(\lambda)$ of (u, u_y) as $M(\lambda) = \left(\begin{smallmatrix} a(\lambda) & b(\lambda) \\ c(\lambda) & d(\lambda) \end{smallmatrix}\right)$, and put $\tau := \mathrm{e}^{-(u(0)+u(1))/4}$. Then the hypotheses of Proposition 8.1 are satisfied for $c^{[1]} = c$, $a^{[1]} = a$ and $c^{[2]} = c_0$, $a^{[2]} = a_0$ by Theorem 5.4.

By Proposition 8.1(1) there exists $C_1 > 0$ and a sequence $(\varepsilon_k)_{k\in\mathbb{Z}}$ with $\varepsilon_k \to 0$ for $k \to \pm\infty$ such that we have for $k > 0$

$$\begin{aligned}&\left|(\lambda_k - \lambda_{k,0}) - 8\,(-1)^k\,\left(\tau^{-1}\,c(\lambda_{k,0}) - c_0(\lambda_{k,0})\right)\right|\\ &\qquad \le \varepsilon_k^{[1]}\cdot|\lambda_k - \lambda_{k,0}| + |\lambda_k - \lambda_{k,0}|\cdot\frac{C_1}{k}\cdot\max_{\lambda\in\overline{U_{k,\delta}}}\left|c_0(\lambda) - \tau^{-1}\,c(\lambda)\right|\end{aligned}$$

and

$$\begin{aligned}&\left|(\lambda_{-k} - \lambda_{-k,0}) - 8\,(-1)^{k+1}\,\lambda_{-k,0}\left(\tau\,c(\lambda_{-k,0}) - c_0(\lambda_{-k,0})\right)\right|\\ &\qquad \le \varepsilon_{-k}^{[1]}\cdot|\lambda_{-k} - \lambda_{-k,0}| + |\lambda_{-k} - \lambda_{-k,0}|\cdot C_1\cdot k\cdot\max_{\lambda\in\overline{U_{-k,\delta}}}|c_0(\lambda) - \tau\,c(\lambda)|\,.\end{aligned}$$

For $k \to \infty$,

$$\text{both}\quad \frac{C_1}{k}\cdot\max_{\lambda\in\overline{U_{k,\delta}}}\left|c_0(\lambda) - \tau^{-1}\,c(\lambda)\right| \quad\text{and}\quad C_1\cdot k\cdot\max_{\lambda\in\overline{U_{-k,\delta}}}|c_0(\lambda) - \tau\,c(\lambda)|$$

converge to zero for $k \to \infty$ by Theorem 5.4, and therefore it follows that there exists $C_2 > 0$ so that

$$\begin{aligned}|\lambda_k - \lambda_{k,0}| &\le C_2\cdot|\tau^{-1}\,c(\lambda_{k,0}) - c_0(\lambda_{k,0})|\\ \text{and}\quad |\lambda_{-k} - \lambda_{-k,0}| &\le \frac{C_2}{k^2}\cdot|\tau\,c(\lambda_{-k,0}) - c_0(\lambda_{-k,0})|\end{aligned}\tag{8.8}$$

holds. It follows from Theorem 7.1 that

$$|\tau^{-1}\,c(\lambda_{k,0}) - c_0(\lambda_{k,0})| \in \ell^2_{-1}(k>0) \quad\text{and}\quad |\tau\,c(\lambda_{-k,0}) - c_0(\lambda_{-k,0})| \in \ell^2_1(k>0)$$

holds, and therefore we obtain from (8.8) that

$$|\lambda_k - \lambda_{k,0}| \in \ell^2_{-1}(k>0) \quad\text{and}\quad |\lambda_{-k} - \lambda_{-k,0}| \in \ell^2_3(k>0)$$

and therefore $\lambda_k - \lambda_{k,0} \in \ell^2_{-1,3}(k)$ holds.

Moreover, by Proposition 8.1(2) there exist $C_2, C_3 > 0$ so that we have for $k > 0$

$$\begin{aligned}&\left|(\mu_{k,0} - \upsilon^{-1}\,\mu_k) - \left(a_0(\lambda_{k,0}) - \upsilon^{-1}\,a(\lambda_{k,0})\right)\right|\\ &\qquad \le C_2\cdot\frac{|\lambda_k - \lambda_{k,0}|}{k}\cdot\max_{\lambda\in\overline{U_{k,\delta}}}\left|a_0(\lambda) - \upsilon^{-1}\,a(\lambda)\right| + C_3\cdot\frac{|\lambda_k - \lambda_{k,0}|}{k}\cdot 1\end{aligned}$$

and

$$\begin{aligned}\big|(\mu_{-k,0} - \upsilon\,\mu_{-k}) \;-\; \big(a_0(\lambda_{-k,0}) - \upsilon\,a(\lambda_{-k,0})\big)\big| \\ \leq C_2 \cdot |\lambda_{-k} - \lambda_{-k,0}| \cdot k^3 \cdot \max_{\lambda \in \overline{U_{-k,\delta}}} |a_0(\lambda) - \upsilon\,a(\lambda)| \\ + C_3 \cdot |\lambda_{-k,0} - \lambda_{-k}| \cdot k^3 \cdot 1.\end{aligned}$$

Because we have previously shown $\lambda_k - \lambda_{k,0} \in \ell^2_{-1,3}(k)$, and for $k \to \infty$

$$\text{both} \quad \max_{\lambda \in \overline{U_{k,\delta}}} \left|a_0(\lambda) - \upsilon^{-1}\,a(\lambda)\right| \quad \text{and} \quad \max_{\lambda \in \overline{U_{-k,\delta}}} |a_0(\lambda) - \upsilon\,a(\lambda)|$$

converge to zero by Theorem 5.4, and we have

$$\left|a_0(\lambda_{k,0}) - \upsilon^{-1}\,a(\lambda_{k,0})\right| \in \ell^2_0(k > 0) \quad \text{and} \quad \left|a_0(\lambda_{-k,0}) - \upsilon\,a(\lambda_{-k,0})\right| \in \ell^2_0(k > 0)$$

by Theorem 7.1, it follows that

$$\mu_k - \upsilon\,\mu_{k,0} \in \ell^2_0(k > 0) \quad \text{and} \quad \mu_{-k} - \upsilon^{-1}\,\mu_{-k,0} \in \ell^2_0(k > 0)$$

holds. □

Definition 8.3 We call a positive generalized divisor $\mathscr{D}$ or the underlying classical divisor D on a spectral curve $\Sigma \subset \mathbb{C}^* \times \mathbb{C}$ (or a discrete multi-set D of points in $\mathbb{C}^* \times \mathbb{C}$) *non-periodic asymptotic*, if the support of D is enumerated by a sequence $(\lambda_k, \mu_k)_{k \in \mathbb{Z}}$, and if there exists a number $\upsilon \in \mathbb{C}^*$ so that we have

$$\lambda_k - \lambda_{k,0} \in \ell^2_{-1,3}(k) \quad \text{and} \quad \left\{\begin{matrix} \mu_k - \upsilon\,\mu_{k,0} & \text{if } k \geq 0 \\ \mu_k - \upsilon^{-1}\,\mu_{k,0} & \text{if } k < 0 \end{matrix}\right\} \in \ell^2_{0,0}(k).$$

We call a non-periodic asymptotic divisor $\mathscr{D}$ resp. D *asymptotic*, if the above holds with $\upsilon = 1$ i.e. if we have

$$\lambda_k - \lambda_{k,0} \in \ell^2_{-1,3}(k) \quad \text{and} \quad \mu_k - \mu_{k,0} \in \ell^2_{0,0}(k).$$

We denote the *space of asymptotic classical divisors* (on any spectral curve, regarded as point multi-sets in $\mathbb{C}^* \times \mathbb{C}^*$) by Div.

Corollary 8.2 shows that if $(\Sigma, \mathscr{D})$ are the spectral data belonging to a non-periodic potential $(u, u_y) \in \mathsf{Pot}_{np}$ (or to a monodromy $M(\lambda)$ satisfying the asymptotic properties of Theorems 5.4 and 7.1) then $\mathscr{D}$ is an non-periodic asymptotic divisor on Σ. If $(u, u_y) \in \mathsf{Pot}$ holds, i.e. if the potential is periodic, or if $\upsilon = 1$ holds in the setting of the present chapter, then $\mathscr{D}$ is in fact an asymptotic divisor on Σ.

In view of Definition 8.3 it is tempting to identify the space Div of asymptotic classical divisors with the Banach space $\ell^2_{-1,3} \oplus \ell^2_{0,0}$. But we need to be careful, because for the points of an asymptotic classical divisor D that lie in the "compact part" of Σ (i.e. for those finitely many points of D which do not need to lie in their excluded domains, see Proposition 6.5(2),(3)) there is no canonical enumeration. Consequently, the space of asymptotic divisors has a different structure from $\ell^2_{-1,3} \oplus \ell^2_{0,0}$ near those asymptotic divisors D which contain a point of higher multiplicity (which can occur only for the finitely many divisor points in the "compact part" of Σ).

To describe the structure of the space of asymptotic divisors, we consider the group $P(\mathbb{Z})$ of finite permutations of $\mathbb{Z}$, i.e. of permutations $\sigma : \mathbb{Z} \to \mathbb{Z}$ for which there exists $N \in \mathbb{N}$ (dependent on σ) with $\sigma(k) = k$ for all $k \in \mathbb{Z}$ with $|k| > N$. $P(\mathbb{Z})$ acts on $\ell^2_{-1,3} \oplus \ell^2_{0,0}$ by permuting the elements of sequences, and the quotient space $(\ell^2_{-1,3} \oplus \ell^2_{0,0})/P(\mathbb{Z})$ is isomorphic to the space Div of asymptotic divisors.

Letting $D^{[1]}, D^{[2]} \in \mathsf{Div}$ be two asymptotic divisors, represented in the usual form $D^{[\nu]} = \{(\lambda_k^{[\nu]}, \mu_k^{[\nu]})\}_{k\in\mathbb{Z}}$ for $\nu \in \{1, 2\}$, we define a distance between them by

$$\|D^{[1]} - D^{[2]}\|_{\mathsf{Div}} := \inf_{\sigma_1,\sigma_2 \in P(\mathbb{Z})} \left(\left\| \lambda^{[1]}_{\sigma_1(k)} - \lambda^{[2]}_{\sigma_2(k)} \right\|^2_{\ell^2_{-1,3}} + \left\| \mu^{[1]}_{\sigma_1(k)} - \mu^{[2]}_{\sigma_2(k)} \right\|^2_{\ell^2_{0,0}} \right)^{1/2}.$$

Then the topology induced by this distance is the quotient topology of $(\ell^2_{-1,3} \oplus \ell^2_{0,0})/P(\mathbb{Z})$; in the sequel we will regard Div with this topology and this distance function.

Near any divisor $D \in \mathsf{Div}$ that does not contain any points of higher multiplicity, the space Div is locally isomorphic to $\ell^2_{-1,3} \oplus \ell^2_{0,0}$ as Hilbert space. However, if D does contain points of higher multiplicity, this is no longer the case. Coordinates of Div near such a point are obtained in the following way: Suppose $D = \{(\lambda_k, \mu_k)\}_{k\in\mathbb{Z}}$, where points (λ_k, μ_k) occur more than once according to multiplicity. Choose $N \in \mathbb{N}$ such that $(\lambda_k, \mu_k) \in \widehat{U}_{k,\delta}$ holds for all k with $|k| > N$ (in particular, none of these (λ_k, μ_k) have multiplicity > 1). Then the $2N + 1$ coefficients of a polynomial with zeros in all the λ_k, $|k| \geq N$ (these coefficients are the elementary symmetric polynomials in λ_k, $|k| \leq N$), together with the λ_k with $|k| > N$ provide coordinates for Div near D.

Part III
The Inverse Problem for the Monodromy

Chapter 9
Asymptotic Spaces of Holomorphic Functions

We now introduce "asymptotic spaces" of holomorphic functions on $\mathbb{C}^*$ or on Σ which have prescribed descent to 0 for $\lambda \to \infty$ and/or for $\lambda \to 0$; for this purpose we will cover $\mathbb{C}^*$ by a sequence $(S_k)_{k\in\mathbb{Z}}$ of annuli, the descent of the functions described by these spaces will be uniform on each of these annuli S_k (up to a factor $w(\lambda)^s$).

The purpose of these spaces is to describe the asymptotic behavior of the monodromy of a potential $(u, u_y) \in \mathsf{Pot}$; more specifically we will see in Chaps. 10 and 11 that the differences between the functions comprising the monodromy matrix and the corresponding functions for the vacuum are members of such asymptotic spaces as we are about to introduce. This formulation of the asymptotic behaviour of the monodromy (which we will regard as the final asymptotics) will liberate us from the special role that the points $\lambda_{k,0}$ play in the Fourier asymptotic of Theorem 7.1.

It is a remarkable fact that we will be able to control the asymptotic behavior of the monodromy by way of these asymptotic spaces not only on the excluded domains, but on all of $\mathbb{C}^*$. This is similar to the basic asymptotics of Theorem 5.4, but it is very interesting that the addition of the Fourier asymptotics of Theorem 7.1 which states only the ℓ^2-summability of the asymptotic difference at the single sequence of points $(\lambda_{k,0})$ already implies control of an ℓ^2-summability-type not only in excluded domains (i.e. small neighborhoods of $\lambda_{k,0}$) but on entire annuli in $\mathbb{C}^*$ via the asymptotic spaces defined below.

The underlying reason why this control is possible, and the fact that inspired the definition of the asymptotic spaces, is that the holomorphic functions comprising the monodromy can be described by infinite sums resp. products (as we will see in Chap. 10), and these infinite sums and products can be estimated on the S_k essentially by the statements of Proposition A.1(2),(3) resp. Proposition A.4(2).

In this chapter, we suppose that a spectral curve Σ is given only where we define or investigate objects involving Σ.

S. Klein, *A Spectral Theory for Simply Periodic Solutions of the Sinh-Gordon Equation*, Lecture Notes in Mathematics 2229,
https://doi.org/10.1007/978-3-030-01276-2_9

For $k \in \mathbb{Z}$ we put

$$S_k := \begin{cases} \{\, \lambda \in \mathbb{C}^* \mid (k - \frac{1}{2})\pi \le |\zeta(\lambda)| \le (k + \frac{1}{2})\pi, \, |\lambda| > 1 \,\} & \text{for } k > 0 \\ \{\, \lambda \in \mathbb{C}^* \mid |\zeta(\lambda)| \le \frac{\pi}{2} \,\} & \text{for } k = 0 \\ \{\, \lambda \in \mathbb{C}^* \mid (-k - \frac{1}{2})\pi \le |\zeta(\lambda)| \le (-k + \frac{1}{2})\pi, \, |\lambda| < 1 \,\} & \text{for } k < 0 \end{cases} \tag{9.1}$$

and

$$\widehat{S}_k := \{\, (\lambda, \mu) \in \Sigma \mid \lambda \in S_k \,\}. \tag{9.2}$$

We call S_k resp. $\widehat{S}_k$ the *annulus through* $\lambda_{k,0}$ in $\mathbb{C}^*$ resp. in Σ. Note that for $0 < \delta < \pi - \frac{1}{2}$, $U_{k,\delta} \subset S_k$ holds, that different S_k resp. $\widehat{S}_k$ intersect at most at their boundary, and that we have $\bigcup_{k \in \mathbb{Z}} S_k = \mathbb{C}^*$ resp. $\bigcup_{k \in \mathbb{Z}} \widehat{S}_k = \Sigma$.

The reader is reminded of the function $w(\lambda)$ introduced in Eq. (5.4).

We now define the asymptotic spaces which we will use in Chap. 11 to describe the asymptotic behavior of the holomorphic functions constituting the monodromy $M(\lambda)$ of a potential $(u, u_y) \in \mathsf{Pot}$, and of many other functions in the course of this work. We introduce three versions of the asymptotic spaces: One that controls the function both near $\lambda = \infty$ and near $\lambda = 0$, and one each where only one of the "ends" near $\lambda = \infty$ resp. near $\lambda = 0$ is controlled.

Definition 9.1 Let $0 < p \le \infty$, $n, m \in \mathbb{Z}$, $s \ge 0$ and a domain $G \subset \mathbb{C}^*$ resp. $\widehat{G} \subset \Sigma$ be given.[1]

Then we say that a holomorphic function $f : G \to \mathbb{C}$ resp. $f : \widehat{G} \to \mathbb{C}$ has *$\ell^p_{n,m}$-asymptotic of type s* if there exists a sequence $(a_k)_{k \in \mathbb{Z}} \in \ell^p_{n,m}(k)$ of non-negative numbers such that

$$\forall k \in \mathbb{Z} \, \forall \lambda \in G \cap S_k \; : \; |f(\lambda)| \le a_k \cdot w(\lambda)^s$$

resp.

$$\forall k \in \mathbb{Z} \, \forall \lambda \in \widehat{G} \cap \widehat{S}_k \; : \; |f(\lambda)| \le a_k \cdot w(\lambda)^s$$

holds. We call any such sequence (a_k) a *bounding sequence* for f. We denote the space of all $\ell^p_{n,m}$-asymptotic functions $f : G \to \mathbb{C}$ resp. $f : \widehat{G} \to \mathbb{C}$ of type s by $\mathrm{As}(G, \ell^p_{n,m}, s)$ resp. $\mathrm{As}(\widehat{G}, \ell^p_{n,m}, s)$. These spaces become Banach spaces via the norm

$$\|f\|_{\mathrm{As}(G, \ell^p_{n,m}, s)} := \inf \|a_k\|_{\ell^p_{n,m}} \quad \text{resp.} \quad \|f\|_{\mathrm{As}(\widehat{G}, \ell^p_{n,m}, s)} := \inf \|a_k\|_{\ell^p_{n,m}} \, ,$$

where the infimum is taken over all bounding sequences (a_k) for f.

---

[1] In most of our applications, we will have $G \in \{\mathbb{C}^*, V_\delta\}$ resp. $\widehat{G} \in \{\Sigma, \widehat{V}_\delta\}$, $p \in \{2, \infty\}$ and $s \in \{0, 1\}$.

Definition 9.2 Let $0 < p \le \infty$, $n, m \in \mathbb{Z}$, $s \ge 0$ and a domain $G \subset \mathbb{C}^*$ resp. $\widehat{G} \subset \Sigma$ be given. We put

$$G_\infty := \{\,\lambda \in \mathbb{C}^* \mid |\lambda| > 1\,\} \cap G \quad \text{and} \quad G_0 := \{\,\lambda \in \mathbb{C}^* \mid |\lambda| < 1\,\} \cap G$$

resp.

$$\widehat{G}_\infty := \{\,(\lambda,\mu) \in \Sigma \mid |\lambda| > 1\,\} \cap \widehat{G} \quad \text{and} \quad \widehat{G}_0 := \{\,(\lambda,\mu) \in \Sigma \mid |\lambda| < 1\,\} \cap \widehat{G}.$$

(1) We say that a holomorphic function $f : G \to \mathbb{C}$ resp. $f : \widehat{G} \to \mathbb{C}$ has *ℓ^p_n-asymptotic of type s near $\lambda = \infty$*, if

$$f|G_\infty \in \mathrm{As}(G_\infty, \ell^p_{n,0}, s) \quad \text{resp.} \quad f|\widehat{G}_\infty \in \mathrm{As}(\widehat{G}_\infty, \ell^p_{n,0}, s)$$

holds. We denote this space of functions by $\mathrm{As}_\infty(G, \ell^p_n, s)$ resp. $\mathrm{As}_\infty(\widehat{G}, \ell^p_n, s)$. It becomes a Banach space via the norm

$$\|f\|_{\mathrm{As}_\infty(G,\ell^p_n,s)} := \|\, f|G_\infty \,\|_{\mathrm{As}(G_\infty,\ell^p_{n,0},s)} \quad \text{resp.}$$

$$\|f\|_{\mathrm{As}_\infty(\widehat{G},\ell^p_n,s)} := \|\, f|\widehat{G}_\infty \,\|_{\mathrm{As}(\widehat{G}_\infty,\ell^p_{n,0},s)}.$$

(2) We say that a holomorphic function $f : G \to \mathbb{C}$ resp. $f : \widehat{G} \to \mathbb{C}$ has *ℓ^p_m-asymptotic of type s near $\lambda = 0$*, if

$$f|G_0 \in \mathrm{As}(G_0, \ell^p_{0,m}, s) \quad \text{resp.} \quad f|\widehat{G}_0 \in \mathrm{As}(\widehat{G}_0, \ell^p_{0,m}, s)$$

holds. We denote this space of functions by $\mathrm{As}_0(G, \ell^p_m, s)$ resp. $\mathrm{As}_0(\widehat{G}, \ell^p_m, s)$. It becomes a Banach space via the norm

$$\|f\|_{\mathrm{As}_0(G,\ell^p_m,s)} := \|\, f|G_0 \,\|_{\mathrm{As}(G_0,\ell^p_{0,m},s)} \quad \text{resp.}$$

$$\|f\|_{\mathrm{As}_0(\widehat{G},\ell^p_m,s)} := \|\, f|\widehat{G}_0 \,\|_{\mathrm{As}(\widehat{G}_0,\ell^p_{0,m},s)}.$$

Remark 9.3 An entire function $f \in \mathrm{As}_\infty(\mathbb{C}, \ell^\infty_0, s)$ satisfies

$$|f(\lambda)| \le C_1 \cdot \exp(s \cdot |\operatorname{Im}(\zeta(\lambda))|) \le C_2 \cdot \exp(\tfrac{s}{4} \cdot \sqrt{|\lambda|}) \quad \text{for } \lambda \in \mathbb{C} \text{ with } |\lambda| \text{ large,}$$

with constants $C_1, C_2 > 0$ and therefore is an entire function of order $\varrho = \frac{1}{2}$ and type $\sigma = \frac{s}{4}$.

The following proposition states several important facts on asymptotic functions. Proposition 9.4(1) shows that a holomorphic function on $\mathbb{C}^*$ which is asymptotic on some V_δ, i.e. on the area outside the excluded domains, is in fact asymptotic on all of $\mathbb{C}^*$; this will be especially useful in several instances.

Proposition 9.4 *Let* $0 < p \leq \infty$, $n, m \in \mathbb{Z}$ *and* $s \geq 0$.

(1) Let $f : \mathbb{C}^* \to \mathbb{C}$ *be a holomorphic function so that* $f|V_\delta \in \mathrm{As}(V_\delta, \ell^p_{n,m}, s)$ *holds for some* $\delta > 0$*, and let* $(a_k) \in \ell^p_{n,m}(k)$ *be a bounding sequence for* f*.*
Then already $f \in \mathrm{As}(\mathbb{C}^*, \ell^p_{n,m}, s)$ *holds and* $((4\mathrm{e}^\delta)^s \cdot a_k)$ *is a bounding sequence for* f*, in particular we have*

$$\|f\|_{\mathrm{As}(\mathbb{C}^*,\ell^p_{n,m},s)} \leq (4\mathrm{e}^\delta)^s \cdot \|\, f|V_\delta \,\|_{\mathrm{As}(V_\delta,\ell^p_{n,m},s)}.$$

(2) Let $G \subset \mathbb{C}^*$ *be a domain,* $f \in \mathrm{As}(G, \ell^p_{n,m}, s)$ *and* $j \in \mathbb{Z}$*. Then we have* $\lambda^{j/2} \cdot f \in \mathrm{As}(G, \ell^p_{n-j,m+j}, s)$*, and for any bounding sequence* (a_k) *for* f*,*

$$b_k := \begin{cases} (4\pi(k+1))^j \cdot a_k & \text{if } k > 0 \\ (4\pi)^j \cdot a_0 & \text{if } k = 0 \\ \left(\frac{1}{4\pi(|k|-1)}\right)^j \cdot a_k & \text{if } k < 0 \end{cases}$$

is a bounding sequence for $\lambda^{j/2} \cdot f$*. In particular, we have*

$$\|\lambda^{j/2} \cdot f\|_{\mathrm{As}(G,\ell^p_{n-j,m+j},s)} \leq (8\pi)^{|j|} \cdot \|f\|_{\mathrm{As}(G,\ell^p_{n,m},s)}.$$

(3) Let $f \in \mathrm{As}(\mathbb{C}^*, \ell^p_{n,m}, s)$ *and* $j \in \mathbb{N}$*. Then for the* j*-th derivative* $f^{(j)}$ *of* f*, we have* $f^{(j)} \in \mathrm{As}(\mathbb{C}^*, \ell^p_{n+j,m-3j}, s)$*, and if* (a_k) *is a bounding sequence for* f*, then* $b_k := \frac{12^s \cdot j!}{r_k^j} \cdot \max\{a_{k-1}, a_k, a_{k+1}\}$ *is a bounding sequence for* $f^{(j)}$*, where*

$$r_k := \begin{cases} k & \text{if } k > 0 \\ 1 & \text{if } k = 0 \\ \frac{1}{16\pi^2 |k|^3} & \text{if } k < 0 \end{cases}.$$

In particular, we have

$$\|f^{(j)}\|_{\mathrm{As}(\mathbb{C}^*,\ell^p_{n+j,m-3j},s)} \leq C \cdot \|f\|_{\mathrm{As}(\mathbb{C}^*,\ell^p_{n,m},s)}$$

with a constant $C > 0$*.*

In (1)–(3), analogous statements hold for the asymptotic spaces $\mathrm{As}_\infty(\mathbb{C}^*, \ell^p_n, s)$ *and* $\mathrm{As}_0(\mathbb{C}^*, \ell^p_m, s)$*.*

Proof For (1). For $k \in \mathbb{Z}$ and $\lambda \in S_k \cap V_\delta$ we have $|f(\lambda)| \leq a_k\, w(\lambda)^s$ by definition. We need to show that for $\lambda \in U_{k,\delta} \subset S_k$,

$$|f(\lambda)| \leq (4\,\mathrm{e}^\delta)^s\, a_k \cdot w(\lambda)^s$$

holds. Indeed we have by Proposition 6.3(1) for any $\lambda' \in \overline{U_{k,\delta}}$

$$\frac{1}{2} \leq w(\lambda') \leq 2\,\mathrm{e}^{\delta}$$

and therefore for $\lambda \in U_{k,\delta}$ by the maximum principle for holomorphic functions

$$|f(\lambda)| \leq \max_{\lambda' \in \partial U_{k,\delta}} |f(\lambda')| \leq a_k \cdot \left(\max_{\lambda' \in \partial U_{k,\delta}} w(\lambda') \right)^s \leq a_k \cdot (2\,\mathrm{e}^{\delta})^s \leq (4\,\mathrm{e}^{\delta})^s\, a_k \cdot w(\lambda)^s .$$

For (2). We let (a_k) be a bounding sequence for f, and define (b_k) as in the Proposition. Note that $(a_k) \in \ell^p_{n,m}(k)$ implies $(b_k) \in \ell^p_{n-j,m+j}(k)$. For $k \in \mathbb{Z}$ and $\lambda \in S_k$, we have

$$\pi \cdot (k - \tfrac{1}{2}) \leq |\zeta(\lambda)| \leq \pi \cdot (k + \tfrac{1}{2})$$

and therefore

$$\begin{cases} 16\,\pi^2\,(k-1)^2 \leq |\lambda| \leq 16\,\pi^2\,(k+1)^2 & \text{if } k > 0 \\ 4\,\pi^2 \leq |\lambda| \leq 16\,\pi^2 & \text{if } k = 0 \\ \frac{1}{16\,\pi^2\,(|k|+1)^2} \leq |\lambda| \leq \frac{1}{16\,\pi^2\,(|k|-1)^2} & \text{if } k < 0 \end{cases},$$

hence

$$\begin{cases} (4\,\pi\,(k-1))^j \leq |\lambda|^{j/2} \leq (4\,\pi\,(k+1))^j & \text{if } k > 0 \\ (2\pi)^j \leq |\lambda|^{j/2} \leq (4\pi)^j & \text{if } k = 0 \\ \left(\frac{1}{4\,\pi\,(|k|+1)}\right)^j \leq |\lambda|^{j/2} \leq \left(\frac{1}{4\,\pi\,(|k|-1)}\right)^j & \text{if } k < 0 \end{cases}.$$

For $\lambda \in S_k \cap G$, we therefore obtain

$$|\lambda^{j/2} \cdot f(\lambda)| \leq |\lambda|^{j/2} \cdot a_k \cdot w(\lambda)^s \leq b_k \cdot w(\lambda)^s ,$$

whence the claimed statements follow.

For (3). We let (a_k) be a bounding sequence for f, and define (r_k) and (b_k) as in the Proposition. Note that $(a_k) \in \ell^p_{n,m}(k)$ implies $(b_k) \in \ell^p_{n+j,m-3j}(k)$. Let $k \in \mathbb{Z}$ be given. r_k is chosen such that for any $\lambda \in S_k$ and $\lambda' \in \mathbb{C}^*$ with $|\lambda' - \lambda| = r_k$, we have

$$|\zeta(\lambda') - \zeta(\lambda)| \leq 1. \tag{9.3}$$

It follows from (9.3) that we have

$$\lambda' \in S_{k-1} \cup S_k \cup S_{k+1}. \tag{9.4}$$

We also obtain from (9.3) by Proposition 5.3(1):

$$\begin{aligned} w(\lambda') &\leq 2\,\mathrm{e}^{|\operatorname{Im}(\zeta(\lambda'))|} \leq 2\,\mathrm{e}^{|\operatorname{Im}(\zeta(\lambda))|}\,\mathrm{e}^{|\operatorname{Im}(\zeta(\lambda')-\zeta(\lambda))|} \leq 2\,\mathrm{e}^{|\operatorname{Im}(\zeta(\lambda))|}\,\mathrm{e}^{|\zeta(\lambda')-\zeta(\lambda)|} \\ &\leq 2 \cdot 2\,w(\lambda) \cdot \mathrm{e} \leq 12\,w(\lambda). \end{aligned}$$

We now obtain by Cauchy's inequality, applied to the holomorphic function f on the disk $\overline{B(\lambda, r_k)}$:

$$\begin{aligned} |f^{(j)}(\lambda)| &\leq \frac{j!}{r_k^j} \cdot \max_{|\lambda'-\lambda|=r_k} |f(\lambda')| \leq \frac{j!}{r_k^j} \cdot \max\{a_{k-1}, a_k, a_{k+1}\} \cdot \left(\max_{|\lambda'-\lambda|=r_k} w(\lambda') \right)^s \\ &\leq \frac{j!}{r_k^j} \cdot \max\{a_{k-1}, a_k, a_{k+1}\} \cdot (12\,w(\lambda))^s = b_k \cdot w(\lambda)^s, \end{aligned}$$

whence the claimed statements follow. □

Chapter 10
Interpolating Holomorphic Functions

One of the main results of this text is that the monodromy $M(\lambda)$ can be reconstructed uniquely (up to a sign in the off-diagonal entries) from the spectral data $(\Sigma, \mathscr{D})$. More specifically, the given spectral data $(\Sigma, \mathscr{D})$ provide the following information on the holomorphic functions comprising the monodromy

$$M(\lambda) = \begin{pmatrix} a(\lambda) & b(\lambda) \\ c(\lambda) & d(\lambda) \end{pmatrix}$$

to which they belong: The components λ_* of the points (λ_*, μ_*) in the support of $\mathscr{D}$ give all the zeros of c (with multiplicity), and moreover the components μ_* provide information on function values of a resp. d by the equations $a(\lambda_*) = \mu_*$, $d(\lambda_*) = \mu_*^{-1}$. (If (λ_*, μ_*) is a point of degree $m \geq 2$ in the divisor $\mathscr{D}$, then the generalized spectral divisor $\mathscr{D}$ also determines the derivatives $d'(\lambda_*), \dots, d^{(m-1)}(\lambda_*)$ in a manner that will be explained in detail in Lemma 12.2.)

So the problem at hand is to reconstruct a holomorphic function on $\mathbb{C}^*$ from either the knowledge of its zeros, or from its function values at a sequence of points, in both cases together with the knowledge of the asymptotic behavior of the function near $\lambda = \infty$ and near $\lambda = 0$. We will address these two problems in the present chapter.

Proposition 10.1 is concerned with the reconstruction of a holomorphic function on $\mathbb{C}^*$ "with the asymptotic behavior of c" from the knowledge of its zeros. In a way, this proposition is an adaption of Hadamard's Factorization Theorem (see for example [Co, Theorem XI.3.4, p. 289]) to our specific situation. The most significant difference between our situation and the classical Theorem is that whereas the classical Theorem concerns entire functions with zeros accumulating near $\lambda = \infty$, we are interested in holomorphic functions on $\mathbb{C}^*$, whose zeros accumulate both near $\lambda = \infty$ and near $\lambda = 0$. Notice that similarly to Hadamard's Theorem, we obtain an explicit representation of c as an infinite product.

S. Klein, *A Spectral Theory for Simply Periodic Solutions of the Sinh-Gordon Equation*, Lecture Notes in Mathematics 2229,
https://doi.org/10.1007/978-3-030-01276-2_10

Thereafter we study in Corollary 10.3 the behavior of the function $\frac{c(\lambda)}{\lambda-\lambda_k}$ on $U_{k,\delta}$; here λ_k is a root of c and $U_{k,\delta}$ is the excluded domain associated to this root. We need to understand the behavior of such functions both for the proof of Proposition 10.4 and on several further occasions in the course of this work.

Finally, Proposition 10.4 concerns the reconstruction of a holomorphic function on $\mathbb{C}^*$ "with the asymptotic behavior of a or d" from the knowledge of its function values at the zeros of c; if c has zeros of higher order, then also values of the derivatives of a resp. d at these points need to be known. We obtain an explicit description of a resp. d as an infinite series.

In Chap. 12, we will use Propositions 10.1 and 10.4 to reconstruct the monodromy $M(\lambda)$ from its spectral data $(\Sigma, \mathscr{D})$.

We mention that in the proofs of the present chapter (and nowhere else) we use the results on infinite sums and products collected in Appendix A.

Proposition 10.1 (Interpolation by the Zeros)

(1) Let a sequence $(\lambda_k)_{k\in\mathbb{Z}}$ in $\mathbb{C}^*$ be given, such that we have

$$\lambda_k - \lambda_{k,0} \in \ell^2_{-1,3}(k).$$

Then the infinite product[1]

$$\tau := \left(\prod_{k\in\mathbb{Z}} \frac{\lambda_{k,0}}{\lambda_k}\right)^{1/2} \tag{10.1}$$

converges absolutely in $\mathbb{C}^*$, and the infinite product

$$c(\lambda) = \frac{1}{4}\,\tau\,(\lambda-\lambda_0)\cdot\prod_{k=1}^{\infty}\frac{\lambda_k-\lambda}{16\pi^2k^2}\cdot\prod_{k=1}^{\infty}\frac{\lambda-\lambda_{-k}}{\lambda} \tag{10.2}$$

converges locally uniformly to a holomorphic function $c = c(\lambda) : \mathbb{C}^* \to \mathbb{C}$. c has zeros in all the λ_k (with the appropriate multiplicity, if some of the λ_k coincide) and no others. Moreover, we have

$$c - \tau\, c_0 \in \mathrm{As}_\infty(\mathbb{C}^*, \ell^2_{-1}, 1) \quad \textit{and} \quad c - \tau^{-1}\, c_0 \in \mathrm{As}_0(\mathbb{C}^*, \ell^2_1, 1). \tag{10.3}$$

(2) Let $R_0 > 0$ be given. Then there exists a constant $C > 0$ (depending only on R_0), such that for any pair of sequences $(\lambda_k^{[1]})_{k\in\mathbb{Z}}, (\lambda_k^{[2]})_{k\in\mathbb{Z}} \in \ell^2_{-1,3}(k)$ with

[1]Because of the freedom of choice of the branch of the square root function, τ is determined only up to sign. In view of the fact that the off-diagonal entries of the monodromy are determined by their zeros only up to a sign (see Proposition 6.7(1)), this is to be expected.

$\|\lambda_k^{[\nu]} - \lambda_{k,0}\|_{\ell^2_{-1,3}} \leq R_0$, *the quantities* $\tau^{[\nu]}$ *and* $c^{[\nu]}$ *corresponding to* $(\lambda_k^{[\nu]})$ *as in (1) satisfy:*

(a) $|\tau^{[1]} - \tau^{[2]}|, |(\tau^{[1]})^{-1} - (\tau^{[2]})^{-1}| \leq C \cdot \|\lambda_k^{[1]} - \lambda_k^{[2]}\|_{\ell^2_{-1,3}}$

(b) $\tau^{[2]} c^{[1]} - \tau^{[1]} c^{[2]} \in \mathrm{As}_\infty(\mathbb{C}^*, \ell^2_{-1}, 1)$
$(\tau^{[2]})^{-1} c^{[1]}(\lambda) - (\tau^{[1]})^{-1} c^{[2]}(\lambda) \in \mathrm{As}_0(\mathbb{C}^*, \ell^2_1, 1)$

(c) *With the notation* $0^{-1} := 1$,

$$a_k := \begin{cases} k^{-1} \cdot |\lambda_k^{[1]} - \lambda_k^{[2]}| & \text{for } k \geq 0 \\ |k|^3 \cdot |\lambda_k^{[1]} - \lambda_k^{[2]}| & \text{for } k < 0 \end{cases} \tag{10.4}$$

and

$$r_k := C \cdot \begin{cases} k \cdot \left(a_k * \frac{1}{|k|}\right) & \text{for } k \geq 0 \\ |k|^{-1} \cdot \left(a_k * \frac{1}{|k|}\right) & \text{for } k < 0 \end{cases}, \tag{10.5}$$

we have $r_k \in \ell^2_{-1,1}(k)$, *and* $(r_k)_{k>0}$ *resp.* $(r_k)_{k<0}$ *is a bounding sequence for* $\tau^{[2]} c^{[1]} - \tau^{[1]} c^{[2]}$ *for* $\lambda \to \infty$ *resp. for* $(\tau^{[2]})^{-1} c^{[1]}(\lambda) - (\tau^{[1]})^{-1} c^{[2]}(\lambda)$ *for* $\lambda \to 0$.

(d) *In particular, we have*

$$\left.\begin{array}{r} \left\|\tau^{[2]} c^{[1]} - \tau^{[1]} c^{[2]}\right\|_{\mathrm{As}_\infty(\mathbb{C}^*, \ell^2_{-1}, 1)} \\ \left\|(\tau^{[2]})^{-1} c^{[1]} - (\tau^{[1]})^{-1} c^{[2]}\right\|_{\mathrm{As}_0(\mathbb{C}^*, \ell^2_1, 1)} \end{array}\right\} \leq C \cdot \left\|\lambda_k^{[1]} - \lambda_k^{[2]}\right\|_{\ell^2_{-1,3}}.$$

Proof We begin by looking at the case $\lambda_k = \lambda_{k,0}$. The corresponding number τ defined by Eq. (10.1) is clearly $\tau = 1$, and we show that the function c defined by Eq. (10.2) equals $c_0(\lambda) = \sqrt{\lambda}\,\sin(\zeta(\lambda))$. Indeed, by virtue of the product expansion of the sine function

$$\sin(z) = z \cdot \prod_{k=1}^{\infty} \left(1 - \frac{z^2}{k^2 \pi^2}\right),$$

the following formulas for the $\lambda_{k,0}$

$$\lambda_{0,0} = -1, \quad \lambda_{k,0} + \lambda_{-k,0} = 16\pi^2 k^2 - 2 \quad \text{and} \quad \lambda_{k,0} \cdot \lambda_{-k,0} = 1,$$

and the equation

$$\zeta(\lambda)^2 = \frac{\lambda^2 + 2\lambda + 1}{16\lambda},$$

we have

$$\begin{aligned}
c_0(\lambda) &= \sqrt{\lambda}\,\sin(\zeta(\lambda)) = \sqrt{\lambda}\,\zeta(\lambda)\prod_{k=1}^{\infty}\left(1-\frac{\zeta(\lambda)^2}{k^2\pi^2}\right)\\
&= \frac{1}{4}(\lambda+1)\prod_{k=1}^{\infty}\frac{16\pi^2k^2\lambda-(\lambda^2+2\lambda+1)}{16\pi^2k^2\cdot\lambda}\\
&= \frac{1}{4}(\lambda-\lambda_{0,0})\prod_{k=1}^{\infty}\frac{-\lambda^2+(\lambda_{k,0}+\lambda_{-k,0})\,\lambda-\lambda_{k,0}\,\lambda_{-k,0}}{16\pi^2k^2\cdot\lambda}\\
&= \frac{1}{4}(\lambda-\lambda_{0,0})\prod_{k=1}^{\infty}\frac{(\lambda_{k,0}-\lambda)\cdot(\lambda-\lambda_{-k,0})}{16\pi^2k^2\cdot\lambda}\\
&= \frac{1}{4}(\lambda-\lambda_{0,0})\prod_{k=1}^{\infty}\frac{\lambda_{k,0}-\lambda}{16\,\pi^2\,k^2}\prod_{k=1}^{\infty}\frac{\lambda-\lambda_{-k,0}}{\lambda}. \qquad (10.6)
\end{aligned}$$

Note that the final expression in (10.6) is the infinite product of Eq. (10.2), where $\lambda_k = \lambda_{k,0}$.

To prove the proposition, we will use the results from Appendix A to show the convergence and estimation of the infinite products involved; we will apply the results from Appendix A both for products over $k \geq 1$ and for products over $k \geq 0$, see Remark A.6. The numbers $C_k > 0$ occurring in the following estimates are constants that depend only on R_0. Moreover we choose $0 < \delta_0 < \delta < \pi - \frac{1}{2}$, then all but finitely many of the λ_k resp. $\lambda_k^{[\nu]}$ are in $U_{k,\delta}$.

We note that because of $\lambda_k - \lambda_{k,0} \in \ell^2_{-1,3}(k)$, we have both $\lambda_k - \lambda_{k,0} \in \ell^2_{-1}$ $(k \geq 1)$ and $\lambda_{-k}^{-1} - \lambda_{k,0} \in \ell^2_{-1}(k \geq 0)$, and

$$\left\|\lambda_k - \lambda_{k,0}\right\|_{\ell^2_{-1}(k\geq 1)}, \ \left\|\lambda_{-k}^{-1} - \lambda_{k,0}\right\|_{\ell^2_{-1}(k\geq 0)} \leq C_1 \cdot R_0. \qquad (10.7)$$

Inspired by this "decomposition", we write

$$c(\lambda) = \frac{1}{4}\,\tau\cdot\varrho\cdot c_+(\lambda)\cdot c_-\left(\frac{1}{\lambda}\right)\cdot\lambda$$

with

$$c_+(\lambda) := \prod_{k=1}^{\infty}\left(1-\frac{\lambda}{\lambda_k}\right), \quad c_-(\lambda) := \prod_{k=0}^{\infty}\left(1-\frac{\lambda}{\lambda_{-k}^{-1}}\right), \quad \varrho := \prod_{k=1}^{\infty}\frac{\lambda_k}{16\,\pi^2\,k^2},$$

and

$$\tau = \left(\sigma_+ \cdot \sigma_-^{-1}\right)^{1/2}$$

with

$$\sigma_+ := \prod_{k=1}^{\infty} \frac{\lambda_{k,0}}{\lambda_k} \quad \text{and} \quad \sigma_- := \prod_{k=0}^{\infty} \frac{\lambda_{k,0}}{\lambda_{-k}^{-1}}.$$

Then the infinite products defining $\sigma_\pm$ and ϱ converge absolutely in $\mathbb{C}^*$ by Proposition A.3, and the infinite products defining $c_\pm(\lambda)$ converge locally uniformly to holomorphic functions on $\mathbb{C}$ by Proposition A.4(1). It also follows from Proposition A.3 that

$$|\sigma_\pm|,\ |\sigma_\pm^{-1}|,\ |\varrho|,\ |\varrho^{-1}| \leq C_2$$

and therefore also

$$|\tau|,\ |\tau^{-1}| \leq C_2 \tag{10.8}$$

holds.

It follows that the product defining τ converges absolutely in $\mathbb{C}^*$, and the infinite products in the definition of c in Eq. (10.2) converge absolutely and locally uniformly to a well-defined holomorphic function c. It is clear that c has zeros in all the λ_k (with appropriate multiplicity), and no others.

Before we show the asymptotic behavior of c given in (10.3), we first work in the setting of (2). For this purpose, we let a pair of sequences $(\lambda_k^{[1]})_{k\in\mathbb{Z}}, (\lambda_k^{[2]})_{k\in\mathbb{Z}} \in \ell^2_{-1,3}(k)$ with $\|\lambda_k^{[\nu]} - \lambda_{k,0}\|_{\ell^2_{-1,3}} \leq R_0$ be given, and denote the quantities associated to $(\lambda_k^{[\nu]})$ by the superscript $^{[\nu]}$ (for $\nu \in \{1, 2\}$). In relation to the splitting (10.7), we note that we have

$$\begin{aligned} C_3 \cdot \left\| \lambda_k^{[1]} - \lambda_k^{[2]} \right\|_{\ell^2_{-1,3}} &\leq \left\| \lambda_k^{[1]} - \lambda_k^{[2]} \right\|_{\ell^2_{-1}(k\geq 1)} + \left\| (\lambda_k^{[1]})^{-1} - (\lambda_k^{[2]})^{-1} \right\|_{\ell^2_{-1}(k\geq 0)} \\ &\leq C_4 \cdot \left\| \lambda_k^{[1]} - \lambda_k^{[2]} \right\|_{\ell^2_{-1,3}} . \end{aligned} \tag{10.9}$$

We now show (2)(a). By Proposition A.3 we have

$$\left| \frac{\sigma_\pm^{[1]}}{\sigma_\pm^{[2]}} - 1 \right| \leq C_5 \cdot \left\| \lambda_k^{[1]} - \lambda_k^{[2]} \right\|_{\ell^2_{-1,3}}$$

and also (by exchanging the roles of $(\lambda_k^{[1]})$ and $(\lambda_k^{[2]})$)

$$\left|\frac{\sigma_\pm^{[2]}}{\sigma_\pm^{[1]}} - 1\right| \leq C_6 \cdot \left\|\lambda_k^{[1]} - \lambda_k^{[2]}\right\|_{\ell^2_{-1,3}},$$

and therefore

$$\begin{aligned}\left|\frac{\sigma_+^{[1]}}{\sigma_-^{[1]}} - \frac{\sigma_+^{[2]}}{\sigma_-^{[2]}}\right| &\leq |\sigma_+^{[2]}| \cdot |\sigma_-^{[2]}|^{-1} \cdot \left(\left|\frac{\sigma_-^{[2]}}{\sigma_-^{[1]}}\right| \cdot \left|\frac{\sigma_+^{[1]}}{\sigma_+^{[2]}} - 1\right| + \left|\frac{\sigma_-^{[2]}}{\sigma_-^{[1]}} - 1\right|\right) \\ &\leq C_2^2 \cdot \left(C_2^2 \cdot C_5 \cdot \left\|\lambda_k^{[1]} - \lambda_k^{[2]}\right\|_{\ell^2_{-1,3}} + C_6 \cdot \left\|\lambda_k^{[1]} - \lambda_k^{[2]}\right\|_{\ell^2_{-1,3}}\right) \\ &= C_7 \cdot \left\|\lambda_k^{[1]} - \lambda_k^{[2]}\right\|_{\ell^2_{-1,3}}.\end{aligned}$$

Because we have $|\tau^{[\nu]}|^2 \geq C_2^{-2} > 0$ by Eq. (10.8), and the square root function is Lipschitz continuous on the interval $[C_2^{-2}, \infty)$, it follows that we have

$$|\tau^{[1]} - \tau^{[2]}| = \left|\left(\frac{\sigma_+^{[1]}}{\sigma_-^{[1]}}\right)^{1/2} - \left(\frac{\sigma_+^{[2]}}{\sigma_-^{[2]}}\right)^{1/2}\right| \leq C_8 \cdot \left\|\lambda_k^{[1]} - \lambda_k^{[2]}\right\|_{\ell^2_{-1,3}}.$$

By exchanging the roles of $(\lambda_k^{[1]})$ and $(\lambda_k^{[2]})$ we also obtain

$$|\left(\tau^{[1]}\right)^{-1} - \left(\tau^{[2]}\right)^{-1}| \leq C_9 \cdot \left\|\lambda_k^{[1]} - \lambda_k^{[2]}\right\|_{\ell^2_{-1,3}},$$

completing the proof of (2)(a).

Continuing the proof of the remaining parts of (2) and of (1), we at first consider only the parts of the Proposition concerned with the asymptotic behaviour of the function c for $\lambda \to \infty$. We begin by investigating the function $\frac{\tau^{[2]} c^{[1]}(\lambda)}{\tau^{[1]} c^{[2]}(\lambda)}$. We have

$$\begin{aligned}\frac{\tau^{[2]} c^{[1]}(\lambda)}{\tau^{[1]} c^{[2]}(\lambda)} - 1 &= \frac{\tau^{[2]} \cdot \frac{1}{4} \tau^{[1]} \varrho^{[1]} c_+^{[1]}(\lambda) c_-^{[1]}(\lambda^{-1}) \lambda}{\tau^{[1]} \cdot \frac{1}{4} \tau^{[2]} \varrho^{[2]} c_+^{[2]}(\lambda) c_-^{[2]}(\lambda^{-1}) \lambda} - 1 \\ &= \frac{\varrho^{[1]} c_+^{[1]}(\lambda) c_-^{[1]}(\lambda^{-1})}{\varrho^{[2]} c_+^{[2]}(\lambda) c_-^{[2]}(\lambda^{-1})} - 1 \\ &= \left(\frac{\varrho^{[1]} c_+^{[1]}(\lambda)}{\varrho^{[2]} c_+^{[2]}(\lambda)} - 1\right) \cdot \frac{c_-^{[1]}(\lambda^{-1})}{c_-^{[2]}(\lambda^{-1})} + \left(\frac{c_-^{[1]}(\lambda^{-1})}{c_-^{[2]}(\lambda^{-1})} - 1\right) \\ &= (g(\lambda) - 1) \cdot h(\lambda) + (h(\lambda) - 1) \qquad (10.10)\end{aligned}$$

with

$$g(\lambda) := \frac{\varrho^{[1]}\, c_+^{[1]}(\lambda)}{\varrho^{[2]}\, c_+^{[2]}(\lambda)} = \prod_{k=1}^{\infty} \frac{\lambda_k^{[1]} - \lambda}{\lambda_k^{[2]} - \lambda}$$

and

$$h(\lambda) := \frac{c_-^{[1]}\left(\lambda^{-1}\right)}{c_-^{[2]}\left(\lambda^{-1}\right)} = \prod_{k=0}^{\infty} \frac{\lambda_{-k}^{[1]} - \lambda}{\lambda_{-k}^{[2]} - \lambda}.$$

By Proposition A.4(2) we have for $\lambda \in S_n \cap V_\delta,\ n \geq 1$,

$$|g(\lambda) - 1| \leq r_n^{[g]},$$

where

$$r_n^{[g]} := C_{10} \cdot \left(\begin{Bmatrix} a_k \text{ for } k > 0 \\ 0 \text{ for } k \leq 0 \end{Bmatrix} * \frac{1}{|k|} \right)_n$$

with a constant $C_{10} > 0$, and the sequence (a_k) is the one from Eq. (10.4).

Likewise by Proposition A.4(2), applied to $(\lambda_{-k}^{[\nu]})^{-1}$ in the place of $\lambda_k^{[\nu]}$, we obtain for $\lambda \in S_n \cap V_\delta,\ n \geq 1$,

$$|h(\lambda) - 1| \leq r_n^{[h]},$$

where

$$r_n^{[h]} := C_{11} \cdot \left(\begin{Bmatrix} \frac{|(\lambda_{-k}^{[1]})^{-1} - |(\lambda_{-k}^{[2]})^{-1}|}{|k|} & \text{for } k \geq 0 \\ 0 & \text{for } k < 0 \end{Bmatrix} * \frac{1}{|k|} \right)_n$$
$$\leq C_{12} \cdot \left(\begin{Bmatrix} a_k \text{ for } k \leq 0 \\ 0 \text{ for } k > 0 \end{Bmatrix} * \frac{1}{|k|} \right)_{-n}$$

with constants $C_{11}, C_{12} > 0$, and again with the sequence (a_k) from Eq. (10.4).

We have $r_n^{[g]}, r_n^{[h]} \leq r_n$, where the sequence (r_n) is defined by Eq. (10.5) (and we choose C as the maximum of C_{10} and C_{12}), and thus we have

$$|g(\lambda) - 1|,\ |h(\lambda) - 1| \ \leq\ r_n. \tag{10.11}$$

From this estimate it also follows that there exists $C_{13} > 0$ so that

$$|g(\lambda)|,\ |h(\lambda)| \ \leq\ C_{13}. \tag{10.12}$$

By taking the absolute value in Eq. (10.10), and then applying the estimates (10.11) and (10.12), we obtain

$$\left| \frac{\tau^{[2]}\, c^{[1]}(\lambda)}{\tau^{[1]}\, c^{[2]}(\lambda)} - 1 \right| \leq C_{14} \cdot r_n \tag{10.13}$$

with a constant $C_{14} > 0$.

By applying this estimate in the setting of (1), i.e. for $\lambda_k^{[1]} = \lambda_k$ and $\lambda_k^{[2]} = \lambda_{k,0}$, we obtain in particular

$$\left| \frac{c(\lambda)}{\tau\, c_0(\lambda)} - 1 \right| \leq C_{14} \cdot r_n .$$

and therefore

$$|c(\lambda) - \tau\, c_0(\lambda)| \leq C_{15} \cdot r_n \cdot |c_0(\lambda)| .$$

Because $|c_0(\lambda)| = |\lambda|^{1/2} \cdot |\sin(\zeta(\lambda))|$ is comparable on V_δ to $|\lambda|^{1/2} \cdot w(\lambda)$ by Proposition 6.3(3), it follows that $(c - \tau\, c_0)|V_\delta \in \mathrm{As}_\infty(V_\delta, \ell^2_{-1}, 1)$ holds. By Proposition 9.4(1) we in fact have $c - \tau\, c_0 \in \mathrm{As}_\infty(\mathbb{C}^*, \ell^2_{-1}, 1)$. This shows the half of (10.3) concerned with $\lambda \to \infty$. Moreover, because of $c_0 \in \mathrm{As}_\infty(\mathbb{C}^*, \ell^\infty_{-1}, 1)$, we conclude $c \in \mathrm{As}_\infty(\mathbb{C}^*, \ell^\infty_{-1}, 1)$ and thus

$$|c(\lambda)| \leq C_{15} \cdot n \cdot w(\lambda) \tag{10.14}$$

with a constant $C_{15} > 0$.

We return to the setting of (2). Again for $\lambda \in S_n \cap V_\delta$ with $n \geq 1$, we have

$$|\tau^{[2]}\, c^{[1]}(\lambda) - \tau^{[1]}\, c^{[2]}(\lambda)| = |\tau^{[1]}| \cdot |c^{[2]}(\lambda)| \cdot \left| \frac{\tau^{[2]}\, c^{[1]}(\lambda)}{\tau^{[1]}\, c^{[2]}(\lambda)} - 1 \right| .$$

From the estimates (10.8), (10.14) and (10.13) we thus conclude

$$|\tau^{[2]}\, c^{[1]}(\lambda) - \tau^{[1]}\, c^{[2]}(\lambda)| \leq C_2 \cdot C_{15} \cdot w(\lambda) \cdot C_{14} \cdot r_n .$$

This shows that $(\tau^{[2]}\, c^{[1]} - \tau^{[1]}\, c^{[2]})|V_\delta \in \mathrm{As}_\infty(V_\delta, \ell^2_{-1}, 1)$ and therefore by Proposition 9.4(1) $\tau^{[2]}\, c^{[1]} - \tau^{[1]}\, c^{[2]} \in \mathrm{As}_\infty(\mathbb{C}^*, \ell^2_{-1}, 1)$ holds. Moreover, we see that (r_k) is a bounding sequence for $\tau^{[2]}\, c^{[1]} - \tau^{[1]}\, c^{[2]}$ (where the constant $C > 0$ occurring in the definition (10.5) of r_k is enlarged appropriately). This shows the half of (2)(b) and (2)(c) concerned with $\lambda \to \infty$, and the half of (2)(d) concerned with $\lambda \to \infty$ follows by application of the version of Young's inequality for weakly ℓ^1-sequences (see Eq. (7.4)).

It remains to show the asymptotic statements in (1) and (2)(b)–(d) also for $\lambda \to 0$. We do this by reducing the situation for $\lambda \to 0$ to the previously proven situation for $\lambda \to \infty$. For this purpose, we first note

$$c_0(\lambda^{-1}) = \lambda^{-1} \cdot c_0(\lambda).$$

Then we put $\widetilde{\lambda}_k := \lambda_{-k}^{-1}$, and denote the quantities associated to $(\widetilde{\lambda}_k)$ by a tilde $\widetilde{\ }$. With $\lambda_k - \lambda_{k,0} \in \ell^2_{-1,3}(k)$, we also have $\widetilde{\lambda}_k - \lambda_{k,0} \in \ell^2_{-1,3}(k)$, and for the corresponding sequences (a_k) and (r_k) defined by Eq. (10.4) resp. (10.5), we have that $\widetilde{a}_k$ is comparable to a_{-k}, and therefore $\widetilde{r}_k$ is comparable to r_{-k}. Explicit calculations yield

$$\sigma_+ = -\widetilde{\lambda}_0 \cdot \widetilde{\sigma}_-, \quad \sigma_- = -\widetilde{\lambda}_0 \cdot \widetilde{\sigma}_+, \quad \tau = \widetilde{\tau}^{-1}, \quad \varrho = -\widetilde{\lambda}_0 \cdot \widetilde{\tau}^2 \cdot \widetilde{\varrho}$$

and for $\lambda \in \mathbb{C}^*$

$$c_+(\lambda) = \frac{1}{1 - \widetilde{\lambda}_0 \cdot \lambda} \cdot \widetilde{c}_-(\lambda) \quad \text{and} \quad c_-(\lambda) = \frac{\widetilde{\lambda}_0 - \lambda}{\widetilde{\lambda}_0} \cdot \widetilde{c}_+(\lambda)$$

and therefore

$$\begin{aligned}
c(\lambda^{-1}) &= \frac{1}{4} \tau \cdot \varrho \cdot c_+(\lambda^{-1}) \cdot c_-(\lambda) \cdot \lambda^{-1} \\
&= \frac{1}{4} \widetilde{\tau}^{-1} \cdot (-\widetilde{\lambda}_0 \, \widetilde{\tau}^2 \, \widetilde{\varrho}) \cdot \left(\frac{1}{1 - \widetilde{\lambda}_0 \, \lambda^{-1}} \, \widetilde{c}_-(\lambda^{-1}) \right) \cdot \left(\frac{\widetilde{\lambda}_0 - \lambda}{\widetilde{\lambda}_0} \, \widetilde{c}_+(\lambda) \right) \cdot \lambda^{-1} \\
&= \frac{1}{4} \cdot \widetilde{\tau} \cdot \widetilde{\varrho} \cdot \widetilde{c}_+(\lambda) \cdot \widetilde{c}_-(\lambda^{-1}) \cdot \lambda \cdot \lambda^{-1} \\
&= \widetilde{c}(\lambda) \cdot \lambda^{-1}.
\end{aligned}$$

Using these formulas, the asymptotic estimates for c resp. for $c^{[\nu]}$ for $\lambda \to 0$ follow from the previously shown asymptotic estimates for $\lambda \to \infty$ applied to $\widetilde{c}$ resp. to $\widetilde{c}^{[\nu]}$. □

Corollary 10.2 *Suppose that $c : \mathbb{C}^* \to \mathbb{C}$ is a holomorphic function which satisfies the following two asymptotic properties for some $\tau \in \mathbb{C}^*$:*

(a) For every $\varepsilon > 0$ there exists $R > 0$ such that we have

$$\textit{for all } \lambda \in \mathbb{C}^* \textit{ with } |\lambda| \geq R : \qquad |c(\lambda) - \tau \, c_0(\lambda)| \leq \varepsilon \, |\lambda|^{1/2} \, w(\lambda)$$

and

$$\textit{for all } \lambda \in \mathbb{C}^* \textit{ with } |\lambda| \leq \tfrac{1}{R} : \qquad |c(\lambda) - \tau^{-1} \, c_0(\lambda)| \leq \varepsilon \, |\lambda|^{1/2} \, w(\lambda).$$

(b) $(c(\lambda_{k,0}))_{k \in \mathbb{Z}} \in \ell^2_{-1,1}(k)$.

Then we already have

$$c - \tau\, c_0 \in \mathrm{As}_\infty(\mathbb{C}^*, \ell^2_{-1}, 1) \quad \textit{and} \quad c - \tau^{-1}\, c_0 \in \mathrm{As}_0(\mathbb{C}^*, \ell^2_1, 1) \tag{10.15}$$

and

$$\tau = \pm \left(\prod_{k\in\mathbb{Z}} \frac{\lambda_{k,0}}{\lambda_k} \right)^{1/2},$$

where $(\lambda_k)_{k\in\mathbb{Z}}$ *is the sequence of zeros of* c *as in Proposition 6.5(2).*

Addendum. *If* L *is a set of holomorphic functions on* $\mathbb{C}^*$ *which satisfy (a) and (b) in such a way that* $R = R(\varepsilon) > 0$ *in (a) can be chosen uniformly for all* $c \in L$*, and in (b) there is a uniform bound* $(z_k)_{k\in\mathbb{Z}} \in \ell^2_{-1,1}(k)$ *for the sequences* $(c(\lambda_{k,0}))$ *for all* $c \in L$*, then there exists a uniform bounding sequence for the asymptotics in* (10.15) *that applies for all* $c \in L$.

Proof By Proposition 6.5(2) the zeros of c are enumerated by a sequence $(\lambda_k)_{k\in\mathbb{Z}}$ such that for every $\delta > 0$, there exists $N \in \mathbb{N}$ so that $\lambda_k \in U_{k,\delta}$ holds for $k \in \mathbb{Z}$ with $|k| \geq N$. By Corollary 8.2 we have $\lambda_k - \lambda_{k,0} \in \ell^2_{-1,3}(k)$. Therefore Proposition 10.1(1) shows that there exists a holomorphic function $\widetilde{c} : \mathbb{C}^* \to \mathbb{C}$ that has zeros in all the λ_k (with the appropriate multiplicity) and no others, and which satisfies

$$\widetilde{c} - \widetilde{\tau}\, c_0 \in \mathrm{As}_\infty(\mathbb{C}^*, \ell^2_{-1}, 1) \quad \text{and} \quad \widetilde{c} - \widetilde{\tau}^{-1}\, c_0 \in \mathrm{As}_0(\mathbb{C}^*, \ell^2_1, 1)$$

with

$$\widetilde{\tau} = \left(\prod_{k\in\mathbb{Z}} \frac{\lambda_{k,0}}{\lambda_k} \right)^{1/2}.$$

By Proposition 6.7(1) we have $(c, \tau) = \pm(\widetilde{c}, \widetilde{\tau})$, whence the claimed statement follows.

In the situation of the addendum, the hypotheses ensure that there is a uniform sequence $(w_k)_{k\in\mathbb{Z}} \in \ell^2_{-1,1}(k)$ of non-negative real numbers so that we have $|\lambda_k - \lambda_{k,0}| \leq w_k$ for every sequence $(\lambda_k)_{k\in\mathbb{Z}}$ that is the sequence of zeros of a $c \in L$. Let

$$a_k := \begin{cases} k^{-1} \cdot w_k & \text{for } k \geq 0 \\ |k|^3 \cdot w_k & \text{for } k < 0 \end{cases} \quad \text{and} \quad r_k := C \cdot \begin{cases} k \cdot \left(a_k * \frac{1}{|k|}\right) & \text{for } k \geq 0 \\ |k|^{-1} \cdot \left(a_k * \frac{1}{|k|}\right) & \text{for } k < 0 \end{cases}$$

with the constant $C > 0$ from Proposition 10.1(2)(c). By applying that proposition with $\lambda^{[1]}_k = \lambda_k$, $\lambda^{[2]}_k = \lambda_{k,0}$, we see that with the sequence $(r_k)_{k>0}$ resp. $(r_k)_{k<0}$

is a bounding sequence for $\widetilde{c} - \widetilde{\tau}\, c_0$ resp. for $\widetilde{c} - \widetilde{\tau}^{-1}\, c_0$ (that is independent of $c \in L$). □

Corollary 10.3

(1) *In the setting of Proposition 10.1(1), $\frac{c(\lambda)}{\lambda-\lambda_k}$ is a holomorphic function on $\mathbb{C}^*$ for every $k \in \mathbb{Z}$. There exists a constant $C > 0$ (depending only on R_0) and a sequence $(r_k) \in \ell^2_{0,-2}(k)$ such that for every $k > 0$ and every $\lambda \in U_{k,\delta}$ we have*

$$\left| \frac{c(\lambda)}{\lambda - \lambda_k} - \tau\, \frac{(-1)^k}{8} \right| \leq C\, \frac{|\lambda - \lambda_k|}{k} + r_k$$

and for every $k < 0$ and every $\lambda \in U_{k,\delta}$ we have

$$\left| \frac{c(\lambda)}{\lambda - \lambda_k} - \left(-\tau^{-1}\, \frac{(-1)^k}{8}\, \lambda_{k,0}^{-1} \right) \right| \leq C\, |\lambda - \lambda_k|\, k^5 + r_k .$$

Here we can choose

$$r_k := \frac{1}{8}\, \lambda_{k,0}^{-1} + \begin{cases} C \cdot \left(a_k * \frac{1}{|k|} \right) & \text{for } k > 0 \\ C\, k^2 \cdot \left(a_k * \frac{1}{|k|} \right) & \text{for } k < 0 \end{cases}$$

with

$$a_k := \begin{cases} k^{-1} \cdot |\lambda_k - \lambda_{k,0}| & \text{for } k > 0 \\ k^3 \cdot |\lambda_k - \lambda_{k,0}| & \text{for } k < 0 \end{cases} .$$

(2) *In the setting of Proposition 10.1(2) there exists a constant $C > 0$ (depending only on R_0) and a sequence $r_k \in \ell^2_{0,-2}(k)$ with $\|r_k\|_{\ell^2_{0,-2}} \leq C \cdot \|\lambda_k^{[1]} - \lambda_k^{[2]}\|_{\ell^2_{-1,3}}$ such that we have for all $k \in \mathbb{Z}$ and $\lambda \in S_k$ if $k > 0$*

$$\left| \frac{c^{[1]}(\lambda)}{\tau^{[1]} \cdot (\lambda - \lambda_k^{[1]})} - \frac{c^{[2]}(\lambda)}{\tau^{[2]} \cdot (\lambda - \lambda_k^{[2]})} \right| \leq r_k$$

and if $k < 0$

$$\left| \frac{c^{[1]}(\lambda)}{(\tau^{[1]})^{-1} \cdot (\lambda - \lambda_k^{[1]})} - \frac{c^{[2]}(\lambda)}{(\tau^{[2]})^{-1} \cdot (\lambda - \lambda_k^{[2]})} \right| \leq r_k .$$

More specifically, we can choose

$$r_k := \begin{cases} C \cdot \left(a_k * \frac{1}{|k|}\right) & \textit{for } k > 0 \\ C\,k^2 \cdot \left(a_k * \frac{1}{|k|}\right) & \textit{for } k < 0 \end{cases}$$

with

$$a_k := \begin{cases} k^{-1} \cdot |\lambda_k^{[1]} - \lambda_k^{[2]}| & \textit{for } k > 0 \\ k^3 \cdot |\lambda_k^{[1]} - \lambda_k^{[2]}| & \textit{for } k < 0 \end{cases}.$$

Proof We consider only the case $k > 0$, and prove (2) before (1).

For (2). We note that because the holomorphic function $c^{[\nu]}$ has a zero at $\lambda = \lambda_k^{[\nu]}$, $\frac{c^{[\nu]}(\lambda)}{\lambda - \lambda_k^{[\nu]}}$ is a holomorphic function on $\mathbb{C}^*$, and we have

$$\begin{aligned} &\frac{c^{[1]}(\lambda)}{\tau^{[1]} \cdot (\lambda - \lambda_k^{[1]})} - \frac{c^{[2]}(\lambda)}{\tau^{[2]} \cdot (\lambda - \lambda_k^{[2]})} \\ &= \frac{1}{\tau^{[1]}\,\tau^{[2]}} \left(\frac{\tau^{[2]}\,c^{[1]}(\lambda) - \tau^{[1]}\,c^{[2]}(\lambda)}{\lambda - \lambda_k^{[1]}} \right. \\ &\qquad \left. + \tau^{[1]}\,c^{[2]}(\lambda) \cdot \left(\frac{1}{\lambda - \lambda_k^{[1]}} - \frac{1}{\lambda - \lambda_k^{[2]}} \right) \right). \end{aligned} \tag{10.16}$$

We choose $\delta > 0$ so large that $\lambda_k^{[\nu]} \in U_{k,\delta}$ holds for all k, and at first consider $\lambda \in S_k \cap V_{2\delta}$. Thus there exists $C_1 > 0$ with $|\lambda - \lambda_k^{[\nu]}| \geq C_1 \cdot k$ for all $k > 0$, $\nu \in \{1, 2\}$ and all $\lambda \in S_k \cap V_{2\delta}$. We also have

$$\left| \frac{1}{\lambda - \lambda_k^{[1]}} - \frac{1}{\lambda - \lambda_k^{[2]}} \right| = \frac{|\lambda_k^{[1]} - \lambda_k^{[2]}|}{|\lambda - \lambda_k^{[1]}| \cdot |\lambda - \lambda_k^{[2]}|} \leq \frac{|\lambda_k^{[1]} - \lambda_k^{[2]}|}{(C_1\,k)^2}.$$

Moreover, by Proposition 10.1(2) we have

$$(\tau^{[2]}\,c^{[1]}(\lambda) - \tau^{[1]}\,c^{[2]}(\lambda)) \in \mathrm{As}_\infty(\mathbb{C}^*, \ell^2_{-1}, 1) \quad \text{and} \quad c^{[\nu]}(\lambda) \in \mathrm{As}_\infty(\mathbb{C}^*, \ell^\infty_{-1}, 1).$$

By plugging these results into Eq. (10.16), we see that there is a sequence $r_k \in \ell^2_0(k > 0)$ with $\|r_k\|_{\ell^2_0} \leq C \cdot \|\lambda_k^{[1]} - \lambda_k^{[2]}\|_{\ell^2_{-1,3}}$ for some $C > 0$ such that for all $k > 0$ and all $\lambda \in S_k \cap V_{2\delta}$ we have

$$\left| \frac{c^{[1]}(\lambda)}{\tau^{[1]} \cdot (\lambda - \lambda_k^{[1]})} - \frac{c^{[2]}(\lambda)}{\tau^{[2]} \cdot (\lambda - \lambda_k^{[2]})} \right| \leq r_k.$$

Because of the maximum principle for holomorphic functions, this inequality then also holds for $\lambda \in U_{k,2\delta} \subset S_k$.

The "more specific" description of r_k follows from the corresponding description in Proposition 10.1(2).

For (1). We consider the holomorphic functions $g_k(\lambda) := \frac{c(\lambda)}{\lambda-\lambda_k}$ and $g_{k,0}(\lambda) := \frac{c_0(\lambda)}{\lambda-\lambda_{k,0}}$ on $U_{k,\delta}$. For $\lambda \in U_{k,\delta}$, we have

$$\frac{c(\lambda)}{\lambda-\lambda_k} - \tau\,\frac{(-1)^k}{8} = g_k(\lambda) - g_k(\lambda_{k,0}) + g_k(\lambda_{k,0}) - \tau\, g_{k,0}(\lambda_{k,0}) + \tau\left(g_{k,0}(\lambda_{k,0}) - \frac{(-1)^k}{8}\right). \tag{10.17}$$

We estimate the three differences on the right hand side separately.

First, for $\lambda \in U_{k,\delta}$, we have the Taylor approximation of c near its zero $\lambda = \lambda_k$

$$|c(\lambda) - c'(\lambda_k)\cdot(\lambda-\lambda_k)| \leq \frac{1}{2}\,|c''(\xi)|\cdot|\lambda-\lambda_k|^2$$

with some $\xi \in U_{k,\delta}$ (dependent on λ), and therefore

$$\left|g_k'(\lambda)\right| = \left|\frac{c'(\lambda)\,(\lambda-\lambda_k) - c(\lambda)}{(\lambda-\lambda_k)^2}\right| \leq \frac{1}{2}\,|c''(\xi)|.$$

Because of $c'' \in \mathrm{As}(\mathbb{C}^*, \ell^\infty_{1,-5}, 1)$, there exists $C > 0$ with $\left|g_k'|U_{k,\delta}\right| \leq \frac{C}{k}$, and thus

$$|g_k(\lambda) - g_k(\lambda_{k,0})| \leq \frac{C}{k}\,|\lambda-\lambda_{k,0}| \leq \frac{C}{k}\,|\lambda-\lambda_k| + \frac{C}{k}|\lambda_k-\lambda_{k,0}|.$$

Second, we have $g_k(\lambda_{k,0}) - \tau\, g_{k,0}(\lambda_{k,0}) \in \ell^2_0(k)$ by (2) (applied with $\lambda_k^{[1]} = \lambda_k$ and $\lambda_k^{[2]} = \lambda_{k,0}$). And third, we have

$$g_{k,0}(\lambda_{k,0}) - \frac{(-1)^k}{8} = c_0'(\lambda_{k,0}) - \frac{(-1)^k}{8} = -\frac{(-1)^k}{8}\,\lambda_{k,0}^{-1}.$$

By plugging the preceding results into Eq. (10.17), we obtain the claimed statement. □

Proposition 10.4 (Interpolation by the Values) *Suppose that sequences* $(\lambda_k)_{k\in\mathbb{Z}}$ *and* $(\mu_k)_{k\in\mathbb{Z}}$*, and a number* $\upsilon \in \mathbb{C}^*$ *are given, such that we have*

$$\lambda_k - \lambda_{k,0} \in \ell^2_{-1,3}(k) \quad \textit{and} \quad \left\{\begin{matrix} \mu_k - \upsilon\,\mu_{k,0} & \textit{if } k \geq 0 \\ \mu_k - \upsilon^{-1}\,\mu_{k,0} & \textit{if } k < 0 \end{matrix}\right\} \in \ell^2_{0,0}(k).$$

We further require that whenever $\lambda_k = \lambda_{\widetilde{k}}$ *holds for some* $k, \widetilde{k} \in \mathbb{Z}$, *we have* $\mu_k = \mu_{\widetilde{k}}$.

Then let $c : \mathbb{C}^* \to \mathbb{C}$ *be the holomorphic function with zeros at the* λ_k *defined in Proposition 10.1, and put for* $k \in \mathbb{Z}$

$$d_k := \#\{\widetilde{k} \in \mathbb{Z} \mid \lambda_{\widetilde{k}} = \lambda_k\} = \mathrm{ord}_{\lambda_k}(c).$$

(1) There are only finitely many $k \in \mathbb{Z}$ *with* $d_k \geq 2$, *the infinite sum*[2]

$$a(\lambda) := \sum_{k \in \mathbb{Z}} \frac{\mu_k \cdot (d_k - 1)! \cdot c(\lambda)}{c^{(d_k)}(\lambda_k) \cdot (\lambda - \lambda_k)^{d_k}} \tag{10.18}$$

converges absolutely and locally uniformly on $\mathbb{C}^*$, *we have*

$$a - \upsilon\, a_0 \in \mathrm{As}_\infty(\mathbb{C}^*, \ell^2_0, 1) \quad \textit{and} \quad a - \upsilon^{-1} a_0 \in \mathrm{As}_0(\mathbb{C}^*, \ell^2_0, 1), \tag{10.19}$$

and

$$a(\lambda_k) = \mu_k \quad \textit{for all } k \in \mathbb{Z}. \tag{10.20}$$

(2) Let $\Lambda := \{\lambda_k \mid k \in \mathbb{Z},\ d_k \geq 2\}$ *be the set of multiple zeros of* c, *and for each* $\lambda_* \in \Lambda$ *choose* $t_{\lambda_*,1}, \dots, t_{\lambda_*,\mathrm{ord}_{\lambda_*}(c)-1} \in \mathbb{C}$. *With the function* a *defined by Eq.* (10.18), $\widetilde{a} : \mathbb{C}^* \to \mathbb{C}$,

$$\widetilde{a}(\lambda) := a(\lambda) + \sum_{\lambda_* \in \Lambda} \sum_{j=1}^{\mathrm{ord}_{\lambda_*}(c)-1} t_{\lambda_*,j} \cdot \frac{c(\lambda)}{(\lambda - \lambda_*)^j} \tag{10.21}$$

is another holomorphic function with

$$\widetilde{a} - \upsilon\, a_0 \in \mathrm{As}_\infty(\mathbb{C}^*, \ell^2_0, 1) \quad \textit{and} \quad \widetilde{a} - \upsilon^{-1} a_0 \in \mathrm{As}_0(\mathbb{C}^*, \ell^2_0, 1), \tag{10.22}$$

and

$$\widetilde{a}(\lambda_k) = \mu_k \quad \textit{for all } k \in \mathbb{Z}. \tag{10.23}$$

Moreover, any holomorphic function $\widetilde{a}$ *that satisfies* (10.22) *and* (10.23) *is obtained in this way.*

(3) Now let two pairs of sequences $(\lambda_k^{[\nu]})_{k \in \mathbb{Z}}$, $(\mu_k^{[\nu]})_{k \in \mathbb{Z}}$ *and numbers* $\upsilon^{[\nu]} \in \mathbb{C}^*$ *(where* $\nu \in \{1, 2\}$*) that satisfy the hypotheses of (1) be given, and denote the corresponding holomorphic function of (1) with* $a^{[\nu]} : \mathbb{C}^* \to \mathbb{C}$. *Then for every*

---

[2]Note that if $d_k > 1$ holds, then the summand $\frac{\mu_k \cdot (d_k-1)! \cdot c(\lambda)}{c^{(d_k)}(\lambda_k) \cdot (\lambda - \lambda_k)^{d_k}}$ will occur d_k times in the sum defining a. This is the reason why the factor $(d_k - 1)!$ (instead of $d_k!$) occurs in the numerator.

$R_0 > 0$ there exists a constant $C > 0$ (depending only on R_0), such that if we have

$$\left\| \lambda_k^{[\upsilon]} - \lambda_{k,0} \right\|_{\ell^2_{-1,3}} \leq R_0 \quad \textit{and} \quad \left\| \begin{Bmatrix} \mu_k - \upsilon\, \mu_{k,0} & \textit{if } k \geq 0 \\ \mu_k - \upsilon^{-1}\, \mu_{k,0} & \textit{if } k < 0 \end{Bmatrix} \right\|_{\ell^2_{0,0}} \leq R_0,$$

then the following holds:

(a) $\upsilon^{[2]}\, a^{[1]} - \upsilon^{[1]}\, a^{[2]} \in \mathrm{As}_\infty(\mathbb{C}^*, \ell^2_0, 1)$,
$(\upsilon^{[2]})^{-1}\, a^{[1]} - (\upsilon^{[1]})^{-1}\, a^{[2]} \in \mathrm{As}_0(\mathbb{C}^*, \ell^2_0, 1)$

(b) With the notation $\frac{1}{0} := 1$,

$$a_k := \begin{cases} k^{-1} \cdot |\lambda_k^{[1]} - \lambda_k^{[2]}| & \textit{for } k \geq 0 \\ |k|^3 \cdot |\lambda_k^{[1]} - \lambda_k^{[2]}| & \textit{for } k < 0 \end{cases}$$

$$\textit{and} \quad b_k := \begin{cases} |\upsilon^{[2]}\, \mu_k^{[1]} - \upsilon^{[1]}\, \mu_k^{[2]}| & \textit{for } k \geq 0 \\ |(\upsilon^{[2]})^{-1}\, \mu_k^{[1]} - (\upsilon^{[1]})^{-1}\, \mu_k^{[2]}| & \textit{for } k < 0 \end{cases} \tag{10.24}$$

and

$$r_k := C \cdot \Bigg(\left(\left(a_k * \frac{1}{|k|} \right) * \frac{1}{|k|} + \left(b_k + \frac{|\tau^{[1]} - \tau^{[2]}|}{|k|} \right) * \frac{1}{|k|} \right. + \frac{|\upsilon^{[1]} - (\upsilon^{[1]})^{-1}| + |\upsilon^{[2]} - (\upsilon^{[2]})^{-1}|}{|k|} \Bigg), \tag{10.25}$$

we have $r_k \in \ell^2_{0,0}(k)$, and $(r_k)_{k>0}$ resp. $(r_k)_{k<0}$ is a bounding sequence for $\upsilon^{[2]}\, a^{[1]} - \upsilon^{[1]}\, a^{[2]}$ for $\lambda \to \infty$ resp. for $(\upsilon^{[2]})^{-1}\, a^{[1]} - (\upsilon^{[1]})^{-1}\, a^{[2]}$ for $\lambda \to 0$.

(c) In particular, we have

$$\left. \begin{array}{r} \left\| \upsilon^{[2]}\, a^{[1]} - \upsilon^{[1]}\, a^{[2]} \right\|_{\mathrm{As}_\infty(\mathbb{C}^*, \ell^2_{-1}, 1)} \\ \left\| (\upsilon^{[2]})^{-1}\, a^{[1]} - (\upsilon^{[1]})^{-1}\, a^{[2]} \right\|_{\mathrm{As}_0(\mathbb{C}^*, \ell^2_1, 1)} \end{array} \right\}$$
$$\leq C \cdot \left(\left\| \lambda_k^{[1]} - \lambda_k^{[2]} \right\|_{\ell^2_{-1,3}} + \|b_k\|_{\ell^2_{0,0}} + |\tau^{[1]} - \tau^{[2]}| + |\upsilon^{[1]} - (\upsilon^{[1]})^{-1}| + |\upsilon^{[2]} - (\upsilon^{[2]})^{-1}| \right).$$

Remark 10.5 If no two λ_k coincide in the setting of Proposition 10.4, then we have $d_k = 1$ for all k, and therefore

$$a(\lambda) = \sum_{k\in\mathbb{Z}} \frac{\mu_k \cdot c(\lambda)}{c'(\lambda_k)\cdot(\lambda-\lambda_k)} \tag{10.26}$$

is the *only* holomorphic function $a : \mathbb{C}^* \to \mathbb{C}$ that satisfies (10.19) and (10.20).

In the general setting, Proposition 10.4(1),(2) shows that for every λ_k that occurs more than once, i.e. $d_k \geq 2$, we may prescribe the values of $a'(\lambda_k), \dots, a^{(d_k-1)}(\lambda_k)$ arbitrarily; then there exists one and only one holomorphic function $a : \mathbb{C}^* \to \mathbb{C}$ that satisfies (10.19) and (10.20), and has the prescribed values of its derivatives.

Proof (of Proposition 10.4) For (1) and (3). Because of the requirement $\lambda_k - \lambda_{k,0} \in \ell^2_{-1,3}(k)$, every excluded domain $U_{k,\delta}$ contains asymptotically and totally exactly one of the λ_k. It follows that only finitely many λ_k can coincide with one another, and therefore we have $d_k = 1$ for all k with only finitely many exceptions.

Considered separately, every single summand $\frac{\mu_k\cdot(d_k-1)!\cdot c(\lambda)}{c^{(d_k)}(\lambda_k)\cdot(\lambda-\lambda_k)^{d_k}}$ with $d_k \geq 2$ is in $\mathrm{As}(\mathbb{C}^*, \ell^\infty_{2d_k-1,1}, 1)$ and therefore in $\mathrm{As}(\mathbb{C}^*, \ell^2_{0,0}, 1)$. For the proof of the convergence results and the claims on the asymptotic behavior, we may therefore suppose without loss of generality that $d_k = 1$ holds for all k. Then $a(\lambda)$ is given by Eq. (10.26).

Because of the hypotheses, the sequence (μ_k) is bounded, we have $c' - \tau\, c_0' \in \mathrm{As}(\mathbb{C}^*, \ell^2_{0,-2}, 1)$ (where τ is the number defined by Eq. (10.1)) and therefore $\frac{1}{c'(\lambda_k)} \in \ell^\infty_{0,2}(k)$, and $\frac{c(\lambda)}{\lambda-\lambda_k} \in \ell^\infty_{2,0}(k)$ locally uniformly with respect to $\lambda \in \mathbb{C}^*$. It follows that

$$\frac{\mu_k \cdot c(\lambda)}{c'(\lambda_k)\cdot(\lambda-\lambda_k)} \in \ell^\infty_{2,2}(k) \subset \ell^1_{0,0}(k)$$

holds (again uniformly in λ). This shows that the infinite sum defining the function a in Eq. (10.26) indeed converges absolutely and locally uniformly, and thus Eq. (10.26) defines a holomorphic function a. For $k, \ell \in \mathbb{Z}$ we have

$$\left.\frac{(d_k-1)!\cdot c(\lambda)}{c^{(d_k)}(\lambda_k)\cdot(\lambda-\lambda_k)^{d_k}}\right|_{\lambda=\lambda_\ell} = \begin{cases} \frac{1}{d_k} & \text{for } k=\ell \\ 0 & \text{for } k\neq\ell \end{cases}.$$

Because the term $\frac{\mu_k\cdot(d_k-1)!\cdot c(\lambda)}{c^{(d_k)}(\lambda_k)\cdot(\lambda-\lambda_k)^{d_k}}$ occurs d_k times in the infinite sum in Eq. (10.18), this shows that $a(\lambda_k) = \mu_k$ holds.

Before we show the claims in (10.22) on the asymptotic behavior of the function a, we work in the setting of (3). As usual, we denote the objects associated with $(\lambda_k^{[\nu]})$ and $(\mu_k^{[\nu]})$ by the superscript $^{[\nu]}$ (for $\nu \in \{1,2\}$). We will first show the asymptotic statements in (3) for the case $\lambda \to \infty$. For this purpose, we will again use statements from Appendix A. The constants $C_k > 0$ occurring in the sequel only depend on R_0 and upper and lower bounds for $|\upsilon^{[\nu]}|$.

We write

$$a^{[\nu]}(\lambda) = a_+^{[\nu]}(\lambda) + a_-^{[\nu]}(\lambda)$$

with

$$a_+^{[\nu]}(\lambda) = \sum_{k=1}^{\infty} \frac{\mu_k^{[\nu]} \cdot c^{[\nu]}(\lambda)}{(c^{[\nu]})'(\lambda_k^{[\nu]}) \cdot (\lambda - \lambda_k^{[\nu]})}$$

$$\text{and} \quad a_-^{[\nu]}(\lambda) = \sum_{k=0}^{\infty} \frac{\mu_{-k}^{[\nu]} \cdot c^{[\nu]}(\lambda)}{(c^{[\nu]})'(\lambda_{-k}^{[\nu]}) \cdot (\lambda - \lambda_{-k}^{[\nu]})}.$$

To estimate $\upsilon^{[2]}\,a^{[1]} - \upsilon^{[1]}\,a^{[2]}$, we will look at the functions $\upsilon^{[2]}\,a_+^{[1]} - \upsilon^{[1]}\,a_+^{[2]}$ and $\upsilon^{[2]}\,a_-^{[1]} - \upsilon^{[1]}\,a_-^{[2]}$ separately. The handling of the two functions is quite dissimilar, owing to the different asymptotic behavior of $\lambda_k^{[\nu]}$, $\mu_k^{[\nu]}$ and $(c^{[\nu]})'(\lambda_{-k}^{[\nu]})$ for $k \to \infty$ and for $k \to -\infty$.

Let us first look at $\upsilon^{[2]}\,a_+^{[1]} - \upsilon^{[1]}\,a_+^{[2]}$. For $\lambda \in \mathbb{C}^*$ we have

$$\begin{aligned}
&\upsilon^{[2]}\,a_+^{[1]}(\lambda) - \upsilon^{[1]}\,a_+^{[2]}(\lambda) \\
&= \sum_{k=1}^{\infty} \left(\frac{\upsilon^{[2]}\,\mu_k^{[1]} \cdot (\tau^{[1]})^{-1}\,c^{[1]}(\lambda)}{(\tau^{[1]})^{-1}\,(c^{[1]})'(\lambda_k^{[1]}) \cdot (\lambda - \lambda_k^{[1]})} - \frac{\upsilon^{[1]}\,\mu_k^{[2]} \cdot (\tau^{[2]})^{-1}\,c^{[2]}(\lambda)}{(\tau^{[2]})^{-1}\,(c^{[2]})'(\lambda_k^{[2]}) \cdot (\lambda - \lambda_k^{[2]})} \right) \\
&= (\tau^{[1]})^{-1}\,c^{[1]}(\lambda) \cdot \sum_{k=1}^{\infty} \left(\upsilon^{[2]}\,\mu_k^{[1]} - \upsilon^{[1]}\,\mu_k^{[2]}\right) \cdot \frac{1}{(\tau^{[1]})^{-1}\,(c^{[1]})'(\lambda_k^{[1]}) \cdot (\lambda - \lambda_k^{[1]})}
\end{aligned} \tag{10.27}$$

$$+ \left((\tau^{[1]})^{-1}\,c^{[1]}(\lambda) - (\tau^{[2]})^{-1}\,c^{[2]}(\lambda)\right) \cdot \sum_{k=1}^{\infty} \frac{\upsilon^{[1]}\,\mu_k^{[2]}}{(\tau^{[1]})^{-1}\,(c^{[1]})'(\lambda_k^{[1]}) \cdot (\lambda - \lambda_k^{[1]})} \tag{10.28}$$

$$+ (\tau^{[2]})^{-1}\,c^{[2]}(\lambda) \cdot \sum_{k=1}^{\infty} \left(\frac{1}{(\tau^{[1]})^{-1}\,(c^{[1]})'(\lambda_k^{[1]})} - \frac{1}{(\tau^{[2]})^{-1}\,(c^{[2]})'(\lambda_k^{[2]})} \right) \cdot \frac{\upsilon^{[1]}\,\mu_k^{[2]}}{\lambda - \lambda_k^{[1]}} \tag{10.29}$$

$$+ (\tau^{[2]})^{-1}\,c^{[2]}(\lambda) \cdot \sum_{k=1}^{\infty} \left(\frac{1}{\lambda - \lambda_k^{[1]}} - \frac{1}{\lambda - \lambda_k^{[2]}} \right) \cdot \frac{\upsilon^{[1]}\,\mu_k^{[2]}}{(\tau^{[2]})^{-1}\,(c^{[2]})'(\lambda_k^{[2]})}. \tag{10.30}$$

We will estimate the four expressions (10.27)–(10.30) individually. For this purpose, we fix $n \in \mathbb{N}$ and consider $\lambda \in S_n \cap V_\delta$.

For the expression (10.27), we note that $(c^{[1]})'(\lambda_k^{[1]})$ is bounded away from zero, and that we have $b_k \in \ell^2(k \geq 1)$ (where b_k is defined by Eq. (10.24)), whence we obtain

$$\left| \sum_{k=1}^{\infty} \left(\upsilon^{[2]}\,\mu_k^{[1]} - \upsilon^{[1]}\,\mu_k^{[2]}\right) \cdot \frac{1}{(\tau^{[1]})^{-1}\,(c^{[1]})'(\lambda_k^{[1]}) \cdot (\lambda - \lambda_k^{[1]})} \right|$$
$$\leq C_1 \cdot \sum_{k=1}^{\infty} \frac{b_k}{|\lambda - \lambda_k^{[1]}|}$$

with a constant $C_1 > 0$. By Proposition A.1(3), the latter sum is $\leq \frac{C_2}{n} \cdot \left(b_k * \frac{1}{|k|}\right)_n \leq \frac{C_3}{n}\, r_n$ with constants $C_2, C_3 > 0$ and (r_k) defined in Eq. (10.25). Because we have $c^{[1]} \in \mathrm{As}_\infty(\mathbb{C}^*, \ell^\infty_{-1}, 1)$, and hence $|c^{[1]}(\lambda)| \leq C_4 \cdot n \cdot w(\lambda)$, it follows that the expression (10.27) is $\leq C_5 \cdot r_n \cdot w(\lambda)$.

For the expression (10.28), we note that $\upsilon^{[1]}\,\mu_k^{[2]} = \upsilon^{[1]}\,\upsilon^{[2]}\,(-1)^k + \ell^2(k \geq 1)$ and moreover $(\tau^{[1]})^{-1}\,(c^{[1]})'(\lambda_k^{[1]}) = \frac{1}{8}\,(-1)^k + \ell^2(k \geq 1)$ holds, whence it follows that there exists a sequence $t_k \in \ell^2(k \geq 1)$ with

$$\frac{\upsilon^{[1]}\,\mu_k^{[2]}}{(\tau^{[1]})^{-1}\,(c^{[1]})'(\lambda_k^{[1]})} = 8\,\upsilon^{[1]}\,\upsilon^{[2]} + t_k$$

and hence

$$\sum_{k=1}^{\infty} \frac{\upsilon^{[1]}\,\mu_k^{[2]}}{(\tau^{[1]})^{-1}\,(c^{[1]})'(\lambda_k^{[1]}) \cdot (\lambda - \lambda_k^{[1]})} = 8\,\upsilon^{[1]}\,\upsilon^{[2]} \sum_{k=1}^{\infty} \frac{1}{\lambda - \lambda_k^{[1]}} + \sum_{k=1}^{\infty} \frac{t_k}{\lambda - \lambda_k^{[1]}}.$$

By Proposition A.1(1),(3) it follows that there exists a constant $C_6 > 0$ with

$$\left| \sum_{k=1}^{\infty} \frac{\upsilon^{[1]}\,\mu_k^{[2]}}{(\tau^{[1]})^{-1}\,(c^{[1]})'(\lambda_k^{[1]}) \cdot (\lambda - \lambda_k^{[1]})} \right| \leq C_6 \cdot \frac{1}{n}. \tag{10.31}$$

By Proposition 10.1(2) we have $(\tau^{[1]})^{-1}\,c^{[1]}(\lambda) - (\tau^{[2]})^{-1}\,c^{[2]}(\lambda) \in \mathrm{As}_\infty(\mathbb{C}^*, \ell^2_{-1}, 1)$, with $C_7 \cdot k \cdot (a_k * \frac{1}{k})$ being a bounding sequence. We have

$$C_7 \cdot k \cdot (a_k * \tfrac{1}{k}) \leq C_8 \cdot k \cdot (a_k * (\tfrac{1}{|k|} * \tfrac{1}{|k|})) = C_8 \cdot k \cdot ((a_k * \tfrac{1}{|k|}) * \tfrac{1}{|k|}) \leq k \cdot r_k.$$

Together with (10.31), it follows that the expression (10.28) is $\leq C_9 \cdot r_n \cdot w(\lambda)$.

For the expression (10.29), we note that by Proposition 10.1(2), we have

$$(\tau^{[1]})^{-1}\,c^{[1]}(\lambda) - (\tau^{[2]})^{-1}\,c^{[2]}(\lambda) = (\tau^{[1]})^{-1}\,(\tau^{[2]})^{-1}\,(\tau^{[2]}\,c^{[1]}(\lambda) - \tau^{[1]}\,c^{[2]}(\lambda))$$
$$\in \mathrm{As}_\infty(\mathbb{C}^*, \ell^2_{-1}, 1),$$

and $C_{10} \cdot k \cdot (a_k * \frac{1}{|k|})$ is a bounding sequence. By Proposition 9.4(3), therefore

$$(\tau^{[1]})^{-1}\,(c^{[1]})' - (\tau^{[2]})^{-1}\,(c^{[2]})' \in \mathrm{As}_\infty(\mathbb{C}^*, \ell^2_0, 1)$$

holds, and $C_{11} \cdot (a_k * \frac{1}{|k|}) \in \ell^2(k \geq 1)$ is a bounding sequence with a constant $C_{11} > 0$. Moreover, $\upsilon^{[1]}\,\mu_k^{[2]}$ is bounded. Thus we have

$$\begin{aligned}
&\left|\sum_{k=1}^{\infty}\left(\frac{1}{(\tau^{[1]})^{-1}\,(c^{[1]})'(\lambda_k^{[1]})} - \frac{1}{(\tau^{[2]})^{-1}\,(c^{[2]})'(\lambda_k^{[2]})}\right)\cdot\frac{\upsilon^{[1]}\,\mu_k^{[2]}}{\lambda-\lambda_k^{[1]}}\right| \\
&\quad\leq C_{12}\cdot\sum_{k=1}^{\infty}\frac{(a_k * \frac{1}{|k|})}{|\lambda-\lambda_k^{[1]}|}
\end{aligned}$$

with a constant $C_{12} > 0$. By Proposition A.1(3) it follows that the sum to the right in the inequality above is $\leq \frac{C_{13}}{n}\cdot((a_k * \frac{1}{|k|}) * \frac{1}{|k|})_n \leq \frac{C_{13}}{n}\cdot r_n$. Together with the fact that $c^{[2]} \in \mathrm{As}_\infty(\mathbb{C}^*, \ell^\infty_{-1}, 1)$ holds, it follows that the expression (10.29) is $\leq C_{14}\cdot r_n\cdot w(\lambda)$.

Finally, for the expression (10.30) we again note that $\frac{\upsilon^{[1]}\,\mu_k^{[2]}}{(\tau^{[2]})^{-1}\,(c^{[2]})'(\lambda_k^{[2]})}$ is bounded, and thus we have

$$\begin{aligned}
&\left|\sum_{k=1}^{\infty}\left(\frac{1}{\lambda-\lambda_k^{[1]}} - \frac{1}{\lambda-\lambda_k^{[2]}}\right)\cdot\frac{\upsilon^{[1]}\,\mu_k^{[2]}}{(\tau^{[2]})^{-1}\,(c^{[2]})'(\lambda_k^{[2]})}\right| \\
&\quad\leq C_{15}\cdot\sum_{k=1}^{\infty}\frac{|\lambda_k^{[1]}-\lambda_k^{[2]}|}{|\lambda-\lambda_k^{[1]}|\cdot|\lambda-\lambda_k^{[2]}|}.
\end{aligned}$$

By Proposition A.1(4), the sum on the right hand side of the preceding inequality is $\leq \frac{C_{16}}{n}\cdot(a_k * \frac{1}{|k|})_n \leq \frac{C_{16}}{n}\cdot r_n$. Similarly as for the expression expression (10.29) above, it follows that the expression (10.30) is $\leq C_{17}\cdot r_n\cdot w(\lambda)$.

The preceding estimates show that for $\lambda \in S_n \cap V_\delta$,

$$\left(\upsilon^{[2]}\,a_+^{[1]}(\lambda) - \upsilon^{[1]}\,a_+^{[2]}(\lambda)\right) \leq C_{18}\cdot r_n\cdot w(\lambda) \tag{10.32}$$

holds with a constant $C_{18} > 0$.

We now attend to $\upsilon^{[2]}\,a_-^{[1]} - \upsilon^{[1]}\,a_-^{[2]}$. We let $\lambda \in \mathbb{C}^*$ be given and write

$$\begin{aligned}
\upsilon^{[2]}\,a_-^{[1]}(\lambda) - \upsilon^{[1]}\,a_-^{[2]}(\lambda) &= \left(\upsilon^{[2]} - (\upsilon^{[2]})^{-1}\right)a_-^{[1]}(\lambda) \\
&\quad + \left((\upsilon^{[2]})^{-1}\,a_-^{[1]}(\lambda) - (\upsilon^{[1]})^{-1}\,a_-^{[2]}(\lambda)\right) + \left((\upsilon^{[1]})^{-1} - \upsilon^{[1]}\right)a_-^{[2]}(\lambda). \qquad (10.33)
\end{aligned}$$

For estimating the various parts of Eq. (10.33), we keep the following inequalities in mind: We fix $n \in \mathbb{N}$ and let $\lambda \in S_n$. Then we have for $k \geq 0$

$$\frac{1}{|\lambda - \lambda_{-k}^{[\nu]}|} \leq \frac{C_{19}}{|n^2 - k^{-2}|} = \frac{C_{19} \cdot k^2}{|n^2 k^2 - 1|}$$
$$= \frac{C_{19} \cdot k^2}{|nk+1| \cdot |nk-1|} \leq \frac{C_{19} \cdot k^2}{nk \cdot |n+k|} = \frac{C_{19} \cdot k}{n \cdot |n+k|}$$

with a constant $C_{19} > 0$. We may also assume without loss of generality that $|\lambda| \geq 2\,|\lambda_{-k}^{[\nu]}|$ holds for all $k \geq 0$ and $\nu \in \{1, 2\}$, then we have

$$\frac{1}{|\lambda - \lambda_{-k}^{[\nu]}|} \leq \frac{1}{2\,|\lambda|}. \tag{10.34}$$

We also have $c^{[\nu]} \in \mathrm{As}_\infty(\mathbb{C}^*, \ell_{-1}^\infty, 1)$ and therefore

$$|c^{[\nu]}(\lambda)| \leq C_{20} \cdot n \cdot w(\lambda).$$

From the asymptotic for $c^{[\nu]}$ for $\lambda \to 0$, namely $c^{[\nu]} - (\tau^{[\nu]})^{-1}\, c_0 \in \mathrm{As}_0(\mathbb{C}^*, \ell_1^2, 1)$ (Proposition 10.1(2)(b)), it follows by Proposition 9.4(3) that $(c^{[\nu]})' - (\tau^{[\nu]})^{-1}\, c_0' \in \mathrm{As}_0(\mathbb{C}^*, \ell_{-2}^2, 1)$ holds, and therefore there exists a constant $C_{21} > 0$ such that

$$\frac{1}{|(c^{[\nu]})'(\lambda_{-k}^{[\nu]})|} \leq C_{21} \cdot \frac{1}{k^2}$$

holds for $k \geq 0$. We also note that $\mu_{-k}^{[\nu]}$ is bounded.

In the first instance, we obtain using these estimates

$$|a_-^{[\nu]}(\lambda)| = \left| \sum_{k=0}^{\infty} \frac{\mu_{-k}^{[\nu]} \cdot c^{[\nu]}(\lambda)}{(c^{[\nu]})'(\lambda_{-k}^{[\nu]}) \cdot (\lambda - \lambda_{-k}^{[\nu]})} \right| \leq C_{22} \cdot \frac{|c(\lambda)|}{|\lambda|} \cdot \sum_{k=0}^{\infty} \frac{1}{k^2} \leq C_{23} \cdot \frac{|c(\lambda)|}{|\lambda|},$$

and therefore

$$a_-^{[\nu]} \in \mathrm{As}_\infty(\mathbb{C}^*, \ell_1^\infty, 1). \tag{10.35}$$

This shows that two of the terms occurring in (10.33), namely $\left(\upsilon^{[2]} - (\upsilon^{[2]})^{-1}\right) a_-^{[1]}$ and $\left((\upsilon^{[1]})^{-1} - \upsilon^{[1]}\right) a_-^{[2]}$ are in $\mathrm{As}_\infty(\mathbb{C}^*, \ell_1^\infty, 1)$ with $C_{23} \cdot \left|\upsilon^{[2]} - (\upsilon^{[2]})^{-1}\right| \cdot \frac{1}{k} \leq C_{23} \cdot r_k$ resp. $C_{23} \cdot \left|\upsilon^{[1]} - (\upsilon^{[1]})^{-1}\right| \cdot \frac{1}{k} \leq C_{23} \cdot r_k$ being a bounding sequence.

We now estimate the remaining term in (10.33), $(\upsilon^{[2]})^{-1}\,a_-^{[1]}(\lambda) - (\upsilon^{[1]})^{-1}\,a_-^{[2]}$. To do so, we split this expression similarly as we did for $\upsilon^{[2]}\,a_+^{[1]} - \upsilon^{[1]}\,a_+^{[2]}$:

$$(\upsilon^{[2]})^{-1}\,a_-^{[1]}(\lambda) - (\upsilon^{[1]})^{-1}\,a_-^{[2]}$$

$$= \sum_{k=0}^{\infty} \left(\frac{(\upsilon^{[2]})^{-1}\,\mu_{-k}^{[1]} \cdot (\tau^{[1]})^{-1}\,c^{[1]}(\lambda)}{(\tau^{[1]})^{-1}\,(c^{[1]})'(\lambda_{-k}^{[1]}) \cdot (\lambda - \lambda_{-k}^{[1]})} - \frac{(\upsilon^{[1]})^{-1}\,\mu_{-k}^{[2]} \cdot (\tau^{[2]})^{-1}\,c^{[2]}(\lambda)}{(\tau^{[2]})^{-1}\,(c^{[2]})'(\lambda_{-k}^{[2]}) \cdot (\lambda - \lambda_{-k}^{[2]})} \right)$$

$$= (\tau^{[1]})^{-1}\,c^{[1]}(\lambda) \cdot \sum_{k=0}^{\infty} \left((\upsilon^{[2]})^{-1}\,\mu_{-k}^{[1]} - (\upsilon^{[1]})^{-1}\,\mu_{-k}^{[2]} \right) \cdot \frac{1}{(\tau^{[1]})^{-1}\,(c^{[1]})'(\lambda_{-k}^{[1]}) \cdot (\lambda - \lambda_{-k}^{[1]})} \tag{10.36}$$

$$+ \left((\tau^{[1]})^{-1}\,c^{[1]}(\lambda) - (\tau^{[2]})^{-1}\,c^{[2]}(\lambda) \right) \cdot \sum_{k=0}^{\infty} \frac{(\upsilon^{[1]})^{-1}\,\mu_{-k}^{[2]}}{(\tau^{[1]})^{-1}\,(c^{[1]})'(\lambda_{-k}^{[1]}) \cdot (\lambda - \lambda_{-k}^{[1]})} \tag{10.37}$$

$$+ (\tau^{[2]})^{-1}\,c^{[2]}(\lambda) \sum_{k=0}^{\infty} \left(\frac{1}{(\tau^{[1]})^{-1}\,(c^{[1]})'(\lambda_{-k}^{[1]})} - \frac{1}{(\tau^{[2]})^{-1}\,(c^{[2]})'(\lambda_{-k}^{[2]})} \right) \cdot \frac{(\upsilon^{[1]})^{-1}\,\mu_{-k}^{[2]}}{\lambda - \lambda_{-k}^{[1]}} \tag{10.38}$$

$$+ (\tau^{[2]})^{-1}\,c^{[2]}(\lambda) \cdot \sum_{k=0}^{\infty} \left(\frac{1}{\lambda - \lambda_{-k}^{[1]}} - \frac{1}{\lambda - \lambda_{-k}^{[2]}} \right) \cdot \frac{(\upsilon^{[1]})^{-1}\,\mu_{-k}^{[2]}}{(\tau^{[2]})^{-1}\,(c^{[2]})'(\lambda_{-k}^{[2]})}. \tag{10.39}$$

We again treat the expressions (10.36)–(10.39) separately:

The expression (10.36) satisfies

$$|(10.36)| \leq C_{24} \cdot n \cdot w(\lambda) \cdot \sum_{k=0}^{\infty} \frac{b_{-k}}{k^2} \cdot \frac{k}{n \cdot |n+k|} \leq C_{24} \cdot w(\lambda) \cdot \sum_{k=0}^{\infty} b_{-k} \cdot \frac{1}{|n+k|}$$

$$\leq C_{24} \cdot w(\lambda) \cdot \left(b_k * \frac{1}{|k|} \right)_n \leq C_{24} \cdot r_n \cdot w(\lambda).$$

The expression (10.37) satisfies

$$|(10.37)| \leq C_{26} \cdot \left| (\tau^{[1]})^{-1}\,c^{[1]}(\lambda) - (\tau^{[2]})^{-1}\,c^{[2]}(\lambda) \right| \cdot \frac{1}{|\lambda|} \cdot \sum_{k=0}^{\infty} \frac{1}{k^2}$$

$$\leq C_{27} \cdot \left| (\tau^{[1]})^{-1}\,c^{[1]}(\lambda) - (\tau^{[2]})^{-1}\,c^{[2]}(\lambda) \right| \cdot \frac{1}{|\lambda|}.$$

We have $(\tau^{[1]})^{-1}\,c^{[1]} - (\tau^{[2]})^{-1}\,c^{[2]} \in \mathrm{As}_\infty(\mathbb{C}^*, \ell^2_{-1}, 1)$ with bounding sequence $C_{28} \cdot k \cdot (a_k * \frac{1}{|k|}) \leq C_{29} \cdot k \cdot r_k$ by Proposition 10.1(2)(b),(c) and therefore it follows that

$$|(10.37)| \leq C_{29} \cdot r_n \cdot w(\lambda).$$

For the expression (10.38) we have

$$
\begin{aligned}
|(10.38)| &\le C_{30}\cdot n\cdot w(\lambda)\cdot\sum_{k=0}^{\infty}\left|\frac{1}{(\tau^{[1]})^{-1}\,(c^{[1]})'(\lambda_{-k}^{[1]})}-\frac{1}{(\tau^{[2]})^{-1}\,(c^{[2]})'(\lambda_{-k}^{[2]})}\right|\cdot\frac{k}{n\cdot|n+k|}\\
&\le C_{31}\cdot w(\lambda)\cdot\sum_{k=0}^{\infty}\frac{\left|(\tau^{[2]})^{-1}\,(c^{[2]})'(\lambda_{-k}^{[2]})-(\tau^{[1]})^{-1}\,(c^{[1]})'(\lambda_{-k}^{[1]})\right|}{k^4}\cdot\frac{k}{|n+k|}.
\end{aligned}
\tag{10.40}
$$

We now note that we have

$$
\begin{aligned}
&(\tau^{[2]})^{-1}\,(c^{[2]})'(\lambda_{-k}^{[2]})-(\tau^{[1]})^{-1}\,(c^{[1]})'(\lambda_{-k}^{[1]})\\
&\quad=\left((\tau^{[2]})^{-1}-(\tau^{[1]})^{-1}\right)(c^{[2]})'(\lambda_{-k}^{[2]})\\
&\qquad+\left((\tau^{[1]})^{-1}\,(c^{[2]})'(\lambda_{-k}^{[2]})-(\tau^{[2]})^{-1}\,(c^{[1]})'(\lambda_{-k}^{[1]})\right)\\
&\qquad+\left((\tau^{[2]})^{-1}-(\tau^{[1]})^{-1}\right)(c^{[1]})'(\lambda_{-k}^{[1]})\\
&\quad=\tau^{[1]}\,\tau^{[2]}\,(\tau^{[1]}-\tau^{[2]})\,((c^{[1]})'(\lambda_{-k}^{[1]})+(c^{[2]})'(\lambda_{-k}^{[2]}))\\
&\qquad+\left((\tau^{[1]})^{-1}\,(c^{[2]})'(\lambda_{-k}^{[2]})-(\tau^{[2]})^{-1}\,(c^{[1]})'(\lambda_{-k}^{[1]})\right),
\end{aligned}
$$

and because of $(c^{[\nu]})'\in \mathrm{As}_0(\mathbb{C}^*,\ell^\infty_{-2},1)$

$$
\left|\tau^{[1]}\,\tau^{[2]}\,(\tau^{[1]}-\tau^{[2]})\,((c^{[1]})'(\lambda_{-k}^{[1]})+(c^{[2]})'(\lambda_{-k}^{[2]}))\right|\le C_{32}\cdot k^2\cdot\left|\tau^{[1]}-\tau^{[2]}\right|.
$$

By Proposition 10.1(2)(b),(c) we have $(\tau^{[1]})^{-1}\,c^{[2]}-(\tau^{[2]})^{-1}\,c^{[1]}\in \mathrm{As}_0(\mathbb{C}^*,\ell^2_1,1)$ with $C_{33}\cdot\frac{1}{k}(a_k*\frac{1}{|k|})_{-k}$ a bounding sequence, and from this it follows by Proposition 9.4(3) that we have $(\tau^{[1]})^{-1}\,(c^{[2]})'-(\tau^{[2]})^{-1}\,(c^{[1]})'\in \mathrm{As}_0(\mathbb{C}^*,\ell^2_{-2},1)$ with $C_{34}\cdot k^2\,(a_k*\frac{1}{|k|})_{-k}$ a bounding sequence. Thus we have

$$
\left|(\tau^{[1]})^{-1}\,(c^{[2]})'(\lambda_{-k}^{[2]})-(\tau^{[2]})^{-1}\,(c^{[1]})'(\lambda_{-k}^{[1]})\right|\le C_{34}\cdot k^2\cdot\left(a_k*\frac{1}{|k|}\right)_{-k},
$$

and therefore

$$
\begin{aligned}
&\left|\tau^{[1]}\,\tau^{[2]}\,(\tau^{[1]}-\tau^{[2]})\,((c^{[1]})'(\lambda_{-k}^{[1]})+(c^{[2]})'(\lambda_{-k}^{[2]}))\right|\\
&\quad\le k^2\cdot\left(C_{32}\left|\tau^{[1]}-\tau^{[2]}\right|+C_{34}\left(a_k*\frac{1}{|k|}\right)_{-k}\right).
\end{aligned}
$$

By plugging this result into (10.40), we obtain

$$
\begin{aligned}
|(10.38)| &\le C_{35} \cdot w(\lambda) \cdot \sum_{k=0}^{\infty} \frac{\left|\tau^{[1]} - \tau^{[2]}\right| + \left(a_k * \frac{1}{|k|}\right)_{-k}}{k} \cdot \frac{1}{|n+k|} \\
&\le C_{35} \cdot w(\lambda) \cdot \left(\left(\frac{\left|\tau^{[1]} - \tau^{[2]}\right|}{|k|} + \left(a_k * \frac{1}{|k|}\right) \right) * \frac{1}{|k|} \right) \\
&\le C_{36} \cdot r_n \cdot w(\lambda).
\end{aligned}
$$

Finally, for the expression (10.39) we have

$$
\begin{aligned}
|(10.39)| &\le C_{37} \cdot n \cdot w(\lambda) \cdot \sum_{k=0}^{\infty} \left| \frac{1}{\lambda - \lambda_{-k}^{[1]}} - \frac{1}{\lambda - \lambda_{-k}^{[2]}} \right| \cdot \frac{1}{k^2} \\
&\le C_{37} \cdot n \cdot w(\lambda) \cdot \sum_{k=0}^{\infty} \frac{|\lambda_{-k}^{[1]} - \lambda_{-k}^{[2]}|}{|\lambda - \lambda_{-k}^{[1]}| \cdot |\lambda - \lambda_{-k}^{[2]}|} \cdot \frac{1}{k^2} \\
&\le C_{38} \cdot n \cdot w(\lambda) \cdot \sum_{k=0}^{\infty} |\lambda_{-k}^{[1]} - \lambda_{-k}^{[2]}| \cdot \frac{k^2}{n^2 \cdot |n+k|^2} \cdot \frac{1}{k^2} \\
&\le C_{38} \cdot \frac{1}{n} \cdot w(\lambda) \cdot \sum_{k=0}^{\infty} \frac{k^{-3} \cdot a_{-k}}{|n+k|^2} \\
&\le C_{38} \cdot w(\lambda) \cdot \sum_{k=0}^{\infty} a_{-k} \cdot \frac{1}{|n+k|} \le C_{38} \cdot w(\lambda) \cdot \left(a_k * \frac{1}{|k|}\right)_n \\
&\le C_{39} \cdot r_n \cdot w(\lambda).
\end{aligned}
$$

From the preceding estimates, we obtain that for all $n \in \mathbb{N}$ and $\lambda \in S_n$ we have

$$
\left(\upsilon^{[2]} a_-^{[1]}(\lambda) - \upsilon^{[1]} a_-^{[2]}(\lambda)\right) \le C_{40} \cdot r_n \cdot w(\lambda) \tag{10.41}
$$

holds with a constant $C_{40} > 0$.

By combining (10.32) and (10.41) we obtain that

$$
\left(\upsilon^{[2]} a^{[1]}(\lambda) - \upsilon^{[1]} a^{[2]}(\lambda)\right) \le C_{41} \cdot r_n \cdot w(\lambda)
$$

holds with a constant $C_{41} > 0$, and thus we have $\upsilon^{[2]} a^{[1]} - \upsilon^{[1]} a^{[2]} \in \mathrm{As}_\infty(\mathbb{C}^*, \ell^2, 1)$ and that r_n is a bounding sequence if the constant $C > 0$ in Eq. (10.25) is chosen appropriately. This shows the parts of (3)(a) and (3)(b) concerning $\lambda \to \infty$. The part of (3)(c) concerning $\lambda \to \infty$ follows by an application of the version (7.4) of Young's inequality.

We next show the part of (10.19) in (1) concerning $\lambda \to \infty$, i.e. we show that in the situation of (1), $a - \upsilon\, a_0 \in \mathrm{As}_\infty(\mathbb{C}^*, \ell^2_0, 1)$ holds. To do so, we apply the situation of (3) with $\lambda_k^{[1]} = \lambda_k$, $\mu_k^{[1]} = \mu_k$ and $\lambda_k^{[2]} = \lambda_{k,0}$, $\mu_k^{[2]} = \mu_{k,0}$. Then $\upsilon^{[1]} = \upsilon$ and $\upsilon^{[2]} = 1$ holds. By (3)(a) we then have $a^{[1]} - \upsilon\, a^{[2]} \in \mathrm{As}_\infty(\mathbb{C}^*, \ell^2_0, 1)$. We therefore only need to show that $a^{[2]}$ (i.e. the function a obtained from Eq. (10.26) by setting $\lambda_k = \lambda_{k,0}$ and $\mu_k = \mu_{k,0}$) equals $a_0 = \cos(\zeta(\lambda))$.

We base our proof of this fact on the partial fraction expansion of the cotangent function: For $z \in \mathbb{C} \setminus \{\, k\pi \mid k \in \mathbb{Z}\,\}$,

$$\cot(z) = \frac{1}{z} + \sum_{k=1}^{\infty} \frac{2z}{z^2 - (k\pi)^2} \tag{10.42}$$

holds. We also remember

$$c_0'(\lambda_{k,0}) = \frac{\lambda_{k,0} - 1}{8\,\lambda_{k,0}}\,\mu_{k,0}. \tag{10.43}$$

Therefore we have for $\lambda \in \mathbb{C}^* \setminus \{\, \lambda_{k,0} \mid k \in \mathbb{Z}\,\}$

$$\begin{aligned}
a_0(\lambda) &= \cos(\zeta(\lambda)) = \sin(\zeta(\lambda)) \cdot \cot(\zeta(\lambda)) = \tfrac{c_0(\lambda)}{\sqrt{\lambda}} \cdot \cot(\zeta(\lambda)) \\
&\overset{(10.42)}{=} \tfrac{c_0(\lambda)}{\sqrt{\lambda}} \cdot \left(\tfrac{1}{\zeta(\lambda)} + \sum_{k=1}^{\infty} \tfrac{2\,\zeta(\lambda)}{\zeta(\lambda)^2 - (k\pi)^2} \right) \\
&= c_0(\lambda) \cdot \left(\tfrac{4}{\lambda+1} + \sum_{k=1}^{\infty} \tfrac{\lambda+1}{2\lambda\,(\zeta(\lambda)^2 - \zeta(\lambda_{k,0})^2)} \right) \\
&= c_0(\lambda) \cdot \left(\tfrac{4}{\lambda+1} + \sum_{k=1}^{\infty} \tfrac{8\,(\lambda+1)\,\lambda_{k,0}}{(\lambda+1)^2\,\lambda_{k,0} - (\lambda_{k,0}+1)^2\,\lambda} \right) \\
&= c_0(\lambda) \cdot \left(\tfrac{4}{\lambda+1} + \sum_{k=1}^{\infty} \tfrac{8\,(\lambda+1)}{(\lambda-\lambda_{-k,0})\,(\lambda-\lambda_{k,0})} \right) \\
&= c_0(\lambda) \cdot \left(\tfrac{4}{\lambda+1} + \sum_{k=1}^{\infty} \tfrac{8}{\lambda_{k,0}-\lambda_{-k,0}} \left(\tfrac{\lambda_{k,0}+1}{\lambda-\lambda_{k,0}} - \tfrac{\lambda_{-k,0}+1}{\lambda-\lambda_{-k,0}} \right) \right) \\
&= c_0(\lambda) \cdot \left(\tfrac{4}{\lambda+1} + \sum_{k \in \mathbb{Z}\setminus\{0\}} \tfrac{8}{\lambda_{k,0}-\lambda_{-k,0}}\, \tfrac{\lambda_{k,0}+1}{\lambda-\lambda_{k,0}} \right)
\end{aligned}$$

$$\overset{(10.43)}{=} c_0(\lambda) \cdot \left(\frac{4}{\lambda+1} \frac{(\lambda_{0,0}-1)\,\mu_{0,0}}{8\,\lambda_{0,0}\, f_0'(\lambda_{k,0})} + \sum_{k\in\mathbb{Z}\setminus\{0\}} \frac{8}{\lambda_{k,0}-\lambda_{-k,0}} \frac{\lambda_{k,0}+1}{\lambda-\lambda_{k,0}} \frac{(\lambda_{k,0}-1)\,\mu_{k,0}}{8\,\lambda_{k,0}\, c_0'(\lambda_{k,0})} \right)$$

$$\overset{\lambda_{0,0}=-1}{=} c_0(\lambda) \cdot \sum_{k\in\mathbb{Z}} \frac{\mu_{k,0}}{c_0'(\lambda_{k,0})\cdot(\lambda-\lambda_{k,0})}.$$

Therefore the desired equality holds for all $\lambda \in \mathbb{C}^* \setminus \{\, \lambda_{k,0} \,|\, k \in \mathbb{Z}\,\}$. Because its both sides are holomorphic in λ, that equality in fact holds for $\lambda \in \mathbb{C}^*$.

Next we show the parts of (1) and (3) concerning $\lambda \to 0$. Similarly as we did in the proof of Proposition 10.1, we do so by reducing the situation for $\lambda \to 0$ to the previously proven situation for $\lambda \to \infty$. For this purpose, we put $\widetilde{\lambda}_k := \lambda_{-k}^{-1}$ and $\widetilde{\mu}_k := \mu_{-k}$. Mutatis mutandis, these sequences again satisfy the hypotheses of part (3) of the proposition. Again we denote the quantities associated to $(\widetilde{\lambda}_k)$ and $\widetilde{\mu}_k$ by a tilde $\widetilde{\ }$. We note that $\widetilde{a}_k$, $\widetilde{b}_k$ and $\widetilde{r}_k$ is comparable to a_{-k}, b_{-k} and r_{-k}, respectively. We may again suppose without loss of generality that $d_k = 1$ holds for all $k \in \mathbb{Z}$. Then we obtain by explicit calculation

$$\tau = \widetilde{\tau}^{-1} \quad \text{and} \quad \upsilon = \widetilde{\upsilon}^{-1},$$

and

$$c(\lambda^{-1}) = \widetilde{c}(\lambda)\cdot\lambda^{-1} \quad \text{and} \quad c'(\lambda_k) = -\widetilde{\lambda}_{-k}\cdot\widetilde{c}'(\widetilde{\lambda}_{-k})$$

and therefore

$$\begin{aligned} a(\lambda^{-1}) &= \sum_{k\in\mathbb{Z}} \frac{\mu_k\cdot c(\lambda^{-1})}{c'(\lambda_k)\cdot(\lambda^{-1}-\lambda_k)} = \sum_{k\in\mathbb{Z}} \frac{\widetilde{\mu}_{-k}\cdot\widetilde{c}(\lambda)\cdot\lambda^{-1}}{(-\widetilde{\lambda}_{-k})\cdot c'(\widetilde{\lambda}_{-k})\cdot(\lambda^{-1}-\widetilde{\lambda}_{-k}^{-1})} \\ &= \sum_{k\in\mathbb{Z}} \frac{\widetilde{\mu}_{-k}\cdot\widetilde{c}(\lambda)}{c'(\widetilde{\lambda}_{-k})\cdot(\lambda-\widetilde{\lambda}_{-k})} = \sum_{k\in\mathbb{Z}} \frac{\widetilde{\mu}_k\cdot\widetilde{c}(\lambda)}{c'(\widetilde{\lambda}_k)\cdot(\lambda-\widetilde{\lambda}_k)} = \widetilde{a}(\lambda). \end{aligned}$$

Using these formulas, one derives the asymptotic estimates for $\lambda \to 0$ in (1) and (3) from the corresponding estimates for $\lambda \to \infty$ that have been shown before.

For (2). For any $\lambda_* \in \Lambda$ we put $d(\lambda_*) := \mathrm{ord}_{\lambda_*}(c) \geq 2$. For $1 \leq j \leq d(\lambda_*) - 1$, we have $\varphi_{\lambda_*,j}(\lambda) := \frac{c(\lambda)}{(\lambda-\lambda_*)^j} \in \mathrm{As}(\mathbb{C}^*, \ell^\infty_{2j-1,1}, 1) \subset \mathrm{As}(\mathbb{C}^*, \ell^2_{0,0}, 1)$, and therefore (10.19) implies (10.22). Moreover, we have $\varphi_{\lambda_*,j}(\lambda_k) = 0$ for all k (including the case $\lambda_k = \lambda_*$), and therefore $\widetilde{a}(\lambda_k) = a(\lambda_k) = \mu_k$.

Now suppose that $\widehat{a}$ is another holomorphic function that satisfies (10.19) and (10.20). Because $\varphi_{\lambda_*,j}$ has at every λ_k with $\lambda_k \neq \lambda_*$ a zero of order d_k, and because the i-th derivative of $\varphi_{\lambda_*,j}$ at λ_* satisfies

$$\varphi^{(i)}_{\lambda_*,j}(\lambda_*) = 0 \quad \text{for } 0 \leq i \leq d(\lambda_*) - j - 1$$

and

$$\varphi_{\lambda_*,j}^{(d(\lambda_*)-j)}(\lambda_*) \neq 0,$$

the numbers $t_{\lambda_*,j}$ can be chosen such that for every $\lambda_* \in \Lambda$ and every $1 \leq j \leq d(\lambda_*)-1$, $\widetilde{a}^{(j)}(\lambda_*) = \widehat{a}^{(j)}(\lambda_*)$ holds, i.e. $\widetilde{a}-\widehat{a}$ has at λ_* a zero of order (at least) $d(\lambda_*)$. By Proposition 6.7(2), $\widetilde{a} = \widehat{a}$ follows. □

Corollary 10.6 *Suppose that* $a : \mathbb{C}^* \to \mathbb{C}$ *is a holomorphic function which satisfies the following two asymptotic properties for some* $\upsilon \in \mathbb{C}^*$*:*

(a) For every $\varepsilon > 0$ *there exists* $R > 0$ *such that we have*

$$\text{for all } \lambda \in \mathbb{C}^* \text{ with } |\lambda| \geq R: \qquad |a(\lambda) - \upsilon\, a_0(\lambda)| \leq \varepsilon\, w(\lambda)$$

and

$$\text{for all } \lambda \in \mathbb{C}^* \text{ with } |\lambda| \leq \tfrac{1}{R}: \qquad |a(\lambda) - \upsilon^{-1}\, a_0(\lambda)| \leq \varepsilon\, w(\lambda).$$

(b) $\begin{Bmatrix} a(\lambda_{k,0}) - \upsilon\, a_0(\lambda_{k,0}) & \text{if } k \geq 0 \\ a(\lambda_{k,0}) - \upsilon^{-1}\, a_0(\lambda_{k,0}) & \text{if } k < 0 \end{Bmatrix} \in \ell^2_{0,0}(k).$

Then we already have

$$a - \upsilon\, a_0 \in \mathrm{As}_\infty(\mathbb{C}^*, \ell^2_0, 1) \quad \text{and} \quad a - \upsilon^{-1}\, a_0 \in \mathrm{As}_0(\mathbb{C}^*, \ell^2_0, 1). \tag{10.44}$$

Addendum. *If* L *is a set of holomorphic functions which satisfy (a) and (b) in such a way that the values of* υ *corresponding to* $a \in L$ *are bounded and bounded away from zero, that* $R = R(\varepsilon) > 0$ *in (a) can be chosen uniformly for all* $a \in L$*, and that in (b) there is a uniform bound* $(z_k)_{k\in\mathbb{Z}} \in \ell^2_{0,0}(k)$ *for the sequences in (b) for all* $a \in L$*, then there exists a uniform bounding sequence for the asymptotics in* (10.44) *that applies for all* $a \in L$*.*

Proof Let $\lambda_k := \lambda_{k,0}$ and $\mu_k := a(\lambda_{k,0})$. Then we have by hypothesis (b)

$$\lambda_k - \lambda_{k,0} \in \ell^2_{-1,3}(k) \quad \text{and} \quad \begin{Bmatrix} \mu_k - \upsilon\, \mu_{k,0} & \text{if } k \geq 0 \\ \mu_k - \upsilon^{-1}\, \mu_{k,0} & \text{if } k < 0 \end{Bmatrix} \in \ell^2_{0,0}(k).$$

By Proposition 10.4(1) it follows that there exists a holomorphic function $\widetilde{a} : \mathbb{C}^* \to \mathbb{C}$ that satisfies

$$\widetilde{a} - \upsilon\, a_0 \in \mathrm{As}_\infty(\mathbb{C}^*, \ell^2_0, 1) \quad \text{and} \quad \widetilde{a} - \upsilon^{-1}\, a_0 \in \mathrm{As}_0(\mathbb{C}^*, \ell^2_0, 1)$$

and $\widetilde{a}(\lambda_{k,0}) = \mu_k$ for all $k \in \mathbb{Z}$. Because no two of the $\lambda_{k,0}$ coincide (i.e. the function c_0 with zeros at the $\lambda_{k,0}$ has no zeros of higher order), it follows from Proposition 6.7(2) that $\widetilde{a} = a$ holds, and thus the claimed statement follows.

In the setting of the addendum, the hypothesis that the numbers υ corresponding to $a \in L$ are bounded and bounded away from zero ensures that there exists a constant $C_1 > 0$ so that $|\upsilon - \upsilon^{-1}| \leq C_1$ holds for all these υ. Let

$$r_k := C \cdot \left(z_k * \frac{1}{|k|} + \frac{C_1}{|k|} \right)$$

with the constant $C > 0$ from Proposition 10.4(3)(b). By applying that proposition with $\lambda_k^{[1]} = \lambda_k^{[2]} = \lambda_{k,0}$, $\mu_k^{[1]} = a(\lambda_{k,0})$ and $\mu_k^{[2]} = \mu_{k,0}$, we see that with the sequence $(r_k)_{k>0}$ resp. $(r_k)_{k<0}$ is a bounding sequence for $a - \upsilon\, a_0$ resp. for $a - \upsilon^{-1}\, a_0$ (that is independent of $a \in L$). □

Chapter 11
Final Description of the Asymptotic of the Monodromy

Resulting from the interpolation theorems from Chap. 10 and by using the asymptotic spaces introduced in Chap. 9, we can now describe the asymptotics of the monodromy, its discriminant $\Delta^2 - 4$ and also of the extended frame in their final form. Moreover, we now prove the refined asymptotics for the branch points $\varkappa_{k,\nu}$ of the spectral curve, which are analogous to Corollary 8.2 for the spectral divisor, and which had been postponed in Chap. 8.

Definition 11.1 We say that a (2×2)-matrix of holomorphic functions $\mathbb{C}^* \to \mathbb{C}$

$$M(\lambda) = \begin{pmatrix} a(\lambda) & b(\lambda) \\ c(\lambda) & d(\lambda) \end{pmatrix}$$

with $\det(M(\lambda)) = 1$ has *non-periodic monodromy asymptotic* or is an *non-periodic (asymptotic) monodromy* with respect to numbers $\tau, \upsilon \in \mathbb{C}^*$ if we have

(a)

$$\begin{aligned} a(\lambda) - \upsilon\, a_0(\lambda) &\in \mathrm{As}_\infty(\mathbb{C}^*, \ell^2_0, 1) \\ b(\lambda) - \tau^{-1}\, b_0(\lambda) &\in \mathrm{As}_\infty(\mathbb{C}^*, \ell^2_1, 1) \\ c(\lambda) - \tau\, c_0(\lambda) &\in \mathrm{As}_\infty(\mathbb{C}^*, \ell^2_{-1}, 1) \\ d(\lambda) - \upsilon^{-1}\, d_0(\lambda) &\in \mathrm{As}_\infty(\mathbb{C}^*, \ell^2_0, 1) \end{aligned}$$

(b)

$$\begin{aligned} a(\lambda) - \upsilon^{-1}\, a_0(\lambda) &\in \mathrm{As}_0(\mathbb{C}^*, \ell^2_0, 1) \\ b(\lambda) - \tau\, b_0(\lambda) &\in \mathrm{As}_0(\mathbb{C}^*, \ell^2_{-1}, 1) \end{aligned}$$

S. Klein, *A Spectral Theory for Simply Periodic Solutions of the Sinh-Gordon Equation*, Lecture Notes in Mathematics 2229,
https://doi.org/10.1007/978-3-030-01276-2_11

$$c(\lambda) - \tau^{-1}\, c_0(\lambda) \in \mathrm{As}_0(\mathbb{C}^*, \ell_1^2, 1)$$
$$d(\lambda) - \upsilon\, d_0(\lambda) \in \mathrm{As}_0(\mathbb{C}^*, \ell_0^2, 1).$$

If this is the case with $\upsilon = 1$, then we say that $M(\lambda)$ has *monodromy asymptotic* or is an *(asymptotic) monodromy* with respect to $\tau \in \mathbb{C}^*$.

We denote the space of non-periodic monodromy asymptotic matrices with respect to τ, υ by $\mathsf{Mon}_{\tau,\upsilon}$, and put $\mathsf{Mon}_\tau := \mathsf{Mon}_{\tau,\upsilon=1}$. $\mathsf{Mon}_{np} := \bigcup_{\tau,\upsilon\in\mathbb{C}^*} \mathsf{Mon}_{\tau,\upsilon}$ is the space of all non-periodic monodromy asymptotic matrices, and $\mathsf{Mon} := \bigcup_{\tau\in\mathbb{C}^*} \mathsf{Mon}_\tau$ is the space of all monodromy asymptotic matrices.

Note that a monodromy $M(\lambda) \in \mathsf{Mon}_{np}$ in particular has the properties shown for spectral monodromies in Theorems 5.4 and 7.1. Therefore Corollary 8.2 concerning the asymptotic behavior of a spectral divisor is applicable to the spectral divisor D associated to a monodromy $M(\lambda) \in \mathsf{Mon}_{np}$.

We next show that the monodromy $M(\lambda)$ associated to a (non-periodic) potential $(u, u_y) \in \mathsf{Pot}_{np}$ indeed has monodromy asymptotic in the sense of Definition 11.1. This is our final form of the description of the asymptotic behavior of the monodromy.

Theorem 11.2 *Let* $(u, u_y) \in \mathsf{Pot}_{np}$, $M(\lambda)$ *be the monodromy associated to* (u, u_y), $\Delta(\lambda)$ *its trace function, and* $\tau := \mathrm{e}^{-(u(0)+u(1))/4}$ *and* $\upsilon := \mathrm{e}^{(u(1)-u(0))/4}$. *Then we have:*

(1) $M(\lambda) \in \mathsf{Mon}_{\tau,\upsilon}$.
Moreover, if P *is a relatively compact subset of* Pot_{np}, *then there exist uniform bounding sequences for the entries of the monodromy that apply to the monodromies of all the potentials* $(u, u_y) \in P$.
(2) If $(u, u_y) \in \mathsf{Pot}$ *holds (i.e.* (u, u_y) *is periodic), then we in fact have* $M(\lambda) \in \mathsf{Mon}_\tau$.
(3) We have the following trace formula*: If* $D = \{(\lambda_k, \mu_k)\}$ *is the classical spectral divisor corresponding to* $M(\lambda)$, *then we have*

$$\mathrm{e}^{(u(0)+u(1))/2} = \prod_{k\in\mathbb{Z}} \frac{\lambda_k}{\lambda_{k,0}}.$$

Remark 11.3 It is tempting to try to replace the condition "P is a relatively compact subset of Pot_{np}" in Theorem 11.2(1) by "P is a small closed ball in Pot_{np}". However with this variant of the condition, the theorem would not be true, as the condition that P needs to be relatively compact is ultimately derived from the requirement that the set $N \subset L^1([a,b])$ is compact in the version of the Riemann-Lebesgue Lemma described in Lemma 5.6 (which had been used to prove the basic asymptotic for the monodromy, Theorem 5.4). It is clear that one does not obtain a uniform estimate in the Riemann-Lebesgue Lemma if N is replaced by a small ball in $L^1([a,b])$, and thus we also cannot relax the condition on P in Theorem 11.2(1) in a similar way.

Proof (of Theorem 11.2) We write

$$M(\lambda) = \begin{pmatrix} a(\lambda) & b(\lambda) \\ c(\lambda) & d(\lambda) \end{pmatrix} \quad \text{and} \quad \Delta(\lambda) = a(\lambda) + d(\lambda).$$

For (1). We note the facts we already know from Theorems 5.4 and 7.1 about the asymptotics of the monodromy of (u, u_y). From them, the statements on the functions a and d follow from Corollary 10.6, and the statements on c follow from Corollary 10.2. The statements on b follow by applying Corollary 10.2 with $-\lambda \cdot b(\lambda)$ in the place of $c(\lambda)$.

If $P \subset \mathsf{Pot}_{np}$ is a relatively compact set, then Theorems 5.4 and 7.1 show that the entries of the monodromies of $(u, u_y) \in P$ satisfy the hypotheses of the Addendum in Corollary 10.2 resp. Corollary 10.6. These Addenda imply the existence of uniform bounding sequences in the sense described in the statement.

For (2). This follows from (1) via the fact that the periodicity of (u, u_y) is equivalent to $\upsilon = 1$.

For (3). The trace formula follows from calculating the parameter τ of the asymptotic of the monodromy in two different ways: On one hand, we have

$$\tau = \mathrm{e}^{-(u(0)+u(1))/4}$$

by Theorem 5.4. On the other hand, we have

$$\tau = \pm \left(\prod_{k \in \mathbb{Z}} \frac{\lambda_{k,0}}{\lambda_k} \right)^{1/2}$$

by Proposition 10.1(1). By combining these two equations, we obtain

$$\mathrm{e}^{(u(0)+u(1))/2} = \tau^{-2} = \prod_{k \in \mathbb{Z}} \frac{\lambda_k}{\lambda_{k,0}}.$$

□

Corollary 11.4 *Let $M(\lambda) \in \mathsf{Mon}_{\tau,\upsilon}$ with $\tau, \upsilon \in \mathbb{C}^*$ be given; this applies for example in the setting of Theorem 11.2 (where $M(\lambda)$ is the monodromy of a non-periodic potential $(u, u_y) \in \mathsf{Pot}_{np}$). We write $M(\lambda) = \left(\begin{smallmatrix} a(\lambda) & b(\lambda) \\ c(\lambda) & d(\lambda) \end{smallmatrix}\right)$ and $\Delta(\lambda) = \operatorname{tr}(M(\lambda)) = a(\lambda) + d(\lambda)$. Then we have*

(1) $a, d, \Delta \in \mathrm{As}(\mathbb{C}^*, \ell^\infty_{0,0}, 1)$, $b \in \mathrm{As}(\mathbb{C}^*, \ell^\infty_{1,-1}, 1)$ *and* $c \in \mathrm{As}(\mathbb{C}^*, \ell^\infty_{-1,1}, 1)$.

(2) $\Delta(\lambda) - \frac{\upsilon+\upsilon^{-1}}{2}\,\Delta_0(\lambda) \in \mathrm{As}(\mathbb{C}^*, \ell^2_{0,0}, 1)$.

(3) $\Delta^2 - \left(\frac{\upsilon+\upsilon^{-1}}{2}\,\Delta_0\right)^2, bc - b_0 c_0 \in \mathrm{As}(\mathbb{C}^*, \ell^2_{0,0}, 2)$.

We now suppose that $\upsilon = 1$ *holds, and let* Σ *be the spectral curve associated to* $M(\lambda)$. *Then we also have*

(4) $\mu - \mu_0, \mu^{-1} - \mu_0^{-1} \in \mathrm{As}(\Sigma, \ell^2_{0,0}, 1)$, *where we set* $\mu_0 := \mathrm{e}^{\mathrm{i}\,\zeta(\lambda)}$.

(5) $\mu, \mu^{-1} \in \mathrm{As}(\Sigma, \ell^\infty_{0,0}, 1)$.

(6) $\left(\frac{\mu-d}{c} - \frac{\mathrm{i}}{\tau\sqrt{\lambda}}\right)|\widehat{V}_\delta \in \mathrm{As}_\infty(\widehat{V}_\delta, \ell^2_1, 0)$ *and* $\left(\frac{\mu-d}{c} - \frac{\mathrm{i}}{\tau^{-1}\sqrt{\lambda}}\right)|\widehat{V}_\delta \in \mathrm{As}_0(\widehat{V}_\delta, \ell^2_{-1}, 0)$ *where* $\sqrt{\lambda}$ *is chosen as the holomorphic function on* $\widehat{V}_\delta$ *with the correct sign.*

Proof For (1). These statements follow from Definition 11.1 together with the fact that

$$a_0, d_0, \Delta_0 \in \mathrm{As}(\mathbb{C}^*, \ell^\infty_{0,0}, 1)\,, \quad b_0 \in \mathrm{As}(\mathbb{C}^*, \ell^\infty_{1,-1}, 1) \quad \text{and} \quad c_0 \in \mathrm{As}(\mathbb{C}^*, \ell^\infty_{-1,1}, 1)$$

holds.

For (2). Because of $\Delta = a + d$ and $\frac{1}{2}\Delta_0 = a_0 = d_0$ we have

$$\begin{aligned} \Delta(\lambda) - \tfrac{\upsilon+\upsilon^{-1}}{2}\,\Delta_0(\lambda) &= \big(a(\lambda) - \upsilon\, a_0(\lambda)\big) + \big(d(\lambda) - \upsilon^{-1}\, d_0(\lambda)\big) \\ &= \big(a(\lambda) - \upsilon^{-1}\, a_0(\lambda)\big) + \big(d(\lambda) - \upsilon\, d_0(\lambda)\big). \end{aligned}$$

The claimed statement follows from this equation together with the asymptotics for a and d in Definition 11.1.

For (3). This follows from the preceding results by the equations

$$\Delta^2 - \left(\tfrac{\upsilon+\upsilon^{-1}}{2}\,\Delta_0\right)^2 = \left(\Delta + \tfrac{\upsilon+\upsilon^{-1}}{2}\,\Delta_0\right)\cdot\left(\Delta - \tfrac{\upsilon+\upsilon^{-1}}{2}\,\Delta_0\right)$$

and

$$\begin{aligned} b\,c - b_0\,c_0 &= (b - \tau^{-1}\,b_0)\cdot c + \tau^{-1}\,b_0\cdot(c - \tau\,c_0) \\ &= (b - \tau\,b_0)\cdot c + \tau\,b_0\cdot(c - \tau^{-1}\,c_0). \end{aligned}$$

For (4). We have $\mu = \frac{1}{2}(\Delta + \sqrt{\Delta^2 - 4})$ and $\mu_0 = \frac{1}{2}(\Delta_0 + \sqrt{\Delta_0^2 - 4})$. Because we have $\Delta - \Delta_0 \in \mathrm{As}(\mathbb{C}^*, \ell^2_{0,0}, 1)$ by (2), it remains to show that

$$\left(\sqrt{\Delta^2 - 4} - \sqrt{\Delta_0^2 - 4}\right) \in \mathrm{As}(\Sigma, \ell^2_{0,0}, 1) \tag{11.1}$$

holds. To show this, we use the formula

$$\sqrt{\Delta^2 - 4} - \sqrt{\Delta_0^2 - 4} = \frac{\Delta^2 - \Delta_0^2}{\sqrt{\Delta^2 - 4} + \sqrt{\Delta_0^2 - 4}}.$$

Because both $\sqrt{\Delta^2-4}$ and $\sqrt{\Delta_0^2-4}$ are comparable to $w(\lambda)$ on $\widehat{V}_\delta$, it follows from (3) that

$$\left(\sqrt{\Delta^2-4}-\sqrt{\Delta_0^2-4}\right)\Big|\,\widehat{V}_\delta \in \mathrm{As}(\widehat{V}_\delta, \ell^2_{0,0}, 1)$$

holds, whence (11.1) follows by Proposition 9.4(1). Therefore we have $\mu-\mu_0 \in \mathrm{As}(\Sigma, \ell^2_{0,0}, 1)$.

Similarly we obtain $\mu^{-1}-\mu_0^{-1} \in \mathrm{As}(\Sigma, \ell^2_{0,0}, 1)$ from the equations $\mu^{-1} = \frac{1}{2}(\Delta-\sqrt{\Delta^2-4})$ and $\mu_0^{-1} = \frac{1}{2}(\Delta_0-\sqrt{\Delta_0^2-4})$,

For (5). Because of $\mu_0, \mu_0^{-1} \in \mathrm{As}(\Sigma, \ell^\infty_{0,0}, 1)$, it follows from (4) that $\mu, \mu^{-1} \in \mathrm{As}(\Sigma, \ell^\infty_{0,0}, 1)$ holds.

For (6). We have $\frac{\mathrm{i}}{\sqrt{\lambda}} = \frac{\mu_0-d_0}{c_0}$ and therefore

$$\frac{\mu-d}{c}-\frac{\mathrm{i}}{\tau\sqrt{\lambda}} = \frac{1}{\tau\,c_0}\big((\mu-\mu_0)+(d-d_0)\big)-\frac{\mu-d}{c\cdot\tau\,c_0}\,(c-\tau\,c_0). \tag{11.2}$$

Because both c and c_0 are comparable to $\sqrt{|\lambda|}\cdot w(\lambda)$ on $\widehat{V}_\delta$, and we have $(\mu-\mu_0)|\widehat{V}_\delta \in \mathrm{As}(\widehat{V}_\delta, \ell^2_{0,0}, 1)$, $d-d_0 \in \mathrm{As}(\mathbb{C}^*, \ell^2_{0,0}, 1)$, $\mu, d \in \mathrm{As}(\Sigma, \ell^\infty_{0,0}, 1)$ and $c-\tau\,c_0 \in \mathrm{As}_\infty(\mathbb{C}^*, \ell^2_{-1,1}, 1)$, it follows that $\left(\frac{\mu-d}{c}-\frac{\mathrm{i}}{\tau\sqrt{\lambda}}\right)|\widehat{V}_\delta \in \mathrm{As}_\infty(\widehat{V}_\delta, \ell^2_1, 0)$ holds.

The other asymptotic assessment $\left(\frac{\mu-d}{c}-\frac{\mathrm{i}}{\tau^{-1}\sqrt{\lambda}}\right)|\widehat{V}_\delta \in \mathrm{As}_0(\widehat{V}_\delta, \ell^2_{-1}, 0)$ is similarly obtained by replacing τ with τ^{-1} in Eq. (11.2). □

In the sequel, we will use the asymptotic behavior of the monodromy described in Definition 11.1 resp. Theorem 11.2 as well as the consequences in Corollary 11.4 without referencing these statements explicitly every time.

The following proposition concerns the zeros of the discriminant Δ^2-4 of an asymptotic monodromy $M(\lambda) \in \mathsf{Mon}$. It shows that the zeros of Δ^2-4 (corresponding to the branch points and singularities of the associated spectral curve Σ) have the same kind of asymptotic behavior as was shown for the divisor points in Corollary 8.2. Note that this statement is true only for periodic asymptotic monodromies (i.e. for the case $\upsilon = 1$).

Proposition 11.5 *Let $M(\lambda) \in \mathsf{Mon}$ be given (or let $M(\lambda)$ be the monodromy of a potential $(u, u_y) \in \mathsf{Pot}$), and let $\Delta(\lambda)$ be the trace function of $M(\lambda)$. We consider the two sequences $(\varkappa_{k,1})$ and $(\varkappa_{k,2})$ of zeros of Δ^2-4 (i.e. of branch points and singularities of the spectral curve Σ associated to Δ) defined in Proposition 6.5(1).*

Then we have $\varkappa_{k,\nu}-\lambda_{k,0} \in \ell^2_{-1,3}(k)$ for $\nu \in \{1,2\}$.

Proof By Corollary 11.4(2) we have $\Delta-\Delta_0 \in \mathrm{As}(\mathbb{C}^*, \ell^2_{0,0}, 1)$. We let $(\Sigma, \mathscr{D})$ be the spectral data of $M(\lambda)$. By Proposition 6.5(2),(3) and Corollary 8.2, the points in the support of $\mathscr{D}$ are enumerated by a sequence $(\lambda_k, \mu_k)_{k\in\mathbb{Z}}$ on Σ, such that $\lambda_k-\lambda_{k,0} \in \ell^2_{-1,3}(k)$ and $\mu_k-\mu_{k,0} \in \ell^2_{0,0}(k)$ holds.

Using the fact that $\Delta_0(\lambda_{k,0}) = 2(-1)^k = \Delta(\varkappa_{k,\nu})$ holds, we write

$$\begin{aligned}\Delta_0(\varkappa_{k,\nu}) - \Delta_0(\lambda_{k,0}) &= \Delta_0(\varkappa_{k,\nu}) - \Delta(\varkappa_{k,\nu})\\ &= \big(\Delta_0(\lambda_k) - \Delta_0(\lambda_{k,0})\big) + \big(\Delta_0(\lambda_{k,0}) - \Delta(\lambda_k)\big)\\ &+ \int_{\lambda_k}^{\varkappa_{k,\nu}} (\Delta_0' - \Delta')(\lambda)\,\mathrm{d}\lambda. \end{aligned} \tag{11.3}$$

Here the integration path is the straight line from λ_k to $\varkappa_{k,\nu}$, this line is contained in $U_{k,\delta}$ for $|k|$ sufficiently large. To evaluate the terms involved in Eq. (11.3), we first note the following Taylor expansion for $\Delta_0(\lambda) = 2\cos(\zeta(\lambda))$ for $\lambda \in U_{k,\delta}$:

$$\Delta_0(\lambda) = \Delta_0(\lambda_{k,0}) + (-1)^k\,(\zeta(\lambda) - \zeta(\lambda_{k,0}))^2 + O\big((\zeta(\lambda) - \zeta(\lambda_{k,0}))^4\big). \tag{11.4}$$

As a consequence, we see that we have for the left hand side of (11.3)

$$\begin{aligned}\Delta_0(\varkappa_{k,\nu}) - \Delta_0(\lambda_{k,0}) &= (-1)^k \cdot (\zeta(\varkappa_{k,\nu}) - \zeta(\lambda_{k,0}))^2\\ &\quad \cdot (1 + O((\zeta(\varkappa_{k,\nu}) - \zeta(\lambda_{k,0}))^2)). \end{aligned} \tag{11.5}$$

We now investigate the three summands on the right hand side of (11.3) individually.

First, we note that because of $\lambda_k - \lambda_{k,0} \in \ell^2_{-1,3}(k)$ we have $\zeta(\lambda_k) - \zeta(\lambda_{k,0}) \in \ell^2_{0,0}(k)$ by Proposition 6.2(1), whence it follows by the Taylor expansion equation (11.4) that

$$\Delta_0(\lambda_k) - \Delta_0(\lambda_{k,0}) \in \ell^1_{0,0}(k) \tag{11.6}$$

holds.

Second, we have

$$\begin{aligned}\Delta(\lambda_k) - \Delta_0(\lambda_{k,0}) &= \mu_k + \mu_k^{-1} - 2\,(-1)^k = (\mu_k^{1/2} - (-1)^k\,\mu_k^{-1/2})^2\\ &= \left(\frac{\mu_k - \mu_k^{-1}}{\mu_k^{1/2} + (-1)^k\,\mu_k^{-1/2}}\right)^2 = \left(\frac{(\mu_k - \mu_{k,0}) - (\mu_k^{-1} - \mu_{k,0}^{-1})}{\mu_k^{1/2} + (-1)^k\,\mu_k^{-1/2}}\right)^2. \end{aligned}$$

We have $\mu_k - \mu_{k,0} \in \ell^2_{0,0}(k)$, and therefore also $\mu_k^{-1} - \mu_{k,0}^{-1} \in \ell^2_{0,0}(k)$ (because $z \mapsto z^{-1}$ is locally Lipschitz continuous near $z = \pm 1$). Because the denominator $\mu_k^{1/2} + (-1)^k\,\mu_k^{-1/2}$ in the expression above is bounded away from zero, it follows that we have

$$\Delta(\lambda_k) - \Delta_0(\lambda_{k,0}) \in \ell^1_{0,0}(k). \tag{11.7}$$

Third, because of $\Delta - \Delta_0 \in \mathrm{As}(\mathbb{C}^*, \ell^2_{0,0}, 1)$ (Corollary 11.4(2)), we have $\Delta' - \Delta'_0 \in \mathrm{As}(\mathbb{C}^*, \ell^2_{1,-3}, 1)$ by Proposition 9.4(3), and thus we obtain

$$\begin{aligned}\left|\int_{\lambda_k}^{\varkappa_{k,\nu}} (\Delta'_0 - \Delta')(\lambda)\,\mathrm{d}\lambda\right| &\leq |\varkappa_{k,\nu} - \lambda_k| \cdot \ell^2_{1,-3}(k) = |\zeta(\varkappa_{k,\nu}) - \zeta(\lambda_k)| \cdot \ell^2_{0,0}(k) \\ &\leq \big(|\zeta(\varkappa_{k,\nu}) - \zeta(\lambda_{k,0})| + \underbrace{|\zeta(\lambda_k) - \zeta(\lambda_{k,0})|}_{\in \ell^2_{0,0}(k)}\big) \cdot \ell^2_{0,0}(k) \\ &= |\zeta(\varkappa_{k,\nu}) - \zeta(\lambda_{k,0})| \cdot \ell^2_{0,0}(k) + \ell^1_{0,0}(k). \qquad (11.8)\end{aligned}$$

By plugging Eqs. (11.5)–(11.8) into Eq. (11.3), we obtain that there exist sequences $a_k \in \ell^1_{0,0}(k)$ and $b_k \in \ell^2_{0,0}(k)$ so that with $r_k := |\zeta(\varkappa_{k,\nu}) - \zeta(\lambda_{k,0})|$ we have

$$r_k^2 = a_k + b_k \cdot r_k$$

and therefore

$$r_k = \frac{b_k}{2} \pm \sqrt{\frac{b_k^2}{4} - a_k} \in \ell^2_{0,0}(k).$$

It follows that $\varkappa_{k,\nu} - \lambda_{k,0} \in \ell^2_{-1,3}(k)$ holds. □

In the next proposition, the discriminant function $\Delta^2 - 4$ of a monodromy $M(\lambda) \in \mathsf{Mon}$ is studied more closely. Proposition 11.6(1),(2) gives a representation of $\Delta^2 - 4$ analogous to the one of the monodromy function c in Proposition 10.1; more specifically, the infinite product in Proposition 11.6(1) is analogous to Eq. (10.2), and the formula in Proposition 11.6(2) is analogous to the trace formula in Eq. (10.1). Proposition 11.6(3) gives a more detailed asymptotic description of $\Delta^2 - 4$ on the excluded domains than is provided by Corollary 11.4(3); we will use this asymptotic description in Chap. 16 to prove certain estimates in preparation for the construction of the Jacobi variety for the spectral curve Σ.

Proposition 11.6 *Let $M(\lambda) \in \mathsf{Mon}$ and $\Delta := \mathrm{tr}(M(\lambda))$; we enumerate the zeros $\varkappa_{k,\nu}$ of $\Delta^2 - 4$ as in Proposition 6.5(1).*

(1) We have

$$\Delta(\lambda)^2 - 4 = -\frac{(\lambda-\varkappa_{0,1})(\lambda-\varkappa_{0,2})}{4\lambda} \prod_{k=1}^{\infty} \frac{(\lambda-\varkappa_{k,1})\,(\lambda-\varkappa_{k,2})}{(16\pi^2 k^2)^2} \cdot \prod_{k=1}^{\infty} \frac{(\lambda-\varkappa_{-k,1})\,(\lambda-\varkappa_{-k,2})}{\lambda^2}.$$

(2) We have

$$\prod_{k\in\mathbb{Z}} \frac{\varkappa_{k,1} \cdot \varkappa_{k,2}}{\lambda_{k,0}^2} = 1.$$

(3) There exists a $C > 0$ and a sequence $a_k \in \ell^2_{0,0}(k)$ such that for all $k > 0$ and all $\lambda \in U_{k,\delta}$ we have

$$\left| \lambda_{k,0} \cdot \frac{\Delta(\lambda)^2 - 4}{(\lambda - \varkappa_{k,1}) \cdot (\lambda - \varkappa_{k,2})} - \left(-\frac{1}{16}\right) \right| \leq \frac{C}{k} \cdot \big(|\lambda - \varkappa_{k,1}| + |\lambda - \varkappa_{k,2}|\big) + a_k \, ,$$

whereas for $k < 0$ and $\lambda \in U_{k,\delta}$ we have

$$\left| \lambda^3_{k,0} \cdot \frac{\Delta(\lambda)^2 - 4}{(\lambda - \varkappa_{k,1}) \cdot (\lambda - \varkappa_{k,2})} - \left(-\frac{1}{16}\right) \right|$$
$$\leq C\, k^3 \cdot \big(|\lambda - \varkappa_{k,1}| + |\lambda - \varkappa_{k,2}|\big) + a_k .$$

Proof For (1) and (2). Let us put $f_0(\lambda) := \Delta_0(\lambda)^2 - 4 = -4 \sin(\zeta(\lambda))^2$ and $f(\lambda) := \Delta(\lambda)^2 - 4$. Then we have $f - f_0 \in \mathrm{As}(\mathbb{C}^*, \ell^2_{0,0}, 2)$ by Corollary 11.4(3).

Moreover we put

$$\widetilde{f}(\lambda) := -\frac{(\lambda - \varkappa_{0,1})(\lambda - \varkappa_{0,2})}{4\lambda} \cdot \prod_{k=1}^{\infty} \frac{(\lambda - \varkappa_{k,1})\,(\lambda - \varkappa_{k,2})}{(16\,\pi^2\,k^2)^2} \cdot \prod_{k=1}^{\infty} \frac{(\lambda - \varkappa_{-k,1})\,(\lambda - \varkappa_{-k,2})}{\lambda^2}$$
$$= -\frac{4}{\lambda} \cdot \frac{1}{\tau_1 \cdot \tau_2} \cdot c_1(\lambda) \cdot c_2(\lambda)$$

with

$$c_\nu(\lambda) := \frac{1}{4}\, \tau_\nu\, (\lambda - \varkappa_{0,\nu}) \prod_{k=1}^{\infty} \frac{\varkappa_{k,\nu} - \lambda}{16\,\pi^2\,k^2} \prod_{k=1}^{\infty} \frac{\lambda - \varkappa_{-k,\nu}}{\lambda}$$

and

$$\tau_\nu := \left(\prod_{k \in \mathbb{Z}} \frac{\lambda_{k,0}}{\varkappa_{k,\nu}} \right)^{1/2}$$

for $\nu \in \{1, 2\}$.

By Proposition 10.1(1) we have

$$c_\nu - \tau_\nu\, c_0 \in \mathrm{As}_\infty(\mathbb{C}^*, \ell^2_{-1}, 1) \quad \text{and} \quad c_\nu - \tau_\nu^{-1}\, c_0 \in \mathrm{As}_0(\mathbb{C}^*, \ell^2_1, 1)$$

and therefore

$$\widetilde{f} - f_0 \in \mathrm{As}_\infty(\mathbb{C}^*, \ell^2_0, 2) \quad \text{and} \quad \widetilde{f} - \tau_1^{-2}\, \tau_2^{-2}\, f_0 \in \mathrm{As}_0(\mathbb{C}^*, \ell^2_0, 2).$$

The holomorphic functions f and $\widetilde{f}$ have the same zeros with multiplicities, and by the preceding asymptotic estimates they satisfy the hypotheses of Proposition 6.7(3) with $\tau_\pm = 1 = \widetilde{\tau}_+$ and $\widetilde{\tau}_- = \tau_1^{-2}\,\tau_2^{-2}$. It follows from that proposition that there exists a constant $A \in \mathbb{C}^*$ with $f = A\cdot\widetilde{f}$, $1 = A\cdot 1$ and $1 = A\cdot\tau_1^{-1}\,\tau_2^{-2}$. Thus we have $\widetilde{f} = f$ and

$$1 = A = \tau_1^2\,\tau_2^2 = \prod_{k\in\mathbb{Z}} \frac{\lambda_{k,0}^2}{\varkappa_{k,1}\,\varkappa_{k,2}}.$$

For (3). We first consider the case $k > 0$. By Corollary 10.3(1), there exists for $\nu \in \{1,2\}$ a constant $C_\nu > 0$ and a sequence $a_k^{[\nu]} \in \ell^2_{0,-2}(k)$ such that we have for all $k > 0$ and all $\lambda \in U_{k,\delta}$

$$\left|\frac{c_\nu(\lambda)}{\tau_\nu^{-1}\cdot(\lambda-\varkappa_{k,\nu})} - \frac{(-1)^k}{8}\right| \le C_\nu \frac{|\lambda-\varkappa_{k,\nu}|}{k} + a_k^{[\nu]},$$

and therefore

$$\begin{aligned}
&\left|\frac{c_1(\lambda)\cdot c_2(\lambda)}{\tau_1\,\tau_2\,(\lambda-\varkappa_{k,1})\,(\lambda-\varkappa_{k,2})} - \frac{1}{64}\right| \\
&\quad\le \left|\frac{c_1(\lambda)}{\tau_1\,(\lambda-\varkappa_{k,1})}\right|\cdot\left|\frac{c_2(\lambda)}{\tau_2\,(\lambda-\varkappa_{k,2})} - \frac{(-1)^k}{8}\right| \\
&\qquad + \left|\frac{c_1(\lambda)}{\tau_1\,(\lambda-\varkappa_{k,1})} - \frac{(-1)^k}{8}\right|\cdot\left|\frac{(-1)^k}{8}\right| \\
&\quad\le C\cdot\left(\frac{|\lambda-\varkappa_{k,1}|}{k} + \frac{|\lambda-\varkappa_{k,2}|}{k}\right) + a_k
\end{aligned} \tag{11.9}$$

with $C > 0$ and $a_k \in \ell^2_0(k > 0)$. Furthermore, we have

$$\left|\frac{\lambda_{k,0}}{\lambda} - 1\right| = \frac{|\lambda_{k,0}-\lambda|}{|\lambda|} = O(k^{-1}). \tag{11.10}$$

By (1), we have

$$\Delta(\lambda)^2 - 4 = -\frac{4}{\lambda}\cdot\frac{c_1(\lambda)\cdot c_2(\lambda)}{\tau_1\,\tau_2}$$

and therefore by the estimates Eqs. (11.9) and (11.10) we obtain

$$\begin{aligned}
&\left|\lambda_{k,0}\cdot\frac{\Delta(\lambda)^2-4}{(\lambda-\varkappa_{k,1})\cdot(\lambda-\varkappa_{k,2})} - \left(-\frac{1}{16}\right)\right| \\
&\quad= \left|4\,\frac{\lambda_{k,0}}{\lambda}\,\frac{c_1(\lambda)\cdot c_2(\lambda)}{\tau_1\,\tau_2\,(\lambda-\varkappa_{k,1})\,(\lambda-\varkappa_{k,2})} - \frac{1}{16}\right|
\end{aligned}$$

$$\leq \underbrace{\left| \frac{\lambda_{k,0}}{\lambda} - 1 \right|}_{=O(k^{-1})} \cdot \underbrace{\left| 4 \frac{c_1(\lambda) \cdot c_2(\lambda)}{\tau_1 \, \tau_2 \, (\lambda - \varkappa_{k,1}) \, (\lambda - \varkappa_{k,2})} \right|}_{=O(1)}$$

$$+ 4 \underbrace{\left| \frac{c_1(\lambda) \cdot c_2(\lambda)}{\tau_1 \, \tau_2 \, (\lambda - \varkappa_{k,1}) \, (\lambda - \varkappa_{k,2})} - \frac{1}{64} \right|}_{\leq C \cdot \left(\frac{|\lambda - \varkappa_{k,1}|}{k} + \frac{|\lambda - \varkappa_{k,2}|}{k} \right) + a_k}$$

$$\leq \frac{C'}{k} \cdot \big(|\lambda - \varkappa_{k,1}| + |\lambda - \varkappa_{k,2}| \big) + a'_k$$

with a new $C' > 0$ and a new sequence $a'_k \in \ell^2_0(k > 0)$. This proves (3) for the case $k > 0$.

In the case $k < 0$ we proceed similarly. The estimate analogous to Eq. (11.9) is

$$\left| \frac{c_1(\lambda) \cdot c_2(\lambda)}{\tau_1^{-1} \, \tau_2^{-1} \, (\lambda - \varkappa_{k,1}) \, (\lambda - \varkappa_{k,2})} - \frac{1}{64} \, \lambda_{k,0}^{-2} \right|$$
$$\leq C \cdot \big(|\lambda - \varkappa_{k,1}| + |\lambda - \varkappa_{k,2}| \big) \cdot k^7 + a_k$$

with $C > 0$ and $a_k \in \ell^2_{-4}(k)$, and the estimate Eq. (11.10) also holds in the case $k < 0$.

We thus obtain (note that we have $\tau_1 \, \tau_2 = \tau_1^{-1} \, \tau_2^{-1}$ by (2))

$$\left| \lambda_{k,0}^3 \cdot \frac{\Delta(\lambda)^2 - 4}{(\lambda - \varkappa_{k,1}) \cdot (\lambda - \varkappa_{k,2})} - \left(-\frac{1}{16} \right) \right| = \left| 4 \, \frac{\lambda_{k,0}^3}{\lambda} \, \frac{c_1(\lambda) \cdot c_2(\lambda)}{\tau_1 \, \tau_2 \, (\lambda - \varkappa_{k,1}) \, (\lambda - \varkappa_{k,2})} - \frac{1}{16} \right|$$

$$\leq \underbrace{\left| \frac{\lambda_{k,0}}{\lambda} - 1 \right|}_{=O(k^{-1})} \cdot \underbrace{\left| 4 \, \lambda_{k,0}^2 \, \frac{c_1(\lambda) \cdot c_2(\lambda)}{\tau_1 \, \tau_2 \, (\lambda - \varkappa_{k,1}) \, (\lambda - \varkappa_{k,2})} \right|}_{=O(1)}$$

$$+ \underbrace{4 \, \lambda_{k,0}^2}_{=O(k^{-4})} \underbrace{\left| \frac{c_1(\lambda) \cdot c_2(\lambda)}{\tau_1^{-1} \, \tau_2^{-1} \, (\lambda - \varkappa_{k,1}) \, (\lambda - \varkappa_{k,2})} - \frac{1}{64} \, \lambda_{k,0}^{-2} \right|}_{\leq C \cdot (|\lambda - \varkappa_{k,1}| + |\lambda - \varkappa_{k,2}|) \cdot k^7 + a_k}$$

$$\leq C' \, k^3 \cdot \big(|\lambda - \varkappa_{k,1}| + |\lambda - \varkappa_{k,2}| \big) + a'_k$$

with new constants $C' > 0$ and $a'_k \in \ell^2_0(k < 0)$. This proves also the case $k < 0$.

□

Finally we consider the extended frame $F(x, \lambda) := F_\lambda(x)$ associated to a potential $(u, u_y) \in \mathsf{Pot}_{np}$, see Chap. 2. For the construction of the Darboux coordinates on the space of (periodic) potentials in Chap. 14, we need an asymptotic estimate for $F(x, \lambda)$ that is uniform in $x \in [0, 1]$, and which is analogous to the asymptotics we have for the monodromy $M(\lambda) = F(1, \lambda)$. The following proposition provides the necessary result:

Proposition 11.7 *Let* $(u, u_y) \in \mathsf{Pot}_{np}$ *and* $F(x,\lambda) := \begin{pmatrix} a(x,\lambda) & b(x,\lambda) \\ c(x,\lambda) & d(x,\lambda) \end{pmatrix}$ *be the extended frame associated to* (u, u_y)*. We also consider the extended frame of the vacuum*

$$F_0(x,\lambda) := \begin{pmatrix} a_0(x,\lambda) & b_0(x,\lambda) \\ c_0(x,\lambda) & d_0(x,\lambda) \end{pmatrix} := \begin{pmatrix} \cos(x\,\zeta(\lambda)) & -\lambda^{-1/2}\sin(x\,\zeta(\lambda)) \\ \lambda^{1/2}\sin(x\,\zeta(\lambda)) & \cos(x\,\zeta(\lambda)) \end{pmatrix}$$

and put $\tau(x) := \mathrm{e}^{-(u(0)+u(x))/4}$, $\upsilon(x) := \mathrm{e}^{(u(x)-u(0))/4}$.

Then for every $x \in [0,1]$ *we have*

(1)

$$a(x,\lambda) - \upsilon(x)\,a_0(x,\lambda) \in \mathrm{As}_\infty(\mathbb{C}^*, \ell^2_0, 1)$$
$$b(x,\lambda) - \tau(x)^{-1}\,b_0(x,\lambda) \in \mathrm{As}_\infty(\mathbb{C}^*, \ell^2_1, 1)$$
$$c(x,\lambda) - \tau(x)\,c_0(x,\lambda) \in \mathrm{As}_\infty(\mathbb{C}^*, \ell^2_{-1}, 1)$$
$$d(x,\lambda) - \upsilon(x)^{-1}\,d_0(x,\lambda) \in \mathrm{As}_\infty(\mathbb{C}^*, \ell^2_0, 1)$$

(2)

$$a(x,\lambda) - \upsilon(x)^{-1}\,a_0(x,\lambda) \in \mathrm{As}_0(\mathbb{C}^*, \ell^2_0, 1)$$
$$b(x,\lambda) - \tau(x)\,b_0(x,\lambda) \in \mathrm{As}_0(\mathbb{C}^*, \ell^2_{-1}, 1)$$
$$c(x,\lambda) - \tau(x)^{-1}\,c_0(x,\lambda) \in \mathrm{As}_0(\mathbb{C}^*, \ell^2_1, 1)$$
$$d(x,\lambda) - \upsilon(x)\,d_0(x,\lambda) \in \mathrm{As}_0(\mathbb{C}^*, \ell^2_0, 1)\,,$$

and these asymptotic estimates are uniform in the sense that there exist bounding sequences that are suitable for all $x \in [0,1]$.

Proof For given $x_0 \in [0,1]$, we consider the functions $\widetilde{u}, \widetilde{u}_y$ defined by

$$\widetilde{u}(t) = \begin{cases} u(t) & \text{for } 0 \le t \le x_0 \\ u(x_0) & \text{for } x_0 < t \le 1 \end{cases} \quad \text{and} \quad \widetilde{u}_y(t) = \begin{cases} u_y(t) & \text{for } 0 \le t \le x_0 \\ 0 & \text{for } x_0 < t \le 1 \end{cases}$$

for $t \in [0,1]$. Then we have $(\widetilde{u}, \widetilde{u}_y) \in \mathsf{Pot}_{np}$.

Let us denote by $\widetilde{F}(t,\lambda)$ the extended frame of the potential $(\widetilde{u}, \widetilde{u}_y)$, and let us denote by $\widehat{F}(t,\lambda)$ the extended frame of the constant potential $(\widehat{u} = u(x_0), \widehat{u}_y = 0) \in \mathsf{Pot}$. (For the geometric meaning of this constant potential, refer to Remark 6.6.) Then we have for $t \in [0,1]$

$$\widetilde{F}(t,\lambda) = \begin{cases} F(t,\lambda) & \text{for } 0 \le t \le x_0 \\ \widehat{F}(t - x_0, \lambda) \cdot F(x_0,\lambda) & \text{for } x_0 < t \le 1 \end{cases},$$

in particular

$$F(x_0, \lambda) = \widehat{F}(1 - x_0, \lambda)^{-1} \cdot \widetilde{F}(1, \lambda).$$

By Theorem 11.2, we have $\widetilde{F}(1, \lambda) = \widetilde{M}(\lambda) \in \mathsf{Mon}_{\tau = \mathrm{e}^{-(u(0)+u(x_0))/4}, \upsilon = \mathrm{e}^{(u(x_0)-u(0))/4}}$, and because the set P of potentials $(\widetilde{u}, \widetilde{u}_y)$, where x_0 runs through $[0, 1]$, is relatively compact in Pot, there exists bounding sequences corresponding to this asymptotic that are independent of $x_0 \in [0, 1]$.

Concerning the extended frame $\widehat{F}(t, \lambda)$ of the constant potential, we note that the corresponding 1-form

$$\widehat{\alpha}_\lambda = \frac{1}{4} \begin{pmatrix} 0 & -\mathrm{e}^{u(x_0)/2} - \lambda^{-1} \mathrm{e}^{-u(x_0)/2} \\ \mathrm{e}^{u(x_0)/2} + \lambda \, \mathrm{e}^{-u(x_0)/2} & 0 \end{pmatrix} \mathrm{d}x$$

is independent of t, and therefore can be calculated explicitly as

$$\widehat{F}(t, \lambda) = \exp(t \cdot \widehat{\alpha}_\lambda) = \begin{pmatrix} \cos(t \cdot \xi(\lambda)) & -\beta(\lambda)^{-1} \cdot \sin(t \cdot \xi(\lambda)) \\ \beta(\lambda) \cdot \sin(t \cdot \xi(\lambda)) & \cos(t \cdot \xi(\lambda)) \end{pmatrix}$$

with

$$\xi(\lambda) := \frac{1}{4} \sqrt{(\lambda \, \mathrm{e}^{-u(x_0)/2} + \mathrm{e}^{u(x_0)/2}) \cdot (\lambda^{-1} \mathrm{e}^{-u(x_0)/2} + \mathrm{e}^{u(x_0)/2})}$$

$$\text{and} \quad \beta(\lambda) := \sqrt{\frac{\mathrm{e}^{u(x_0)/2} + \lambda \, \mathrm{e}^{-u(x_0)/2}}{\mathrm{e}^{u(x_0)/2} + \lambda^{-1} \mathrm{e}^{-u(x_0)/2}}}$$

(compare the proof of Proposition 6.5(2)). An explicit calculation shows that $\widetilde{F}(t, \lambda)$ is asymptotically close to

$$\begin{pmatrix} a_0(t,\lambda) & \mathrm{e}^{u(x_0)/2} \, b_0(t,\lambda) \\ \mathrm{e}^{-u(x_0)/2} \, c_0(t,\lambda) & d_0(t,\lambda) \end{pmatrix} \quad \text{resp. to} \quad \begin{pmatrix} a_0(t,\lambda) & \mathrm{e}^{-u(x_0)/2} \, b_0(t,\lambda) \\ \mathrm{e}^{u(x_0)/2} \, c_0(t,\lambda) & d_0(t,\lambda) \end{pmatrix}$$

for $\lambda \to \infty$ resp. for $\lambda \to 0$.

It follows that for $\lambda \to \infty$, $F(x_0, \lambda) = \widehat{F}(1 - x_0, \lambda)^{-1} \cdot \widetilde{F}(1, \lambda)$ is for $\lambda \to \infty$ asymptotically close to (where we abbreviate $\zeta := \zeta(\lambda)$)

$$\begin{pmatrix} a_0(1 - x_0, \lambda) & \mathrm{e}^{u(x_0)/2} \, b_0(1 - x_0, \lambda) \\ \mathrm{e}^{-u(x_0)/2} \, c_0(1 - x_0, \lambda) & d_0(1 - x_0, \lambda) \end{pmatrix}^{-1} \cdot \begin{pmatrix} \mathrm{e}^{(u(x_0)-u(0))/4} \, a_0(1, \lambda) & \mathrm{e}^{(u(0)+u(x_0))/4} \, b_0(1, \lambda) \\ \mathrm{e}^{-(u(0)+u(x_0))/4} \, c_0(1, \lambda) & \mathrm{e}^{-(u(x_0)-u(0))/4} \, d_0(1, \lambda) \end{pmatrix}$$

$$= \begin{pmatrix} d_0(1 - x_0, \lambda) & -\mathrm{e}^{u(x_0)/2} \, b_0(1 - x_0, \lambda) \\ -\mathrm{e}^{-u(x_0)/2} \, c_0(1 - x_0, \lambda) & a_0(1 - x_0, \lambda) \end{pmatrix} \cdot \begin{pmatrix} \mathrm{e}^{(u(x_0)-u(0))/4} \, a_0(1, \lambda) & \mathrm{e}^{(u(0)+u(x_0))/4} \, b_0(1, \lambda) \\ \mathrm{e}^{-(u(0)+u(x_0))/4} \, c_0(1, \lambda) & \mathrm{e}^{-(u(x_0)-u(0))/4} \, d_0(1, \lambda) \end{pmatrix}$$

$$= \left(\begin{matrix} \mathrm{e}^{(u(x_0)-u(0))/4} \cos((1 - x_0)\zeta) \cos(\zeta) + \mathrm{e}^{(u(x_0)-u(0))/4} \sin((1 - x_0)\zeta) \sin(\zeta) \\ -\mathrm{e}^{-(u(0)+u(x_0))/4} \lambda^{1/2} \sin((1 - x_0)\zeta) \cos(\zeta) + \mathrm{e}^{-(u(0)+u(x_0))/4} \cos((1 - x_0)\zeta) \lambda^{1/2} \sin(\zeta) \end{matrix} \right.$$

$$\left. \begin{matrix} -\mathrm{e}^{(u(0)+u(x_0))/4} \cos((1 - x_0)\zeta) \lambda^{-1/2} \sin(\zeta) + \mathrm{e}^{(u(x_0)+u(0))/4} \lambda^{-1/2} \sin((1 - x_0)\zeta) \cos(\zeta) \\ \mathrm{e}^{(u(0)-u(x_0))/4} \lambda^{1/2} \sin((1 - x_0)\zeta) \lambda^{-1/2} \sin(\zeta) + \mathrm{e}^{-(u(x_0)-u(0))/4} \cos((1 - x_0)\zeta) \cos(\zeta) \end{matrix} \right)$$

$$= \begin{pmatrix} e^{(u(x_0)-u(0))/4}\cos(x_0\,\zeta) & -e^{(u(0)+u(x_0))/4}\lambda^{-1/2}\sin(x_0\,\zeta) \\ e^{-(u(0)+u(x_0))/4}\lambda^{1/2}\sin(x_0\,\zeta) & e^{-(u(x_0)-u(0))/4}\cos(x_0\,\zeta) \end{pmatrix}$$

$$= \begin{pmatrix} \upsilon(x_0)\,a_0(x_0,\lambda) & \tau(x_0)^{-1}\,b_0(x_0,\lambda) \\ \tau(x_0)\,c_0(x_0,\lambda) & \upsilon(x_0)^{-1}\,d_0(x_0,\lambda) \end{pmatrix}.$$

An analogous calculation shows that for $x \to 0$, $F(x_0,\lambda)$ is asymptotically close to

$$\begin{pmatrix} \upsilon(x_0)^{-1}\,a_0(x_0,\lambda) & \tau(x_0)\,b_0(x_0,\lambda) \\ \tau(x_0)^{-1}\,c_0(x_0,\lambda) & \upsilon(x_0)\,d_0(x_0,\lambda) \end{pmatrix}.$$

□

The main reason why we considered non-periodic potentials and monodromies was to facilitate the proof of Proposition 11.7. Therefore we will no longer consider non-periodically asymptotic objects from here on, i.e. we will always suppose $\upsilon = 1$ from now on.

Chapter 12
Non-special Divisors and the Inverse Problem for the Monodromy

In this chapter we characterize the pairs $(\Sigma, \mathscr{D})$ which uniquely define a monodromy $M(\lambda) \in \mathsf{Mon}$ such that $(\Sigma, \mathscr{D})$ is the spectral data of $M(\lambda)$. Remember that in Chap. 3, we associated to any $M(\lambda) \in \mathsf{Mon}$ the trace function $\Delta(\lambda) := \operatorname{tr}(M(\lambda))$ and thereby the spectral curve

$$\Sigma := \{ (\lambda, \mu) \in \mathbb{C}^* \times \mathbb{C} \mid \mu^2 - \Delta(\lambda) \cdot \mu + 1 = 0 \}, \tag{12.1}$$

and moreover the spectral divisor $\mathscr{D}$, which is positive and asymptotic.

We now consider the inverse problem for the construction of the spectral data $(\Sigma, \mathscr{D})$ from the monodromy $M(\lambda)$. We will see that the view of the spectral data from Chap. 3, namely taking the *generalized* divisor of the eigenline bundle of $M(\lambda)$ as the spectral divisor, will be fundamentally important for solving this inverse problem.

More explicitly, we let a pair $(\Sigma, \mathscr{D})$ be given, and ask if there exists a monodromy $M(\lambda) \in \mathsf{Mon}$ such that $(\Sigma, \mathscr{D})$ is the spectral data of $M(\lambda)$. Here we suppose that Σ is a spectral curve "of the type we consider", i.e. that there exists a holomorphic function $\Delta : \mathbb{C}^* \to \mathbb{C}$ with $\Delta - \Delta_0 \in \mathrm{As}(\mathbb{C}^*, \ell^2_{0,0}, 1)$ so that the complex curve Σ is given by Eq. (12.1). We further suppose that $\mathscr{D}$ is a generalized divisor on Σ, i.e. a subsheaf of the sheaf of meromorphic functions on Σ that is finitely generated over the sheaf $\mathscr{O}$ of holomorphic functions on Σ; moreover we require $\mathscr{D}$ to be positive (i.e. $\mathscr{O} \subset \mathscr{D}$) and asymptotic in the sense of Definition 8.3.

However it turns out that two further conditions need to be imposed on $\mathscr{D}$ so that $(\Sigma, \mathscr{D})$ is the spectral data of a monodromy $M(\lambda) \in \mathsf{Mon}$. The first of these conditions, which we call "$\mathscr{D}$ is *compatible*" below, is a condition concerning the local structure of spectral divisors at singularities of Σ, namely precisely the property that has been shown for spectral divisors in Proposition 3.7(2). Thus every spectral divisor $\mathscr{D}$ also is compatible.

S. Klein, *A Spectral Theory for Simply Periodic Solutions of the Sinh-Gordon Equation*, Lecture Notes in Mathematics 2229,
https://doi.org/10.1007/978-3-030-01276-2_12

The second condition that we need to impose concerns also the asymptotic behavior of $\mathscr{D}$, namely that $\mathscr{D}$ is non-special (which is also required in the case of finite type divisors). We have not yet shown that all spectral divisors are non-special (because the definition of specialty involves the asymptotic behavior near $\lambda = \infty$ resp. $\lambda = 0$, we could not have done so in Chap. 3), but this follows from Theorem 12.3 below, which states that a compatible, asymptotic, positive generalized divisor $\mathscr{D}$ is a spectral divisor corresponding to some monodromy $M(\lambda)$ if and only if it is non-special.

As in Chap. 3, we let $\mathscr{O}$ be the sheaf of holomorphic functions on Σ, $\widehat{\Sigma}$ be the normalization of Σ, and $\widehat{\mathscr{O}}$ be the direct image onto Σ of the sheaf of holomorphic functions on $\widehat{\Sigma}$. For a positive generalized divisor $\mathscr{D}$ we call the classical divisor

$$D : \Sigma \to \mathbb{N}_0,\ (\lambda, \mu) \mapsto \dim(\mathscr{D}_{(\lambda,\mu)}/\mathscr{O}_{(\lambda,\mu)})$$

the underlying classical divisor of Σ, and in Definition 8.3 we described what it means for $\mathscr{D}$ resp. D to be asymptotic. In the sequel, we will also use other notations from Chap. 3.

Definition 12.1 Let $\mathscr{D}$ be an asymptotic, positive generalized divisor on Σ.

(1) We call $\mathscr{D}$ *compatible*, if for every singular point $(\lambda_*, \mu_*) \in \Sigma$ that is in the support of $\mathscr{D}$, say with degree $m := \dim(\mathscr{D}_{(\lambda_*,\mu_*)}/\mathscr{O}_{(\lambda_*,\mu_*)}) \geq 1$, there exists $g \in \mathscr{D}_{(\lambda_*,\mu_*)}$ so that $\eta := g - \frac{\mu-\mu^{-1}}{(\lambda-\lambda_*)^m}$ is a meromorphic function germ in λ alone, with $\operatorname{polord}^{\Sigma}(\eta) \leq \operatorname{polord}^{\Sigma}(g)$.
(2) We call $\mathscr{D}$ *special*, if $\mathscr{D}$ contains a section f that is a non-constant meromorphic function on Σ which is bounded on $\widehat{V}_\delta$ for some $\delta > 0$. Otherwise we say that $\mathscr{D}$ is *non-special*.

We prepare the proof of the theorem on the reconstruction of the monodromy from the spectral data (Theorem 12.3) with the following lemma:

Lemma 12.2 *Let $\mathscr{D}$ be a compatible, positive generalized divisor on Σ, and let $(\lambda_*, \mu_*) \in \Sigma$ be in the support of $\mathscr{D}$, i.e. $m := \dim(\mathscr{D}_{(\lambda_*,\mu_*)}/\mathscr{O}_{(\lambda_*,\mu_*)}) \geq 1$ holds. We further suppose that $(\lambda - \lambda_*)^{-1}$ is* not *a section of $\mathscr{D}$.*

Moreover, we suppose that holomorphic function germs c, d in λ at $\lambda = \lambda_*$ are given with $\operatorname{ord}^{\mathbb{C}}_{\lambda_*}(c) = m$. We give a characterization of the condition that $\frac{\mu-d}{c} \in \mathscr{D}_{(\lambda_*,\mu)}$ holds for "all" (one or two) values of $\mu \in \mathbb{C}^*$ with $(\lambda_*, \mu) \in \Sigma$; for this we distinguish between regular and singular points (λ_*, μ_*) of Σ.

(1) Suppose that (λ_*, μ_*) is a regular point of Σ. If (λ_*, μ_*) is not a branch point (i.e. $\Delta(\lambda_*)^2 - 4 \neq 0$ holds), then the holomorphic function μ^{-1} on Σ can near (λ_*, μ_*) be represented as a holomorphic function $\vartheta(\lambda)$ in λ, and we have $\frac{\mu-d}{c} \in \mathscr{D}_{(\lambda_*,\mu_*)}$ and $\frac{\mu-d}{c} \in \mathscr{D}_{(\lambda_*,\mu_*^{-1})}$ if and only if we have

$$d^{(k)}(\lambda_*) = \vartheta^{(k)}(\lambda_*) \quad \textit{for } k \in \{0, \ldots, m-1\}. \tag{12.2}$$

If (λ_*, μ_*) *is a branch point (i.e.* $\Delta^2 - 4$ *has a simple zero at* λ_**) then we have* $m = 1$*, and* $\frac{\mu-d}{c} \in \mathscr{D}_{(\lambda_*,\mu_*)}$ *holds if and only if we have*

$$d(\lambda_*) = \mu_*^{-1}. \tag{12.3}$$

(2) Suppose that (λ_*, μ_*) *is a singular point of* Σ *(i.e.* $\Delta^2 - 4$ *has a zero of order* $\widehat{n} \geq 2$ *at* λ_**); we let* $g \in \mathscr{D}_{(\lambda_*,\mu_*)}$ *be so that* $\eta := g - \frac{\mu-\mu^{-1}}{(\lambda-\lambda_*)^m}$ *is a meromorphic function germ in* λ *alone, with* $\operatorname{polord}^{\Sigma}(\eta) \leq \operatorname{polord}^{\Sigma}(g)$*. Then* $m \leq \widehat{n} - j_0$ *holds,* $\widetilde{\eta} := \frac{1}{2}\big(\Delta - (\lambda - \lambda_*)^m \cdot \eta\big)$ *is a holomorphic function germ in* λ*, and* $\frac{\mu-d}{c} \in \mathscr{D}_{(\lambda_*,\mu_*)}$ *holds if and only if we have*

$$d^{(k)}(\lambda_*) = \widetilde{\eta}^{(k)}(\lambda_*) \quad \textit{for } k \in \{0, \ldots, m-1\}. \tag{12.4}$$

Addendum. *If the condition of Eq.* (12.2), (12.3) *or* (12.4) *holds in the setting of (1) resp. (2), then the function germ* $b := \frac{(\Delta-d)\cdot d - 1}{c}$ *is holomorphic at* λ_**.*

Proof For (1). If (λ_*, μ_*) is a regular point of Σ, then $\mathscr{D}_{(\lambda_*,\mu_*)}$ is completely characterized by the underlying classical divisor, therefore a meromorphic function germ f is in $\mathscr{D}_{(\lambda_*,\mu_*)}$ if and only if it has at (λ_*, μ_*) at most a pole of order m. Let us now first suppose that (λ_*, μ_*) is not a branch point of Σ. We then have $\mu_*^{-1} \neq \mu_*$, and because of the hypothesis that $(\lambda - \lambda_*)^{-1}$ is not a section of $\mathscr{D}$, we see that (λ_*, μ_*^{-1}) is not in the support of $\mathscr{D}$. Therefore $f \in \mathscr{D}_{(\lambda_*,\mu_*^{-1})}$ holds if and only if f is holomorphic at (λ_*, μ_*). Moreover, $\mathscr{O}_{(\lambda_*,\mu_*)}$ is generated by λ as a ring, and therefore the holomorphic function μ^{-1} can near (λ_*, μ_*) be represented as a holomorphic function $\vartheta(\lambda)$ in λ. In any event, $\frac{\mu-d}{c}$ has at (λ_*, μ_*) a pole of order at most m, and therefore $\frac{\mu-d(\lambda)}{c(\lambda)} \in \mathscr{D}_{(\lambda_*,\mu_*)}$ holds. In this situation, Eq. (12.2) is equivalent to the fact that the holomorphic function germ $\mu - d(\lambda)$ has at (λ_*, μ_*^{-1}) a zero of order at least m, and this is in turn equivalent to the fact that $\frac{\mu-d(\lambda)}{c(\lambda)}$ is holomorphic at (λ_*, μ_*^{-1}), which is equivalent to $\frac{\mu-d(\lambda)}{c(\lambda)} \in \mathscr{D}_{(\lambda_*,\mu_*)}$. Moreover, we have

$$(\Delta - d)\cdot d - 1 = (\mu + \mu^{-1} - d)\cdot d - \mu \cdot \mu^{-1} = -(\mu - d)\cdot(\mu^{-1} - d), \tag{12.5}$$

therefore the function $(\Delta - d)\cdot d - 1$ has a zero of order at least m at λ_*. Because c also has a zero of order m at λ_*, it follows that b is holomorphic at λ_*.

Now suppose that (λ_*, μ_*) is a regular branch point of Σ. Then $\mu_* = \mu_*^{-1}$ holds, and because $\mathrm{d}\lambda$ has a simple zero at λ_*, the function $(\lambda - \lambda_*)^{-1}$ has a pole of order 2 when regarded as a meromorphic function on Σ. Therefore $m \geq 2$ would imply that $(\lambda - \lambda_*)^{-1}$ is a section of $\mathscr{D}$, in contradiction to the hypothesis. Hence $m = 1$ holds. Thus c has a simple zero as a function in λ, but a zero of order 2 as a holomorphic function germ on Σ. In this situation, the equation $d(\lambda_*) = \mu_*^{-1} = \mu_*$ is equivalent to the fact that $\mu - d(\lambda)$ has (at least) a simple zero at (λ_*, μ_*) (because μ is a coordinate on Σ near (λ_*, μ_*)); this is equivalent to

$\frac{\mu-d}{c}$ having at most a pole of order 1 at (λ_*, μ_*) and therefore to $\frac{\mu-d}{c} \in \mathscr{D}_{(\lambda_*,\mu_*)}$. Moreover, because $\mu^{-1} - d$ also has a zero at (λ_*, μ_*), Eq. (12.5) shows that the function $(\Delta - d) \cdot d - 1$ has as function on Σ a zero of order at least 2, and therefore as function in λ a zero of order at least 1. Because c also has a zero of order $m = 1$ as function in λ, it follows again that b is holomorphic.

For (2). Because $\mathscr{D}$ is compatible, there exists $g \in \mathscr{D}_{(\lambda_*,\mu_*)}$ so that $\eta := g - \frac{\mu-\mu^{-1}}{(\lambda-\lambda_*)^m}$ is a meromorphic function germ in λ alone, with $\operatorname{polord}^\Sigma(\eta) \leq \operatorname{polord}^\Sigma(g)$. Here the pole order of η (as a function of λ) is at most m, and therefore $(\lambda - \lambda_*)^m \cdot \eta$ and hence also $\widetilde{\eta}$ is holomorphic. Moreover, because c has at λ_* a zero of order m, there exists an invertible $\gamma \in \mathscr{O}_{\lambda_*}(\lambda)$ with $c = \gamma \cdot (\lambda - \lambda_*)^m$. We now calculate

$$\begin{aligned}
\frac{\mu - d}{c} - \frac{1}{2\gamma}\, g &= \frac{\mu - d}{c} - \frac{1}{2\gamma}\left(\eta + \frac{\mu - \mu^{-1}}{(\lambda - \lambda_*)^m}\right) \\
&= \frac{\mu - d}{c} - \frac{\mu - \mu^{-1}}{2\gamma\,(\lambda - \lambda_*)^m} - \frac{(\lambda - \lambda_*)^m \cdot \eta}{2\gamma\,(\lambda - \lambda_*)^m} \\
&= \frac{2\mu - 2d - \mu + \mu^{-1} - (\lambda - \lambda_*)^m \cdot \eta}{2\,c} \\
&= \frac{\mu + \mu^{-1} - 2d - (\lambda - \lambda_*)^m \cdot \eta}{2\,c} \\
&= \frac{\Delta - 2d - (\lambda - \lambda_*)^m \cdot \eta}{2\,c} = \frac{\widetilde{\eta} - d}{c}.
\end{aligned}$$

Because we have $\frac{1}{2\gamma} g \in \mathscr{D}_{(\lambda_*,\mu_*)}$, this calculation shows that $\frac{\mu-d}{c} \in \mathscr{D}_{(\lambda_*,\mu_*)}$ holds if and only if $\frac{\widetilde{\eta}-d}{c} \in \mathscr{D}_{(\lambda_*,\mu_*)}$ holds. Because the latter germ is meromorphic in λ alone, and we have $(\lambda - \lambda_*)^{-1} \notin \mathscr{D}_{(\lambda_*,\mu_*)}$ by hypothesis, $\frac{\widetilde{\eta}-d}{c} \in \mathscr{D}_{(\lambda_*,\mu_*)}$ holds if and only if $\frac{\widetilde{\eta}-d}{c}$ is holomorphic, and this is true if and only if $\widetilde{\eta} - d$ has a zero of order at least m (with respect to λ). The latter condition is equivalent to (12.4).

For the proof of the Addendum, we let $\widehat{n}$ be the order of the singularity (λ_*, μ_*) of Σ, i.e. $\widehat{n}$ is the order of the zero of $\Delta^2 - 4$ at λ_*. In the sequel, we will distinguish between the cases that $\widehat{n} = 2n + 1$ is odd, and that $\widehat{n} = 2n$ is even. Moreover let $s \geq 0$ be the maximal pole order that occurs in $\mathscr{D}_{(\lambda_*,\mu_*)}$. As a first stage for the proof, we will show that

$$s \leq \widehat{n} - 2j_0 \tag{12.6}$$

holds, where $j_0 \in \{0, \ldots, n\}$ is the number from Proposition 3.3. Let $f \in \mathscr{D}_{(\lambda_*,\mu_*)}$ be of pole order s. By the fact that $\mathscr{D}$ is a $\mathscr{O}$-module and by Proposition 3.3 it then follows that

$$(\lambda - \lambda_*)^{n-j_0} \cdot f,\ \frac{\mu - \mu^{-1}}{(\lambda - \lambda_*)^{j_0}} \cdot f \in \mathscr{D}_{(\lambda_*,\mu_*)}. \tag{12.7}$$

If $\widehat{n} = 2n+1$ is odd, then (12.7) means that

$$\sqrt{\lambda-\lambda_*}^{\,2n-2j_0}\cdot f,\ \sqrt{\lambda-\lambda_*}^{\,2n-2j_0+1}\cdot f\ \in \mathscr{D}_{(\lambda_*,\mu_*)}$$

and therefore

$$\sqrt{\lambda-\lambda_*}^{\,-s+2n-2j_0}\cdot \widehat{\mathscr{O}}_{(\lambda_*,\mu_*)} \subset \mathscr{D}_{(\lambda_*,\mu_*)}$$

holds, where $\widehat{\mathscr{O}}$ denotes the direct image of the sheaf of holomorphic functions on the normalization of Σ. If $-s+2n-2j_0 \leq -2$ were true, this would imply $(\lambda-\lambda_*)^{-1} \in \mathscr{D}_{(\lambda_*,\mu_*)}$, in contradiction to the hypothesis. Thus we have $-s+2n-2j_0 \geq -1$ and therefore (12.6) holds in this case.

If $\widehat{n} = 2n$ is even, then we represent $f = (f_1, f_2)$ as a pair of germs of meromorphic functions in λ at the two points above (λ_*, μ_*) in the normalization of Σ, and denote the pole order of f_ν by s_ν ($\nu \in \{1,2\}$), ordered such that $s_1 \geq s_2$ holds. Then we have $s_1 + s_2 = s$, and (12.7) means that

$$(\lambda-\lambda_*)^{-s_1+n-j_0}\cdot \widehat{\mathscr{O}}_{(\lambda_*,\mu_*)} \subset \mathscr{D}_{(\lambda_*,\mu_*)}$$

holds. If $-s_1+n-j_0 \leq -1$ were true, this would imply $(\lambda-\lambda_*)^{-1} \in \mathscr{D}_{(\lambda_*,\mu_*)}$, in contradiction to the hypothesis. Therefore we have $-s_1+n-j_0 \geq 0$, whence (12.6) follows also in this case.

It follows from (12.6) by Proposition 3.5 that

$$m = s + j_0 \leq \widehat{n} - j_0 \tag{12.8}$$

holds.

To show that b is holomorphic in λ, we note that by (12.4), $d - \widetilde{\eta}$ has at λ_* a zero of order m (with respect to λ), so we can represent d in the form $d = \widetilde{\eta} + (\lambda-\lambda_*)^m \cdot \zeta$ with a holomorphic function germ ζ in λ. We now calculate:

$$\begin{aligned} b = \frac{(\varDelta - d)\cdot d - 1}{c} &= \frac{(\varDelta - \widetilde{\eta} - (\lambda-\lambda_*)^m\cdot\zeta)\cdot(\widetilde{\eta} + (\lambda-\lambda_*)^m\cdot\zeta) - 1}{\gamma\cdot(\lambda-\lambda_*)^m} \\ &= \frac{1}{\gamma}\cdot\left(\frac{(\varDelta-\widetilde{\eta})\cdot\widetilde{\eta} - 1}{(\lambda-\lambda_*)^m} + \varDelta\,\zeta - 2\,\widetilde{\eta}\,\zeta - (\lambda-\lambda_*)^m\,\zeta^2\right). \end{aligned} \tag{12.9}$$

Because $\varDelta\,\zeta - 2\,\widetilde{\eta}\,\zeta - (\lambda-\lambda_*)^m\,\zeta^2$ is obviously holomorphic, it remains to show that $\frac{(\varDelta-\widetilde{\eta})\cdot\widetilde{\eta}-1}{(\lambda-\lambda_*)^m}$ is holomorphic. For this purpose, we continue to calculate:

$$\begin{aligned} \frac{(\varDelta-\widetilde{\eta})\cdot\widetilde{\eta}-1}{(\lambda-\lambda_*)^m} &= \frac{\big(\varDelta - \tfrac{1}{2}(\varDelta - (\lambda-\lambda_*)^m\eta)\big)\cdot\tfrac{1}{2}\big(\varDelta - (\lambda-\lambda_*)^m\eta\big) - 1}{(\lambda-\lambda_*)^m} \\ &= \frac{1}{4}\,\frac{\big(\varDelta + (\lambda-\lambda_*)^m\eta\big)\cdot\big(\varDelta - (\lambda-\lambda_*)^m\eta\big) - 4}{(\lambda-\lambda_*)^m} \end{aligned}$$

$$= \frac{1}{4} \frac{\Delta^2 - \left((\lambda - \lambda_*)^m \, \eta\right)^2 - 4}{(\lambda - \lambda_*)^m}$$

$$= \frac{1}{4} \left(\frac{\Delta^2 - 4}{(\lambda - \lambda_*)^m} - (\lambda - \lambda_*)^m \cdot \eta^2 \right) . \tag{12.10}$$

$\Delta^2 - 4$ has (with respect to λ) a zero of order $\widehat{n}$, whereas $(\lambda - \lambda_*)^m$ has by (12.8) a zero of order at most $\widehat{n} - j_0 \leq \widehat{n}$. Therefore $\frac{\Delta^2-4}{(\lambda-\lambda_*)^m}$ is holomorphic in λ. Moreover, we have

$$\operatorname{polord}^{\mathbb{C}}(\eta^2) = \operatorname{polord}^{\Sigma}(\eta) \leq \operatorname{polord}^{\Sigma}(g) \leq s = m - j_0 \leq m$$

(where the last equals sign again follows from Proposition 3.5), and hence $(\lambda - \lambda_*)^m \cdot \eta^2$ is also holomorphic. Therefore it follows from Eq. (12.10) that $\frac{(\Delta - \widetilde{\eta}) \cdot \widetilde{\eta} - 1}{(\lambda - \lambda_*)^m}$ is holomorphic, and thus we see from Eq. (12.9) that b is holomorphic. □

Now we are ready to prove that any non-special, compatible, asymptotic, positive generalized divisor on Σ is the spectral divisor of an asymptotic monodromy $M(\lambda) \in \mathsf{Mon}$. This is the content of the implication (a) ⇒ (c) of the following theorem. Also note the equivalence (a) ⇔ (b) of the theorem, which shows that for a compatible, asymptotic, positive divisor $\mathscr{D}$ the property that $\mathscr{D}$ is non-special (which is by its definition related to the behavior of $\mathscr{D}$ near $\lambda = 0$ and $\lambda = \infty$) can be characterized in terms of the local behavior of $\mathscr{D}$ at the points of its support.

Theorem 12.3 *Let $\mathscr{D}$ be a compatible, asymptotic, positive generalized divisor on Σ. Then the following statements are equivalent:*

(a) $\mathscr{D}$ is non-special.
(b) There does not exist any $\lambda_* \in \mathbb{C}^*$ so that the meromorphic function $(\lambda - \lambda_*)^{-1}$ on Σ is a section of $\mathscr{D}$.
(c) There exists a monodromy $M(\lambda) = \begin{pmatrix} a(\lambda) & b(\lambda) \\ c(\lambda) & d(\lambda) \end{pmatrix} \in \mathsf{Mon}$ with $a + d = \Delta$, such that $\mathscr{D}$ is the spectral divisor of $M(\lambda)$, i.e. $\mathscr{D}$ is generated by 1 and $\frac{\mu - d}{c}$ over $\mathscr{O}$. Moreover, $M(\lambda)$ is determined uniquely up to a joint change of sign of the functions b and c.
(d) There is one and only one global meromorphic function f on Σ that is a section of $\mathscr{D}$ and such that $\left(f - \frac{\mathrm{i}}{\tau\sqrt{\lambda}}\right) | \widehat{V}_\delta \in \mathrm{As}_\infty(\widehat{V}_\delta, \ell^2_1, 0)$ and $\left(f - \frac{\mathrm{i}}{\tau^{-1}\sqrt{\lambda}}\right) | \widehat{V}_\delta \in \mathrm{As}_0(\widehat{V}_\delta, \ell^2_{-1}, 0)$ holds for some $\delta > 0$. (Here $\sqrt{\lambda}$ is a fixed holomorphic branch of the square root function on $\widehat{V}_\delta$, and $\tau \in \mathbb{C}^*$ is defined by Eq. (10.1) via the underlying classical divisor $D = \{(\lambda_k, \mu_k)\}$ of $\mathscr{D}$.)

Addendum. *If $\mathscr{D}$ is non-special, then the following statements hold concerning the monodromy $M(\lambda)$ in (c), where we let $D = \{(\lambda_k, \mu_k)\}_{k\in\mathbb{Z}}$ be the underlying classical divisor of $\mathscr{D}$:*

(1) c is characterized up to sign by the fact that it has zeros in all the λ_k (with the correct multiplicity) and no others; this function is given by Eq. (10.2) *in Proposition 10.1.*
(2) The functions a resp. d satisfy $a(\lambda_k) = \mu_k$ resp. $d(\lambda_k) = \mu_k^{-1}$ for all $k \in \mathbb{Z}$; these functions are given by Eqs. (10.18) *and* (10.21) *in Proposition 10.4 (where the $t_{\lambda_*,j}$ are uniquely determined from the spectral data $(\Sigma, \mathscr{D})$).*
(3) The function b is characterized by $\det(M) = ad - bc = 1$.
(4) We have $M(\lambda) \in \mathsf{Mon}_\tau$ with $\tau = \pm\left(\prod_{k\in\mathbb{Z}} \frac{\lambda_{k,0}}{\lambda_k}\right)^{1/2}$.
(5) $M(\lambda)$ is uniquely determined by the spectral data $(\Sigma, \mathscr{D})$ up to a joint change of sign of b, c and τ.

Moreover, the uniquely determined function f from (d) then equals $\pm\frac{\mu-d}{c}$. If we choose a sign for the holomorphic function $\sqrt{\lambda}$ on $\widehat{V}_\delta$, the condition that $f = \frac{\mu-d}{c}$ holds in (d) determines the sign of c in (1) and the sign of τ in (4), and therefore determines $M(\lambda)$ in (5).

Proof For (a) $\Rightarrow$ (b). Assume to the contrary that there exists $\lambda_* \in \mathbb{C}^*$ so that $f := (\lambda - \lambda_*)^{-1}$ is a section of $\mathscr{D}$. Then we can choose $\delta > 0$ so large that $\lambda_* \notin V_\delta$ holds, and with this choice of δ, the non-constant meromorphic function f is bounded on $\widehat{V}_\delta$. Therefore $\mathscr{D}$ would be special, in contradiction to the hypothesis (a).

For (b) $\Rightarrow$ (c). Because $\mathscr{D}$ is asymptotic, the underlying classical divisor D is of the form $D = \{(\lambda_k, \mu_k)\}_{k\in\mathbb{Z}}$ with $\lambda_k - \lambda_{k,0} \in \ell^2_{-1,3}(k)$ and $\mu_k - \mu_{k,0} \in \ell^2_{0,0}(k)$.

By Proposition 10.1(1),

$$\tau := \pm\left(\prod_{k\in\mathbb{Z}} \frac{\lambda_{k,0}}{\lambda_k}\right)^{1/2}$$

then converges in $\mathbb{C}^*$, and after fixing the sign of τ, there exists a holomorphic function $c(\lambda)$ on $\mathbb{C}^*$ with zeros in all the λ_k (with multiplicity) and no others, and which satisfies the asymptotic property

$$c - \tau\, c_0 \in \mathrm{As}_\infty(\mathbb{C}^*, \ell^2_{-1}, 1) \quad \text{and} \quad c - \tau^{-1}\, c_0 \in \mathrm{As}_0(\mathbb{C}^*, \ell^2_1, 1);$$

c is unique by Proposition 6.7(1), and $c(\lambda)$ is given explicitly by Eq. (10.2).

The next step of the construction is to obtain a holomorphic function d with $d - d_0 \in \mathrm{As}(\mathbb{C}^*, \ell^2_{0,0}, 1)$ such that $\frac{\mu-d}{c}$ is a global section of $\mathscr{D}$. For this purpose, we use Lemma 12.2, which is applicable because of the hypothesis (b). This lemma shows that the condition that $\frac{\mu-d}{c}$ be a section of $\mathscr{D}$ is equivalent to the prescription of the values of $d^{(\nu)}(\lambda_k)$ for every $k \in \mathbb{Z}$ and $\nu \in \{0, \ldots, m-1\}$, where m is the degree of the point (λ_k, μ_k) in $\mathscr{D}$, and m also equals the multiplicity of the zero λ_k of c. Therefore Proposition 10.4(1), (2) shows (see Remark 10.5) that there exists one and only one holomorphic function d with $d - d_0 \in \mathrm{As}(\mathbb{C}^*, \ell^2_{0,0}, 1)$ so that the conditions from Lemma 12.2 are satisfied for all points in the support of $\mathscr{D}$, and therefore $\frac{\mu - d}{c}$ is a global section of $\mathscr{D}$.

We let $a := \Delta - d$, this is also a holomorphic function in λ and we have $a - a_0 \in \mathrm{As}(\mathbb{C}^*, \ell^2_{0,0}, 1)$. Moreover, $b := \frac{ad-1}{c}$ is holomorphic by the Addendum of Lemma 12.2. From the equation

$$b - \tau^{-1} b_0 = \frac{ad - 1}{c} - \frac{a_0 d_0 - 1}{\tau c_0} = \frac{1}{c} \cdot \big((a - a_0) d + a_0 (d - d_0) \big) - \frac{b_0}{\tau c} \cdot (c - \tau c_0)$$

one obtains $(b - \tau^{-1} b_0)|V_\delta \in \mathrm{As}_\infty(\mathbb{C}^*, \ell^2_1, 1)$ and then by Proposition 9.4(1) $b - \tau^{-1} b_0 \in \mathrm{As}_\infty(\mathbb{C}^*, \ell^2_1, 1)$. Similarly one also shows $b - \tau b_0 \in \mathrm{As}_0(\mathbb{C}^*, \ell^2_{-1}, 1)$.

We now consider

$$M(\lambda) := \begin{pmatrix} a(\lambda) & b(\lambda) \\ c(\lambda) & d(\lambda) \end{pmatrix}.$$

By the preceding construction, we have $\det(M(\lambda)) = 1$, $M(\lambda) \in \mathsf{Mon}_\tau$ and $a + d = \Delta$. Moreover, both 1 and $\frac{\mu - d}{c}$ are global sections of $\mathscr{D}$, so the spectral divisor $\mathscr{D}_M$ of $M(\lambda)$ (which is generated by 1 and $\frac{\mu - d}{c}$) is contained in $\mathscr{D}$. Both $\mathscr{D}$ and $\mathscr{D}_M$ are asymptotic divisors, i.e. they have asymptotically and totally the same degree. Therefore $\mathscr{D}_M \subset \mathscr{D}$ in fact implies $\mathscr{D}_M = \mathscr{D}$.

For (c) ⇒ (d). We let $M(\lambda) = \left(\begin{smallmatrix} a(\lambda) & b(\lambda) \\ c(\lambda) & d(\lambda) \end{smallmatrix}\right)$ be the monodromy from (c). Then $f := \frac{\mu - d}{c}$ is a section of $\mathscr{D}$, and by Corollary 11.4(6) we have $\left(f - \frac{\mathrm{i}}{\tau\sqrt{\lambda}}\right)|\widehat{V}_\delta \in \mathrm{As}_\infty(\widehat{V}_\delta, \ell^2_1, 0)$ and $\left(f - \frac{\mathrm{i}}{\tau^{-1}\sqrt{\lambda}}\right)|\widehat{V}_\delta \in \mathrm{As}_0(\widehat{V}_\delta, \ell^2_{-1}, 0)$.

For the proof of the uniqueness of f, we suppose that another section $\widetilde{f}$ of $\mathscr{D}$ with $\left(\widetilde{f} - \frac{\mathrm{i}}{\tau\sqrt{\lambda}}\right)|\widehat{V}_\delta \in \mathrm{As}_\infty(\widehat{V}_\delta, \ell^2_1, 0)$ and $\left(\widetilde{f} - \frac{\mathrm{i}}{\tau^{-1}\sqrt{\lambda}}\right)|\widehat{V}_\delta \in \mathrm{As}_0(\widehat{V}_\delta, \ell^2_{-1}, 0)$ is given. We first show that the odd parts of f and $\widetilde{f}$ are equal. Because f and $\widetilde{f}$ are sections in $\mathscr{D}$, $g := c \cdot (f - \widetilde{f})$ is a holomorphic function on Σ, and we have $(f - \widetilde{f})|\widehat{V}_\delta \in \mathrm{As}(\widehat{V}_\delta, \ell^2_{1,-1}, 0)$, and therefore $g|\widehat{V}_\delta \in \mathrm{As}(\widehat{V}_\delta, \ell^2_{0,0}, 1)$. We now represent $g = g_+ + (\mu - \mu^{-1}) \cdot g_-$ with holomorphic functions g_+, g_- in $\lambda \in \mathbb{C}^*$. Along with g, we also have $((\mu - \mu^{-1}) \cdot g_-)|\widehat{V}_\delta \in \mathrm{As}(\widehat{V}_\delta, \ell^2_{0,0}, 1)$. Because $\mu - \mu^{-1}$ is comparable to $w(\lambda)$ on $\widehat{V}_\delta$, this implies $g_-|V_\delta \in \mathrm{As}(V_\delta, \ell^2_{0,0}, 0)$ and therefore by Proposition 9.4(1) $g_- \in \mathrm{As}(\mathbb{C}^*, \ell^2_{0,0}, 0)$. This shows that the holomorphic function g_- on $\mathbb{C}^*$ can be extended to a holomorphic function on $\mathbb{P}^1$ by setting $g_-(0) = g_-(\infty) = 0$. It follows that $g_- = 0$ holds, and thus the odd parts of f and $\widetilde{f}$ are equal.

Because we have

$$f = \frac{\mu - d}{c} = \frac{(\Delta - 2d) + (\mu - \mu^{-1})}{2c},$$

it follows that there exists a meromorphic function $\widetilde{d}$ in λ so that

$$\widetilde{f} = \frac{(\Delta - 2\widetilde{d}) + (\mu - \mu^{-1})}{2c} = \frac{\mu - \widetilde{d}}{c}$$

holds. Because $g = c \cdot (\widetilde{f} - f)$ is holomorphic on Σ, $\widetilde{d}$ is in fact holomorphic in λ. Moreover, $(\widetilde{f} - f)|\widehat{V_\delta} \in \mathrm{As}(\widehat{V_\delta}, \ell^2_{1,-1}, 0)$ implies

$$-\frac{d - \widetilde{d}}{c}\bigg|_{V_\delta} = \left(\frac{\Delta - 2\widetilde{d}}{2c} - \frac{\Delta - 2d}{2c}\right)\bigg|_{V_\delta} \in \mathrm{As}(V_\delta, \ell^2_{1,-1}, 0)$$

and therefore $(d - \widetilde{d})|V_\delta \in \mathrm{As}(V_\delta, \ell^2_{0,0}, 1)$, hence $d - \widetilde{d} \in \mathrm{As}(\mathbb{C}^*, \ell^2_{0,0}, 1)$ by Proposition 9.4(1). Lemma 12.2 shows that the condition $\frac{\mu - d}{c} \in H^0(\Sigma, \mathscr{D})$ prescribes for any (λ_*, μ_*) in the support of $\mathscr{D}$, say of degree m, the values of $d(\lambda_*)$, $d'(\lambda_*)$, $\ldots$, $d^{(m-1)}(\lambda_*)$, and therefore we have $\mathrm{ord}_{\lambda_*}(d - \widetilde{d}) \geq \mathrm{ord}_{\lambda_*}(c)$ for every $\lambda_* \in \mathbb{C}$ with $c(\lambda_*) = 0$. Proposition 6.7(2) shows that $d = \widetilde{d}$ and therefore $f = \widetilde{f}$ holds.

For (d) ⇒ (a). Suppose that (d) holds, i.e. there exists one and only one section f of $\mathscr{D}$ such that $\left(f - \frac{\mathrm{i}}{\tau\sqrt{\lambda}}\right)|\widehat{V_\delta} \in \mathrm{As}_\infty(\widehat{V_\delta}, \ell^2_1, 0)$ and $\left(f - \frac{\mathrm{i}}{\tau^{-1}\sqrt{\lambda}}\right)|\widehat{V_\delta} \in \mathrm{As}_0(\widehat{V_\delta}, \ell^2_{-1}, 0)$ holds for some $\delta > 0$.

Assume to the contrary that $\mathscr{D}$ is special. Thus there exists a section g of $\mathscr{D}$ which is non-constant and bounded on $\widehat{V_\delta}$ for some $\delta > 0$. Then we have in particular $\frac{1}{\sqrt{\lambda}} \cdot g|\widehat{V_\delta} \in \mathrm{As}(\widehat{V_\delta}, \ell^2_{1,-1}, 0)$, and therefore for every $s \in \mathbb{C}^*$, $f_s := f + s \cdot \frac{1}{\sqrt{\lambda}} g$ is a section of $\mathscr{D}$ different from f, for which $\left(f_s - \frac{\mathrm{i}}{\tau\sqrt{\lambda}}\right)|\widehat{V_\delta} \in \mathrm{As}_\infty(\widehat{V_\delta}, \ell^2_1, 0)$ and $\left(f_s - \frac{\mathrm{i}}{\tau^{-1}\sqrt{\lambda}}\right)|\widehat{V_\delta} \in \mathrm{As}_0(\widehat{V_\delta}, \ell^2_{-1}, 0)$ holds. This contradicts the uniqueness statement of (d). □

With this theorem, we have clarified the equivalence between monodromies $M(\lambda) \in \mathsf{Mon}$ and spectral data $(\Sigma, \mathscr{D})$.

From this point onward, we would like to avoid technical complications by restricting the class of divisors in such a way that the generalized divisor $\mathscr{D}$ is uniquely determined by the underlying classical divisor D. We will achieve this by requiring that the degree of $\mathscr{D}$ above any $\lambda_* \in \mathbb{C}^*$ is at most 1; we will call such divisors *tame*. The set of tame divisors is open and dense in Div; also note that the divisor $\mathscr{D}_0$ of the vacuum is tame (see Chap. 4). Because tame generalized divisors are uniquely characterized by their underlying classical divisors, we will work mostly with the classical divisors in the case of tame divisors.

The concept is made precise by the following definition and proposition:

Definition 12.4

(1) We say that an asymptotic classical divisor $D = \{(\lambda_k, \mu_k)\} \in \mathsf{Div}$ (viewed as a multi-set of points in $\mathbb{C}^* \times \mathbb{C}^*$) is *tame* if the λ_k are pairwise unequal. We denote the subset of tame divisors in Div by Div_{tame}.

(2) We say that a generalized divisor $\mathcal{D}$ on the spectral curve Σ is *tame*, if it is compatible, asymptotic and positive, and if the total degree of $\mathcal{D}$ above any $\lambda_* \in \mathbb{C}^*$ is at most 1.

(3) We say that a potential $(u, u_y) \in \mathsf{Pot}$ is *tame*, if its associated spectral divisor is tame. We denote the subset of tame potentials in Pot by Pot_{tame}.

At the end of Chap. 8 we identified the space Div of asymptotic classical divisors with the quotient space $(\ell^2_{-1,3} \oplus \ell^2_{0,0})/P(\mathbb{Z})$, and viewed in this way, Div_{tame} is an open and dense subset of Div. The singular points of Div are those $D \in \mathsf{Div}$ which contain a point of multiplicity ≥ 2, and therefore the points of Div_{tame} are regular points of Div. Thus Div_{tame} has the structure of a smooth Banach manifold of type $\ell^2_{-1,3} \oplus \ell^2_{0,0}$.

The following Proposition 12.5(1) shows that every tame generalized divisor $\mathcal{D}$ is non-special, and therefore is the spectral divisor of a monodromy $M(\lambda)$. It also shows that if a point $(\lambda_k, \mu_k) \in \operatorname{supp}(\mathcal{D})$ with $|k|$ large is a singular point of Σ, then near this point, $\mathcal{D}$ looks like the spectral divisor of the vacuum, see Chap. 4. Proposition 12.5(2) shows that the classical tame divisors D (regarded as multi-sets of points in $\mathbb{C}^* \times \mathbb{C}^*$) are in one-to-one correspondence with the spectral data $(\Sigma, \mathcal{D})$, where $\mathcal{D}$ is a tame generalized divisor on the spectral curve Σ. The latter fact justifies our approach of investigating tame classical (rather than generalized) divisors in the sequel.

Proposition 12.5

(1) Let $\mathcal{D}$ be a tame generalized divisor on Σ. Then the following holds:

(a) The underlying classical divisor D of $\mathcal{D}$ is tame.

(b) $\mathcal{D}$ is non-special.

(c) We write $D = \{(\lambda_k, \mu_k)\}$ as in Proposition 6.5(3). Then there exists $N \in \mathbb{N}$ such that for every $k \in \mathbb{Z}$ with $|k| > N$ where (λ_k, μ_k) is a singular point of Σ, this singular point is an ordinary double point and we have $\mathcal{D}_{(\lambda_k,\mu_k)} = \widehat{\mathcal{O}}_{(\lambda_k,\mu_k)}$. Here $\widehat{\mathcal{O}}$ denotes the direct image in Σ of the sheaf of holomorphic functions on the normalization of Σ.

(2) Let $D = \{(\lambda_k, \mu_k)\} \in \mathsf{Div}_{tame}$. Then there exists one and only one holomorphic function $\Delta : \mathbb{C}^* \to \mathbb{C}$ with $\Delta - \Delta_0 \in \mathrm{As}(\mathbb{C}^*, \ell^2_{0,0}, 1)$ so that the hyperelliptic complex curve

$$\Sigma := \{\, (\lambda, \mu) \in \mathbb{C}^* \times \mathbb{C}^* \,\big|\, \mu^2 - \Delta(\lambda) \cdot \mu + 1 = 0 \,\} \tag{12.11}$$

contains D, and there exists one and only one tame generalized divisor $\mathcal{D}$ on Σ so that D is the support of $\mathcal{D}$.

Proof For (1)(a). This is obvious.

For (1)(b). For any $\lambda_* \in \mathbb{C}^*$, the meromorphic function $(\lambda - \lambda_*)^{-1}$ on Σ has total pole order 2 above λ_* (regardless of whether Σ is regular or singular above λ_*), and therefore cannot be a section of the tame divisor $\mathscr{D}$. Thus $\mathscr{D}$ is non-special by Theorem 12.3(b)⇒(a).

For (1)(c). We enumerate the zeros of the discriminant function $\Delta^2 - 4$ by two sequences $(\varkappa_{k,1})$ and $(\varkappa_{k,2})$ as in Proposition 6.5(1), then there exists $N \in \mathbb{N}$ so that $\varkappa_{k,\nu} \in U_{k,\delta}$ and $(\lambda_k, \mu_k) \in \widehat{U}_{k,\delta}$ holds for all $k \in \mathbb{Z}$ with $|k| > N$ and $\nu \in \{1, 2\}$. Because the excluded domains $U_{k,\delta}$ are disjoint, it follows that $\Delta^2 - 4$ can have only zeros of order ≤ 2 on $U_{k,\delta}$ with $|k| > N$, and therefore any singularities of Σ in $\widehat{U}_{k,\delta}$ are then of order 2, i.e. ordinary double points.

Now let $k \in \mathbb{Z}$ be given with $|k| > N$ so that the divisor point (λ_k, μ_k) is a singular point of Σ. By the preceding argument, this singularity of Σ is then an ordinary double point, and hence the δ-invariant of Σ at (λ_k, μ_k) is 1. Moreover, $\mathscr{D}$ is non-special by (1)(b), and therefore is the spectral divisor of a monodromy $M(\lambda) \in \mathsf{Mon}$ by Theorem 12.3(a)⇒(c). Here the degree of (λ_k, μ_k) in $\mathscr{D}$ is 1, because $\mathscr{D}$ is tame, and therefore it follows from Proposition 3.7(3) that $j_0 = 1$ and $s = 0$ holds, where j_0 is the number from Proposition 3.3 and s is the maximal pole order occurring in $\mathscr{D}_{(\lambda_k,\mu_k)}$. Because j_0 thus equals the δ-invariant of (λ_k, μ_k), we have $\mathscr{R}_{(\lambda_k,\mu_k)} = \widehat{\mathscr{O}}_{(\lambda_k,\mu_k)}$ for the ring $\mathscr{R}_{(\lambda_k,\mu_k)}$ over which $\mathscr{D}_{(\lambda_k,\mu_k)}$ is locally free (see Proposition 3.3), and because of $s = 0$ we have $\mathscr{D}_{(\lambda_k,\mu_k)} = \mathscr{R}_{(\lambda_k,\mu_k)}$. Thus $\mathscr{D}_{(\lambda_k,\mu_k)} = \widehat{\mathscr{O}}_{(\lambda_k,\mu_k)}$ holds.

For (2). We have

$$\frac{1}{2}\left(\mu_k + \mu_k^{-1}\right) - \mu_{k,0} = \frac{1}{2}\left(1 - \frac{1}{\mu_k\,\mu_{k,0}}\right) \cdot (\mu_k - \mu_{k,0}).$$

$\mu_k - \mu_{k,0} \in \ell^2_{0,0}(k)$ and $\mu_{k,0} = (-1)^k$ implies $\lim_{k\to\pm\infty} \frac{1}{2}\big(1 - \frac{1}{\mu_k\,\mu_{k,0}}\big) = 0$ and therefore by the above calculation $\frac{1}{2}\left(\mu_k + \mu_k^{-1}\right) - \mu_{k,0} \in \ell^2_{0,0}(k)$. Because D is tame, the λ_k are pairwise different, therefore Proposition 10.4(1),(2) shows that there exists one and only one holomorphic function $\frac{1}{2}\Delta : \mathbb{C}^* \to \mathbb{C}$ with $\frac{1}{2}\Delta - \frac{1}{2}\Delta_0 \in \mathrm{As}(\mathbb{C}^*, \ell^2_{0,0}, 1)$ and

$$\frac{1}{2}\,\Delta(\lambda_k) = \frac{1}{2}\left(\mu_k + \mu_k^{-1}\right) \quad \text{for all } k \in \mathbb{Z}.$$

If we now define the complex curve Σ by Eq. (12.11), then we have $D \subset \Sigma$, and there is no other way to choose Δ so that this inclusion holds. We let $\mathscr{O}$ be the sheaf of holomorphic functions on Σ, and $\widehat{\mathscr{O}}$ be the direct image of the sheaf of holomorphic functions on the normalization $\widehat{\Sigma}$ of Σ.

By Proposition 10.1(1) there exists (up to sign) one and only one holomorphic function $c : \mathbb{C}^* \to \mathbb{C}$ with $c - \tau\, c_0 \in \mathrm{As}_\infty(\mathbb{C}^*, \ell^2_{-1}, 1)$ and $c - \tau^{-1}\, c_0 \in \mathrm{As}_0(\mathbb{C}^*, \ell^2_1, 1)$ for some $\tau \in \mathbb{C}^*$, which has zeros at all the λ_k, $k \in \mathbb{Z}$ and no others. And again because the λ_k are pairwise different, there exists one and only one holomorphic function $d : \mathbb{C}^* \to \mathbb{C}$ with $d - d_0 \in \mathrm{As}(\mathbb{C}^*, \ell^2_{0,0}, 1)$ such that $d(\lambda_k) = \mu_k^{-1}$ holds for all $k \in \mathbb{Z}$, by Proposition 10.4(1),(2).

We let $\mathscr{D}$ be the generalized divisor on Σ generated by 1 and $\frac{\mu - d}{c}$ over $\mathscr{O}$. Clearly $\mathscr{D}$ is positive, and the support of $\mathscr{D}$ is D, hence $\mathscr{D}$ is asymptotic. Because D is tame, the total degree of $\mathscr{D}$ above any $\lambda_* \in \mathbb{C}^*$ is at most 1. Moreover $\mathscr{D}$ is compatible: If $(\lambda_*, \mu_*) \in D$ is a singular point of Σ, let $g := 2\,\frac{\mu - d}{\lambda - \lambda_*}$, then $g \in H^0(\Sigma, \mathscr{D})$ holds and $\eta := g - \frac{\mu - \mu^{-1}}{\lambda - \lambda_*} = \frac{\Delta - 2d}{\lambda - \lambda_*}$ is a meromorphic function in λ with a pole of order at most 1. Therefore it follows that $\mathscr{D}$ is tame. □

By virtue of Proposition 12.5, the tame classical divisors are in one-to-one correspondence with spectral data $(\Sigma, \mathscr{D})$, where Σ is a spectral curve and $\mathscr{D}$ is a tame, generalized divisor on Σ. In the sequel, we will therefore speak of the spectral curve Σ and trace function $\Delta = \mu + \mu^{-1}$, and of the tame generalized divisor $\mathscr{D}$ on Σ associated to a tame classical divisor $D \in \mathsf{Div}_{tame}$ (regarded as a multi-set of points in $\mathbb{C}^* \times \mathbb{C}^*$). In this manner, any tame classical divisor D gives rise to a monodromy $M(\lambda)$ with spectral divisor D by Theorem 12.3.

Part IV
The Inverse Problem for Periodic Potentials (Cauchy Data)

Chapter 13
Divisors of Finite Type

An asymptotic divisor $\mathscr{D}$ on a spectral curve Σ is said to be *of finite type*, if the following conditions hold:

1. The spectral curve Σ has finite geometric genus (i.e. only finitely many of the double points of the spectral curve of the vacuum have "opened up" into a pair of branch points with positive distance).
2. All but finitely many of the points (λ_*, μ_*) in the support of $\mathscr{D}$ lie in double points of Σ, and we have $\mathscr{D}_{(\lambda_*,\mu_*)} = \widehat{\mathscr{O}}_{(\lambda_*,\mu_*)}$, where $\widehat{\mathscr{O}}$ denotes the direct image in Σ of the sheaf of holomorphic functions on the normalization of Σ.

Spectral data $(\Sigma, \mathscr{D})$ of finite type thus look like the spectral data of the vacuum near all but finitely many of the divisor points, see Chap. 4. In particular the equation $\mathscr{D}_{(\lambda_*,\mu_*)} = \widehat{\mathscr{O}}_{(\lambda_*,\mu_*)}$ implies that the ring $\mathscr{R}_{(\lambda_*,\mu_*)}$ from Proposition 3.3 over which $\mathscr{D}_{(\lambda_*,\mu_*)}$ is locally free equals $\widehat{\mathscr{O}}_{(\lambda_*,\mu_*)}$; this shows that a generalized divisor $\mathscr{D}$ of finite type is locally free in the normalization $\widehat{\Sigma}$ of Σ with the exception of finitely many points.

Among the asymptotic divisors, those that are of finite type play a special role, because the spectral data associated to solutions u of the sinh-Gordon equation that are doubly periodic, meaning that there is a number $\tau \in \mathbb{C}$ with $\mathrm{Im}(\tau) > 0$ such that $u(z+1) = u(z+\tau) = u(z)$ holds for all $z \in \mathbb{C}$, are of finite type. (Conversely though, not every solution u associated to spectral data of finite type is doubly periodic.) Doubly periodic solutions of the sinh-Gordon equation that satisfy an additional closing condition correspond to constant mean curvature tori in $\mathbb{R}^3$, or to minimal tori in S^3. The finite type solutions u of the sinh-Gordon equation, and consequently the minimal tori, have been classified by Pinkall and Sterling [Pi-S], and independently by Hitchin [Hi] as was described in more detail in the Introduction.

S. Klein, *A Spectral Theory for Simply Periodic Solutions of the Sinh-Gordon Equation*, Lecture Notes in Mathematics 2229,
https://doi.org/10.1007/978-3-030-01276-2_13

We are interested in finite type divisors here, because for them, the inverse problem for the potential, i.e. the reconstruction of the potential from the (finite type) spectral divisor, has already been solved. In Chap. 15 we will use this result to solve the inverse problem also for tame spectral divisors that are not of finite type essentially by a limit argument. To make this argument work, it is essential to know that the finite type potentials are dense in the space of all potentials.

For many other integrable systems, where one can similarly define the concept of a finite type potential, the finite type potentials are known to be dense in the space of all potentials. For example, this is true for the integrable system associated to the KdV equation (where the "finite type potentials" are called finite-gap solutions), see [Kapp-P, Section 11]. Therefore we expect the finite type potentials to be dense in the space of all potentials also in our setting for the sinh-Gordon equation, and likewise that the set of divisors of finite type is dense in the space of all asymptotic potentials.

It is the objective of the present chapter to prove the latter statement, i.e. that the divisors of finite type are indeed dense in Div. Later, in Corollary 15.6(1), we will see that also the set of finite type potentials in Pot_{tame} is dense in Pot_{tame}.

We will see that it suffices to consider divisors of finite type that are tame, because the tame divisors themselves comprise an open and dense set in Div. Proposition 12.5(1)(c) shows that for a tame generalized divisor $\mathscr{D}$ and a point (λ_k, μ_k) in the support of $\mathscr{D}$ with $|k|$ large that is a singular point of the associated spectral curve Σ, we already have that the singularity (λ_k, μ_k) is an ordinary double point of Σ and that $\mathscr{D}_{(\lambda_k,\mu_k)} = \widehat{\mathscr{O}}_{(\lambda_k,\mu_k)}$ holds. Moreover by Proposition 12.5 the generalized tame divisors are in one-to-one correspondence with classical tame divisors. Therefore the definition of classical finite type divisors given below is for tame divisors equivalent to the definition of finite type above.

Definition 13.1

(1) Let $D = \{ (\lambda_k, \mu_k) \,|\, k \in \mathbb{Z} \}$ be an asymptotic divisor on a spectral curve Σ with trace function $\Delta = \mu + \mu^{-1}$, and let $\varkappa_{k,\nu}$ be the zeros of $\Delta^2 - 4$ (see Proposition 6.5(1)). We say that D is *of finite type*, if $\varkappa_{k,1} = \varkappa_{k,2} = \lambda_k$ holds for all $k \in \mathbb{Z}$ with at most finitely many exceptions.

(2) We say that a potential $(u, u_y) \in \mathsf{Pot}$ is *of finite type*, if its associated spectral divisor D is of finite type.

As promised we will show in the present chapter that the set of finite type divisors is dense in Div. In fact, the following theorem states that the set of tame divisors of finite type is dense in Div_{tame}. Because Div_{tame} is open and dense in Div, it follows from this theorem that the finite type divisors are dense in Div.

Theorem 13.2 *Let* $D = \{(\lambda_k, \mu_k)\} \in \mathsf{Div}_{tame}$ *be given. Then for every* $\varepsilon > 0$ *there exists a divisor of finite type* $D^* = \{(\lambda_k^*, \mu_k^*)\} \in \mathsf{Div}_{tame}$ *with*

$$\|D^* - D\|_{\mathsf{Div}} \leq \varepsilon.$$

Moreover for given $N \in \mathbb{N}$, D^* can be chosen such that

$$\forall k \in \mathbb{Z},\ |k| \leq N\ :\ \lambda_k^* = \lambda_k,\quad \mu_k^* = \mu_k \tag{13.1}$$

holds.

We prepare the proof of Theorem 13.2 with two lemmas. For a given holomorphic function Δ with $\Delta - \Delta_0 \in \mathrm{As}(\mathbb{C}^*, \ell^2_{0,0}, 1)$ (intended to be the trace function of a spectral curve Σ), the first lemma (Lemma 13.3) studies the asymptotic behavior of the sequence (η_n) of zeros of Δ'. These zeros are of interest in relation to potentials of finite type because if Σ has an ordinary double point (or, in fact, any other singularity) above some $\lambda_* \in \mathbb{C}^*$, then $\Delta'(\lambda_*) = 0$ holds. The second lemma (Lemma 13.4) provides a Lipschitz estimate for the quantity $\Delta(\eta_n) - \Delta(\lambda_n)$ where (λ_k) is a fixed sequence, but Δ and correspondingly (η_n) varies. Both lemmas will be used in the proof of Theorem 13.2 to set up a fixed point equation to describe the property of a divisor being of finite type, and to show that the Banach fixed point theorem applies to this equation.

Lemma 13.3

(1) *Let $\Delta : \mathbb{C}^* \to \mathbb{C}$ be a holomorphic function with $\Delta - \Delta_0 \in \mathrm{As}(\mathbb{C}^*, \ell^2_{0,0}, 1)$. Then Δ' has asymptotically and totally a zero in each excluded domain $U_{k,\delta}$, and besides these exactly one additional zero.*

If $\eta_* \in \mathbb{C}^*$ denotes a zero of Δ' for which $|\eta_* - 1|$ becomes minimal, and if we denote by $(\eta_k)_{k\in\mathbb{Z}}$ the sequence of the remaining zeros of Δ', where $\eta_k \in U_{k,\delta}$ holds for $|k|$ large, then $\eta_k - \lambda_{k,0} \in \ell^2_{-1,3}(k)$ holds.

In the sequel, we will denote the sequence of all zeros of Δ' by $(\eta_k)_{k\in\mathbb{Z}\cup\{*\}}$.

(2) *Let $R_0 > 0$, let $\Delta^{[1]}, \Delta^{[2]} : \mathbb{C}^* \to \mathbb{C}$ be two holomorphic functions with $\Delta^{[\nu]} - \Delta_0 \in \mathrm{As}(\mathbb{C}^*, \ell^2_{0,0}, 1)$ and $\|\Delta^{[\nu]} - \Delta_0\|_{\mathrm{As}(\mathbb{C}^*, \ell^2_{0,0}, 1)} \leq R_0$ for $\nu \in \{1, 2\}$.*

We denote the zeros of $\Delta^{[\nu]\prime}$ as in (1) by $(\eta_k^{[\nu]})_{k\in\mathbb{Z}\cup\{*\}}$. Then there exists a constant $C > 0$, dependent only on R_0, such that

$$|\eta_k^{[1]} - \eta_k^{[2]}| \leq C \cdot \begin{Bmatrix} k & \text{if } k > 0 \\ |k|^{-3} & \text{if } k < 0 \end{Bmatrix} \cdot \max\{b_{k-1}, b_k, b_{k+1}\}$$

holds, where $(b_k)_{k\in\mathbb{Z}} \in \ell^2_{0,0}(k)$ is a bounding sequence for the function $\Delta^{[1]} - \Delta^{[2]}$ in $\mathrm{As}(\mathbb{C}^*, \ell^2_{0,0}, 1)$.

Proof For (1). Because of $\Delta - \Delta_0 \in \mathrm{As}(\mathbb{C}^*, \ell^2_{0,0}, 1)$, we have by Proposition 9.4(3)

$$\Delta' - \Delta_0' \in \mathrm{As}(\mathbb{C}^*, \ell^2_{1,-3}, 1) \quad \text{and} \quad \Delta'' - \Delta_0'' \in \mathrm{As}(\mathbb{C}^*, \ell^2_{2,-6}, 1) \tag{13.2}$$

Moreover, we have

$$\Delta_0' \in \mathrm{As}(\mathbb{C}^*, \ell^\infty_{1,-3}, 1) \quad \text{and} \quad \Delta_0'' \in \mathrm{As}(\mathbb{C}^*, \ell^\infty_{2,-6}, 1) \tag{13.3}$$

and therefore also

$$\Delta' \in \mathrm{As}(\mathbb{C}^*, \ell^\infty_{1,-3}, 1) \quad \text{and} \quad \Delta'' \in \mathrm{As}(\mathbb{C}^*, \ell^\infty_{2,-6}, 1). \tag{13.4}$$

We begin by showing that Δ' has at least one zero η_*. It follows from (13.2), (13.3) and (13.4) that we have

$$\left(\frac{\Delta''}{\Delta'} - \frac{\Delta''_0}{\Delta'_0} \right) \bigg|_{V_\delta} = \frac{(\Delta'' - \Delta''_0) \cdot \Delta'_0 + \Delta''_0 \cdot (\Delta'_0 - \Delta')}{\Delta' \cdot \Delta'_0} \bigg|_{V_\delta} \in \mathrm{As}(\mathbb{C}^*, \ell^2_{1,-3}, 1);$$

because the sequence of circumferences of $U_{k,\delta}$ is in $\ell^\infty_{-1,3}(k)$, this implies

$$\int_{\partial U_{k,\delta}} \left(\frac{\Delta''}{\Delta'} - \frac{\Delta''_0}{\Delta'_0} \right) \mathrm{d}\lambda \in \ell^2_{0,0}(k).$$

Because $\int_{\partial U_{k,\delta}} \left(\frac{\Delta''}{\Delta'} - \frac{\Delta''_0}{\Delta'_0} \right) \mathrm{d}\lambda$ is the difference of the "zero counting integrals" for Δ' and for Δ'_0, this integral is an integer multiple of $2\pi \mathrm{i}$, and therefore equals zero for large $|k|$. Thus we see that for $|k|$ large, Δ' and Δ'_0 have the same number of zeros on $U_{k,\delta}$, i.e. one. In particular, Δ' has at least one zero, and we let $\eta_* \in \mathbb{C}^*$ be the zero of Δ' for which $|\eta_* - 1|$ becomes minimal. (If there are several such zeros, we make an arbitrary choice.)

We now put

$$\tau := \sqrt{\eta_*} \in \mathbb{C}^* \quad \text{and} \quad c(\lambda) := -8\tau \, \frac{\lambda^2}{\lambda - \eta_*} \, \Delta'(\lambda). \tag{13.5}$$

Because η_* is a zero of Δ', $c(\lambda)$ is a holomorphic function on $\mathbb{C}^*$, and we claim that we have

$$c - \tau\, c_0 \in \mathrm{As}_\infty(\mathbb{C}^*, \ell^2_{-1}, 1) \quad \text{and} \quad c - \tau^{-1}\, c_0 \in \mathrm{As}_0(\mathbb{C}^*, \ell^2_1, 1). \tag{13.6}$$

Indeed, because of $\Delta_0(\lambda) = \cos(\zeta(\lambda))$ and $c_0(\lambda) = \sqrt{\lambda}\, \sin(\zeta(\lambda))$, we have

$$\Delta'_0(\lambda) = -\frac{1}{8} \left(\lambda^{-1/2} - \lambda^{-3/2} \right) \sin(\zeta(\lambda)) = -\frac{\lambda - 1}{8\lambda^2}\, c_0(\lambda)$$

and therefore

$$c_0(\lambda) = -\frac{8\lambda^2}{\lambda - 1}\, \Delta'_0(\lambda).$$

We thus obtain via (13.2) and (13.3)

$$c - \tau\, c_0 = \frac{-8\,\tau\,\lambda^2}{\lambda - \eta_*} \cdot (\Delta' - \Delta'_0) + \frac{-8\,\tau\,(\eta_* - 1)\,\lambda^2}{(\lambda - \eta_*) \cdot (\lambda - 1)} \cdot \Delta'_0 \in \mathrm{As}_\infty(\mathbb{C}^*, \ell^2_{-1}, 1)$$

and

$$c-\tau^{-1}\,c_0 = \frac{-8\,\tau^{-1}\,\eta_*\,\lambda^2}{\lambda-\eta_*}\cdot(\Delta'-\Delta_0')+\frac{-8\,\tau^{-1}\,(\eta_*-1)\,\lambda^3}{(\lambda-\eta_*)\cdot(\lambda-1)}\cdot\Delta_0' \;\in\; \mathrm{As}_0(\mathbb{C}^*,\ell^2_1,1).$$

By Proposition 6.5(2) it follows from (13.6) that the holomorphic function $c(\lambda)$ has asymptotically and totally exactly one zero in every excluded domain $U_{k,\delta}$, and that the sequence $(\eta_k)_{k\in\mathbb{Z}}$ of these zeros satisfies $\eta_k - \lambda_{k,0} \in \ell^2_{-1,3}(k)$. Because of $\Delta'(\lambda) = \frac{\lambda-\eta_*}{-8\,\tau\,\lambda^2}\cdot c(\lambda)$, it follows that η_* and the $(\eta_k)_{k\in\mathbb{Z}}$ are all the zeros of Δ'.

For (2). In the setting of (2), we denote the objects defined above in relation to $\Delta^{[\nu]}$ by the superscript $^{[\nu]}$ $(\nu \in \{1,2\})$. If (b_k) is a bounding sequence for $\Delta^{[1]} - \Delta^{[2]} \in \mathrm{As}(\mathbb{C}^*,\ell^2_{0,0},1)$, then by Proposition 9.4(3) there exists a constant $C_1 > 0$ so that

$$\begin{cases} \frac{C_1}{k}\cdot\max\{b_{k-1},b_k,b_{k+1}\} & \text{for } k>0 \\ C_1\cdot k^3\cdot\max\{b_{k-1},b_k,b_{k+1}\} & \text{for } k<0 \end{cases}$$

is a bounding sequence for $\Delta^{[1]\prime} - \Delta^{[2]\prime} \in \mathrm{As}(\mathbb{C}^*,\ell^2_{1,-3},1)$, and therefore with another constant $C_2 > 0$ and the sequence

$$\widetilde{b}_k := \begin{cases} C_2\cdot k\cdot\max\{b_{k-1},b_k,b_{k+1}\} & \text{for } k>0 \\ \frac{C_2}{k}\cdot\max\{b_{k-1},b_k,b_{k+1}\} & \text{for } k<0 \end{cases},$$

$(\widetilde{b}_k)_{k>0}$ is a bounding sequence for $\tau^{[2]}\,c^{[1]} - \tau^{[1]}\,c^{[2]} \in \mathrm{As}_\infty(\mathbb{C}^*,\ell^2_{-1},1)$ and $(\widetilde{b}_k)_{k<0}$ is a bounding sequence for $\tau^{[1]}\,c^{[1]} - \tau^{[2]}\,c^{[2]} \in \mathrm{As}_0(\mathbb{C}^*,\ell^2_1,1)$. By application of Proposition 8.1(1), the claimed statement follows. □

Lemma 13.4 *Let $R_0 > 0$, and let $(\lambda_k)_{k\in\mathbb{Z}}$ with $\lambda_k - \lambda_{k,0} \in \ell^2_{-1,3}(k)$ and $\|\lambda_k - \lambda_{k,0}\|_{\ell^2_{-1,3}} \le R_0$ be given. Further suppose that for $\nu \in \{1,2\}$, $\Delta^{[\nu]} : \mathbb{C}^* \to \mathbb{C}$ is a holomorphic function with $\Delta^{[\nu]} - \Delta_0 \in \mathrm{As}(\mathbb{C}^*,\ell^2_{0,0},1)$ and $\|\Delta^{[\nu]} - \Delta_0\|_{\mathrm{As}(\mathbb{C}^*,\ell^2_{0,0},1)} \le R_0$. We put $w_k^{[\nu]} := \Delta^{[\nu]}(\lambda_k)$ and let $(\eta_k^{[\nu]})_{k\in\mathbb{Z}\cup\{*\}}$ be the sequence of zeros of $(\Delta^{[\nu]})'$ as in Lemma 13.3(1). Then there exists a constant $C > 0$, depending only on R_0, so that we have for $n \in \mathbb{Z}$*

$$\begin{aligned}&\left|\left(\Delta^{[1]}(\eta_n^{[1]}) - \Delta^{[1]}(\lambda_n)\right) - \left(\Delta^{[2]}(\eta_n^{[2]}) - \Delta^{[2]}(\lambda_n)\right)\right| \\ &\quad\le C\cdot\left(|\zeta(\eta_n^{[1]}) - \zeta(\lambda_n)|\cdot\left(|w_k^{[1]} - w_k^{[2]}| * \frac{1}{|k|}\right)_n + |\zeta(\eta_n^{[1]}) - \zeta(\eta_n^{[2]})|^2\right).\end{aligned}$$

Proof In the sequel, we always have $\nu \in \{1, 2\}$, and all C_k are constants > 0 that depend only on R_0. We have

$$\begin{aligned}
&\left(\Delta^{[1]}(\eta_n^{[1]}) - \Delta^{[1]}(\lambda_n)\right) - \left(\Delta^{[2]}(\eta_n^{[2]}) - \Delta^{[2]}(\lambda_n)\right) \\
&= \int_{\lambda_n}^{\eta_n^{[1]}} (\Delta^{[1]})'(\lambda)\,\mathrm{d}\lambda - \int_{\lambda_n}^{\eta_n^{[2]}} (\Delta^{[2]})'(\lambda)\,\mathrm{d}\lambda \\
&= \int_{\lambda_n}^{\eta_n^{[1]}} (\Delta^{[1]} - \Delta^{[2]})'(\lambda)\,\mathrm{d}\lambda + \int_{\eta_n^{[2]}}^{\eta_n^{[1]}} (\Delta^{[2]})'(\lambda)\,\mathrm{d}\lambda.
\end{aligned} \tag{13.7}$$

We will estimate the two summands in the last expression individually.

For the first term in (13.7), we have

$$\begin{aligned}
\left|\int_{\lambda_n}^{\eta_n^{[1]}} (\Delta^{[1]} - \Delta^{[2]})'(\lambda)\,\mathrm{d}\lambda\right| &\leq |\eta_n^{[1]} - \lambda_n| \cdot \max_{\lambda \in U_{n,\delta}} \left|\left(\Delta^{[1]} - \Delta^{[2]}\right)'(\lambda)\right| \\
&\overset{(*)}{\leq} |\eta_n^{[1]} - \lambda_n| \cdot \ell^\infty_{1,-3}(n) \cdot \max_{\lambda \in U_{n,\delta}} \left|\left(\Delta^{[1]} - \Delta^{[2]}\right)(\lambda)\right| \\
&\overset{(\dagger)}{\leq} C_1 \cdot |\zeta(\eta_n^{[1]}) - \zeta(\lambda_n)| \cdot \max_{\lambda \in U_{n,\delta}} \left|\left(\Delta^{[1]} - \Delta^{[2]}\right)(\lambda)\right|, v
\end{aligned} \tag{13.8}$$

where the estimate marked $(*)$ follows from Cauchy's inequality, and the estimate marked $(\dagger)$ follows from Proposition 6.2(1).

To estimate $\left(\Delta^{[1]} - \Delta^{[2]}\right)(\lambda)$, we apply Proposition 10.4(3): In the setting described there, we choose $\lambda_k^{[1]} = \lambda_k^{[2]} = \lambda_k$ and $\mu_k^{[\nu]} = \frac{1}{2} w_k^{[\nu]}$. Then we have $\tau^{[1]} = \tau^{[2]}$, $\upsilon^{[1]} = \upsilon^{[2]} = 1$, $a_k = 0$ and $b_k = \frac{1}{2}\,|w_k^{[1]} - w_k^{[2]}|$. It follows from Proposition 10.4(3)(b) that there exists $C_2 > 0$ so that

$$r_n := C_2 \cdot \left(|w_k^{[1]} - w_k^{[2]}| * \frac{1}{|k|}\right)_n \in \ell^2_{0,0}(n)$$

is a bounding sequence for $\Delta^{[1]} - \Delta^{[2]}$ in $\mathrm{As}(\mathbb{C}^*, \ell^2_{0,0}, 1)$; here we set $\frac{1}{0} := 1$ as usual. Because $w(\lambda)$ is bounded on U_δ, there exists $C_3 > 0$ so that

$$\max_{\lambda \in U_{n,\delta}} \left|\left(\Delta^{[1]} - \Delta^{[2]}\right)(\lambda)\right| \leq C_3 \cdot r_n.$$

By plugging this estimate into (13.8) we obtain

$$\left|\int_{\lambda_n}^{\eta_n^{[1]}} (\Delta^{[1]} - \Delta^{[2]})'(\lambda)\,\mathrm{d}\lambda\right| \leq C_4 \cdot |\zeta(\eta_n^{[1]}) - \zeta(\lambda_n)| \cdot \left(|w_k^{[1]} - w_k^{[2]}| * \frac{1}{|k|}\right)_n \tag{13.9}$$

with $C_4 := C_3 \cdot C_2 \cdot C_1$.

We now turn our attention to the second summand in (13.7). We have

$$\int_{\eta_n^{[2]}}^{\eta_n^{[1]}} (\Delta^{[2]})'(\lambda)\,\mathrm{d}\lambda = \int_{\eta_n^{[2]}}^{\eta_n^{[1]}} g_n(\lambda)\cdot(\lambda-\eta_n^{[2]})\,\mathrm{d}\lambda, \tag{13.10}$$

where the function

$$g_n(\lambda) := \frac{(\Delta^{[2]})'(\lambda)}{\lambda-\eta_n^{[2]}}$$

is holomorphic because $(\Delta^{[2]})'$ has a zero at $\eta_n^{[2]}$. We now define (compare the proof of Lemma 13.3(1))

$$\tau := \sqrt{\eta_*^{[2]}} \in \mathbb{C}^* \quad \text{and} \quad c(\lambda) := -8\tau\,\frac{\lambda^2}{\lambda-\eta_*^{[2]}}\,(\Delta^{[2]})'(\lambda),$$

then we have $c - \tau\, c_0 \in \mathrm{As}_\infty(\mathbb{C}^*, \ell^2_{-1}, 1)$, $c - \tau^{-1}\, c_0 \in \mathrm{As}_0(\mathbb{C}^*, \ell^2_1, 1)$ and

$$g_n(\lambda) = -\frac{1}{8\tau}\,\frac{\lambda-\eta_*^{[2]}}{\lambda^2}\,\frac{c(\lambda)}{\lambda-\eta_n^{[2]}}.$$

For $\lambda \in U_{n,\delta}$ we have by Corollary 10.3(1): $\left|\frac{c(\lambda)}{\lambda-\eta_n^{[2]}}\right| \in \ell^\infty_{0,-2}(n)$, and we also have $\left|\frac{\lambda-\eta_*^{[2]}}{\lambda^2}\right| \in \ell^\infty_{2,-4}(n)$. Thus we obtain

$$|g_n(\lambda)| \in \ell^\infty_{2,-6}(n) \quad \text{for } \lambda \in U_{n,\delta} \quad ,$$

and therefore by Eq. (13.10)

$$\begin{aligned}\left|\int_{\eta_n^{[2]}}^{\eta_n^{[1]}} (\Delta^{[2]})'(\lambda)\,\mathrm{d}\lambda\right| &\le |\eta_n^{[2]}-\eta_n^{[1]}|\cdot \ell^\infty_{2,-6}(n)\cdot \max_{\lambda\in[\eta_n^{[1]},\eta_n^{[2]}]} |\lambda-\eta_n^{[2]}| \\ &\le \ell^\infty_{2,-6}(n)\cdot|\eta_n^{[2]}-\eta_n^{[1]}|^2 \overset{(*)}{\le} C_5\cdot|\zeta(\eta_n^{[2]})-\zeta(\eta_n^{[1]})|^2,\end{aligned} \tag{13.11}$$

where $(*)$ again follows from Proposition 6.2(1).

By taking the absolute value in Eq. (13.7) and then applying the estimates (13.9) and (13.11), we obtain the claimed statement. □

Proof (of Theorem 13.2) The following construction depends on a number $N \in \mathbb{N}$. We will see that if N is chosen large enough, then this construction will yield a divisor of finite type D^* so that (13.1) holds.

The idea of the proof is as follows: We need to construct the trace function $\varDelta(\lambda)$ (with the asymptotic $\varDelta - \varDelta_0 \in \mathrm{As}(\mathbb{C}^*, \ell^2_{0,0}, 1))$ for a spectral curve Σ such that $(\lambda_k, \mu_k) \in \Sigma$ holds for $|k| \leq N$ and so that Σ has a double point in each excluded domain $\widehat{U}_{k,\delta}$ with $|k| > N$. The latter condition means: We have $\varDelta(\eta_k) = 2(-1)^k$ for $|k| > N$, where $(\eta_k)_{k \in \mathbb{Z} \cup \{*\}}$ is the sequence of zeros of $\varDelta'$ as in Lemma 13.3(1). Because the λ_k are pairwise unequal (D being tame), Proposition 10.4 shows that we can uniquely determine a trace function $\varDelta$ with $\varDelta - \varDelta_0 \in \mathrm{As}(\mathbb{C}^*, \ell^2_{0,0}, 1)$ by prescribing the values of $\varDelta$ at the points λ_k as

$$\varDelta(\lambda_k) = 2(-1)^k + z_k$$

with a sequence $(z_k) \in \ell^2_{0,0}(k)$. $\varDelta'$ has a zero in each excluded domain $U_{k,\delta}$ with $|k|$ large, and therefore $\varDelta$ is approximately constant on $U_{k,\delta}$. If some $\varDelta$ defined via any sequence $(z_k) \in \ell^2_{0,0}(k)$ is given, we can therefore expect to decrease $|\varDelta(\eta_k) - 2(-1)^k|$ by passing from (z_k) to $(\widetilde{z}_k)$ defined by

$$\widetilde{z}_k = z_k - \left(\varDelta(\eta_k) - 2(-1)^k\right) = \varDelta(\lambda_k) - \varDelta(\eta_k).$$

In particular the desired equality $\varDelta(\eta_k) = 2(-1)^k$ is equivalent to the fixed point equation $\widetilde{z}_k = z_k$. The following proof shows that (for N chosen suitably large), the iteration map $(z_k) \mapsto (\widetilde{z}_k)$ defines a contraction with respect to the ℓ^2-norm, and thus has a unique fixed point by Banach's Fixed Point Theorem.

To carry out the described idea, we consider the Banach space $\mathfrak{B}_N$ of ℓ^2-sequences (z_k), where the index k runs through all the integers with $|k| > N$, equipped with the ℓ^2-norm. To each member $(z_k) \in \mathfrak{B}_N$ we associate the holomorphic function $\varDelta : \mathbb{C}^* \to \mathbb{C}$ with $\varDelta - \varDelta_0 \in \mathrm{As}(\mathbb{C}^*, \ell^2_{0,0}, 1)$ and

$$\varDelta(\lambda_k) = \begin{cases} \mu_k + \mu_k^{-1} & \text{for } |k| \leq N \\ 2(-1)^k + z_k & \text{for } |k| > N \end{cases}$$

for all $k \in \mathbb{Z}$; existence of $\varDelta$ follows from Proposition 10.4(1) (applied to $\frac{1}{2}\varDelta$ with $\upsilon = 1$), and $\varDelta$ is uniquely determined by these equations because D is tame, and hence the λ_k are pairwise unequal (Proposition 10.4(2)). In this situation, we also have the sequence $(\eta_k)_{k \in \mathbb{Z} \cup \{*\}}$ of the zeros of $\varDelta'$ as in Lemma 13.3(1); the sequence satisfies $\eta_k - \lambda_{k,0} \in \ell^2_{-1,3}(k)$.

We now define for given $(z_k) \in \mathfrak{B}_N$ and the associated objects $\varDelta$, (η_k) a new sequence $(\widetilde{z}_k)_{|k|>N}$ by

$$\widetilde{z}_k := z_k - \left(\varDelta(\eta_k) - 2\,(-1)^k\right) = \varDelta(\lambda_k) - \varDelta(\eta_k).$$

We then have for $|k| > N$

$$|\widetilde{z}_k| = |\varDelta(\lambda_k) - \varDelta(\eta_k)| \leq \int_{\eta_k}^{\lambda_k} \underbrace{|\varDelta'(\lambda)|}_{\in \mathrm{As}(\mathbb{C}^*, \ell^\infty_{1,-3}, 1)} \, d\lambda \leq \ell^\infty_{1,-3}(k) \cdot \underbrace{|\lambda_k - \eta_k|}_{\in \ell^2_{-1,3}(k)} \in \ell^2_{0,0}(k),$$

and therefore we have $(\widetilde{z}_k) \in \mathfrak{B}_N$. Hence, the map

$$\Phi : \mathfrak{B}_N \to \mathfrak{B}_N, \ (z_k) \mapsto (\widetilde{z}_k)$$

is well-defined.

Before we show that for sufficiently large N, Φ is a contraction on some small closed ball in $\mathfrak{B}_N$, we look at what happens if we apply Φ to the sequence $z_k = 0 \in \mathfrak{B}_N$. It should be noted that the function $\Delta^{[0]}$ associated to this sequence depends on the choice of N, as do the associated sequences $(\eta_k^{[0]})$ and $(\widetilde{z}_n) := \Phi(0) \in \mathfrak{B}_N$. First we note that $\Delta^{[0]}$ is characterized by

$$\frac{1}{2}\,\Delta^{[0]}(\lambda_k) = \mu_k^{[0]} := \begin{cases} \frac{1}{2}(\mu_k + \mu_k^{-1}) & \text{for } |k| \leq N \\ \frac{1}{2}(\mu_{k,0} + \mu_{k,0}^{-1}) = (-1)^k & \text{for } |k| > N \end{cases},$$

and therefore, if we put $R_0 := \max\{\|\lambda_k - \lambda_{k,0}\|_{\ell^2_{-1,3}}, \ \|\frac{1}{2}(\mu_k + \mu_k^{-1}) - \mu_{k,0}\|_{\ell^2_{0,0}}\}$ (the value of this constant depends on the divisor D, but not on N), we have

$$\|\lambda_k - \lambda_{k,0}\|_{\ell^2_{-1,3}} \leq R_0 \quad \text{and} \quad \|\mu_k^{[0]} - \mu_{k,0}\|_{\ell^2_{0,0}} \leq R_0,$$

therefore it follows from Proposition 10.4(3)(c) that there exists a constant $C_1 > 0$ (again depending on D, but not on N) so that

$$\|\Delta^{[0]} - \Delta_0\|_{\mathrm{As}(\mathbb{C}^*, \ell^2_{0,0}, 1)} \leq C_1 \cdot R_0$$

holds, moreover by Proposition 10.4(3)(b) (note that we have $\upsilon^{[\nu]} = 1$ and $\tau^{[2]} = 1$ in the application of that proposition), the $\ell^2_{0,0}$-sequence

$$C_1 \cdot \left(\left(a_k * \frac{1}{|k|} \right) * \frac{1}{|k|} + \left(|\mu_k^{[0]} - \mu_{k,0}| + \frac{|\tau - 1|}{|k|} \right) * \frac{1}{|k|} \right)$$

with

$$a_k := \begin{cases} k^{-1} \cdot |\lambda_k - \lambda_{k,0}| & \text{for } k \geq 0 \\ |k|^3 \cdot |\lambda_k - \lambda_{k,0}| & \text{for } k < 0 \end{cases}$$

is a bounding sequence for $\Delta^{[0]} - \Delta_0$. Because we have $|\mu_k^{[0]} - \mu_{k,0}| \leq |\frac{1}{2}(\mu_k + \mu_k^{-1}) - \mu_{k,0}|$ for all k, in fact

$$b_k := C_1 \cdot \left(\left(a_k * \frac{1}{|k|} \right) * \frac{1}{|k|} + \left(\left|\tfrac{1}{2}(\mu_k + \mu_k^{-1}) - \mu_{k,0}\right| + \frac{|\tau - 1|}{|k|} \right) * \frac{1}{|k|} \right)$$

is another bounding sequence in $\ell^2_{0,0}(k)$ for $\Delta^{[0]} - \Delta_0$; this sequence does not depend on N. Next we estimate $\|\eta_k^{[0]} - \lambda_{k,0}\|_{\ell^2_{-1,3}(|k|>N)}$ by applying Lemma 13.3(2) in the setting $\Delta^{[1]} = \Delta^{[0]}$, $\Delta^{[2]} = \Delta_0$: Because $\|\Delta^{[0]} - \Delta_0\|_{\mathrm{As}(\mathbb{C}^*,\ell^2_{0,0},1)}$ is bounded independently of N, there exists a constant $C_2 > 0$ which is independent of N, so that we have for $k \in \mathbb{Z}$

$$|\eta_k^{[0]} - \lambda_{k,0}| \leq \tfrac{1}{3}\, C_2 \cdot \begin{Bmatrix} k & \text{if } k > 0 \\ |k|^{-3} & \text{if } k < 0 \end{Bmatrix} \cdot \max\{b_{k-1}, b_k, b_{k+1}\}.$$

Therefore we have

$$\|\eta_k^{[0]} - \lambda_{k,0}\|_{\ell^2_{-1,3}(|k|>N)} \leq C_2 \cdot \|b_k\|_{\ell^2_{0,0}(|k|>N)}. \tag{13.12}$$

Finally, we have

$$\left|\tilde{z}_k^{[0]}\right| = \left|\Delta^{[0]}(\lambda_k) - \Delta^{[0]}(\eta_k)\right| \leq \left(|\eta_k^{[0]} - \lambda_{k,0}| + |\lambda_k - \lambda_{k,0}|\right) \cdot \max_{\lambda \in U_{k,\delta}} |\Delta^{[0]\prime}(\lambda)|.$$

Because we have $\Delta^{[0]} - \Delta_0 \in \mathrm{As}(\mathbb{C}^*, \ell^2_{0,0}, 1)$, where (b_k) is a bounding sequence independent of N, we also have $\Delta^{[0]\prime} - \Delta_0' \in \mathrm{As}(\mathbb{C}^*, \ell^2_{1,-3}, 1)$ with a bounding sequence independent of N by Proposition 9.4(3). Because we also have $\Delta_0' \in \mathrm{As}(\mathbb{C}^*, \ell^\infty_{1,-3}, 1)$, it follows that we have $\Delta^{[0]\prime} \in \mathrm{As}(\mathbb{C}^*, \ell^\infty_{1,-3}, 1)$ with a bounding sequence $(c_k) \in \ell^\infty_{1,-3}(k)$ that is independent of N. With this sequence, we have

$$\left|\tilde{z}_k^{[0]}\right| \leq \left(|\eta_k^{[0]} - \lambda_{k,0}| + |\lambda_k - \lambda_{k,0}|\right) \cdot c_k.$$

Hence there exist constants $C_3, C_4 > 0$, independent of N, so that

$$\begin{aligned} \left\|\tilde{z}_k^{[0]}\right\|_{\mathfrak{B}_N} &\leq C_3 \cdot \left(\|\eta_k^{[0]} - \lambda_{k,0}\|_{\ell^2_{-1,3}} + \|\lambda_k - \lambda_{k,0}\|_{\ell^2_{-1,3}(|k|>N)}\right) \\ &\overset{(13.12)}{\leq} C_3 \cdot \left(C_2 \cdot \|b_k\|_{\ell^2_{0,0}(|k|>N)} + \|\lambda_k - \lambda_{k,0}\|_{\ell^2_{-1,3}(|k|>N)}\right) \\ &\leq C_4 \cdot \left(\|b_k\|_{\ell^2_{0,0}(|k|>N)} + \|\lambda_k - \lambda_{k,0}\|_{\ell^2_{-1,3}(|k|>N)}\right). \end{aligned} \tag{13.13}$$

We now fix besides $N \in \mathbb{N}$ also $\delta > 0$, and consider the closed ball

$$\mathfrak{B}_{N,\delta} := \left\{ (z_k) \in \mathfrak{B}_N \;\middle|\; \|z_k\|_{\mathfrak{B}_N} \leq \delta \right\}$$

in $\mathfrak{B}_N$. We will show that if we choose δ small enough and N large enough, then Φ is a contracting self-mapping on $\mathfrak{B}_{N,\delta}$.

For this purpose, we let two sequences $(z_k^{[1]}), (z_k^{[2]}) \in \mathfrak{B}_N$ be given, and we denote the objects associated to $(z_k^{[\nu]})$ by $\Delta^{[\nu]}$, $(\eta_k^{[\nu]})$ and $(\tilde{z}_k^{[\nu]})$. Moreover, we denote the objects associated to the zero sequence $(z_k = 0)$ by $\Delta^{[0]}$, $(\eta_k^{[0]})$ and $(\tilde{z}_k^{[0]})$ as before. By Lemma 13.4 (for its application, note that for the $w_k^{[\nu]}$ from the Lemma we have $w_k^{[1]} - w_k^{[2]} = \Delta^{[1]}(\lambda_k) - \Delta^{[2]}(\lambda_k) = z_k^{[1]} - z_k^{[2]}$) we have for $|n| > N$

$$|\tilde{z}_n^{[1]} - \tilde{z}_n^{[2]}| = \Big|\Big(\Delta^{[1]}(\eta_n^{[1]}) - \Delta^{[1]}(\lambda_n)\Big) - \Big(\Delta^{[2]}(\eta_n^{[2]}) - \Delta^{[2]}(\lambda_n)\Big)\Big|$$
$$\leq C_5 \cdot \left(|\zeta(\eta_n^{[1]}) - \zeta(\lambda_n)| \cdot \left(\frac{1}{|k|} * |z_k^{[1]} - z_k^{[2]}|\right)_n + |\zeta(\eta_n^{[1]}) - \zeta(\eta_n^{[2]})|^2\right),$$

where C_5 and all C_k $(k > 5)$ occurring in the sequel are positive constants, which apply uniformly for all $(z_k) \in \mathfrak{B}_{N,\delta}$, all sufficiently large $N \in \mathbb{N}$, and all $\delta > 0$ which are smaller than some arbitrarily fixed upper bound. From this estimate, we obtain by Cauchy-Schwarz's inequality, the variant (7.4) of Young's inequality for weakly ℓ^1-sequences and Proposition 6.2(1)

$$\begin{aligned}
\|\tilde{z}_n^{[1]} - \tilde{z}_n^{[2]}\|_{\mathfrak{B}_N} &\leq C_6 \cdot \|\tilde{z}_n^{[1]} - \tilde{z}_n^{[2]}\|_{\ell^1_{0,0}(|n|>N)} \\
&\leq C_7 \cdot \Big(\|\zeta(\eta_n^{[1]}) - \zeta(\lambda_n)\|_{\ell^2_{0,0}(|n|>N)} \cdot \|z_n^{[1]} - z_n^{[2]}\|_{\ell^2_{0,0}(|n|>N)} \\
&\quad + \|\zeta(\eta_n^{[1]}) - \zeta(\eta_n^{[2]})\|^2_{\ell^2_{0,0}(|n|>N)}\Big) \\
&\leq C_8 \cdot \Big(\|\eta_n^{[1]} - \lambda_n\|_{\ell^2_{-1,3}(|n|>N)} \cdot \|z_n^{[1]} - z_n^{[2]}\|_{\ell^2_{0,0}(|n|>N)} \\
&\quad + \|\eta_n^{[1]} - \eta_n^{[2]}\|^2_{\ell^2_{-1,3}(|n|>N)}\Big).
\end{aligned} \tag{13.14}$$

We now note that we have by Lemma 13.3(2) and Proposition 10.4(2)(c) (in the application of the latter proposition, we have $\lambda_k^{[1]} = \lambda_k^{[2]}$, $\tau^{[1]} = \tau^{[2]}$ and $\upsilon^{[\nu]} = 1$)

$$\|\eta_n^{[1]} - \eta_n^{[2]}\|_{\ell^2_{-1,3}(|n|>N)} \leq C_9 \cdot \|\Delta^{[1]} - \Delta^{[2]}\|_{\mathrm{As}(\mathbb{C}^*,\ell^2_{0,0},1)} \leq C_{10} \cdot \|z_n^{[1]} - z_n^{[2]}\|_{\mathfrak{B}_N}, \tag{13.15}$$

and for $\nu \in \{1, 2\}$ we also have

$$\begin{aligned}
\|\eta_n^{[\nu]} - \lambda_n\|_{\ell^2_{-1,3}(|n|>N)} &\leq \|\eta_n^{[\nu]} - \eta_n^{[0]}\|_{\ell^2_{-1,3}} + \|\eta_n^{[0]} - \lambda_{n,0}\|_{\ell^2_{-1,3}(|n|>N)} \\
&\quad + \|\lambda_n - \lambda_{n,0}\|_{\ell^2_{-1,3}(|n|>N)} \\
&\overset{(*)}{\leq} C_{10} \cdot \|z_n^{[\nu]}\|_{\mathfrak{B}_N} + C_2 \cdot \|b_n\|_{\ell^2_{0,0}(|n|>N)} + \|\lambda_n - \lambda_{n,0}\|_{\ell^2_{-1,3}(|n|>N)},
\end{aligned} \tag{13.16}$$

where $(*)$ follows from (13.12) and (13.15). By applying this to (13.14), we obtain

$$
\begin{aligned}
&\|\widetilde{z}_n^{[1]} - \widetilde{z}_n^{[2]}\|_{\mathfrak{B}_N} \\
&\quad \le \Big(C_{10} \cdot \|z_n^{[1]}\|_{\mathfrak{B}_N} + C_2 \cdot \|b_k\|_{\ell^2_{0,0}(|n|>N)} \\
&\qquad + \|\lambda_n - \lambda_{n,0}\|_{\ell^2_{-1,3}(|n|>N)} + C_{10}^2 \cdot \|z_n^{[1]} - z_n^{[2]}\|_{\mathfrak{B}_N} \Big) \cdot \|z_n^{[1]} - z_n^{[2]}\|_{\mathfrak{B}_N} \\
&\quad \le C_{11} \cdot \Big(\|z_n^{[1]}\|_{\mathfrak{B}_N} + \|z_n^{[2]}\|_{\mathfrak{B}_N} + \|b_k\|_{\ell^2_{0,0}(|n|>N)} \\
&\qquad + \|\lambda_n - \lambda_{n,0}\|_{\ell^2_{-1,3}(|n|>N)} \Big) \cdot \|z_n^{[1]} - z_n^{[2]}\|_{\mathfrak{B}_N} .
\end{aligned} \tag{13.17}
$$

Inequality (13.17) shows that Φ is Lipschitz continuous on $\mathfrak{B}_{N,\delta}$.

If we now choose

$$\delta := \min\left\{ \frac{1}{8\,C_{11}}, \frac{\varepsilon}{C_{10} + C_2 + 2} \right\}$$

and then $N \in \mathbb{N}$ so large that the following inequalities hold:

$$
\begin{aligned}
\|b_n\|_{\ell^2_{0,0}(|n|>N)} &\le \min\left\{ \delta, \frac{\delta}{4\,C_4} \right\} \\
\|\lambda_n - \lambda_{n,0}\|_{\ell^2_{-1,3}(|n|>N)} &\le \min\left\{ \delta, \frac{\delta}{4\,C_4} \right\} \\
\|\mu_n - \mu_{n,0}\|_{\ell^2_{0,0}(|n|>N)} &\le \delta,
\end{aligned}
$$

then it follows from Eq. (13.17) that for $(z_k^{[\nu]}) \in \mathfrak{B}_{N,\delta}$ we have

$$\|\widetilde{z}_n^{[1]} - \widetilde{z}_n^{[2]}\|_{\mathfrak{B}_N} \le C_{11} \cdot (\delta + \delta + \delta + \delta) \cdot \|z_n^{[1]} - z_n^{[2]}\|_{\mathfrak{B}_N} \le \tfrac{1}{2} \|z_n^{[1]} - z_n^{[2]}\|_{\mathfrak{B}_N} . \tag{13.18}$$

Moreover, by setting $z_k^{[1]} = z_k \in \mathfrak{B}_{N,\delta}$ and $z_k^{[2]} = 0$ in this inequality, we also see

$$\|\widetilde{z}_n - \widetilde{z}_n^{[0]}\|_{\mathfrak{B}_N} \le \tfrac{1}{2} \|z_n\|_{\mathfrak{B}_N}$$

and therefore

$$
\begin{aligned}
\|\widetilde{z}_n\|_{\mathfrak{B}_N} &\le \tfrac{1}{2}\, \|z_n\|_{\mathfrak{B}_N} + \|\widetilde{z}_n^{[0]}\|_{\mathfrak{B}_N} \\
&\overset{(13.13)}{\le} \tfrac{1}{2}\, \|z_n\|_{\mathfrak{B}_N} + C_4 \cdot \Big(\|b_n\|_{\ell^2_{0,0}(|n|>N)} + \|\lambda_n - \lambda_{n,0}\|_{\ell^2_{-1,3}(|n|>N)} \Big) \\
&\le \tfrac{1}{2}\, \delta + C_4 \cdot \Big(\tfrac{\delta}{4\,C_4} + \tfrac{\delta}{4\,C_4} \Big) = \delta .
\end{aligned} \tag{13.19}
$$

The inequality (13.19) shows that Φ maps the complete metric space $\mathfrak{B}_{N,\delta}$ into itself, and (13.18) shows that Φ is a contraction (with Lipschitz constant $\frac{1}{2}$) on this space.

The Banach Fixed Point Theorem therefore implies that Φ has exactly one fixed point (z_n^*) in $\mathfrak{B}_{N,\delta}$. We let Δ^* and (η_k^*) be the objects associated to (z_n^*). Then we define the divisor $D^* = \{(\lambda_k^*, \mu_k^*)\}$ by

$$\lambda_k^* := \begin{cases} \lambda_k & \text{for } |k| \leq N \\ \eta_k^* & \text{for } |k| > N \end{cases} \quad \text{and} \quad \mu_k^* := \begin{cases} \mu_k & \text{for } |k| \leq N \\ (-1)^k & \text{for } |k| > N \end{cases}.$$

By construction, D^* is an asymptotic divisor on the spectral curve Σ^* corresponding to Δ^*. Because (z_k^*) is a fixed point of Φ, we have for $|k| > N$

$$z_k^* = z_k^* - (\Delta^*(\eta_k^*) - 2(-1)^k)$$

and therefore $\Delta^*(\eta_k^*) = 2(-1)^k$. Because we have $\Delta^{*\prime}(\eta_k^*) = 0$ by the definition of η_k^*, it follows that the spectral curve Σ^* associated to Δ^* has a double point at $(\eta_k^*, 2(-1)^k) = (\lambda_k^*, \mu_k^*)$. Hence D^* is of finite type.

Moreover we have

$$\begin{aligned} \|D^* - D\|_{\mathsf{Div}} &\leq \|\lambda_n^* - \lambda_n\|_{\ell^2_{-1,3}(n)} + \|\mu_n^* - \mu_n\|_{\ell^2_{0,0}(n)} \\ &= \|\eta_n^* - \lambda_n\|_{\ell^2_{-1,3}(|n|>N)} + \|(-1)^n - \mu_n\|_{\ell^2_{0,0}(|n|>N)} \\ &\overset{(13.16)}{\leq} C_{10} \cdot \|z_n^*\|_{\mathfrak{B}_N} + C_2 \cdot \|b_n\|_{\ell^2_{0,0}(|n|>N)} \\ &\quad + \|\lambda_n - \lambda_{n,0}\|_{\ell^2_{-1,3}(|n|>N)} + \|(-1)^n - \mu_n\|_{\ell^2_{0,0}(|n|>N)} \\ &\leq C_{10} \cdot \delta + C_2 \cdot \delta + \delta + \delta = (C_{10} + C_2 + 2) \cdot \delta \leq \varepsilon. \end{aligned}$$

Finally, if we choose N large enough that $\lambda_k \in U_{k,\delta}$ holds for $|k| > N$, and suppose that ε is small enough that the condition $\|D^* - D\|_{\mathsf{Div}} < \varepsilon$ implies that then also $\lambda_k^* \in U_{k,\delta}$ holds for $|k| > N$, then the λ_k^* are necessarily pairwise unequal. Therefore D^* is then tame. □

Chapter 14
Darboux Coordinates for the Space of Potentials

In the present chapter we will equip the space of potentials Pot with the structure of an (infinite-dimensional) symplectic manifold, and construct coordinates for this manifold which are adapted to the symplectic structure. By analogy with the finite-dimensional situation, we will call these coordinates *Darboux coordinates*.

More specifically, we consider the Hilbert space Pot (a hyperplane in $\mathsf{Pot}_{np} = W^{1,2}([0,1]) \times L^2([0,1])$). For $(u, u_y) \in \mathsf{Pot}$, the tangent space $T_{(u,u_y)}\mathsf{Pot}$ is canonically isomorphic to Pot, and we will typically denote elements of $T_{(u,u_y)}\mathsf{Pot}$ by $(\delta u, \delta u_y)$ and $(\widetilde{\delta} u, \widetilde{\delta} u_y)$. For any function f defined on Pot, we let $\delta f := \frac{\partial f}{\partial(u,u_y)} \cdot (\delta u, \delta u_y)$ be the variation of f in the direction of $(\delta u, \delta u_y) \in T_{(u,u_y)}\mathsf{Pot}$.

In this setting we define a complex bilinear form Ω on each tangent space $T_{(u,u_y)}\mathsf{Pot}$ by setting

$$\Omega\big((\delta u, \delta u_y),\, (\widetilde{\delta} u, \widetilde{\delta} u_y) \big) = \int_0^1 \big(\delta u \cdot \widetilde{\delta} u_y - \widetilde{\delta} u \cdot \delta u_y\big)\, \mathrm{d}x. \tag{14.1}$$

Ω defines a non-degenerate symplectic form on Pot.

We recall a theorem for the *finite-dimensional* situation due to Darboux:

Theorem 14.1 (Darboux [Da]) *Any symplectic manifold (M, Ω) of dimension $2n < \infty$ is* locally *symplectomorphic to an open subset of $(\mathbb{R}^{2n}, \Omega_0)$, where Ω_0 is the canonical symplectic form on $\mathbb{R}^{2n}$.*

In the sequel, we will obtain coordinates on the symplectic space Pot near any tame potential (u, u_y), which are analogous to the coordinates promised by Darboux's theorem for the finite-dimensional case. These coordinates are an excellent and important instrument for understanding the structure of Pot; we will base our proof (in Chap. 15) that the map $\mathsf{Pot}_{tame} \to \mathsf{Div}_{tame}$ is a diffeomorphism on the use of these coordinates. Concerning the analogous coordinates for the space

S. Klein, *A Spectral Theory for Simply Periodic Solutions of the Sinh-Gordon Equation*, Lecture Notes in Mathematics 2229,
https://doi.org/10.1007/978-3-030-01276-2_14

of potentials for the 1-dimensional Schrödinger equation (see [Pö-T, Theorem 2.8, p. 44]), Pöschel and Trubowitz write: "It is only a slight exaggeration to say that [these coordinates are] the basis of almost everything else we are going to do.". In our situation, we will not use the Darboux coordinates quite as intensely, but they will still prove to be very useful. The application of the Darboux coordinates for the 1-dimensional Schrödinger equation to the spectral theory for the KdV equation (for which the 1-dimensional Schrödinger operator is the Lax operator) is described in [Kapp-P, Section 8].

In our situation concerning the sinh-Gordon equation, Knopf [Kn-2] has constructed a symplectic basis (with respect to Ω) for a certain linear subspace U of $T_{(u,u_y)}\mathsf{Pot}$, and thereby effectively Darboux coordinates for this subspace. Knopf did not show that this subspace is actually dense in $T_{(u,u_y)}\mathsf{Pot}$, but we will combine his representation with the asymptotic estimates of the present work (especially the asymptotic description of the extended frame given in Proposition 11.7, giving rise to the asymptotic descriptions of $\delta M(\lambda)$ in Lemma 14.6 and of $\delta\lambda_k$, $\delta\mu_k$ in Lemma 14.8 below) to prove that in fact $\overline{U} = T_{(u,u_y)}\mathsf{Pot}$ holds. Because of this, we will see that Knopf [Kn-2] essentially provides Darboux coordinates for Pot (near tame potentials).

Because we will base our construction of Darboux coordinates on [Kn-2], we now report on the results of that paper, while using the notations of the present book. When adapting the results from [Kn-2], one should note that the conventions of that paper differ from ours in several important points: (1) The norming of the potential u differ, as is evidenced by Knopf's sinh-Gordon equation $\Delta u + 2\sinh(2u) = 0$ ([Kn-2, Equation (1.1)]) in comparison to our sinh-Gordon equation $\Delta u + \sinh(u) = 0$ (Eq. (2.1)). (2) Knopf's flat connection form α_λ differs from ours by a factor of 2, and by an additional conjugation with a constant matrix (compare Eq. (3.2) to [Kn-2, Equations (2.1)]). (3) Knopf normalizes the eigenvector field of the monodromy as $(1, \frac{\mu-a}{b})$, whereas we use $(\frac{\mu-d}{c}, 1)$. In the definition of the (classical) spectral divisor, the pair of holomorphic functions (c, a) is therefore replaced by (b, d).

We fix $(u, u_y) \in \mathsf{Pot}_{tame}$ and let $D = \{(\lambda_k, \mu_k)\}$ be the spectral divisor of (u, u_y). By hypothesis, D is tame, i.e. D does not contain any multiple points. Therefore λ_k and μ_k can be interpreted as well-defined smooth functions on Pot near (u, u_y), and $c'(\lambda_k) \neq 0$ holds for all $k \in \mathbb{Z}$.

In the sequel we consider the extended frame $F(x, \lambda)$ corresponding to the potential (u, u_y), we write it in the form

$$F(x,\lambda) = \begin{pmatrix} a(x,\lambda) & b(x,\lambda) \\ c(x,\lambda) & d(x,\lambda) \end{pmatrix} \tag{14.2}$$

like we already did in Proposition 11.7, and also consider the extended frame of the vacuum

$$F_0(x,\lambda) := \begin{pmatrix} a_0(x,\lambda) & b_0(x,\lambda) \\ c_0(x,\lambda) & d_0(x,\lambda) \end{pmatrix} := \begin{pmatrix} \cos(x\,\zeta(\lambda)) & -\lambda^{-1/2}\sin(x\,\zeta(\lambda)) \\ \lambda^{1/2}\sin(x\,\zeta(\lambda)) & \cos(x\,\zeta(\lambda)) \end{pmatrix}.$$

We now define for $k \in \mathbb{Z}$

$$v_k := (v_{k,1}, v_{k,2}) \quad \text{and} \quad w_k := (w_{k,1}, w_{k,2}) \tag{14.3}$$

with

$$\begin{aligned}
v_{k,1}(x) &:= a(x,\lambda_k)\, c(x,\lambda_k) \\
v_{k,2}(x) &:= \tfrac{i}{4}\Big((e^{u(x)/2} - \lambda_k\, e^{-u(x)/2})\, a(x,\lambda_k)^2 + (e^{u(x)/2} - \lambda_k^{-1}\, e^{-u(x)/2})\, c(x,\lambda_k)^2 \Big) \\
w_{k,1}(x) &:= a(x,\lambda_k)\, d(x,\lambda_k) + b(x,\lambda_k)\, c(x,\lambda_k) \\
w_{k,2}(x) &:= \tfrac{i}{2}\big((e^{u(x)/2} - \lambda_k\, e^{-u(x)/2})\, a(x,\lambda_k)\, b(x,\lambda_k) \\
&\qquad + (e^{u(x)/2} - \lambda_k^{-1}\, e^{-u(x)/2})\, c(x,\lambda_k)\, d(x,\lambda_k) \big)
\end{aligned}$$

and we also put

$$\vartheta_k := \int_0^1 \Big(\lambda_k\, a(x,\lambda_k)^2 + \lambda_k^{-1}\, c(x,\lambda_k)^2 \Big) \cdot e^{u(x)/2}\, dx. \tag{14.4}$$

In [Kn-2], it is shown in the proof of Theorem 5.5 that $c'(\lambda_k) = -i\, \frac{\mu_k}{2\lambda_k} \cdot \vartheta_k$ holds; because we have $c'(\lambda_k) \neq 0$ it follows that

$$\vartheta_k \neq 0$$

holds.

We can now state the main result of [Kn-2]:

Theorem 14.2 (Knopf [Kn-2]) *Let* $(u, u_y) \in \mathsf{Pot}_{tame}$.

(1) For $k \in \mathbb{Z}$ *we have* $v_k, w_k \in T_{(u,u_y)}\mathsf{Pot}$, *and up to the factor* ϑ_k, (v_k, w_k) *is a system of symplectic vectors with respect to* Ω, *i.e. we have for all* $k, \ell \in \mathbb{Z}$

$$\begin{aligned}
\Omega(v_k, v_\ell) &= 0 \\
\Omega(v_k, w_\ell) &= \vartheta_k \cdot \delta_{k\ell} \quad (\delta_{k\ell} : \textit{Kronecker delta}) \\
\Omega(w_k, w_\ell) &= 0.
\end{aligned}$$

(2) For all $(\delta u, \delta u_y), (\widetilde{\delta} u, \widetilde{\delta} u_y) \in T_{(u,u_y)}\mathsf{Pot}$ *that are finite linear combinations of the* v_k *and* w_k, *we have*

$$\Omega\big((\delta u, \delta u_y), (\widetilde{\delta} u, \widetilde{\delta} u_y) \big) = \frac{i}{2} \sum_{k \in \mathbb{Z}} \left(\frac{\delta \lambda_k}{\lambda_k} \cdot \frac{\widetilde{\delta} \mu_k}{\mu_k} - \frac{\widetilde{\delta} \lambda_k}{\lambda_k} \cdot \frac{\delta \mu_k}{\mu_k} \right). \tag{14.5}$$

In the sequel, we liberate Knopf's preceding result from the stated restrictions. More specifically, we prove the following two statements:

1. The symplectic system (v_k, w_k) defined by Eqs. (14.3) spans $T_{(u,u_y)}\mathsf{Pot}$ (in the Hilbert space sense, i.e. $\overline{\mathrm{span}\{\, v_k, w_k \mid k \in \mathbb{Z}\,\}} = T_{(u,u_y)}\mathsf{Pot}$).
2. The infinite sum on the right-hand side of Eq. (14.5) converges absolutely for arbitrary $(\delta u, \delta u_y)$, $(\widetilde{\delta u}, \widetilde{\delta u}_y) \in T_{(u,u_y)}\mathsf{Pot}$.

In this way we will generalize Knopf's Theorem 14.2 to the result of the following theorem. This result shows that the (v_k, w_k) essentially define Darboux coordinates on Pot (near any tame potential), and the functions $\lambda_k^{-1}\,\delta\lambda_k$ and $\mu_k^{-1}\,\delta\mu_k$ define Darboux coordinates on Div (near any tame divisor, where Div is equipped with the symplectic structure induced by the map $\mathsf{Pot} \to \mathsf{Div}$ associating to each potential its associated spectral divisor).

Theorem 14.3 *Let* $(u, u_y) \in \mathsf{Pot}_{tame}$.

(1) The symplectic system (v_k, w_k) *defined by Eqs.* (14.3) *is a symplectic basis of* $T_{(u,u_y)}\mathsf{Pot}$ *(as before, up to the factor* ϑ_k*).*
(2) For all $(\delta u, \delta u_y)$, $(\widetilde{\delta u}, \widetilde{\delta u}_y) \in T_{(u,u_y)}\mathsf{Pot}$,

$$\Omega\big(\,(\delta u, \delta u_y),\ (\widetilde{\delta u}, \widetilde{\delta u}_y)\,\big) = \frac{\mathrm{i}}{2}\sum_{k\in\mathbb{Z}}\left(\frac{\delta\lambda_k}{\lambda_k}\cdot\frac{\widetilde{\delta\mu}_k}{\mu_k} - \frac{\widetilde{\delta\lambda}_k}{\lambda_k}\cdot\frac{\delta\mu_k}{\mu_k}\right) \tag{14.6}$$

holds, and the sum on the right hand side of this equation converges.

Remark 14.4 In Theorem 14.3 we are still restricted to tame potentials resp. divisors, i.e. spectral divisors without multiple points. To lift this restriction, we would need to deal with the complication that if $(\lambda_k, \mu_k) = (\lambda_{k'}, \mu_{k'})$ holds for $k \neq k'$, then also $v_k = v_{k'}$ and $w_k = w_{k'}$ holds, and therefore we do not get "enough" independent coordinates by the (v_k, w_k) in this case. To obtain the "missing" coordinates, we would need to consider appropriate coordinates on Div in such points; they are given by the elementary symmetric polynomials in the λ_k resp. μ_k, see the discussion at the end of Chap. 8.

The remainder of the chapter is dedicated to the proof of Theorem 14.3.

Lemma 14.5 *Let* $(u, u_y) \in \mathsf{Pot}_{tame}$ *be given. The symplectic system* (v_k, w_k) *defined by Eqs.* (14.3) *is a basis of* $T_{(u,u_y)}\mathsf{Pot}$.

Proof By Theorem 14.2(1) (=[Kn-2, Theorem 5.4]), (v_k, w_k) is a symplectic system (up to the factor ϑ_k) with respect to the non-degenerate symplectic form Ω, and therefore these vectors are linear independent. It remains to show that they span $T_{(u,u_y)}\mathsf{Pot} \cong \{\,(u, u_y) \in W^{1,2}([0,1]) \times L^2([0,1]) \,\big|\, u(0) = u(1)\,\}$ in the Hilbert space sense, i.e. that the set $\mathrm{span}\{\, v_k, w_k \mid k \in \mathbb{Z}\,\}$ is dense in $T_{(u,u_y)}\mathsf{Pot}$. Because $\{\, u \in W^{1,2}([0,1]) \,\big|\, u(0) = u(1)\,\}$ is dense in $L^2([0,1])$, the preceding claim is equivalent to the statement that the set $\mathrm{span}\{\, v_k, w_k \mid k \in \mathbb{Z}\,\}$ is dense in $L^2([0,1]) \times L^2([0,1])$. We will prove the latter statement by applying the following fact from functional analysis, which we cite from [Pö-T, Theorem D.3, p. 163f.] (where its proof can also be found):

> Let $(h_i)_{i\in I}$ be an orthonormal basis of a Hilbert space H, and suppose that $(\widetilde{h}_i)_{i\in I}$ is another sequence of vectors in H that is linear independent. If in addition
>
> $$\sum_{i\in I} \|h_i - \widetilde{h}_i\|^2 < \infty,$$
>
> then $(\widetilde{h}_i)_{i\in I}$ is also a basis of H.

To apply this theorem to the present situation (with $H := L^2([0,1]) \times L^2([0,1])$), we need to investigate the asymptotic behavior of the v_k and the w_k. For this purpose, we denote for $1 \leq p \leq \infty$ and $n \in \mathbb{Z}$ by $\ell^p_n(L^2([0,1]))$ the space of sequences $(f_k)_{k\geq 1}$ in $L^2([0,1])$ with $(\|f_k\|_{L^2([0,1])}) \in \ell^p_n \, (k \geq 1)$.

As an exemplar case we look at $v_{k,1}(x) = a(x,\lambda_k)\cdot c(x,\lambda_k)$ for $k > 0$. By Proposition 11.7 we have

$$\begin{aligned} &a - \upsilon\, a_0 \in \mathrm{As}_\infty(\mathbb{C}^*, \ell^2_0, 1), \quad c - \tau\, c_0 \in \mathrm{As}_\infty(\mathbb{C}^*, \ell^2_{-1}, 1), \\ &c \in \mathrm{As}_\infty(\mathbb{C}^*, \ell^\infty_{-1}, 1) \quad \text{and} \quad \upsilon\, a_0 \in \mathrm{As}_\infty(\mathbb{C}^*, \ell^\infty_0, 1), \end{aligned}$$

where all these asymptotic assessments are uniform in $x \in [0,1]$ and where we put $\tau(x) := \mathrm{e}^{-(u(0)+u(x))/4}$, $\upsilon(x) := \mathrm{e}^{(u(x)-u(0))/4}$. As a consequence we have

$$\begin{aligned} a(x,\lambda_k) - \upsilon(x)\, a_0(x,\lambda_k) &\in \ell^2_0(L^2([0,1])) \\ c(x,\lambda_k) - \tau(x)\, c_0(x,\lambda_k) &\in \ell^2_{-1}(L^2([0,1])) \\ c(x,\lambda_k) &\in \ell^\infty_{-1}(L^2([0,1])) \\ \upsilon(x)\, a_0(x,\lambda_k) &\in \ell^\infty_0(L^2([0,1])) \end{aligned}$$

and therefore

$$\begin{aligned} &v_{k,1}(x) - \frac{1}{2}\sqrt{\lambda_k}\, \mathrm{e}^{-u(0)/2} \sin(2\zeta(\lambda_k)x) \\ &\quad = a(x,\lambda_k)\cdot c(x,\lambda_k) - \upsilon(x)\, a_0(x,\lambda_k)\, \tau(x)\, c_0(x,\lambda_k) \in \ell^2_{-1}(L^2([0,1])). \end{aligned}$$

Moreover, it is easy to check that

$$\sqrt{\lambda_k}\, \sin(2\zeta(\lambda_k)x) - 4\pi k\, \sin(2k\pi\, x) \in \ell^2_{-1}(L^2([0,1]))$$

holds, and thus we obtain

$$v_{k,1}(x) - 2\pi k\, \mathrm{e}^{-u(0)/2} \sin(2k\pi\, x) \in \ell^2_{-1}(L^2([0,1])).$$

By carrying out analogous calculations for $v_{k,2}$ and $w_{k,\nu}$, one obtains for $k > 0$

$$\begin{aligned}
&v_{k,1}(x) - k \cdot 2\pi\, \mathrm{e}^{-u(0)/2} \cdot \sin(2k\pi\, x) && \in \ell^2_{-1}(L^2([0,1])) \\
&v_{k,2}(x) - k^2 \cdot (-4)\pi^2\, \mathrm{i}\, \mathrm{e}^{-u(0)/2} \cdot \cos(2k\pi\, x) && \in \ell^2_{-2}(L^2([0,1])) \\
&w_{k,1}(x) - \cos(2k\pi\, x) && \in \ell^2_{0}(L^2([0,1])) \\
&w_{k,2}(x) - k \cdot 2\pi \mathrm{i} \cdot \sin(2k\pi\, x) && \in \ell^2_{-1}(L^2([0,1]))
\end{aligned}$$

and also

$$\begin{aligned}
&v_{-k,1}(x) - k^{-1} \cdot \tfrac{1}{8\pi}\, \mathrm{e}^{u(0)/2} \cdot \sin(2k\pi\, x) && \in \ell^2_{1}(L^2([0,1])) \\
&v_{-k,2}(x) - \tfrac{\mathrm{i}}{4}\, \mathrm{e}^{u(0)/2} \cdot \cos(2k\pi\, x) && \in \ell^2_{0}(L^2([0,1])) \\
&w_{-k,1}(x) - \cos(2k\pi\, x) && \in \ell^2_{0}(L^2([0,1])) \\
&w_{-k,2}(x) - k \cdot (-2)\pi \mathrm{i} \cdot \sin(2k\pi\, x) && \in \ell^2_{-1}(L^2([0,1]))
\end{aligned}$$

We now define elements of $L^2([0,1]) \times L^2([0,1])$ for $k \geq 1$ by:

$$\begin{aligned}
\widetilde{h}_{s1,k} &:= \sqrt{2} \cdot \left(\tfrac{1}{4\pi}\, \mathrm{e}^{u(0)/2} \cdot k^{-1} \cdot v_k + 4\pi\, \mathrm{e}^{-u(0)/2} \cdot k \cdot v_{-k} \right) \\
\widetilde{h}_{s2,k} &:= \tfrac{\sqrt{2}}{4\pi \mathrm{i}} \cdot k^{-1} \cdot (w_k - w_{-k}) \\
\widetilde{h}_{c1,k} &:= \tfrac{\sqrt{2}}{2} \cdot (w_k + w_{-k}) \\
\widetilde{h}_{c2,k} &:= \sqrt{2} \cdot \left(\tfrac{1}{8\pi^2}\, \mathrm{i}\, \mathrm{e}^{u(0)/2} \cdot k^{-2} \cdot v_k - 2\mathrm{i}\, \mathrm{e}^{-u(0)/2} \cdot v_{-k} \right)
\end{aligned} \tag{14.7}$$

Moreover, we put

$$\widetilde{h}_{c1,0} := v_0 \quad \text{and} \quad \widetilde{h}_{c2,0} := w_0 .$$

Because the vectors in $\{\, v_k, w_k \,|\, k \in \mathbb{Z} \,\}$ are linear independent, the above definitions show that the vectors in $\{\, \widetilde{h}_{c1,k}, \widetilde{h}_{c2,k} \,|\, k \geq 0 \,\} \cup \{\, \widetilde{h}_{s1,k}, \widetilde{h}_{s2,k} \,|\, k \geq 1 \,\}$ are also linear independent. Moreover, from the preceding asymptotic assessments of the $v_{\pm k,\nu}$ and the $w_{\pm k,\nu}$ one can calculate that

$$\begin{aligned}
&\widetilde{h}_{c1,k} - h_{c1,k},\ \widetilde{h}_{c2,k} - h_{c2,k} \ \in\ \ell^2_0(L^2([0,1])) \quad \text{for } k \geq 0 \\
\text{and}\quad &\widetilde{h}_{s1,k} - h_{s1,k},\ \widetilde{h}_{s2,k} - h_{s2,k} \ \in\ \ell^2_0(L^2([0,1])) \quad \text{for } k \geq 1
\end{aligned} \tag{14.8}$$

holds, where we put for $k \geq 0$

$$h_{c1,k}(x) := \left(\sqrt{2}\cos(2k\pi\, x),\, 0\right) \quad \text{and} \quad h_{c2,k}(x) := \left(0,\, \sqrt{2}\cos(2k\pi\, x)\right)$$

and for $k \geq 1$

$$h_{s1,k}(x) := \big(\sqrt{2}\,\sin(2k\pi\,x),\,0\big) \quad\text{and}\quad h_{s2,k}(x) := \big(0,\,\sqrt{2}\,\sin(2k\pi\,x)\big).$$

The set $\{\,h_{c1,k}, h_{c2,k}\,|\,k \geq 0\,\} \cup \{\,h_{s1,k}, h_{s2,k}\,|\,k \geq 1\,\}$ is an orthonormal basis of the Hilbert space $H := L^2([0,1]) \times L^2([0,1])$, the vectors in the set $\{\,\widetilde{h}_{c1,k}, \widetilde{h}_{c2,k}\,|\,k \geq 0\,\} \cup \{\,\widetilde{h}_{s1,k}, \widetilde{h}_{s2,k}\,|\,k \geq 1\,\}$ are linear independent, and by (14.8) we have

$$\sum_{k=0}^{\infty} \|\widetilde{h}_{c1,k} - h_{c1,k}\|_H^2 + \sum_{k=0}^{\infty} \|\widetilde{h}_{c2,k} - h_{c2,k}\|_H^2 + \sum_{k=1}^{\infty} \|\widetilde{h}_{s1,k} - h_{s1,k}\|_H^2 + \sum_{k=1}^{\infty} \|\widetilde{h}_{s2,k} - h_{s2,k}\|_H^2 < \infty.$$

It follows by the theorem cited at the beginning of the proof that $\{\,\widetilde{h}_{c1,k}, \widetilde{h}_{c2,k}\,|\,k \geq 0\,\} \cup \{\,\widetilde{h}_{s1,k}, \widetilde{h}_{s2,k}\,|\,k \geq 1\,\}$ is a basis of H. By solving the Eqs. (14.7) defining the $\widetilde{h}_{\dots}$ for $v_{\pm k}$ and $w_{\pm k}$ it follows that $\{\,v_k, w_k\,|\,k \in \mathbb{Z}\,\}$ also is a basis of H. This completes the proof. □

After the question of Theorem 14.3(1) has been settled with the preceding Lemma, we next need to show the convergence of the infinite sum occurring in Theorem 14.3(2). To do so, we will have to describe the asymptotic behavior of the summand $\frac{\delta\lambda_k}{\lambda_k} \cdot \frac{\widetilde{\delta}\mu_k}{\mu_k} - \frac{\widetilde{\delta}\lambda_k}{\lambda_k} \cdot \frac{\delta\mu_k}{\mu_k}$ for $k \to \pm\infty$, and thus we will need to find out about the asymptotic behaviour of the functions $\delta\lambda_k$ and $\delta\mu_k$.

As preparation for the latter task, the following Lemma describes the asymptotic behaviour of the variation $\delta M(\lambda)$ of the monodromy $M(\lambda)$.

Lemma 14.6 *Let $(u, u_y) \in \mathsf{Pot}_{tame}$ be given. We denote the monodromy of (u, u_y) by $M(\lambda) = \begin{pmatrix} a(\lambda) & b(\lambda) \\ c(\lambda) & d(\lambda) \end{pmatrix}$ as usual.*

(1) (The case $\lambda \to \infty$.) *There exist functions depending on $\lambda \in \mathbb{C}^*$ and $t \in [0,1]$ with*

$$\begin{aligned} f_{1,a} &\in \mathrm{As}_\infty(\mathbb{C}^*, \ell^2_0, 2), & f_{2,a} &\in \mathrm{As}_\infty(\mathbb{C}^*, \ell^\infty_1, 2),\\ f_{1,b} &\in \mathrm{As}_\infty(\mathbb{C}^*, \ell^2_1, 2), & f_{2,b} &\in \mathrm{As}_\infty(\mathbb{C}^*, \ell^\infty_2, 2),\\ f_{1,c} &\in \mathrm{As}_\infty(\mathbb{C}^*, \ell^2_{-1}, 2), & f_{2,c} &\in \mathrm{As}_\infty(\mathbb{C}^*, \ell^\infty_0, 2),\\ f_{1,d} &\in \mathrm{As}_\infty(\mathbb{C}^*, \ell^2_0, 2), & f_{2,d} &\in \mathrm{As}_\infty(\mathbb{C}^*, \ell^\infty_1, 2), \end{aligned}$$

where all these memberships in asymptotic spaces are uniform in $t \in [0,1]$, so that we have

$$\begin{aligned} \delta a(\lambda) = {}& -\tfrac{1}{2}\,a(\lambda) \int_0^1 \delta u_z(t)\,\cos(2\zeta(\lambda)t)\,\mathrm{d}t + \tfrac{1}{2}\,\tau\,\lambda^{1/2}\,b(\lambda) \int_0^1 \delta u_z(t)\,\sin(2\zeta(\lambda)t)\,\mathrm{d}t\\ & + \int_0^1 \delta u_z(t)\,f_{1,a}(t)\,\mathrm{d}t + \int_0^1 \delta u(t)\,f_{2,a}(t)\,\mathrm{d}t \end{aligned}$$

$$\delta b(\lambda) = \tfrac{1}{2}\,b(\lambda) \left(\int_0^1 \delta u_z(t)\,\cos(2\zeta(\lambda)t)\,\mathrm{d}t - \delta u(0) \right) + \cdots$$

$$\cdots + \tfrac{1}{2}\,\tau^{-1}\,\lambda^{-1/2}\,a(\lambda)\,\int_0^1 \delta u_z(t)\,\sin(2\zeta(\lambda)t)\,\mathrm{d}t$$

$$+ \int_0^1 \delta u_z(t)\,f_{1,b}(t)\,\mathrm{d}t + \int_0^1 \delta u(t)\,f_{2,b}(t)\,\mathrm{d}t$$

$$\delta c(\lambda) = -\tfrac{1}{2}\,c(\lambda)\left(\int_0^1 \delta u_z(t)\,\cos(2\zeta(\lambda)t)\,\mathrm{d}t - \delta u(0)\right)$$

$$+ \tfrac{1}{2}\,\tau\,\lambda^{1/2}\,d(\lambda)\,\int_0^1 \delta u_z(t)\,\sin(2\zeta(\lambda)t)\,\mathrm{d}t$$

$$+ \int_0^1 \delta u_z(t)\,f_{1,c}(t)\,\mathrm{d}t + \int_0^1 \delta u(t)\,f_{2,c}(t)\,\mathrm{d}t$$

$$\delta d(\lambda) = \tfrac{1}{2}\,d(\lambda)\,\int_0^1 \delta u_z(t)\,\cos(2\zeta(\lambda)t)\,\mathrm{d}t + \tfrac{1}{2}\,\tau^{-1}\,\lambda^{-1/2}\,c(\lambda)\,\int_0^1 \delta u_z(t)\,\sin(2\zeta(\lambda)t)\,\mathrm{d}t$$

$$+ \int_0^1 \delta u_z(t)\,f_{1,d}(t)\,\mathrm{d}t + \int_0^1 \delta u(t)\,f_{2,d}(t)\,\mathrm{d}t.$$

(2) (The case $\lambda \to 0$.) *There exist functions depending on* $\lambda \in \mathbb{C}^*$ *and* $t \in [0,1]$ *with*

$$g_{1,a} \in \mathrm{As}_0(\mathbb{C}^*, \ell^2_0, 2), \qquad g_{2,a} \in \mathrm{As}_0(\mathbb{C}^*, \ell^\infty_1, 2),$$
$$g_{1,b} \in \mathrm{As}_0(\mathbb{C}^*, \ell^2_{-1}, 2), \qquad g_{2,b} \in \mathrm{As}_0(\mathbb{C}^*, \ell^\infty_0, 2),$$
$$g_{1,c} \in \mathrm{As}_0(\mathbb{C}^*, \ell^2_1, 2) \qquad g_{2,c} \in \mathrm{As}_0(\mathbb{C}^*, \ell^\infty_2, 2)$$
$$g_{1,d} \in \mathrm{As}_0(\mathbb{C}^*, \ell^2_0, 2), \qquad g_{2,d} \in \mathrm{As}_0(\mathbb{C}^*, \ell^\infty_1, 2),$$

where all these memberships in asymptotic spaces are uniform in $t \in [0,1]$, *so that we have*

$$\delta a(\lambda) = \tfrac{1}{2}\,a(\lambda)\,\int_0^1 \delta u_{\bar z}(t)\,\cos(2\zeta(\lambda)t)\,\mathrm{d}t - \tfrac{1}{2}\,\tau^{-1}\,\lambda^{1/2}\,b(\lambda)\,\int_0^1 \delta u_{\bar z}(t)\,\sin(2\zeta(\lambda)t)\,\mathrm{d}t$$

$$+ \int_0^1 \delta u_{\bar z}(t)\,g_{1,a}(t)\,\mathrm{d}t + \int_0^1 \delta u(t)\,g_{2,a}(t)\,\mathrm{d}t$$

$$\delta b(\lambda) = \tfrac{1}{2}\,b(\lambda)\left(-\int_0^1 \delta u_{\bar z}(t)\,\cos(2\zeta(\lambda)t)\,\mathrm{d}t + \delta u(0)\right)$$

$$- \tfrac{1}{2}\,\tau\,\lambda^{-1/2}\,a(\lambda)\,\int_0^1 \delta u_{\bar z}(t)\,\sin(2\zeta(\lambda)t)\,\mathrm{d}t$$

$$+ \int_0^1 \delta u_{\bar z}(t)\,g_{1,b}(t)\,\mathrm{d}t + \int_0^1 \delta u(t)\,g_{2,b}\,\mathrm{d}t$$

$$\delta c(\lambda) = -\tfrac{1}{2}c(\lambda)\left(-\int_0^1 \delta u_{\bar z}(t)\cos(2\zeta(\lambda)t)\mathrm{d}t + \delta u(0)\right)$$

$$- \tfrac{1}{2}\tau^{-1}\lambda^{1/2}d(\lambda)\,\int_0^1 \delta u_{\bar z}(t)\sin(2\zeta(\lambda)t)\mathrm{d}t + \cdots$$

$$\cdots + \int_0^1 \delta u_{\overline{z}}(t)\, g_{1,c}(t)\,\mathrm{d}t + \int_0^1 \delta u(t)\, g_{2,c}(t)\,\mathrm{d}t$$

$$\delta d(\lambda) = -\tfrac{1}{2}\, d(\lambda) \int_0^1 \delta u_{\overline{z}}(t)\, \cos(2\zeta(\lambda)t)\,\mathrm{d}t - \tfrac{1}{2}\,\tau\,\lambda^{-1/2}\, c(\lambda) \int_0^1 \delta u_{\overline{z}}(t)\, \sin(2\zeta(\lambda)t)\,\mathrm{d}t$$

$$+ \int_0^1 \delta u_{\overline{z}}(t)\, g_{1,d}(t)\,\mathrm{d}t + \int_0^1 \delta u(t)\, g_{2,d}(t)\,\mathrm{d}t.$$

Remark 14.7 If we apply Lemma 14.6 to the vacuum potential, i.e. $(u, u_y) = (0,0)$, then all the functions $f_{\dots}$ and $g_{\dots}$ vanish (see the proof below). It follows that up to factors which do not depend on $(\delta u, \delta u_y)$, the variations $\delta a_0(\lambda_{k,0}) \cdot (\delta u, \delta u_y), \dots, \delta d_0(\lambda_{k,0}) \cdot (\delta u, \delta u_y)$ are the $|k|$-th Fourier coefficients of δu_z (for $k > 0$) or of $\delta u_{\overline{z}}$ (for $k < 0$). This is a linearised version of the observation that for the monodromy $M(\lambda)$ corresponding to a potential $(u, u_y) \in \mathsf{Pot}$, a first order approximation of $M(\lambda_{k,0})$ is given by the $|k|$-th Fourier coefficients of u_z resp. of $u_{\overline{z}}$, see Theorem 7.1.

Proof (of Lemma 14.6) For (1). Let $(u, u_y) \in \mathsf{Pot}_{tame}$ and $(\delta u, \delta u_y) \in T_{(u,u_y)}\mathsf{Pot}$ be given.

We once again need to use the regauging of the extended frame which we described in the proof of Theorem 5.4. In the sequel, we will use the notations taken from that proof, especially with regard to the regauging map g defined in Eq. (5.14), the regauged 1-form $\widetilde{\alpha} = \widetilde{\alpha}_0 + \beta + \gamma$ from Eq. (5.16), the regauged extended frame resp. monodromy $\widetilde{F}(x)$ resp. $\widetilde{M}$ defined by Eq. (5.12), and the corresponding vacuum solution $\widetilde{F}_0$ given by Eq. (5.21).

We begin by calculating $\delta\widetilde{F}(x)$. From the fact that $\widetilde{F}$ satisfies the homogeneous linear initial value problem

$$\widetilde{F}'(x) = \widetilde{\alpha}\,\widetilde{F}(x) \quad \text{with} \quad \widetilde{F}(0) = \mathbb{1},$$

it follows that $\delta\widetilde{F}$ solves the inhomogeneous linear initial value problem

$$(\delta\widetilde{F})'(x) = \widetilde{\alpha}\cdot\delta\widetilde{F} + \delta\widetilde{\alpha}\cdot\widetilde{F} \quad \text{with} \quad \delta\widetilde{F}(0) = 0,$$

and therefore $\delta\widetilde{F}$ is given by

$$\delta\widetilde{F}(x) = \widetilde{F}(x) \int_0^x \widetilde{F}(t)^{-1}\cdot\delta\widetilde{\alpha}(t)\cdot\widetilde{F}(t)\,\mathrm{d}t. \tag{14.9}$$

Now we have

$$\widetilde{\alpha} = \widetilde{\alpha}_0 - \frac{1}{2}u_z\, L + \gamma \quad \text{with} \quad \gamma = \frac{1}{4}\lambda^{-1/2}\begin{pmatrix} 0 & -e^{u}+1 \\ e^{-u}-1 & 0 \end{pmatrix} \quad \text{and} \quad L := \begin{pmatrix} 1 & 0 \\ 0 & -1 \end{pmatrix}$$

and therefore

$$\delta\widetilde{\alpha} = -\frac{1}{2}\,\delta u_z\, L + \delta\gamma \quad \text{with} \quad \delta\gamma = \frac{1}{4}\,\lambda^{-1/2}\,\delta u \begin{pmatrix} 0 & -\mathrm{e}^{u} \\ -\mathrm{e}^{-u} & 0 \end{pmatrix}.$$

It therefore follows from Eq. (14.9) that we have

$$\delta\widetilde{F}(x) = \widetilde{F}(x) \int_0^x \big(\delta u_z(t)\cdot \widetilde{X}(t) + \delta u_z(t)\cdot \widetilde{R}_1(t) + \delta u(t)\cdot \widetilde{R}_2(t)\big)\,\mathrm{d}t \tag{14.10}$$

with

$$\widetilde{X}(t) := -\frac{1}{2}\,\widetilde{F}_0(t)^{-1}\,L\,\widetilde{F}_0(t) \overset{(5.21)}{=} \frac{1}{2}\begin{pmatrix} -\cos(2\,\zeta(\lambda)\,t) & \sin(2\,\zeta(\lambda)\,t) \\ \sin(2\,\zeta(\lambda)\,t) & \cos(2\,\zeta(\lambda)\,t) \end{pmatrix} \tag{14.11}$$

$$\widetilde{R}_1(t) := -\frac{1}{2}\,\big((\widetilde{F}(t)^{-1} - \widetilde{F}_0(t)^{-1})\,L\,\widetilde{F}_0(t) + \widetilde{F}(t)^{-1}\,L\,(\widetilde{F}(t) - \widetilde{F}_0(t))\big) \tag{14.12}$$

$$\widetilde{R}_2(t) := \frac{1}{4}\,\lambda^{-1/2}\,\widetilde{F}(t)^{-1}\begin{pmatrix} 0 & -\mathrm{e}^{u(t)} \\ -\mathrm{e}^{-u(t)} & 0 \end{pmatrix}\widetilde{F}(t). \tag{14.13}$$

We now undo the regauging by g, to obtain δF from $\delta\widetilde{F}$: By Eq. (5.12) we have

$$F(x) = g(x)\cdot\widetilde{F}(x)\cdot g(0)^{-1}$$

and therefore

$$\begin{aligned}
&\delta F(x) \\
&\quad = g(x)\cdot\delta\widetilde{F}(x)\cdot g(0)^{-1} + \delta g(x)\cdot\widetilde{F}(x)\cdot g(0)^{-1} + g(x)\cdot\widetilde{F}(x)\cdot\delta(g(0)^{-1}) \\
&\quad = g(x)\cdot\delta\widetilde{F}(x)\cdot g(0)^{-1} + \delta g(x)\cdot\widetilde{F}(x)\cdot g(0)^{-1} \\
&\qquad - g(x)\cdot\widetilde{F}(x)\cdot g(0)^{-1}\cdot\delta g(0)\cdot g(0)^{-1} \\
&\quad = g(x)\cdot\delta\widetilde{F}(x)\cdot g(0)^{-1} + \delta g(x)\cdot g(x)^{-1}\cdot F(x) - F(x)\cdot\delta g(0)\cdot g(0)^{-1}.
\end{aligned}$$

When we set $x = 1$ in the above equation, we obtain (remember that the potential $(u, u_y) \in \mathsf{Pot}_{tame}$ is periodic, and therefore $g(1) = g(0)$ holds)

$$\begin{aligned}
\delta F(1) &= g(0)\cdot\delta\widetilde{F}(1)\cdot g(0)^{-1} + \delta g(0)\cdot g(0)^{-1}\cdot F(1) - F(1)\cdot\delta g(0)\cdot g(0)^{-1} \\
&= g(0)\cdot\delta\widetilde{F}(1)\cdot g(0)^{-1} + [g(0)^{-1}\cdot\delta g(0), F(1)] \\
&\overset{(*)}{=} g(0)\cdot\delta\widetilde{F}(1)\cdot g(0)^{-1} + \frac{1}{4}\,\delta u(0)\,[L, F(1)]
\end{aligned}$$

$$= g(0) \cdot \delta\widetilde{F}(1) \cdot g(0)^{-1} + \frac{1}{2}\,\delta u(0) \begin{pmatrix} 0 & -b(1,\lambda) \\ c(1,\lambda) & 0 \end{pmatrix}$$

$$= F(1) \int_0^1 \left(\delta u_z(t) \cdot X(t) + \delta u_z(t) \cdot R_1(t) + \delta u(t) \cdot R_2(t)\right) \mathrm{d}t$$

$$+ \frac{1}{2}\,\delta u(0) \begin{pmatrix} 0 & -b(1,\lambda) \\ c(1,\lambda) & 0 \end{pmatrix} \tag{14.14}$$

with (compare Eqs. (14.11)–(14.13))

$$X(t) := g(0)\,\widetilde{X}(t)\,g(0)^{-1} = \frac{1}{2} \begin{pmatrix} -\cos(2\,\zeta(\lambda)\,t) & \tau^{-1}\,\lambda^{-1/2}\,\sin(2\,\zeta(\lambda)\,t) \\ \tau\,\lambda^{1/2}\,\sin(2\,\zeta(\lambda)\,t) & \cos(2\,\zeta(\lambda)\,t) \end{pmatrix}$$

$$R_1(t) := g(0)\,\widetilde{R}_1(t)\,g(0)^{-1}$$

$$= -\frac{1}{2}\left((F(t)^{-1} - \widehat{F}_0(t)^{-1})\,L\,\widehat{F}_0(t) + F(t)^{-1}\,L\,(F(t) - \widehat{F}_0(t))\right)$$

$$R_2(t) := g(0)\,\widetilde{R}_2(t)\,g(0)^{-1}$$

$$= \frac{1}{4}\,\lambda^{-1/2}\,F(t)^{-1} \begin{pmatrix} 0 & -\lambda^{-1/2}\,\tau^{-1}\,\mathrm{e}^{u(t)} \\ -\lambda^{1/2}\,\tau\,\mathrm{e}^{-u(t)} & 0 \end{pmatrix} F(t)$$

$$\widehat{F}_0(t) := g(0)\,\widetilde{F}_0(t)\,g(0)^{-1} = \begin{pmatrix} a_0 & \tau^{-1}\,b_0 \\ \tau\,c_0 & d_0 \end{pmatrix};$$

for the equals sign marked $(*)$ we note that we obtain from Eq. (5.14)

$$\delta g(x) = \frac{1}{4}\,\delta u(x) \begin{pmatrix} \mathrm{e}^{u/4} & 0 \\ 0 & -\lambda^{1/2}\,\mathrm{e}^{-u/4} \end{pmatrix} \quad \text{and therefore} \quad g(x)^{-1} \cdot \delta g(x) = \frac{1}{4}\,\delta u(x)\,L.$$

We now write the extended frame F in the form

$$F = \begin{pmatrix} a & b \\ c & d \end{pmatrix}$$

(note that the functions $a, \dots, d$ here depend not only on $\lambda \in \mathbb{C}^*$, but on $x \in [0,1]$ as well). Because of $\det(F) = \det(\widehat{F}_0) = 1$ we then have

$$F^{-1} = \begin{pmatrix} d & -b \\ -c & a \end{pmatrix} \quad \text{and} \quad \widehat{F}_0^{-1} = \begin{pmatrix} d_0 & -\tau^{-1}\,b_0 \\ -\tau\,c_0 & a_0 \end{pmatrix}.$$

By carrying out the matrix multiplications in Eq. (14.14), and using these formulae, we see that the component functions of $\delta M = \delta F(1)$ are indeed of the form claimed in part (1) of the Lemma, where

$$\begin{aligned} f_{a,1} &= -\tfrac{1}{2}\big((d-d_0)a_0 + (b-\tau^{-1}b_0)c_0\big), & f_{a,2} &= \tfrac{1}{4}\big(\tau\, a\, b\, \mathrm{e}^{-u(t)} - \tau^{-1}\, c\, d\, \lambda^{-1}\, \mathrm{e}^{u(t)}\big),\\ f_{b,1} &= -\tfrac{1}{2}\big((d-d_0)b_0 + (b-\tau^{-1}b_0)d_0\big), & f_{b,2} &= \tfrac{1}{4}\big(\tau\, b^2\, \mathrm{e}^{-u(t)} - \tau^{-1}\, d^2\, \lambda^{-1}\, \mathrm{e}^{u(t)}\big),\\ f_{c,1} &= -\tfrac{1}{2}\big(-(c-\tau\, c_0)a_0 + (a-a_0)c_0\big), & f_{c,2} &= \tfrac{1}{4}\big(\tau\, a^2\, \mathrm{e}^{-u(t)} + \tau^{-1}\, c^2\, \lambda^{-1}\, \mathrm{e}^{u(t)}\big),\\ f_{d,1} &= -\tfrac{1}{2}\big(-(c-\tau\, c_0)b_0 + (a-a_0)d_0\big), & f_{d,2} &= \tfrac{1}{4}\big(-\tau\, a\, b\, \mathrm{e}^{-u(t)} + \tau^{-1}\, c\, d\, \lambda^{-1}\, \mathrm{e}^{u(t)}\big). \end{aligned}$$

It remains to show that these functions have the asymptotic behavior stated in part (1) of the Lemma, and this follows from the asymptotic behavior of the extended frame F described in Proposition 11.7, together with the fact that the functions $\mathrm{e}^{\pm u(t)}$ are (continuous and therefore) bounded for $t \in [0,1]$.

For (2). We consider besides the given potential (u, u_y) also the potential $(\widetilde{u}, \widetilde{u}_y) := (-u, u_y)$, and denote the quantities associated to the latter potential by a tilde. We will reduce (2) for (u, u_y) to the statement of (1) applied to $(\widetilde{u}, \widetilde{u}_y)$.

We have $\widetilde{\tau} = \tau^{-1}$,

$$(\delta\widetilde{u}, \delta\widetilde{u}_y) = (-\delta u, \delta u_y)$$

and therefore

$$\delta\widetilde{u} = -\delta u \quad \text{and} \quad \delta\widetilde{u}_z = -\delta u_{\overline{z}}.$$

By Proposition 2.2(1) we have

$$M(\lambda^{-1}) = g^{-1} \cdot \widetilde{M}(\lambda) \cdot g \quad \text{with} \quad g := \begin{pmatrix} 1 & 0 \\ 0 & \lambda \end{pmatrix},$$

i.e.

$$\begin{pmatrix} a(\lambda^{-1}) & b(\lambda^{-1}) \\ c(\lambda^{-1}) & d(\lambda^{-1}) \end{pmatrix} = \begin{pmatrix} \widetilde{a}(\lambda) & \lambda \cdot \widetilde{b}(\lambda) \\ \lambda^{-1} \cdot \widetilde{c}(\lambda) & \widetilde{d}(\lambda) \end{pmatrix}.$$

By differentiation of this equation with respect to (u, u_y) it follows:

$$\begin{pmatrix} \delta a(\lambda^{-1}) & \delta b(\lambda^{-1}) \\ \delta c(\lambda^{-1}) & \delta d(\lambda^{-1}) \end{pmatrix} = \begin{pmatrix} \delta\widetilde{a}(\lambda) & \lambda \cdot \delta\widetilde{b}(\lambda) \\ \lambda^{-1} \cdot \delta\widetilde{c}(\lambda) & \delta\widetilde{d}(\lambda) \end{pmatrix}.$$

Thus we obtain via the result of (1)

$$\begin{aligned} \delta a(\lambda^{-1}) &= \delta\widetilde{a}(\lambda)\\ &= -\tfrac{1}{2}\,\widetilde{a}(\lambda) \int_0^1 \delta\widetilde{u}_z(t)\, \cos(2\zeta(\lambda)t)\,\mathrm{d}t\\ &\qquad + \tfrac{1}{2}\,\widetilde{\tau}\,\lambda^{1/2}\,\widetilde{b}(\lambda) \int_0^1 \delta\widetilde{u}_z(t)\, \sin(2\zeta(\lambda)t)\,\mathrm{d}t + \cdots \end{aligned}$$

$$+\int_0^1 \delta\widetilde{u}_z(t)\,\widetilde{f}_{1,a}(t)\,\mathrm{d}t + \int_0^1 \delta\widetilde{u}(t)\,\widetilde{f}_{2,a}(t)\,\mathrm{d}t$$

$$= \tfrac{1}{2}\,a(\lambda^{-1}) \int_0^1 \delta u_{\bar z}(t)\,\cos(2\zeta(\lambda^{-1})t)\,\mathrm{d}t$$

$$-\tfrac{1}{2}\,\tau^{-1}\,\lambda^{1/2}\,\lambda^{-1}\,b(\lambda^{-1}) \int_0^1 \delta u_{\bar z}(t)\,\sin(2\zeta(\lambda^{-1})t)\,\mathrm{d}t$$

$$+\int_0^1 \delta u_{\bar z}(t)\,g_{1,a}(t)\big|_{\lambda^{-1}}\,\mathrm{d}t + \int_0^1 \delta u(t)\,g_{2,a}(t)\big|_{\lambda^{-1}}\,\mathrm{d}t$$

with the functions $g_{1,a} := -\widetilde{f}_{1,a}\big|_{\lambda^{-1}}$ and $g_{2,a} := -\widetilde{f}_{2,a}\big|_{\lambda^{-1}}$. The uniform asymptotic properties $\widetilde{f}_{1,a} \in \mathrm{As}_\infty(\mathbb{C}^*, \ell^2_0, 2)$ and $\widetilde{f}_{2,a} \in \mathrm{As}_\infty(\mathbb{C}^*, \ell^\infty_1, 2)$ imply that also $g_{1,a} \in \mathrm{As}_0(\mathbb{C}^*, \ell^2_0, 2)$ and $g_{2,a} \in \mathrm{As}_0(\mathbb{C}^*, \ell^\infty_1, 2)$ holds. By substituting λ for λ^{-1} in the above equation we obtain the result for δa in (2). The asymptotic equations for $\delta b, \delta c, \delta d$ are shown in the same way. □

We apply Lemma 14.6 to describe the asymptotic behavior of $\delta\lambda_k$ and $\delta\mu_k$:

Lemma 14.8 *Let $(u, u_y) \in \mathsf{Pot}_{tame}$ be given, and let $D = \{(\lambda_k, \mu_k)\} \in \mathsf{Div}_{tame}$ be the spectral divisor of (u, u_y). Because D does not contain any double points, the functions λ_k and μ_k are well-defined and smooth on neighborhoods of (u, u_y) in Pot. In this setting, there exist sequences $(a_k) \in \ell^2_{-1,3}(k)$ and $(b_k) \in \ell^2_{0,0}(k)$ of non-negative real numbers, so that we have for $(\delta u, \delta u_y) \in T_{(u,u_y)}\mathsf{Pot}$:*

(1) For $k > 0$:

$$\left|\delta\lambda_k - (-4)\,(-1)^k\,\lambda_k^{1/2}\,\mu_k^{-1} \int_0^1 \delta u_z(t)\,\sin(2k\pi t)\,\mathrm{d}t\right| \le a_k \cdot \|(\delta u, \delta u_y)\|_{\mathsf{Pot}}$$

$$\left|\delta\mu_k - \left(-\frac{1}{2}\right)\,\mu_k \int_0^1 \delta u_z(t)\,\cos(2k\pi t)\,\mathrm{d}t\right| \le b_k \cdot \|(\delta u, \delta u_y)\|_{\mathsf{Pot}}$$

(2) For $k < 0$:

$$\left|\delta\lambda_k - 4\,(-1)^k\,\lambda_k^{3/2}\,\mu_k^{-1} \int_0^1 \delta u_{\bar z}(t)\,\sin(2k\pi t)\,\mathrm{d}t\right| \le a_k \cdot \|(\delta u, \delta u_y)\|_{\mathsf{Pot}}$$

$$\left|\delta\mu_k - \frac{1}{2}\,\mu_k \int_0^1 \delta u_{\bar z}(t)\,\cos(2k\pi t)\,\mathrm{d}t\right| \le b_k \cdot \|(\delta u, \delta u_y)\|_{\mathsf{Pot}}$$

Moreover, $\delta\lambda_k$ and $\delta\mu_k$ depend continuously on $(\delta u, \delta u_y)$.

Proof *For (1).* We have $k > 0$ here, and use the notations of Lemma 14.6. Because D is tame, we have $c'(\lambda_k) \neq 0$, and therefore the implicit function theorem shows that the function λ_k is smooth, and that we have

$$\delta\lambda_k = -\frac{1}{c'(\lambda_k)}\,(\delta c)(\lambda_k). \tag{14.15}$$

We have $c'(\lambda_k) = \left.\frac{c(\lambda)}{\lambda-\lambda_k}\right|_{\lambda=\lambda_k}$, and therefore Corollary 10.3(1) shows that

$$c'(\lambda_k) - \tau\,\frac{(-1)^k}{8} \in \ell^2_0(k)$$

holds; because the function $z \mapsto z^{-1}$ is locally Lipschitz continuous near $z = \tau\,\frac{(-1)^k}{8}$, it follows that we also have

$$\frac{1}{c'(\lambda_k)} = \frac{8\,(-1)^k}{\tau} + r_k^{[1]} \tag{14.16}$$

with a sequence $(r_k^{[1]}) \in \ell^2_0(k)$.

Moreover, we estimate $(\delta c)(\lambda_k)$ via Lemma 14.6(1): Because we have $c(\lambda_k) = 0$ and $d(\lambda_k) = \mu_k^{-1}$, we obtain

$$\begin{aligned}(\delta c)(\lambda_k) = {}& \frac{1}{2}\,\tau\,\lambda_k^{1/2}\,\mu_k^{-1}\int_0^1 \delta u_z(t)\,\sin(2\zeta(\lambda_k)t)\,\mathrm{d}t \\ &+ \int_0^1 \delta u_z(t)\,f_{1,c}(t)\,\mathrm{d}t + \int_0^1 \delta u(t)\,f_{2,c}(t)\,\mathrm{d}t\end{aligned} \tag{14.17}$$

with functions $f_{1,c} \in \mathrm{As}_\infty(\mathbb{C}^*, \ell^2_{-1}, 2)$ and $f_{2,c} \in \mathrm{As}_\infty(\mathbb{C}^*, \ell^\infty_0, 2)$ uniformly in $t \in [0,1]$. We let $(r_k^{[f_{c,1}]}) \in \ell^2_{-1}(k)$ resp. $(r_k^{[f_{c,2}]}) \in \ell^\infty_0(k)$ be bounding sequences for $f_{c,1}(t)$ resp. $f_{c,2}(t)$ for all $t \in [0,1]$. Then we have

$$|(\delta c)(\lambda_k)| \leq \left(|\tau|\,|\lambda_k|^{1/2}\,|\mu_k|^{-1} + r_k^{[f_{1,c}]}\right)\cdot \|\delta u_z\|_{L^2([0,1])} + r_k^{[f_{2,c}]}\cdot\|\delta u\|_{L^2([0,1])} \tag{14.18}$$

by Cauchy-Schwarz's inequality, and the fact that $|\sin(2\zeta(\lambda_k)t)| \leq 2$ holds for all $k \in \mathbb{Z}$ and $t \in [0,1]$.

We also note that we have $\lambda_k - \lambda_{k,0} \in \ell^2_{-1}(k > 0)$ and therefore $\zeta(\lambda_k) - \pi k = \zeta(\lambda_k) - \zeta(\lambda_{k,0}) \in \ell^2_0(k > 0)$; because $\sin(z)$ is Lipschitz continuous on any horizontal strip in the complex plane, it follows that there exists a sequence $(r_k^{[2]}) \in \ell^2(k)$ such that

$$|\sin(2\,\zeta(\lambda_k)\,t) - \sin(2\pi k t)| \leq r_k^{[2]} \tag{14.19}$$

holds for all $k > 0$ and $t \in [0,1]$.

We now apply the preceding estimates to Eq. (14.15) to obtain the asymptotic assessment for $\delta\lambda_k$ claimed in the Lemma. Indeed, we have

$$\begin{aligned}
\delta\lambda_k &\overset{(14.15)}{=} -\frac{1}{c'(\lambda_k)}\,(\delta c)(\lambda_k) \\
&= -\frac{8\,(-1)^k}{\tau}\,(\delta c)(\lambda_k) - \left(\frac{1}{c'(\lambda_k)} - \frac{8\,(-1)^k}{\tau}\right)(\delta c)(\lambda_k) \\
&\overset{(14.17)}{=} -4\,(-1)^k\,\lambda_k^{1/2}\,\mu_k^{-1}\int_0^1 \delta u_z(t)\,\sin(2\zeta(\lambda_k)t)\,\mathrm{d}t \\
&\quad - \frac{8\,(-1)^k}{\tau}\left(\int_0^1 \delta u_z(t)\,f_{1,c}(t)\,\mathrm{d}t + \int_0^1 \delta u(t)\,f_{2,c}(t)\,\mathrm{d}t\right) \\
&\quad - \left(\frac{1}{c'(\lambda_k)} - \frac{8\,(-1)^k}{\tau}\right)(\delta c)(\lambda_k)
\end{aligned}$$

and therefore

$$\begin{aligned}
&\delta\lambda_k - (-4)\,(-1)^k\,\lambda_k^{1/2}\,\mu_k^{-1}\int_0^1 \delta u_z(t)\,\sin(2k\pi t)\,\mathrm{d}t \\
&= -4\,(-1)^k\,\lambda_k^{1/2}\,\mu_k^{-1}\int_0^1 \delta u_z(t)\,\big(\sin(2\zeta(\lambda_k)t) - \sin(2k\pi t)\big)\,\mathrm{d}t \\
&\quad - \frac{8\,(-1)^k}{\tau}\left(\int_0^1 \delta u_z(t)\,f_{1,c}(t)\,\mathrm{d}t + \int_0^1 \delta u(t)\,f_{2,c}(t)\,\mathrm{d}t\right) \\
&\quad - \left(\frac{1}{c'(\lambda_k)} - \frac{8\,(-1)^k}{\tau}\right)(\delta c)(\lambda_k).
\end{aligned}$$

By use of the triangle inequality, Cauchy-Schwarz's inequality and the estimates (14.16), (14.18) and (14.19), we obtain

$$\begin{aligned}
&\left|\delta\lambda_k - (-4)\,(-1)^k\,\lambda_k^{1/2}\,\mu_k^{-1}\int_0^1 \delta u_z(t)\,\sin(2k\pi t)\,\mathrm{d}t\right| \\
&\leq a_k^{[1]}\cdot\|\delta u_z\|_{L^2([0,1])} + a_k^{[2]}\cdot\|\delta u\|_{L^2([0,1])} \\
&\leq (a_k^{[1]} + a_k^{[2]})\cdot\|(\delta u, \delta u_y)\|_{\mathsf{Pot}}
\end{aligned}$$

with

$$\begin{aligned}
a_k^{[1]} &:= 4\,|\lambda_k|^{1/2}\,|\mu_k|^{-1}\,r_k^{[2]} + 8\,|\tau|^{-1}\,r_k^{[f_{1,c}]} + r_k^{[1]}\cdot\left(|\tau|\,|\lambda_k|^{1/2}\,|\mu_k|^{-1} + r_k^{[f_{1,c}]}\right), \\
a_k^{[2]} &:= 8\,|\tau|^{-1}\,r_k^{[f_{2,c}]} + r_k^{[1]}\cdot r_k^{[f_{2,c}]}.
\end{aligned}$$

Because we have $a_k^{[1]}, a_k^{[2]} \in \ell^2_{-1}(k)$, the asymptotic estimate claimed for $\delta\lambda_k$ in part (1) of the lemma follows.

For the estimate for $\delta\mu_k$, we first note that $\mu_k = a(\lambda_k)$ is a smooth function near (u, u_y). However, the computation of $\delta\mu_k$ becomes slightly easier if we base our calculation on $\mu_k^{-1} = d(\lambda_k)$ instead. The reason for this is that the expressions for $\delta d(\lambda_k)$ in Lemma 14.6 are somewhat simpler than those for $\delta a(\lambda_k)$, owing to the fact that $c(\lambda_k) = 0$ holds, whereas we do not know anything about $b(\lambda_k)$.

Thus we calculate:

$$\begin{aligned}\delta\mu_k &= \delta\big((\mu_k^{-1})^{-1}\big) = -\mu_k^2 \cdot \delta(\mu_k^{-1}) = -\mu_k^2 \cdot \delta(d(\lambda_k))\\ &= -\mu_k^2 \cdot \big(\delta d(\lambda_k) + d'(\lambda_k) \cdot \delta\lambda_k\big)\,. \end{aligned} \tag{14.20}$$

For $\delta d(\lambda_k)$, we have by Lemma 14.6(1) (again noting that $c(\lambda_k) = 0$ and $d(\lambda_k) = \mu_k^{-1}$ holds):

$$\begin{aligned}\delta d(\lambda_k) &= \tfrac{1}{2}\,\mu_k^{-1} \int_0^1 \delta u_z(t)\, \cos(2\zeta(\lambda_k)t)\,\mathrm{d}t\\ &\quad + \int_0^1 \delta u_z(t)\, f_{1,d}(t)\,\mathrm{d}t + \int_0^1 \delta u(t)\, f_{2,d}(t)\,\mathrm{d}t \end{aligned} \tag{14.21}$$

with functions $f_{1,d} \in \mathrm{As}_\infty(\mathbb{C}^*, \ell^2_0, 2)$ and $f_{2,d} \in \mathrm{As}_\infty(\mathbb{C}^*, \ell^\infty_1, 2)$ uniformly in $t \in [0,1]$. We let $(r_k^{[f_{d,1}]}) \in \ell^2_0(k)$ resp. $(r_k^{[f_{d,2}]}) \in \ell^\infty_1(k)$ be uniform bounding sequences for $f_{d,1}(t)$ resp. $f_{d,2}(t)$ for all $t \in [0,1]$.

Next, we estimate $d'(\lambda_k)$. We have $d_0'(\lambda_{k,0}) = 0$ and therefore

$$d'(\lambda_k) = d'(\lambda_k) - d_0'(\lambda_{k,0}) = d'(\lambda_k) - d_0'(\lambda_k) + \int_{\lambda_{k,0}}^{\lambda_k} d_0''(\lambda)\,\mathrm{d}\lambda\,.$$

We have $d' - d_0' \in \mathrm{As}(\mathbb{C}^*, \ell^2_{1,-3}, 1)$ and therefore $d'(\lambda_k) - d_0'(\lambda_k) \in \ell^2_{1,-3}(k)$. Moreover, we have $d_0'' \in \mathrm{As}(\mathbb{C}^*, \ell^\infty_{2,-6}, 1)$ and $\lambda_k - \lambda_{k,0} \in \ell^2_{-1,3}(k)$ and therefore $\int_{\lambda_{k,0}}^{\lambda_k} d_0''(\lambda)\,\mathrm{d}\lambda \in \ell^2_1(k)$. Thus we obtain

$$|d'(\lambda_k)| =: r_k^{[3]} \in \ell^2_1(k). \tag{14.22}$$

Moreover, it follows from the asymptotic estimate for $\delta\lambda_k$ shown above that we have

$$|\delta\lambda_k| \leq \widetilde{a}_k \cdot \|(\delta u, \delta u_y)\|_{\mathsf{Pot}} \tag{14.23}$$

with the sequence

$$\widetilde{a}_k := 4\,|\lambda_k|^{1/2}\,|\mu_k|^{-1} + a_k \in \ell^\infty_{-1}(k).$$

Similarly as Eq. (14.19), we note moreover that there exists a sequence $(r_k^{[4]}) \in \ell^2(k)$ such that

$$|\cos(2\,\zeta(\lambda_k)\,t) - \cos(2\pi kt)| \le r_k^{[4]} \tag{14.24}$$

holds for all $k > 0$ and $t \in [0,1]$.

We now obtain

$$\begin{aligned}\delta\mu_k &\overset{(14.20)}{=} -\mu_k^2 \cdot \big(\delta d(\lambda_k) + d'(\lambda_k)\cdot\delta\lambda_k\big)\\ &\overset{(14.21)}{=} -\tfrac{1}{2}\,\mu_k \int_0^1 \delta u_z(t)\,\cos(2\zeta(\lambda_k)t)\,\mathrm{d}t - \mu_k^2 \int_0^1 \delta u_z(t)\,f_{1,d}(t)\,\mathrm{d}t\\ &\quad - \mu_k^2 \int_0^1 \delta u(t)\,f_{2,d}(t)\,\mathrm{d}t - \mu_k^2 \cdot d'(\lambda_k)\cdot\delta\lambda_k\end{aligned}$$

and therefore

$$\begin{aligned}&\delta\mu_k - \left(-\tfrac{1}{2}\right)\,\mu_k \int_0^1 \delta u_z(t)\,\cos(2k\pi t)\,\mathrm{d}t\\ &= -\tfrac{1}{2}\,\mu_k^2 \int_0^1 \delta u_z(t)\,\big(\cos(2\zeta(\lambda_k)t) - \cos(2k\pi t)\big)\,\mathrm{d}t\\ &\quad - \mu_k^2 \int_0^1 \delta u_z(t)\,f_{1,d}(t)\,\mathrm{d}t\\ &\quad - \mu_k^2 \int_0^1 \delta u(t)\,f_{2,d}(t)\,\mathrm{d}t - \mu_k^2\cdot d'(\lambda_k)\cdot\delta\lambda_k .\end{aligned}$$

By use of the triangle inequality, Cauchy-Schwarz's inequality and the estimates (14.22), (14.23) and (14.24), we obtain

$$\begin{aligned}&\left|\delta\mu_k - \left(-\tfrac{1}{2}\right)\,\mu_k \int_0^1 \delta u_z(t)\,\cos(2k\pi t)\,\mathrm{d}t\right|\\ &\le\; b_k^{[1]}\cdot\|\delta u_z\|_{L^2([0,1])} + b_k^{[2]}\cdot\|\delta u\|_{L^2([0,1])} + r_k^{[3]}\cdot\widetilde{a}_k\cdot\|(\delta u,\delta u_y)\|_{\mathsf{Pot}}\\ &\le\; (b_k^{[1]} + b_k^{[2]} + r_k^{[3]}\cdot\widetilde{a}_k)\cdot\|(\delta u,\delta u_y)\|_{\mathsf{Pot}}\end{aligned}$$

with

$$b_k^{[1]} := -\frac{1}{2}\,|\mu_k|^2\,r_k^{[4]} + |\mu_k|^2\,r_k^{[f_{1,d}]} \quad\text{and}\quad b_k^{[2]} := |\mu_k|^2\,r_k^{[f_{2,d}]}.$$

Because we have $b_k^{[1]}, b_k^{[2]}, r_k^{[3]}\cdot\widetilde{a}_k \in \ell_0^2(k)$, and moreover $\mu_k - \mu_k^{-3} \in \ell_0^2(k)$ holds, the asymptotic estimate claimed for $\delta\mu_k$ in part (1) of the lemma follows.

For (2). This is shown similarly as (1), using the asymptotic estimates for δM for $\lambda \to 0$ from Lemma 14.6(2).

Equations (14.15) and (14.20) also show that $\delta\lambda_k$ and $\delta\mu_k$ depend continuously on $(\delta u, \delta u_y)$. □

Lemma 14.9 *Let* $(u, u_y) \in \mathsf{Pot}_{tame}$ *and* $(\delta u, \delta u_y), (\widetilde{\delta u}, \widetilde{\delta u}_y) \in T_{(u,u_y)}\mathsf{Pot}$ *be given. Then the sum*

$$\frac{1}{2} \sum_{k\in\mathbb{Z}} \left(\frac{\delta\lambda_k}{\lambda_k} \cdot \frac{\widetilde{\delta\mu}_k}{\mu_k} - \frac{\widetilde{\delta\lambda}_k}{\lambda_k} \cdot \frac{\delta\mu_k}{\mu_k} \right)$$

converges absolutely, and it depends continuously on $(\delta u, \delta u_y)$ *and* $(\widetilde{\delta u}, \widetilde{\delta u}_y)$.

Proof Because of $(\delta u, \delta u_y) \in T_{(u,u_y)}\mathsf{Pot} \subset W^{1,2}([0,1]) \times L^2([0,1])$, we know that $\delta u_z, \delta u_{\overline{z}} \in L^2([0,1])$, and therefore the Fourier coefficients

$$\int_0^1 \delta u_z(t)\, \cos(2k\pi t)\, \mathrm{d}t \quad \text{and} \quad \int_0^1 \delta u_z(t)\, \sin(2k\pi t)\, \mathrm{d}t$$

and

$$\int_0^1 \delta u_{\overline{z}}(t)\, \cos(2k\pi t)\, \mathrm{d}t \quad \text{and} \quad \int_0^1 \delta u_{\overline{z}}(t)\, \sin(2k\pi t)\, \mathrm{d}t$$

are in $\ell^2(k)$. Lemma 14.8 therefore shows that

$$\delta\lambda_k \in \ell^2_{-1,3}(k) \quad \text{and} \quad \delta\mu_k \in \ell^2_{0,0}(k)$$

holds. Moreover, because of $\lambda_k = \lambda_{k,0} + \ell^2_{-1,3}(k)$ we have $\frac{1}{\lambda_k} \in \ell^\infty_{2,-2}(k)$, and because of $\mu_k = \mu_{k,0} + \ell^2_{0,0}(k)$, we have $\frac{1}{\mu_k} \in \ell^\infty_{0,0}(k)$. Thus we obtain

$$\frac{\delta\lambda_k}{\lambda_k} \in \ell^2_{1,1}(k) \quad \text{and} \quad \frac{\delta\mu_k}{\mu_k} \in \ell^2_{0,0}(k).$$

By applying this result both for $(\delta u, \delta u_y)$ and $(\widetilde{\delta u}, \widetilde{\delta u}_y)$ we see that

$$\frac{\delta\lambda_k}{\lambda_k} \cdot \frac{\widetilde{\delta\mu}_k}{\mu_k} - \frac{\widetilde{\delta\lambda}_k}{\lambda_k} \cdot \frac{\delta\mu_k}{\mu_k} \in \ell^1_{1,1}(k) \subset \ell^1(k)$$

holds, which means that the sum under investigation converges absolutely in $\mathbb{C}$.

Moreover, the summands depend continuously on $(\delta u, \delta u_y)$ and $(\widetilde{\delta u}, \widetilde{\delta u}_y)$ by Lemma 14.8. Because the above estimates show that the convergence of the sum is locally uniform, it follows that also the sum depends continuously on $(\delta u, \delta u_y)$ and $(\widetilde{\delta u}, \widetilde{\delta u}_y)$. □

We now have all the pieces in place to complete the proof of Theorem 14.3.

Proof (of Theorem 14.3) For (1). This is Lemma 14.5.

For (2). By Theorem 14.2(2) (=[Kn-2, Theorem 5.5]), Eq. (14.6) holds for all $(\delta u, \delta u_y), (\widetilde{\delta u}, \widetilde{\delta u}_y) \in T_{(u,u_y)}\mathsf{Pot}$ that are *finite* linear combinations of the v_k and the w_k. Because the vectors $(\delta u, \delta u_y)$ with this property are dense in $T_{(u,u_y)}\mathsf{Pot}$ by Lemma 14.5, and both sides of Eq. (14.6) are well-defined and continuous for all $(\delta u, \delta u_y), (\widetilde{\delta u}, \widetilde{\delta u}_y) \in T_{(u,u_y)}\mathsf{Pot}$ (the right-hand side by Lemma 14.9), it follows that Eq. (14.6) holds for all $(\delta u, \delta u_y), (\widetilde{\delta u}, \widetilde{\delta u}_y) \in T_{(u,u_y)}\mathsf{Pot}$. □

Chapter 15
The Inverse Problem for Cauchy Data Along the Real Line

In Chap. 12 we showed that the monodromy $M(\lambda)$ of a tame potential $(u, u_y) \in \mathsf{Pot}$ is uniquely determined by its (tame) spectral divisor D, at least up to a change of sign of the off-diagonal entries. We are now ready to prove the important result that also the potential (u, u_y) itself is uniquely determined by D, at least if D is tame. In fact, we will show that for the map $\varPhi : \mathsf{Pot} \to \mathsf{Div}$ that associates to each potential $(u, u_y) \in \mathsf{Pot}$ its spectral divisor, $\varPhi|\mathsf{Pot}_{tame} : \mathsf{Pot}_{tame} \to \mathsf{Div}_{tame}$ is a diffeomorphism onto an open and dense subset of Div_{tame}. We will obtain this result relatively easily by using the Darboux coordinates for Pot_{tame} constructed in the preceding chapter.

We will prove

Theorem 15.1 $\varPhi|\mathsf{Pot}_{tame} : \mathsf{Pot}_{tame} \to \mathsf{Div}_{tame}$ *is a diffeomorphism onto an open and dense subset of* Div_{tame}.

Remark 15.2 $\varPhi$ is not immersive at potentials $(u, u_y) \in \mathsf{Pot} \setminus \mathsf{Pot}_{tame}$. This is to be expected if the classical spectral divisor $D \in \mathsf{Div}$ corresponding to (u, u_y) contains singular points of the associated spectral curve Σ with multiplicity ≥ 2, because then not even the monodromy $M(\lambda)$ is uniquely determined by D, as we need the additional information contained in the generalized spectral divisor $\mathscr{D}$ to reconstruct $M(\lambda)$ in this case (see Chap. 12). However, even if all the points occurring in D with multiplicity ≥ 2 are regular points of Σ (then the monodromy $M(\lambda)$ is uniquely determined by D), there is an entire family of integral curves of x-translation in Div that intersect in such a point, and therefore $\varPhi$ cannot be immersive. To make $\varPhi$ an immersion (and consequently a local diffeomorphism) near such points, we would therefore need to replace the range Div of $\varPhi$ by a suitable blow-up at its singularities (see e.g. [Har-1, p. 163ff.]). We do not carry out such a construction here.

The reason why the image of $\varPhi|\mathsf{Pot}_{tame}$ is not all of Div_{tame} is that even though any tame divisor $D \in \mathsf{Div}_{tame}$ is non-special, it is possible for D to become

S. Klein, *A Spectral Theory for Simply Periodic Solutions of the Sinh-Gordon Equation*, Lecture Notes in Mathematics 2229,
https://doi.org/10.1007/978-3-030-01276-2_15

special under x-translation. If this occurs, the potential corresponding to D has a singularity for the corresponding value of x, and thus D cannot correspond to a potential $(u, u_y) \in \mathsf{Pot}$ in our sense. The investigation of sinh-Gordon potentials with singularities, corresponding to divisors $D \in \mathsf{Div}$ which become special under x-translation for some value of x, would be extremely interesting for the study of compact constant mean curvatures; we discuss this perspective in the concluding Chap. 21.

For the proof of Theorem 15.1, we will use a well-known fact from the theory of solutions of the sinh-Gordon equation of finite type: For any divisor $D \in \mathsf{Div}$ of finite type (see Definition 13.1), such that the x-translations $D(x)$ are non-special for every $x \in [0, 1]$, there exists one and only one potential $(u, u_y) \in \mathsf{Pot}$ with $\Phi((u, u_y)) = D$.

By the x-translation of a divisor $D \in \mathsf{Div}$ we here mean the following: Suppose that a potential $(u, u_y) \in \mathsf{Pot}$ and $x \in \mathbb{R}$ are given, and denote the translation by x on $\mathbb{R}$ by $L_x(x') := x' + x$. Then $(u \circ L_x, u_y \circ L_x)$ is another (periodic) potential in Pot, and if (Σ, D) are the spectral data of (u, u_y), then the spectral curve of $(u \circ L_x, u_y \circ L_x)$ will also be Σ for all $x \in \mathbb{R}$, whereas we denote the spectral divisor of $(u \circ L_x, u_y \circ L_x)$ by $D(x) \in \mathsf{Div}$; this is the x-translation of the divisor D.

At least if $D \in \mathsf{Div}$ is tame, we can characterize the x-translation of D by differential equations, without recourse to a potential (u, u_y) of which D is the spectral divisor: We write $D = \{(\lambda_k, \mu_k)\}$, and regard λ_k and μ_k as functions in $x \in \mathbb{R}$. Because $D(x)$ is a divisor on the spectral curve Σ associated to D for all $x \in \mathbb{R}$, it suffices to describe the motion of $\lambda_k(x)$ under x. Let us write the monodromy corresponding to $D(x)$ by Theorem 12.3 as $M(\lambda, x) = \left(\begin{smallmatrix} a(\lambda,x) & b(\lambda,x) \\ c(\lambda,x) & d(\lambda,x) \end{smallmatrix}\right)$,[1] then the characteristic equation $c(\lambda_k(x), x) = 0$ together with the differential Equation (3.3) for $\frac{\partial}{\partial x} M(\lambda, x)$ yields a differential equation for λ_k:

$$\frac{\partial \lambda_k}{\partial x} = -\frac{1}{4\,c'(\lambda_k)} \cdot (\lambda\,\tau + \tau^{-1}) \cdot (\mu_k - \mu_k^{-1}) \quad \text{with} \quad \tau := \left(\prod_{k \in \mathbb{Z}} \frac{\lambda_{k,0}}{\lambda_k} \right)^{1/2} \tag{15.1}$$

(see the proof of Proposition 19.2 for the detailed calculation). For given $x \in \mathbb{R}$, if there is a solution λ_k of this differential equation on $[0, x]$ (resp. on $[x, 0]$) for all $k \in \mathbb{Z}$, and $D(x) := \{(\lambda_k(x), \mu_k(x))\} \in \mathsf{Div}$ holds, we call $D(x)$ the x-translation of D; here $\mu_k(x)$ is determined by the condition that $D(x)$ moves smoothly on the spectral curve Σ of D. Note that it is not clear a priori that $D(x)$ exists for any $x \neq 0$ (when one does not know that D stems from a potential $(u, u_y) \in \mathsf{Pot}$). As long as $D(x)$ is defined, it depends smoothly on D and on x.

---

[1]Note that the functions a, b, c, d comprising this monodromy are (for $x \neq 1$) different from the functions comprising the extended frame, which we also denoted by a, b, c, d e.g. in Proposition 11.7 and in Chap. 14.

We will study the x-translation (and the y-translation as well) for general tame divisors D in Chap. 19 by using the Jacobi variety associated to the spectral curve corresponding to D. At the moment however, we are interested in the x-translation only of finite type divisors D. If $D = \{(\lambda_k, \mu_k)\} \in \mathsf{Div}_{tame}$ is given and (λ_k, μ_k) is a double point of the spectral curve Σ corresponding to D for some $k \in \mathbb{Z}$, then the corresponding λ_k will not move under x-translation, i.e. we will have $(\lambda_k(x), \mu_k(x)) = (\lambda_k(0), \mu_k(0))$ for all $x \in \mathbb{R}$. If D is of finite type, the system of differential Equations (15.1) thus contains in fact only finitely many nontrivial equations; it follows that solutions $\lambda_k(x)$ exist and $D(x) = (\lambda_k(x), \mu_k(x))$ remains in Div as long as $D(x)$ does not run into a non-tame (special) divisor, and this is the case at least for small $|x|$. Note that the x-translation of a finite type divisor D is again of finite type.

The following fact is a well-known result from the theory of finite type solutions of the sinh-Gordon equation.

Lemma 15.3 (Bobenko) *For any $D \in \mathsf{Div}$ that is of finite type, and so that the x-translation $D(x)$ exists for all $x \in [0, 1]$ and is non-special, there exists one and only one potential $(u, u_y) \in \mathsf{Pot}$ with $\Phi((u, u_y)) = D$.*

Proof The potential (u, u_y) corresponding to D has been constructed explicitly in terms of theta functions by Bobenko [Bo-1, Theorem 4.1], and also see the construction by Babich [Ba-1, Ba-2]. In the case where D satisfies a condition of reality that is equivalent to the condition that the corresponding potential (u, u_y) is real-valued (this reality condition already implies that $D(x)$ is non-special for all x), a nice explicit construction of the potential (u, u_y) in terms of vector-valued Baker-Akhiezer functions on the spectral curve Σ corresponding to D was described by Knopf [Kn-1, the proof of Proposition 4.34]. □

The instrument for the application of Lemma 15.3 to our situation, where the divisor D is in general of infinite type, is the following lemma:

Lemma 15.4 *The set of divisors $D \in \mathsf{Div}_{tame}$ that are of finite type, and such that $D(x)$ exists for all $x \in [0, 1]$ and is tame (in particular, non-special), is dense in Div_{tame}.*

Proof For $N \in \mathbb{N}$ we let Div_N be the set of those divisors $D = \{(\lambda_k, \mu_k)\} \in \mathsf{Div}_{tame}$ for which (λ_k, μ_k) is a double point of the spectral curve Σ associated to D for all $k \in \mathbb{Z}$ with $|k| > N$; clearly the divisors in Div_N are of finite type. Moreover, we let $\mathsf{Div}_{N,xtame}$ be the set of divisors $D \in \mathsf{Div}_N$ so that $D(x)$ exists and is tame for all $x \in [0, 1]$. We will show that $\mathsf{Div}_{N,xtame}$ is open and dense in Div_N.

As discussed above, the only obstacle for the existence of $D(x)$ for a finite type divisor D is that the x-translation runs into a special divisor. It follows that for $D \in \mathsf{Div}_N \backslash \mathsf{Div}_{N,xtame}$ there exists $x \in (0, 1]$ such that $D(x)$ is non-tame, i.e. such that there exist $k, \ell \in \mathbb{Z}$ with $|k|, |\ell| \leq N$ and $k \neq \ell$ so that $\lambda_k(x) = \lambda_\ell(x)$

holds; this value of x is then the maximal value for which $D(x)$ is defined. This shows that

$$\mathsf{Div}_N \setminus \mathsf{Div}_{N,xtame} = \bigcup_{\substack{|k|,|\ell| \leq N \\ k \neq \ell}} \bigcup_{x \in [0,1]} \{ D = \{(\lambda_k, \mu_k)\} \in \mathsf{Div}_N \mid \lambda_k(x) = \lambda_\ell(x) \}$$

holds. The set $\{ D = \{(\lambda_k, \mu_k)\} \in \mathsf{Div}_N \mid \lambda_k(x) = \lambda_\ell(x) \}$ is a complex hypersurface in Div_N for every fixed value of k, ℓ, x because $D(x)$ depends smoothly on D. Hence $\bigcup_{x \in [0,1]} \{ D = \{(\lambda_k, \mu_k)\} \in \mathsf{Div}_N \mid \lambda_k(x) = \lambda_\ell(x) \}$ is for every fixed value of k, ℓ contained in a real hypersurface in Div_N, because $D(x)$ also depends smoothly on x. Thus $\mathsf{Div}_N \setminus \mathsf{Div}_{N,xtame}$ is contained in a union of finitely many real hypersurfaces in Div_N, and therefore $\mathsf{Div}_{N,xtame}$ is open and dense in Div_N.

Therefore the set $\bigcup_{N \in \mathbb{N}} \mathsf{Div}_{N,xtame}$ of divisors $D \in \mathsf{Div}_{tame}$ that are of finite type and such that $D(x)$ exists and is tame for all $x \in [0, 1]$ is open and dense in the set $\bigcup_{N \in \mathbb{N}} \mathsf{Div}_N$ of all divisors of finite type in Div_{tame}. Because the set of finite type divisors in Div_{tame} is dense in Div_{tame} by Theorem 13.2, we see that the set of divisors $D \in \mathsf{Div}_{tame}$ that are of finite type, and such that $D(x)$ exists for all $x \in [0, 1]$ and is tame, is dense in Div_{tame}. □

To further prepare the proof of Theorem 15.1, we show that the map $\Phi : \mathsf{Pot} \to \mathsf{Div}$ is smooth, and calculate the action of its derivative on the symplectic basis (v_k, w_k) of $T_{(u,u_y)}\mathsf{Pot}$ from Chap. 14:

Lemma 15.5

(1) The map $\Phi : \mathsf{Pot} \to \mathsf{Div}$, $(u, u_y) \mapsto (\lambda_k, \mu_k)_{k \in \mathbb{Z}}$ *is smooth at* $(u, u_y) \in \mathsf{Pot}_{tame}$ *in the "weak" sense that every component function* $\lambda_k : \mathsf{Pot} \to \mathbb{C}$ *or* $\mu_k : \mathsf{Pot} \to \mathbb{C}$ *(with* $k \in \mathbb{Z}$*) depends smoothly on* (u, u_y).[2]

(2) Let $(u, u_y) \in \mathsf{Pot}_{tame}$ *be given, let* $F(x, \lambda)$ *be the extended frame corresponding to* (u, u_y) *written as in Eq.* (14.2), *let* (v_k, w_k) *be the symplectic basis of* $T_{(u,u_y)}\mathsf{Pot}$ *defined in* (14.3) *(see also Theorem 14.3(1)), let* ϑ_k *be the numbers defined in Eq.* (14.4), *and put*

$$\widetilde{\vartheta}_k := \int_0^1 \Big(\lambda_k \, a(x, \lambda_k) \, b(x, \lambda_k) + \lambda_k^{-1} \, c(x, \lambda_k) \, d(x, \lambda_k) \Big) \cdot \mathrm{e}^{u(x)/2} \, \mathrm{d}x .$$

For $k \in \mathbb{Z}$ *we moreover let* $e_k = (\delta_{jk})_{j \in \mathbb{Z}}$ *be the* $\mathbb{Z}$*-sequence whose* k*-th member is* 1 *and all other members are* 0*; then we define elements of* $T_{\Phi((u,u_y))}\mathsf{Div} \cong \ell^2_{-1,3} \oplus \ell^2_{0,0}$ *by* $e_{\lambda,k} := (e_k, 0)$ *and* $e_{\mu,k} := (0, e_k)$*. Clearly* $(e_{\lambda,k}, e_{\mu,k})_{k \in \mathbb{Z}}$ *is a basis of* $T_{\Phi((u,u_y))}\mathsf{Div}$.

---

[2] We will see in the proof of Theorem 15.1 that Φ is in fact smooth in a "stronger" sense, namely as a map into the Banach space Div.

In this setting we have for $k \in \mathbb{Z}$ (where we interpret $\Phi'((u, u_y))$ according to the "weak" differentiability of (1)):

$$\Phi'((u, u_y))v_k = -\mathrm{i}\,\vartheta_k\,\mu_k\,e_{\mu,k},$$
$$\Phi'((u, u_y))w_k = -2\,\lambda_k\,e_{\lambda,k} + \mathrm{i}\,\mu_k\,\widetilde{\vartheta}_k\,e_{\mu,k}.$$

Proof For (1). We begin by showing that Φ is smooth in the "weak" sense described in the lemma: The 1-form α_λ defined by Eq. (3.2) depends smoothly on $(u, u_y) \in \mathsf{Pot}$. Therefore the extended frame $F_\lambda(x)$, characterized by the initial value problem

$$\mathrm{d}F_\lambda = \alpha_\lambda\,F_\lambda \quad \text{with} \quad F_\lambda(0) = \mathbb{1}$$

also depends smoothly on (u, u_y), and hence, the monodromy $M(\lambda) = F_\lambda(1) = \left(\begin{smallmatrix} a(\lambda) & b(\lambda) \\ c(\lambda) & d(\lambda) \end{smallmatrix}\right)$ depends smoothly on (u, u_y); here the entries of the monodromy and the entries of the extended frame are related by $a(\lambda) = a(1, \lambda)$ and likewise for the functions b, c and d.

In the spectral divisor $D = \{(\lambda_k, \mu_k)\}$ corresponding to the monodromy $M(\lambda)$, the λ_k are the zeros of the function $c(\lambda)$, which depends smoothly on (u, u_y). Therefore the λ-coordinates of the divisor points depend smoothly on (u, u_y); this is true even in the event of zeros of c of higher order by virtue of our construction of Div as the quotient $(\ell^2_{-1,3} \oplus \ell^2_{0,0})/P(\mathbb{Z})$, see the end of Chap. 8. Finally, the μ-components of the divisor points are obtained as $\mu_k = a(\lambda_k)$, where the function a depends smoothly on (u, u_y). Thus the μ-components also depend smoothly on (u, u_y). Hence, the map $\Phi : \mathsf{Pot} \to \mathsf{Div}$ is smooth.

For (2). We let $(u, u_y) \in \mathsf{Pot}_{tame}$ be given, and let $D := \Phi((u, u_y)) \in \mathsf{Div}_{tame}$ be the spectral divisor of (u, u_y). We equip $T_{(u,u_y)}\mathsf{Pot}$ with the non-degenerate symplectic form Ω from Eq. (14.1). Moreover, because D is tame, D is a regular point of Div and therefore we can consider the tangent space $T_D\mathsf{Div} \cong \ell^2_{-1,3} \oplus \ell^2_{0,0}$, which we equip with the non-degenerate symplectic form $\widetilde{\Omega} : T_D\mathsf{Div} \times T_D\mathsf{Div} \to \mathbb{C}$ defined by

$$\widetilde{\Omega}\big((\delta\lambda_k, \delta\mu_k), (\widetilde{\delta}\lambda_k, \widetilde{\delta}\mu_k)\big) = \frac{\mathrm{i}}{2} \sum_{k\in\mathbb{Z}} \left(\frac{\delta\lambda_k}{\lambda_k} \cdot \frac{\widetilde{\delta}\mu_k}{\mu_k} - \frac{\widetilde{\delta}\lambda_k}{\lambda_k} \cdot \frac{\delta\mu_k}{\mu_k} \right);$$

the sum converges by Lemma 14.9. By Theorem 14.3(2) the derivative $\Phi'((u, u_y)) : T_{(u,u_y)}\mathsf{Pot} \to T_D\mathsf{Div}$ is a symplectomorphism from $(T_{(u,u_y)}\mathsf{Pot}, \Omega)$ to $(T_D\mathsf{Div}, \widetilde{\Omega})$. Moreover we have $\Phi'((u, u_y))(\delta u, \delta u_y) = (\delta\lambda_k, \delta\mu_k)$; from this equation and the definitions of $\widetilde{\Omega}$, $e_{\lambda,k}$ and $e_{\mu,k}$ it follows that

$$\delta\lambda_k = -\frac{2}{\mathrm{i}}\,\lambda_k\,\mu_k\,\widetilde{\Omega}\left(e_{\mu,k},\,\Phi'((u,u_y))(\delta u,\delta u_y)\right)$$
$$\text{and}\quad \delta\mu_k = \frac{2}{\mathrm{i}}\,\lambda_k\,\mu_k\,\widetilde{\Omega}\left(e_{\lambda,k},\,\Phi'((u,u_y))(\delta u,\delta u_y)\right)$$

holds.

Now let $(\delta u, \delta u_y) \in T_{(u,u_y)}\mathsf{Pot}$ be given. Knopf has shown [Kn-2, the proof of Theorem 5.5] (essentially by evaluating an ungauged analogue to our Eq. (14.9)) that the corresponding variations δa and δc of entries of the extended frame satisfy

$$\delta a(\lambda_k) = -\mathrm{i}\,\mu_k\,\Omega\left(w_k,(\delta u,\delta u_y)\right) \quad\text{and}\quad \delta c(\lambda_k) = -\mathrm{i}\,\mu_k\,\Omega\left(v_k,(\delta u,\delta u_y)\right).$$

By an analogous calculation he also showed that the derivatives $a'(\lambda)$ and $c'(\lambda)$ of a resp. c with respect to λ satisfy

$$a'(\lambda_k) = -\mathrm{i}\frac{\mu_k}{2\,\lambda_k}\,\widetilde{\vartheta}_k \quad\text{and}\quad c'(\lambda_k) = -\mathrm{i}\frac{\mu_k}{2\,\lambda_k}\,\vartheta_k.$$

By the implicit function theorem for λ_k and differentiation of the equality $\mu_k = a(\lambda_k)$ we have

$$\delta\lambda_k = -\frac{\delta c(\lambda_k)}{c'(\lambda_k)} \quad\text{and}\quad \delta\mu_k = \delta a(\lambda_k) + a'(\lambda_k)\cdot\delta\lambda_k.$$

By combining the preceding equations, and the fact that $\Phi'((u,u_y))$ is a symplectomorphism with respect to Ω and $\widetilde{\Omega}$, we obtain

$$\begin{aligned}
-\frac{2}{\mathrm{i}}\,\lambda_k\,\mu_k\,\widetilde{\Omega}\left(e_{\mu,k},\,\Phi'((u,u_y))(\delta u,\delta u_y)\right) &= \delta\lambda_k = -\frac{\delta c(\lambda_k)}{c'(\lambda_k)}\\
&= -\frac{-\mathrm{i}\,\mu_k\,\Omega\left(v_k,(\delta u,\delta u_y)\right)}{-\mathrm{i}\frac{\mu_k}{2\,\lambda_k}\,\vartheta_k} = -2\,\frac{\lambda_k}{\vartheta_k}\,\Omega\left(v_k,(\delta u,\delta u_y)\right)\\
&= -2\,\frac{\lambda_k}{\vartheta_k}\,\widetilde{\Omega}\left(\Phi'((u,u_y))v_k,\,\Phi'((u,u_y))(\delta u,\delta u_y)\right)
\end{aligned}$$

and therefore, because $(\delta u, \delta u_y) \in T_{(u,u_y)}\mathsf{Pot}$ was arbitrary and $\Phi'((u,u_y))$ is a symplectomorphism,

$$-\frac{2}{\mathrm{i}}\,\lambda_k\,\mu_k\,e_{\mu,k} = -2\,\frac{\lambda_k}{\vartheta_k}\,\Phi'((u,u_y))v_k,$$

whence

$$\Phi'((u,u_y))v_k = -\mathrm{i}\,\vartheta_k\,\mu_k\,e_{\mu,k}$$

follows.

Similarly we have

$$\begin{aligned}
&\frac{2}{\mathrm{i}}\,\lambda_k\,\mu_k\,\widetilde{\Omega}\left(e_{\lambda,k},\Phi'((u,u_y))(\delta u,\delta u_y)\right)\\
&= \delta\mu_k = \delta a(\lambda_k) + a'(\lambda_k)\cdot\delta\lambda_k\\
&= -\mathrm{i}\,\mu_k\,\Omega\left(w_k,(\delta u,\delta u_y)\right) + (-\mathrm{i})\frac{\mu_k}{2\,\lambda_k}\,\widetilde{\vartheta}_k\cdot\delta\lambda_k\\
&= -\mathrm{i}\,\mu_k\,\widetilde{\Omega}\left(\Phi'((u,u_y))w_k,\Phi'((u,u_y))(\delta u,\delta u_y)\right)\\
&\quad + \mu_k^2\,\widetilde{\vartheta}_k\,\widetilde{\Omega}\left(e_{\mu,k},\Phi'((u,u_y))(\delta u,\delta u_y)\right)
\end{aligned}$$

and therefore

$$\frac{2}{\mathrm{i}}\,\lambda_k\,\mu_k\,e_{\lambda,k} = -\mathrm{i}\,\mu_k\,\Phi'((u,u_y))w_k + \mu_k^2\,\widetilde{\vartheta}_k\,e_{\mu,k},$$

whence

$$\Phi'((u,u_y))w_k = \mathrm{i}\,\mu_k\,\widetilde{\vartheta}_k\,e_{\mu,k} - 2\,\lambda_k\,e_{\lambda,k}$$

follows. □

Proof (of Theorem 15.1) We let $(u,u_y)\in\mathsf{Pot}_{tame}$ be given, and denote the spectral divisor of (u,u_y) by $D := \Phi((u,u_y))\in\mathsf{Div}_{tame}$. We use the notations from Lemma 15.5 and its proof, in particular we equip $T_{(u,u_y)}\mathsf{Pot}$ and $T_D\mathsf{Div}$ with the symplectic forms Ω resp. $\widetilde{\Omega}$.

To prove that Φ is a local diffeomorphism near (u,u_y), we apply the inverse function theorem. We therefore need to show that $\Phi : \mathsf{Pot}\to\mathsf{Div}$ is smooth at (u,u_y) as a map between Banach spaces, and that the derivative $\Phi'((u,u_y)) : T_{(u,u_y)}\mathsf{Pot}\to T_D\mathsf{Div}$ of Φ is an invertible linear operator of Banach spaces.

Because Φ is smooth in the "weak" sense of Lemma 15.5(1), we have the linear map $\Phi'((u,u_y)) : T_{(u,u_y)}\mathsf{Pot}\to T_D\mathsf{Div}$. This linear map is a symplectomorphism with respect to Ω and $\widetilde{\Omega}$ by Theorem 14.3(2), and is therefore bijective. It remains to show that this linear map is an invertible operator of Banach spaces, i.e. that it is continuous and that its inverse is also continuous.

For this we need to show that there exist constants $C_1, C_2 > 0$ so that

$$\begin{aligned}
&C_1\cdot\|v_k\|_{\mathsf{Pot}} \le \left\|\Phi'((u,u_y))v_k\right\|_{\mathsf{Div}} \le C_2\cdot\|v_k\|_{\mathsf{Pot}}\\
\text{and}\quad &C_1\cdot\|w_k\|_{\mathsf{Pot}} \le \left\|\Phi'((u,u_y))w_k\right\|_{\mathsf{Div}} \le C_2\cdot\|w_k\|_{\mathsf{Pot}}
\end{aligned}$$

holds for all $k \in \mathbb{Z}$, where (v_k, w_k) is the basis of $T_{(u,u_y)}\mathsf{Pot}$ introduced in (14.3), see also Theorem 14.3(1). It follows from the asymptotic assessment for v_k and w_k in the proof of Lemma 14.5 that there exist constants $C_3, \dots, C_6 > 0$ so that

$$\|v_k\|_{\mathsf{Pot}} = \begin{Bmatrix} C_3\,k^2 & \text{for } k \geq 0 \\ C_4 & \text{for } k < 0 \end{Bmatrix} + \ell^2_{-2,0}(k)$$

$$\text{and} \quad \|w_k\|_{\mathsf{Pot}} = \begin{Bmatrix} C_5\,k & \text{for } k \geq 0 \\ C_6\,|k| & \text{for } k < 0 \end{Bmatrix} + \ell^2_{-1,-1}(k)$$

holds. To show that $\Phi'((u, u_y))$ is an invertible operator of Banach spaces it therefore suffices to show that there exist constants $C_7, C_8 > 0$ so that

$$\begin{Bmatrix} C_7\,k^2 & \text{for } k \geq 0 \\ C_7 & \text{for } k < 0 \end{Bmatrix} \leq \left\| \Phi'((u, u_y))v_k \right\|_{\mathsf{Div}} \leq \begin{Bmatrix} C_8\,k^2 & \text{for } k \geq 0 \\ C_8 & \text{for } k < 0 \end{Bmatrix} \tag{15.2}$$

$$\text{and} \quad \begin{Bmatrix} C_7\,k & \text{for } k \geq 0 \\ C_7\,|k| & \text{for } k < 0 \end{Bmatrix} \leq \left\| \Phi'((u, u_y))w_k \right\|_{\mathsf{Div}} \leq \begin{Bmatrix} C_8\,k & \text{for } k \geq 0 \\ C_8\,|k| & \text{for } k < 0 \end{Bmatrix} \tag{15.3}$$

holds for all $k \in \mathbb{Z}$ with $|k|$ large.

For (15.2), we note that $\Phi'((u, u_y))v_k = -i\,\vartheta_k\,\mu_k\,e_{\mu,k}$ holds by Lemma 15.5(2). The asymptotics for the extended frame $F(x, \lambda)$ in Proposition 11.7 show that $\vartheta_k = \vartheta_{k,0} + \ell^2_{-2,0}(k)$ holds, where

$$\vartheta_{k,0} := \int_0^1 \left(\lambda_{k,0}\,\cos(k\pi\,x)^2 + \sin(k\pi\,x)^2 \right)\,\mathrm{d}x = \frac{1}{2}(\lambda_{k,0} + 1)$$

denotes the quantity corresponding to ϑ_k for the vacuum. Moreover $\mu_k = (-1)^k + \ell^2_{0,0}(k)$ and $\|e_{\mu,k}\|_{\mathsf{Div}} = 1$ holds. Thus we obtain

$$\left\| \Phi'((u, u_y))v_k \right\|_{\mathsf{Div}} = |\vartheta_k| \cdot |\mu_k| \cdot \|e_{\mu,k}\|_{\mathsf{Div}} = \frac{1}{2}\,(\lambda_{k,0} + 1) + \ell^2_{-2,0}(k),$$

which implies (15.2).

For (15.3), we similarly note that $\Phi'((u, u_y))w_k = -2\,\lambda_k\,e_{\lambda,k} + \mathrm{i}\,\mu_k\,\widetilde{\vartheta}_k\,e_{\mu,k}$ holds by Lemma 15.5(2). In addition to the information on μ_k and $e_{\mu,k}$ already given, we have $\lambda_k = \lambda_{k,0} + \ell^2_{-1,3}(k)$,

$$\|e_{\lambda,k}\|_{\mathsf{Div}} = \begin{cases} k^{-1} & \text{for } k \geq 0 \\ |k|^3 & \text{for } k < 0 \end{cases},$$

and the asymptotics for the extended frame $F(x, \lambda)$ in Proposition 11.7 show that $\widetilde{\vartheta}_k \in \ell^2_{-1,-1}(k)$ holds, because the quantity $\widetilde{\vartheta}_{k,0}$ corresponding to $\widetilde{\vartheta}_k$ for the vacuum vanishes:

$$\widetilde{\vartheta}_{k,0} = \int_0^1 \left(-\lambda_{k,0}^{1/2} \cos(k\pi\, x)\, \sin(k\pi\, x) + \lambda_{k,0}^{-1/2} \sin(k\pi\, x)\, \cos(k\pi\, x)\right) \mathrm{d}x = 0.$$

Thus we obtain

$$\begin{aligned}\left\| \Phi'((u, u_y))w_k - (-2)\,\lambda_k\, e_{\lambda,k} \right\|_{\mathsf{Div}} &= \left\| \mathrm{i}\, \mu_k\, \widetilde{\vartheta}_k\, e_{\mu,k} \right\|_{\mathsf{Div}} \\ &= (1 + \ell^2_{0,0}(k)) \cdot \ell^2_{-1,-1}(k) \cdot 1 \in \ell^2_{-1,-1}(k)\end{aligned}$$

with

$$\begin{aligned}\left\| (-2)\,\lambda_k\, e_{\lambda,k} \right\|_{\mathsf{Div}} &= 2 \cdot (\lambda_{k,0} + \ell^2_{-1,3}(k)) \cdot \begin{Bmatrix} k^{-1} & \text{for } k \geq 0 \\ |k|^3 & \text{for } k < 0 \end{Bmatrix} \\ &= \begin{Bmatrix} C_9\, k & \text{for } k \geq 0 \\ C_{10}\, |k| & \text{for } k < 0 \end{Bmatrix} + \ell^2_{0,0}(k),\end{aligned}$$

which implies (15.3).

Therefore $\Phi'((u, u_y))$ is an invertible operator of Banach spaces. Hence it follows from the inverse function theorem that Φ is a local diffeomorphism near (u, u_y).

Because $\Phi|\mathsf{Pot}_{tame}$ thus is a local diffeomorphism, its image is an open subset of Div_{tame}. Because all finite type divisors $D \in \mathsf{Div}_{tame}$ for which $D(x)$ exists and is tame for all $x \in [0, 1]$ are contained in this image by Lemma 15.3, and the set of these divisors is dense in Div_{tame} by Lemma 15.4, the image of $\Phi|\mathsf{Pot}_{tame}$ is also dense in Div_{tame}.

It remains to show that $\Phi|\mathsf{Pot}_{tame}$ is injective. For this purpose let $(u^{[1]}, u_y^{[1]})$, $(u^{[2]}, u_y^{[2]}) \in \mathsf{Pot}_{tame}$ be given with $\Phi((u^{[1]}, u_y^{[1]})) = \Phi((u^{[2]}, u_y^{[2]})) =: D \in \mathsf{Div}_{tame}$. Because Φ is a diffeomorphism near $(u^{[\nu]}, u_y^{[\nu]})$ (for $\nu \in \{1, 2\}$), there exist neighborhoods $U^{[\nu]}$ of $(u^{[\nu]}, u_y^{[\nu]})$ in Pot_{tame} and $V^{[\nu]}$ of D in Div_{tame} so that $\Phi|U^{[\nu]} : U^{[\nu]} \to V^{[\nu]}$ is a diffeomorphism.

By Lemma 15.4 there exists a sequence $(D_n)_{n\geq 1}$ of divisors of finite type, such that $D_n(x)$ exists and is tame for all $x \in [0, 1]$, which converges to D. Without loss of generality, we may suppose $D_n \in V^{[1]} \cap V^{[2]}$ for all $n \geq 1$, and then $(u^{[\nu,n]}, u_y^{[\nu,n]}) := (\Phi|U^{[\nu]})^{-1}(D_n)$ is a sequence in $U^{[\nu]}$ which converges to $(\Phi|U^{[\nu]})^{-1}(D_n) = (u^{[\nu]}, u_y^{[\nu]})$. On the other hand, D_n has only one pre-image under Φ by the uniqueness statement in Lemma 15.3. Therefore $(u^{[1,n]}, u_y^{[1,n]}) = (u^{[2,n]}, u_y^{[2,n]})$ holds for all $n \geq 1$, whence $(u^{[1]}, u_y^{[1]}) = (u^{[2]}, u_y^{[2]})$ follows by taking the limit $n \to \infty$. □

Corollary 15.6

(1) The set of potentials of finite type in Pot_{tame} *is dense in* Pot_{tame}.
(2) The set of divisors $D \in \mathsf{Div}_{tame}$ *such that* $D(x)$ *exists and is tame for all* $x \in [0, 1]$ *is open and dense in* Div_{tame}.

Proof For (1). The set Div_{fin} of divisors of finite type in Div_{tame} is dense in Div_{tame} by Theorem 13.2; because $\Phi[\mathsf{Pot}_{tame}]$ is open in Div_{tame} by Theorem 15.1, it follows that $\mathsf{Div}_{fin} \cap \Phi[\mathsf{Pot}_{tame}]$ is dense in $\Phi[\mathsf{Pot}_{tame}]$. Because $\Phi : \mathsf{Pot}_{tame} \to \Phi[\mathsf{Pot}_{tame}]$ is a diffeomorphism again by Theorem 15.1, it follows that $\Phi^{-1}[\mathsf{Div}_{fin} \cap \Phi[\mathsf{Pot}_{tame}]]$ is dense in Pot_{tame}; this is the set of finite type potentials in Pot_{tame}.

For (2). The set of divisors $D \in \mathsf{Div}_{tame}$ such that $D(x)$ exists and is tame for all $x \in [0, 1]$ is open in Div_{tame} because $D(x)$ depends continuously on D and on x. This set is dense in Div_{tame} because already the set of finite type divisors such that $D(x)$ exists and is tame for all $x \in [0, 1]$ is dense in Div_{tame} by Lemma 15.4. □

Part V
The Jacobi Variety of the Spectral Curve

Chapter 16
Estimate of Certain Integrals

Our next big objective is the construction of Jacobi coordinates for the spectral curve Σ corresponding to some potential $(u, u_y) \in \mathsf{Pot}$. Similarly as for the construction of the Jacobi variety for compact Riemann surfaces, our construction of the Jacobi variety for Σ in Chap. 18 will involve path integrals on the spectral curve Σ which are of the form

$$\int_{P_0}^{P_1} \frac{\Phi(\lambda)}{\mu - \mu^{-1}} \, \mathrm{d}\lambda,$$

where Φ is a holomorphic function in λ.

To make the construction tractable, we will need to do two things: In the present chapter, we will estimate $\int_{P_0}^{P_1} \frac{1}{\mu-\mu^{-1}} \, \mathrm{d}\lambda$ and $\int_{P_0}^{P_1} \frac{1}{|\mu-\mu^{-1}|} \, |\mathrm{d}\lambda|$ (Proposition 16.5). In the former integral, the integrand is holomorphic, and therefore the value of the integral depends only on the homology type of the path of integration. However the second integrand is not holomorphic, and thus the value of the second integral depends on the specific choice of the path of integration. We will choose a special class of *admissible* paths of integration, which permit to connect each pair of points contained in an excluded domain $\widehat{U}_{k,\delta} \subset \Sigma$, but which are simple enough (they are composed of lifts of straight lines and circular arcs in the λ-plane) so that the integral in question can be estimated relatively explicitly.

The other preparation needed for the construction of the Jacobi coordinates on Σ is the study of the asymptotic behavior of holomorphic 1-forms on Σ which are square-integrable. This is the topic of Chap. 17.

We suppose that Σ either is the spectral curve associated to some potential $(u, u_y) \in \mathsf{Pot}$, or the spectral curve defined by a holomorphic function $\Delta : \mathbb{C}^* \to \mathbb{C}$ with $\Delta - \Delta_0 \in \mathrm{As}(\mathbb{C}^*, \ell^2_{0,0}, 1)$. As before, we have the holomorphic functions

S. Klein, *A Spectral Theory for Simply Periodic Solutions of the Sinh-Gordon Equation*, Lecture Notes in Mathematics 2229,
https://doi.org/10.1007/978-3-030-01276-2_16

$\lambda, \mu : \Sigma \to \mathbb{C}^*$, and we denote the zeros of $\Delta^2 - 4$ (corresponding to the branch points resp. singularities of Σ) by $\varkappa_{k,\nu}$ as in Proposition 6.5(1); we have $\varkappa_{k,\nu} - \lambda_{k,0} \in \ell^2_{-1,3}(k)$ by Proposition 11.5.

Because $\varkappa_{k,\nu}$ is a zero of $\Delta^2 - 4$ for $k \in \mathbb{Z}$ and $\nu \in \{1, 2\}$, there exists one and only one point in Σ above $\varkappa_{k,\nu}$ (which is a fixed point of the hyperelliptic involution of Σ). In the sequel we will take the liberty of denoting this point of Σ also by $\varkappa_{k,\nu}$.

To prepare for the evaluation of the mentioned integrals, we begin by showing that the function $\frac{1}{\mu-\mu^{-1}}$ behaves on $\widehat{U}_{k,\delta}$ essentially like $\frac{1}{\sqrt{(\lambda-\varkappa_{k,1})\cdot(\lambda-\varkappa_{k,2})}}$, provided that $|k|$ is so large that $\varkappa_{k,1}$ and $\varkappa_{k,2}$ are the only zeros of Δ^2-4 (branch points or singularities of Σ) that occur in $\widehat{U}_{k,\delta}$. This statement is made precise by the following lemma:

Lemma 16.1 *Let $k \in \mathbb{Z}$. There exists $N \in \mathbb{N}$ so that for all $k \in \mathbb{Z}$ with $|k| > N$ there exists a holomorphic function Ψ_k on $\widehat{U}_{k,\delta}$ with*

$$\Psi_k(\lambda, \mu)^2 = (\lambda - \varkappa_{k,1}) \cdot (\lambda - \varkappa_{k,2}). \tag{16.1}$$

We have $\Psi_k \circ \sigma = -\Psi_k$, where σ is the hyperelliptic involution of Σ. Ψ_k has the following asymptotic property, which also fixes the sign of Ψ_k:

There exists $C > 0$ and a sequence $a_k \in \ell^2_{-1,3}(|k| > N)$ such that for all $k \in \mathbb{Z}$ with $|k| > N$ and all $(\lambda, \mu) \in \widehat{U}_{k,\delta} \setminus \{\varkappa_{k,1}, \varkappa_{k,2}\}$ we have if $k > 0$

$$\left| \frac{1}{\mu - \mu^{-1}} - 4\mathrm{i}(-1)^k \lambda_{k,0}^{1/2} \frac{1}{\Psi_k(\lambda, \mu)} \right| \leq \frac{C \cdot \big(|\lambda - \varkappa_{k,1}| + |\lambda - \varkappa_{k,2}|\big) + a_k}{|\Psi_k(\lambda, \mu)|},$$

and if $k < 0$

$$\left| \frac{1}{\mu - \mu^{-1}} - 4\mathrm{i}(-1)^k \lambda_{k,0}^{3/2} \frac{1}{\Psi_k(\lambda, \mu)} \right| \leq \frac{C \cdot \big(|\lambda - \varkappa_{k,1}| + |\lambda - \varkappa_{k,2}|\big) + a_k}{|\Psi_k(\lambda, \mu)|}.$$

Proof By Proposition 6.5(1) there exists $N \in \mathbb{N}$ so that for every k with $|k| > N$, $\varkappa_{k,1}$ and $\varkappa_{k,2}$ are all the zeros of $\Delta^2 - 4$ in $U_{k,\delta}$, and therefore $\varkappa_{k,1}$ and $\varkappa_{k,2}$ are all the branch points of $\widehat{U}_{k,\delta}$. For $|k| > N$ it is therefore clear that there exists a holomorphic function Ψ_k on $\widehat{U}_{k,\delta}$ so that Eq. (16.1) holds; we then also have $\Psi_k \circ \sigma = -\Psi_k$. Equation (16.1) determines Ψ_k only up to sign on each of the (one or two) connected components of $\widehat{U}_{k,\delta} \setminus \{\varkappa_{k,1}, \varkappa_{k,2}\}$. The sign is fixed by the stated asymptotic property for Ψ_k, which we now prove:

We let $\varrho = 1$ if $k > 0$, $\varrho = 3$ if $k < 0$. By Proposition 11.6(3) there exists $C_1 > 0$ such that we have for $\lambda \in U_{k,\delta}$

$$\left|\lambda_{k,0}^{\varrho} \cdot \frac{\Delta(\lambda)^2 - 4}{(\lambda - \varkappa_{k,1}) \cdot (\lambda - \varkappa_{k,2})} - \left(-\frac{1}{16}\right)\right|$$
$$\leq C_1 \cdot \begin{Bmatrix} k^{-1} & \text{if } k > 0 \\ k^3 & \text{if } k < 0 \end{Bmatrix} \cdot \left(|\lambda - \varkappa_{k,1}| + |\lambda - \varkappa_{k,2}|\right) + \ell^2_{0,0}(k). \tag{16.2}$$

Because of the equation $\mu - \mu^{-1} = \sqrt{\Delta(\lambda)^2 - 4}$ and the fact that $z \mapsto z^{-1/2}$ is Lipschitz continuous near $z = -\frac{1}{16}$, there exists $C_2 > 0$ and $\varepsilon \in \{-1, 1\}$ with

$$\left|\lambda_{k,0}^{-\varrho/2} \cdot \frac{\Psi_k(\lambda, \mu)}{\mu - \mu^{-1}} - 4\mathrm{i}\,\varepsilon\right| \leq C_2 \cdot \begin{Bmatrix} k^{-1} & \text{if } k > 0 \\ k^3 & \text{if } k < 0 \end{Bmatrix} \cdot \left(|\lambda - \varkappa_{k,1}| + |\lambda - \varkappa_{k,2}|\right) + \ell^2_{0,0}(k),$$

whence

$$\left|\frac{1}{\sqrt{\mu - \mu^{-1}}} - 4\mathrm{i}\,\varepsilon\,\lambda_{k,0}^{\varrho/2} \frac{1}{\Psi_k(\lambda, \mu)}\right| \leq \frac{C_3 \cdot \left(|\lambda - \varkappa_{k,1}| + |\lambda - \varkappa_{k,2}|\right) + \ell^2_{-1,3}(k)}{|\Psi_k(\lambda, \mu)|}$$

follows. By choosing the proper sign for Ψ_k, we can arrange $\varepsilon = 1$, and therefore obtain the claimed asymptotic property. □

For $k \in \mathbb{Z}$ we let $\varkappa_{k,*} \in \mathbb{C}^*$ be the midpoint between $\varkappa_{k,1}$ and $\varkappa_{k,2}$, i.e.

$$\varkappa_{k,*} := \frac{1}{2}(\varkappa_{k,1} + \varkappa_{k,2}). \tag{16.3}$$

Because of $\varkappa_{k,\nu} - \lambda_{k,0} \in \ell^2_{-1,3}(k)$ we have $\varkappa_{k,*} - \lambda_{k,0} \in \ell^2_{-1,3}(k)$, and for $|k|$ large, we have $\varkappa_{k,*} \in U_{k,\delta}$.

If $k \in \mathbb{Z}$ is such that $\varkappa_{k,1}$ and $\varkappa_{k,2}$ are the only zeros of $\Delta^2 - 4$ in $\widehat{U}_{k,\delta}$ (i.e. $|k| > N$ in the notations of Lemma 16.1), then we put

$$\widehat{U}'_{k,\delta} := \begin{cases} \widehat{U}_{k,\delta} & \text{if } \varkappa_{k,1} \neq \varkappa_{k,2} \\ \widehat{U}_{k,\delta} \setminus \{\varkappa_{k,*}\} & \text{if } \varkappa_{k,1} = \varkappa_{k,2} \end{cases}$$

$$\text{and} \quad U'_{k,\delta} := \begin{cases} U_{k,\delta} & \text{if } \varkappa_{k,1} \neq \varkappa_{k,2} \\ U_{k,\delta} \setminus \{\varkappa_{k,*}\} & \text{if } \varkappa_{k,1} = \varkappa_{k,2} \end{cases}. \tag{16.4}$$

Note that $U'_{k,\delta}$ is connected in any case; whereas $\widehat{U}'_{k,\delta}$ is connected only if $\varkappa_{k,1} \neq \varkappa_{k,2}$, for $\varkappa_{k,1} = \varkappa_{k,2}$ it has two connected components.

In the sequel, we will estimate the integrals of $\frac{1}{\Psi_k}$ and of $\frac{1}{|\Psi_k|}$ along paths in the excluded domain. Because the second integrand $\frac{1}{|\Psi_k|}$ is not holomorphic, we need to specify in detail which paths we permit for estimating the corresponding integral, as was explained above. The admissible paths introduced in the following definition are the paths of integration that we permit in this context:

Definition 16.2 Let $N \in \mathbb{N}$ be as in Lemma 16.1 and $k \in \mathbb{Z}$ with $|k| > N$. We call a path γ in $\widehat{U}'_{k,\delta}$ *admissible*, if the following properties hold, where we denote by (λ_k^o, μ_k^o) and (λ_k, μ_k) the endpoints of γ:

(1) γ is composed of the lifts to Σ of at most three each of the following two types of path segments in $U'_{k,\delta}$:

  (a) A segment of a line through $\varkappa_{k,*}$ and contained in $U'_{k,\delta}$, traversed in either direction,
  (b) A segment of a circle with center $\varkappa_{k,*}$ and contained in $U'_{k,\delta}$, where the circle may be parameterized in either direction, and the angular length is at most 4π (i.e. at most two full revolutions of the circle).

(2) For all (λ, μ) on the trace of γ and for $\nu \in \{1, 2\}$ we have

$$|\lambda - \varkappa_{k,\nu}| \leq \max\left\{2\,|\varkappa_{k,1} - \varkappa_{k,2}|,\ |\lambda_k^o - \varkappa_{k,*}|,\ |\lambda_k - \varkappa_{k,*}|\right\}. \tag{16.5}$$

(3) The trace of γ meets the points $\varkappa_{k,1}$ and $\varkappa_{k,2}$ at most in its endpoints.

Proposition 16.3 *Let $N \in \mathbb{N}$ be as in Lemma 16.1 and $k \in \mathbb{Z}$ with $|k| > N$. For every $(\lambda_k^o, \mu_k^o), (\lambda_k, \mu_k) \in \widehat{U}'_{k,\delta}$ (from the same connected component of $\widehat{U}'_{k,\delta}$ if $\varkappa_{k,1} = \varkappa_{k,2}$) there exists an admissible path γ in $\widehat{U}'_{k,\delta}$ which connects (λ_k^o, μ_k^o) to (λ_k, μ_k).*

Proof We distinguish cases depending on whether $|\lambda_k - \varkappa_{k,*}|$ or $|\lambda_k^o - \varkappa_{k,*}|$ is larger or smaller than $|\varkappa_{k,1} - \varkappa_{k,2}|$. In the sequel, "line" or "ray" always means a line or ray contained in $U_{k,\delta}$ starting in $\varkappa_{k,*}$ and passing through another point, and "circle (segment)" always means a circle (segment) contained in $U_{k,\delta}$ with center $\varkappa_{k,*}$.

If $|\lambda_k^o - \varkappa_{k,*}|, |\lambda_k - \varkappa_{k,*}| \leq |\varkappa_{k,1} - \varkappa_{k,2}|$ holds, we construct γ as composition of lifts in $\widehat{U}'_{k,\delta}$ of the following segments in $U'_{k,\delta}$:

- If the ray through λ_k^o passes through either $\varkappa_{k,1}$ or $\varkappa_{k,2}$, a short circle segment so that the ray through its endpoint passes through neither $\varkappa_{k,1}$ nor $\varkappa_{k,2}$; otherwise nothing.
- The segment of the line through the previous endpoint from that endpoint to a point $\lambda \in U_{k,\delta}$ with $|\lambda - \varkappa_{k,*}| = |\varkappa_{k,1} - \varkappa_{k,2}|$.
- A circle segment from the previous endpoint, which contains a full circle if and only if (λ_k^o, μ_k^o) and (λ_k, μ_k) are on different "leaves" of $\widehat{U}'_{k,\delta}$, and whose endpoint is chosen in the following way: If the ray through λ_k does not pass through either $\varkappa_{k,1}$ or $\varkappa_{k,2}$, then the endpoint is such that the ray through the endpoint passes through λ_k; otherwise the arc is extended beyond that point by a short length so that the ray through its endpoint does not pass through either $\varkappa_{k,1}$ or $\varkappa_{k,2}$.

- A line segment connecting the previous endpoint to a point $\lambda' \in U_{k,\delta}$ with $|\lambda' - \varkappa_{k,*}| = |\lambda_k - \varkappa_{k,*}|$.
- If the lift of the previous endpoint is not (λ_k, μ_k), then a short circle segment connecting the previous endpoint to (λ_k, μ_k).

If either $|\lambda_k^o - \varkappa_{k,*}| > |\varkappa_{k,1} - \varkappa_{k,2}|$ or $|\lambda_k - \varkappa_{k,*}| > |\varkappa_{k,1} - \varkappa_{k,2}|$ holds, we may suppose without loss of generality that $|\lambda_k^o - \varkappa_{k,*}| \leq |\lambda_k - \varkappa_{k,*}|$ and therefore $|\lambda_k - \varkappa_{k,*}| > |\varkappa_{k,1} - \varkappa_{k,2}|$ holds. We make no assumptions on the relation between $|\lambda_k^o - \varkappa_{k,*}|$ and $|\varkappa_{k,1} - \varkappa_{k,2}|$. In this case, we construct γ as composition of lifts in $\widehat{U}'_{k,\delta}$ of the following segments in $U'_{k,\delta}$:

- If the segment of the ray through λ_k^o from λ_k^o to a point $\lambda \in U_{k,\delta}$ with $|\lambda - \varkappa_{k,*}| = |\lambda_k - \varkappa_{k,*}|$ would pass through either $\varkappa_{k,1}$ or $\varkappa_{k,2}$, a short circle segment so that the ray through its endpoint passes through neither $\varkappa_{k,1}$ nor $\varkappa_{k,2}$; otherwise nothing.
- A line segment from the previous endpoint to a point λ with $|\lambda - \varkappa_{k,*}| = |\lambda_k - \varkappa_{k,*}|$.
- A circle segment from the previous endpoint, which contains a full circle if and only if (λ_k^o, μ_k^o) and (λ_k, μ_k) are on different "leaves" of $\widehat{U}'_{k,\delta}$, and which ends in (λ_k, μ_k).

□

We now estimate the path integral along admissible paths for the integrands $\frac{1}{\Psi_k}$, $\frac{1}{|\Psi_k|}$ and $\frac{|\lambda-\xi_k|}{|\Psi_k|}$, and also the area integral of $\left(\frac{|\lambda-\xi_k|}{|\Psi_k|}\right)^2$ over an excluded domain. Note that while the holomorphic function Ψ_k is well-defined only on $\widehat{U}_{k,\delta}$ for k sufficiently large (see Lemma 16.1), the continuous function $|\Psi_k|$ is well-defined on all of Σ and for all $k \in \mathbb{Z}$, namely by

$$|\Psi_k(\lambda)| := \sqrt{|\lambda - \varkappa_{k,1}| \cdot |\lambda - \varkappa_{k,2}|},$$

where $\sqrt{}$ denotes the real square root function, moreover $|\Psi_k|$ depends only on $\lambda \in \mathbb{C}^*$, and not on μ.

Lemma 16.4 *Let $N \in \mathbb{N}$ be as in Lemma 16.1, $k \in \mathbb{Z}$, and $(\lambda_k^o, \mu_k^o), (\lambda_k, \mu_k) \in \widehat{U}'_{k,\delta}$. If $\varkappa_{k,1} \neq \varkappa_{k,2}$ holds, we let $\xi_k \in U_{k,\delta}$ be given, whereas for $\varkappa_{k,1} = \varkappa_{k,2}$ we put $\xi_k := \varkappa_{k,*} \in U_{k,\delta}$ and require that (λ_k^o, μ_k^o) and (λ_k, μ_k) are in the same connected component of $\widehat{U}'_{k,\delta}$.*

(1) For $|k| > N$ we have

$$\int_{(\lambda_k^o,\mu_k^o)}^{(\lambda_k,\mu_k)} \frac{1}{\Psi_k}\, d\lambda = \ln\left(\frac{\lambda_k - \varkappa_{k,*} + \Psi_k(\lambda_k, \mu_k)}{\lambda_k^o - \varkappa_{k,*} + \Psi_k(\lambda_k^o, \mu_k^o)}\right),$$

where we integrate along any path that is entirely contained in $\widehat{U}'_{k,\delta}$. Here $\ln(z)$ is the branch of the complex logarithm function with $\ln(1) = 2\pi i m$, where $m \in \mathbb{Z}$ is the winding number of the path of integration around the pair of branch points $\varkappa_{k,1}, \varkappa_{k,2}$.

(2) *There exists a constant* $C > 0$ *(depending neither on* k *nor on* $\varkappa_{k,\nu}$*) so that for* $|k| > N$ *we have*

$$\int_{(\lambda_k^o,\mu_k^o)}^{(\lambda_k,\mu_k)} \frac{1}{|\Psi_k|}\,|\mathrm{d}\lambda| \leq C\cdot\left|\int_{(\lambda_k^o,\mu_k^o)}^{(\lambda_k,\mu_k)} \frac{1}{\Psi_k}\,\mathrm{d}\lambda\right|, \tag{16.6}$$

where we integrate along an admissible path (Definition 16.2).

(3) *There exist constants* $C_1, C_2 > 0$ *(depending neither on* k *nor on* $\varkappa_{k,\nu}$ *or* ξ_k*) so that we have for any* $k \in \mathbb{Z}$

$$\int_{\varkappa_{k,1}}^{(\lambda_k,\mu_k)} \frac{|\lambda-\xi_k|}{|\Psi_k|}\,|\mathrm{d}\lambda| \leq \left(C_1\left|\frac{\xi_k-\varkappa_{k,*}}{\varkappa_{k,1}-\varkappa_{k,2}}\right| + C_2\right)\cdot|\lambda_k-\varkappa_{k,1}|.$$

Here we integrate along the lift of the straight line from $\varkappa_{k,1}$ *to* λ_k*, and the expression* $\left|\frac{\xi_k-\varkappa_{k,*}}{\varkappa_{k,1}-\varkappa_{k,2}}\right|$ *is to be read as* 0 *in the case* $\varkappa_{k,1}=\varkappa_{k,2}$.

(4) *There exists* $C > 0$ *(depending neither on* k *nor on* $\varkappa_{k,\nu}$ *or* ξ_k*) so that we have for any* $k \in \mathbb{Z}$

$$\int_{U_{k,\delta}} \frac{|\lambda-\xi_k|^2}{|\lambda-\varkappa_{k,1}|\cdot|\lambda-\varkappa_{k,2}|}\,\mathrm{d}^2\lambda \leq C\cdot\left(|\xi_k-\varkappa_{k,*}|^2 + \frac{|\xi_k-\varkappa_{k,*}|^4}{|\varkappa_{k,1}-\varkappa_{k,2}|^2}\right) + \ell^\infty_{-2,6}(k),$$

where $\mathrm{d}^2\lambda$ *denotes the 2-dimensional Lebesgue measure on* $\mathbb{C}$*. In the case* $\varkappa_{k,1}=\varkappa_{k,2}$*, the expression* $\frac{|\xi_k-\varkappa_{k,*}|^4}{|\varkappa_{k,1}-\varkappa_{k,2}|^2}$ *is to be read as* 0.

Proof (of Lemma 16.4) For (1). It is easily checked that $\ln(\lambda-\varkappa_{k,*}+\Psi_k(\lambda,\mu))$ is an anti-derivative of $\frac{1}{\Psi_k(\lambda,\mu)}$, and the claimed formula follows.

For (2). We first show that the inequality to be shown is invariant under change of the $\varkappa_{k,\nu}$. For this purpose, we at first suppose that $\varkappa_{k,1}\neq\varkappa_{k,2}$ holds, let $\widehat{U}$ be the hyperelliptic surface above $\mathbb{C}$ with branch points above $\varkappa_1 := 1$ and $\varkappa_2 := -1$, and let $\Psi(\lambda,\mu) := \sqrt{\lambda^2-1}$ be the corresponding holomorphic function on $\widehat{U}$ with zeros in the branch points. We then have

$$\begin{aligned}\Psi_k(\lambda,\mu) &= \sqrt{(\lambda-\varkappa_{k,1})\cdot(\lambda-\varkappa_{k,2})} = \sqrt{(\lambda-\varkappa_{k,*})^2-\left(\tfrac{1}{2}(\varkappa_{k,1}-\varkappa_{k,2})\right)^2}\\ &= \frac{1}{2}(\varkappa_{k,1}-\varkappa_{k,2})\cdot\Psi\left(\frac{\lambda-\varkappa_{k,*}}{\frac{1}{2}(\varkappa_{k,1}-\varkappa_{k,2})},\mu\right)\end{aligned}$$

and therefore, if we choose points $(\lambda^o,\mu^o), (\lambda,\mu) \in \widehat{U}$ with $\lambda^o = \frac{\lambda_k^o-\varkappa_{k,*}}{\frac{1}{2}(\varkappa_{k,1}-\varkappa_{k,2})}$ resp. $\lambda = \frac{\lambda_k-\varkappa_{k,*}}{\frac{1}{2}(\varkappa_{k,1}-\varkappa_{k,2})}$, we have

$$\int_{(\lambda_k^o,\mu_k^o)}^{(\lambda_k,\mu_k)} \frac{1}{\Psi_k}\,\mathrm{d}\lambda' = \int_{(\lambda^o,\mu^o)}^{(\lambda,\mu)} \frac{1}{\Psi}\,\mathrm{d}\lambda'$$

and similarly

$$\int_{(\lambda_k^o,\mu_k^o)}^{(\lambda_k,\mu_k)} \frac{1}{|\Psi_k|}\,|\mathrm{d}\lambda'| = \int_{(\lambda^o,\mu^o)}^{(\lambda,\mu)} \frac{1}{|\Psi|}\,|\mathrm{d}\lambda'|.$$

Therefore we see that the inequality (16.6) is implied by the claim that there exists $C > 0$ so that we have for all $(\lambda^o, \mu^o), (\lambda, \mu) \in \widehat{U}$

$$\int_{(\lambda^o,\mu^o)}^{(\lambda,\mu)} \frac{1}{|\Psi|}\,|\mathrm{d}\lambda'| \leq C \cdot \left| \int_{(\lambda^o,\mu^o)}^{(\lambda,\mu)} \frac{1}{\Psi}\,\mathrm{d}\lambda' \right|. \tag{16.7}$$

For reasons of continuity this is true even in the case $\varkappa_{k,1} = \varkappa_{k,2}$.

For the proof of (16.7), we note that $\varkappa_1 - \varkappa_2 = 2$ and $\varkappa_* := \frac{1}{2}(\varkappa_1 + \varkappa_2) = 0$ holds. We first consider the case where $|\lambda^o|, |\lambda| \leq 2$ holds. Then the admissible path of integration from (λ^o, μ^o) to (λ, μ) runs entirely in the pre-image of $\{\lambda' \in \mathbb{C} \,|\, |\lambda'| \leq 2\}$. We note that

$$\int_{(\lambda^o,\mu^o)}^{(\lambda,\mu)} \frac{1}{\Psi}\,\mathrm{d}\lambda' = \ln\left(\frac{\lambda - \varkappa_* + \Psi(\lambda,\mu)}{\lambda^o - \varkappa_* + \Psi(\lambda^o,\mu^o)}\right)$$

equals 0 if and only if we choose the branch of the complex logarithm with $\ln(1) = 0$ (corresponding to integration paths that do not wind around the branch points $\varkappa_\nu$), and moreover

$$\lambda - \varkappa_* + \Psi(\lambda,\mu) = \lambda^o - \varkappa_* + \Psi(\lambda^o,\mu^o)$$

and therefore $(\lambda, \mu) = (\lambda^o, \mu^o)$ holds.

On the other hand, the function $\int_{(\lambda^o,\mu^o)}^{(\lambda,\mu)} \frac{1}{|\Psi|}\,|\mathrm{d}\lambda'|$ also equals zero for $(\lambda, \mu) = (\lambda^o, \mu^o)$, and we will show below that

$$\lim_{(\lambda,\mu)\to(\lambda^o,\mu^o)} \frac{\int_{(\lambda^o,\mu^o)}^{(\lambda,\mu)} \frac{1}{|\Psi|}\,|\mathrm{d}\lambda'|}{\left|\int_{(\lambda^o,\mu^o)}^{(\lambda,\mu)} \frac{1}{\Psi}\,\mathrm{d}\lambda'\right|} \leq \sqrt{2} \tag{16.8}$$

holds. Moreover, there exists a constant $C_1 > 0$ so that $\int_{(\lambda^o,\mu^o)}^{(\lambda,\mu)} \frac{1}{|\Psi|}\,|\mathrm{d}\lambda'| \leq C_1$ holds. It follows from these facts by an argument of compactness that (16.7) holds for $|\lambda^o|, |\lambda| \leq 2$.

For the proof of (16.8): If $(\lambda^o, \mu^o) \neq \varkappa_\nu$ holds for $\nu \in \{1, 2\}$, then we have $\Psi(\lambda^o, \mu^o) \neq 0$ and therefore by l'Hospital's rule and the Fundamental Theorem of Calculus

$$\lim_{(\lambda,\mu)\to(\lambda^o,\mu^o)} \frac{\int_{(\lambda^o,\mu^o)}^{(\lambda,\mu)} \frac{1}{|\Psi|}\,|\mathrm{d}\lambda'|}{\left|\int_{(\lambda^o,\mu^o)}^{(\lambda,\mu)} \frac{1}{\Psi}\,\mathrm{d}\lambda'\right|} = \frac{\frac{1}{|\Psi(\lambda^o,\mu^o)|}}{\left|\frac{1}{\Psi(\lambda^o,\mu^o)}\right|} = 1 \leq \sqrt{2}.$$

We now look at the case $(\lambda^o, \mu^o) = \varkappa_\nu$, say with $\nu = 1$. We may suppose that (λ, μ) is close to (λ^o, μ^o), more specifically that

$$|\lambda - \varkappa_1| \leq 1 \quad \text{and therefore} \quad |\lambda - \varkappa_2| \geq 1 \tag{16.9}$$

holds. Moreover, we may suppose that the path of integration in (16.8) is a straight line from $(\lambda^o, \mu^o) = \varkappa_1$ to (λ, μ) (this is an admissible path). From (16.9) it follows that we have

$$\frac{1}{|\Psi(\lambda', \mu')|} \leq \frac{1}{\sqrt{|\lambda' - \varkappa_1|}}$$

for (λ', μ') on the path of integration, and therefore

$$\int_{\varkappa_1}^{(\lambda,\mu)} \frac{1}{|\Psi|}\,|\mathrm{d}\lambda'| \leq \int_{\varkappa_1}^{(\lambda,\mu)} \frac{1}{\sqrt{|\lambda'-\varkappa_1|}}\,|\mathrm{d}\lambda'| \overset{(*)}{=} \left|\int_{\varkappa_1}^{(\lambda,\mu)} \frac{1}{\sqrt{\lambda'-\varkappa_1}}\,\mathrm{d}\lambda'\right| = 2\sqrt{|\lambda - \varkappa_1|},$$

where the equals sign marked $(*)$ follows from our choice of the path of integration, which implies that $\frac{1}{\sqrt{\lambda'-\varkappa_1}}$ has constant argument. On the other hand, we have

$$\int_{\varkappa_1}^{(\lambda,\mu)} \frac{1}{\Psi}\,\mathrm{d}\lambda' = \ln\left(\frac{\lambda - \varkappa_* + \Psi(\lambda, \mu)}{\varkappa_1 - \varkappa_* + \Psi(\varkappa_1)}\right) = \ln\left(\lambda + \sqrt{\lambda^2 - 1}\right) = \operatorname{arcosh}(\lambda).$$

Thus we obtain, again by the rule of l'Hospital:

$$\lim_{(\lambda,\mu)\to\varkappa_1} \frac{\int_{\varkappa_1}^{(\lambda,\mu)} \frac{1}{|\Psi|}\,|\mathrm{d}\lambda'|}{\left|\int_{\varkappa_1}^{(\lambda,\mu)} \frac{1}{\Psi}\,\mathrm{d}\lambda'\right|} \leq \lim_{(\lambda,\mu)\to\varkappa_1} \frac{2\sqrt{|\lambda - \varkappa_1|}}{|\operatorname{arcosh}(\lambda)|} = \lim_{(\lambda,\mu)\to\varkappa_1} \frac{\frac{1}{\sqrt{|\lambda-\varkappa_1|}}}{\left|\frac{1}{\sqrt{\lambda^2-1}}\right|} = \sqrt{2}.$$

This completes the proof of (16.8).

We now prove (16.7) in the case where at least one of the inequalities $|\lambda| > 2$ and $|\lambda^o| > 2$ holds. The parts of the admissible path of integration that run within $\{\lambda' \in \mathbb{C} \,|\, |\lambda'| \leq 2\}$ are handled by the above argument, so we only consider the parts that run within $\{\lambda' \in \mathbb{C} \,|\, |\lambda'| \geq 2\}$ in the sequel. For such λ', we have

$|\lambda' - \varkappa_\upsilon| \geq \frac{1}{2}|\lambda'|$ and therefore

$$\frac{1}{|\Psi(\lambda', \mu')|} \leq \frac{2}{|\lambda'|}. \tag{16.10}$$

We also note that we have

$$\begin{aligned}\int_{(\lambda^o,\mu^o)}^{(\lambda,\mu)} \frac{1}{\Psi}\,\mathrm{d}\lambda' &= \ln\left(\frac{\lambda - \varkappa_* + \Psi(\lambda,\mu)}{\lambda^o - \varkappa_* + \Psi(\lambda^o,\mu^o)}\right)\\ &= \ln\left|\frac{\lambda - \varkappa_* + \Psi(\lambda,\mu)}{\lambda^o - \varkappa_* + \Psi(\lambda^o,\mu^o)}\right| + \mathrm{i}\cdot\arg\left(\frac{\lambda - \varkappa_* + \Psi(\lambda,\mu)}{\lambda^o - \varkappa_* + \Psi(\lambda^o,\mu^o)}\right)\end{aligned}$$

and therefore

$$\left|\int_{(\lambda^o,\mu^o)}^{(\lambda,\mu)} \frac{1}{\Psi}\,\mathrm{d}\lambda'\right| \geq \max\left\{\ln\left|\frac{\lambda-\varkappa_*+\Psi(\lambda,\mu)}{\lambda^o-\varkappa_*+\Psi(\lambda^o,\mu^o)}\right|,\ \left|\arg\left(\frac{\lambda-\varkappa_*+\Psi(\lambda,\mu)}{\lambda^o-\varkappa_*+\Psi(\lambda^o,\mu^o)}\right)\right|\right\}.$$

In the sequel, we will handle the segments of the part of the admissible path of integration contained in $\{|\lambda| \geq 2\}$ individually, treating lines and circles separately. We will estimate $\int \frac{1}{|\Psi|}|\mathrm{d}\lambda'|$ by a multiple of $\ln\left|\frac{\lambda-\varkappa_*+\Psi(\lambda,\mu)}{\lambda^o-\varkappa_*+\Psi(\lambda^o,\mu^o)}\right|$ for lines, and by a multiple of $\left|\arg\left(\frac{\lambda-\varkappa_*+\Psi(\lambda,\mu)}{\lambda^o-\varkappa_*+\Psi(\lambda^o,\mu^o)}\right)\right|$ for circles. For this purpose we denote by γ the relevant path, and by (λ_1, μ_1), (λ_2, μ_2) its endpoints.

For lines, we have by (16.10) with a constant $C_2 > 0$

$$\int_\gamma \frac{1}{|\Psi|}|\mathrm{d}\lambda'| \leq \int_\gamma \frac{2}{|\lambda'|}|\mathrm{d}\lambda'| = 2\ln\left(\frac{|\lambda_2|}{|\lambda_1|}\right) \leq C_2 \ln\left|\frac{\lambda - \varkappa_* + \Psi(\lambda,\mu)}{\lambda^o - \varkappa_* + \Psi(\lambda^o,\mu^o)}\right|.$$

For circles, $|\lambda|$ is constant along the path of integration, and therefore we have again by (16.10) with a constant $C_3 > 0$

$$\int_\gamma \frac{1}{|\Psi|}|\mathrm{d}\lambda'| \leq \int_\gamma \frac{2}{|\lambda'|}|\mathrm{d}\lambda'| = 2\arg\left(\frac{\lambda_2}{\lambda_1}\right) \leq C_3 \left|\arg\left(\frac{\lambda - \varkappa_* + \Psi(\lambda,\mu)}{\lambda^o - \varkappa_* + \Psi(\lambda^o,\mu^o)}\right)\right|.$$

This completes the proof of (16.6).

For (3). In the case $\varkappa_{k,1} = \varkappa_{k,2}$, the integrand equals 1 because of our hypothesis $\xi_k = \varkappa_{k,*}$ for this case, and therefore we then have

$$\int_{\varkappa_{k,1}}^{(\lambda_k,\mu_k)} \frac{|\lambda - \xi_k|}{|\Psi_k|}|\mathrm{d}\lambda| \leq |\lambda_k - \varkappa_{k,1}|.$$

Thus we only need to consider the case $\varkappa_{k,1} \neq \varkappa_{k,2}$ in the sequel. For this case we will show as an intermediate step that there are constants $C_3, C_4 > 0$ so that

$$\begin{aligned}
&\int_{\varkappa_{k,1}}^{(\lambda_k,\mu_k)} \frac{|\lambda - \xi_k|}{|\Psi_k|} \, |d\lambda| \\
&\quad \leq C_3 \cdot |\lambda_k - \varkappa_{k,1}| \\
&\qquad + C_4 \cdot (|\xi_k - \varkappa_{k,*}| + |\varkappa_{k,1} - \varkappa_{k,2}|) \cdot \left| \operatorname{arcosh}\left(\frac{\lambda_k - \varkappa_{k,*}}{\frac{1}{2}(\varkappa_{k,1} - \varkappa_{k,2})} \right) \right|
\end{aligned} \tag{16.11}$$

holds.

Indeed, it follows from an argument of compactness that C_3, C_4 can be chosen such that (16.11) holds for $|k| \leq N$. Thus we only consider $|k| > N$ in the sequel. Then there exists a constant $C_5 > 0$ so that we have

$$\begin{aligned}
\int_{\varkappa_{k,1}}^{(\lambda_k,\mu_k)} \frac{1}{|\Psi_k|} \, |d\lambda| &\overset{(2)}{\leq} C_5 \cdot \left| \int_{\varkappa_{k,1}}^{(\lambda_k,\mu_k)} \frac{1}{\Psi_k} \, d\lambda \right| \\
&\overset{(1)}{=} C_5 \cdot \left| \ln\left(\frac{\lambda_k - \varkappa_{k,*} + \Psi_k(\lambda_k, \mu_k)}{\varkappa_{k,1} - \varkappa_{k,*} + \Psi_k(\varkappa_{k,1})} \right) \right| \\
&\overset{(*)}{=} C_5 \cdot \left| \operatorname{arcosh}\left(\frac{\lambda_k - \varkappa_{k,*}}{\frac{1}{2}(\varkappa_{k,1} - \varkappa_{k,2})} \right) \right| ,
\end{aligned} \tag{16.12}$$

where (2) and (1) refer to the respective parts of the present lemma, and the equality marked $(*)$ follows from the equation $\operatorname{arcosh}(z) = \ln(z + \sqrt{z^2 - 1})$, and therefore

$$\begin{aligned}
&\int_{\varkappa_{k,1}}^{(\lambda_k,\mu_k)} \frac{|\lambda - \xi_k|}{|\Psi_k|} \, |d\lambda| \\
&\leq \int_{\varkappa_{k,1}}^{(\lambda_k,\mu_k)} \frac{|\lambda - \varkappa_{k,*}|}{|\Psi_k|} \, |d\lambda| + |\varkappa_{k,*} - \xi_k| \cdot \int_{\varkappa_{k,1}}^{(\lambda_k,\mu_k)} \frac{1}{|\Psi_k|} \, |d\lambda| \\
&\leq \int_{\varkappa_{k,1}}^{(\lambda_k,\mu_k)} \frac{|\lambda - \varkappa_{k,*}|}{|\Psi_k|} \, |d\lambda| + C_5 \cdot |\varkappa_{k,*} - \xi_k| \cdot \left| \operatorname{arcosh}\left(\frac{\lambda_k - \varkappa_{k,*}}{\frac{1}{2}(\varkappa_{k,1} - \varkappa_{k,2})} \right) \right| .
\end{aligned}$$

This calculation shows that it suffices to consider the case $\xi_k = \varkappa_{k,*}$ in the proof of (16.11). In this case, (16.11) takes the form of the following inequality

$$\int_{\varkappa_{k,1}}^{(\lambda_k,\mu_k)} \frac{|\lambda - \varkappa_{k,*}|}{|\Psi_k|} \, |d\lambda| \leq C_3 \cdot |\lambda_k - \varkappa_{k,1}| + C_4 \cdot |\varkappa_{k,1} - \varkappa_{k,2}| \cdot \left| \operatorname{arcosh}\left(\frac{\lambda_k - \varkappa_{k,*}}{\frac{1}{2}(\varkappa_{k,1} - \varkappa_{k,2})} \right) \right| , \tag{16.13}$$

which we will now prove.

For the proof of (16.13) we first look at the case $|\lambda_k - \varkappa_{k,*}| \leq |\varkappa_{k,1} - \varkappa_{k,2}|$. Then we have for any $\lambda \in [\varkappa_{k,1}, \lambda_k]$

$$\begin{aligned}|\lambda - \varkappa_{k,*}| &\leq |\lambda - \varkappa_{k,1}| + |\varkappa_{k,1} - \varkappa_{k,*}| \leq |\lambda_k - \varkappa_{k,1}| + |\varkappa_{k,1} - \varkappa_{k,*}| \\ &\leq |\lambda_k - \varkappa_{k,*}| + 2\,|\varkappa_{k,1} - \varkappa_{k,*}| \leq 2\,|\varkappa_{k,1} - \varkappa_{k,2}|\end{aligned}$$

and therefore by (16.12)

$$\begin{aligned}\int_{\varkappa_{k,1}}^{(\lambda_k,\mu_k)} \frac{|\lambda - \varkappa_{k,*}|}{|\Psi_k|}\,|d\lambda| &\leq 2\,|\varkappa_{k,1} - \varkappa_{k,2}| \cdot \int_{\varkappa_{k,1}}^{(\lambda_k,\mu_k)} \frac{1}{|\Psi_k|}\,|d\lambda| \\ &\leq 2\,C_5\,|\varkappa_{k,1} - \varkappa_{k,2}| \cdot \left|\operatorname{arcosh}\left(\frac{\lambda_k - \varkappa_{k,*}}{\frac{1}{2}\,(\varkappa_{k,1} - \varkappa_{k,2})}\right)\right|,\end{aligned} \tag{16.14}$$

whence (16.13) follows for this case.

We now consider the case $|\lambda_k - \varkappa_{k,*}| > |\varkappa_{k,1} - \varkappa_{k,2}|$. Because the endpoints of the line segment $[\varkappa_{k,1}, \lambda_k]$ then satisfy

$$|\varkappa_{k,1} - \varkappa_{k,*}| < |\varkappa_{k,1} - \varkappa_{k,2}| \quad \text{and} \quad |\lambda_k - \varkappa_{k,*}| > |\varkappa_{k,1} - \varkappa_{k,2}|,$$

there is a unique intersection point between this line segment and the circle $\{\,\lambda \in \mathbb{C}\,\big|\,|\lambda - \varkappa_{k,*}| = |\varkappa_{k,1} - \varkappa_{k,2}|\,\}$, which we denote by $\lambda_* \in \mathbb{C}^*$. We further choose $\mu_* \in \mathbb{C}$ so that $(\lambda_*, \mu_*) \in \Sigma$ is in the appropriate leaf of Σ. Then we have

$$\int_{\varkappa_{k,1}}^{(\lambda_k,\mu_k)} \frac{|\lambda - \varkappa_{k,*}|}{|\Psi_k|}\,|d\lambda| = \int_{\varkappa_{k,1}}^{(\lambda_*,\mu_*)} \frac{|\lambda - \varkappa_{k,*}|}{|\Psi_k|}\,|d\lambda| + \int_{(\lambda_*,\mu_*)}^{(\lambda_k,\mu_k)} \frac{|\lambda - \varkappa_{k,*}|}{|\Psi_k|}\,|d\lambda|, \tag{16.15}$$

where we integrate along straight lines in all three integrals.

We estimate the two integrals on the right hand side of Eq. (16.15) separately. For the first integral, the estimate (16.14) is applicable because of $|\lambda_* - \varkappa_{k,*}| = |\varkappa_{k,1} - \varkappa_{k,2}|$, and therefore we have

$$\begin{aligned}\int_{\varkappa_{k,1}}^{(\lambda_*,\mu_*)} \frac{|\lambda - \varkappa_{k,*}|}{|\Psi_k|}\,|d\lambda| &\leq 2\,C_5\,|\varkappa_{k,1} - \varkappa_{k,2}| \cdot \left|\operatorname{arcosh}\left(\frac{\lambda_* - \varkappa_{k,*}}{\frac{1}{2}\,(\varkappa_{k,1} - \varkappa_{k,2})}\right)\right| \\ &\leq 2\,C_5\,|\varkappa_{k,1} - \varkappa_{k,2}| \cdot \left|\operatorname{arcosh}\left(\frac{\lambda_k - \varkappa_{k,*}}{\frac{1}{2}\,(\varkappa_{k,1} - \varkappa_{k,2})}\right)\right|.\end{aligned}$$

Concerning the second integral, we note that for $\lambda \in [\lambda_*, \lambda_k]$ and $\nu \in \{1, 2\}$ we have

$$|\lambda - \varkappa_{k,*}| \geq |\varkappa_{k,1} - \varkappa_{k,2}| = 2\,|\varkappa_{k,*} - \varkappa_{k,\nu}|$$

and therefore

$$|\lambda - \varkappa_{k,\nu}| \geq |\lambda - \varkappa_{k,*}| - |\varkappa_{k,*} - \varkappa_{k,\nu}| \geq \frac{1}{2}\,|\lambda - \varkappa_{k,*}|,$$

whence

$$\frac{|\lambda - \varkappa_{k,*}|}{|\Psi_k(\lambda,\mu)|} = \frac{|\lambda - \varkappa_{k,*}|}{\sqrt{|\lambda - \varkappa_{k,1}| \cdot |\lambda - \varkappa_{k,2}|}} \leq 2$$

follows. Thus we obtain

$$\int_{(\lambda_*,\mu_*)}^{(\lambda_k,\mu_k)} \frac{|\lambda - \varkappa_{k,*}|}{|\Psi_k|}\,|\mathrm{d}\lambda| \leq \int_{(\lambda_*,\mu_*)}^{(\lambda_k,\mu_k)} 2\,|\mathrm{d}\lambda| = 2\,|\lambda_k - \lambda_*| \leq 2\,|\lambda_k - \varkappa_{k,1}|.$$

By plugging these estimates into Eq. (16.15), we obtain the estimate (16.13) also for the case of $|\lambda_k - \varkappa_{k,*}| > |\varkappa_{k,1} - \varkappa_{k,2}|$. This concludes the proof of (16.13) and therefore of (16.11).

To derive the estimate claimed in part (3) of the lemma from (16.11), we use the fact that there exists a constant $C_{\text{arcosh}} > 0$ so that for every $z \in \mathbb{C}$ we have

$$|\operatorname{arcosh}(z)| \leq C_{\text{arcosh}} \cdot |z - 1|. \tag{16.16}$$

For the proof of (16.16) we note that $\operatorname{arcosh}(z)$ is differentiable on a suitably doubly slitted plane with $\frac{\mathrm{d}}{\mathrm{d}z}\operatorname{arcosh}(z) = \frac{1}{\sqrt{z^2-1}}$ and $\operatorname{arcosh}(1) = 0$, and therefore we have

$$|\operatorname{arcosh}(z)| = |\operatorname{arcosh}(z) - \operatorname{arcosh}(1)| = \left|\int_1^z \frac{1}{\sqrt{z^2-1}}\,\mathrm{d}z\right| \leq \int_1^z \frac{1}{\sqrt{|z^2-1|}}\,|\mathrm{d}z|,$$

where the integration takes place along a path that is chosen in the slitted plane with a length commensurate with $|z - 1|$. Because the integrand has only finitely many poles of order at most $\frac{1}{2}$, the above integral is finite, and because the integrand is bounded for $|z| \to \infty$, there in fact exists a constant $C_{\text{arcosh}} > 0$ with

$$|\operatorname{arcosh}(z)| \leq \int_1^z \frac{1}{\sqrt{|z^2-1|}}\,|\mathrm{d}z| \leq C_{\text{arcosh}} \cdot |z - 1|.$$

By applying (16.16) to the estimate (16.11), we see

$$\begin{aligned}
&\int_{\varkappa_{k,1}}^{(\lambda_k,\mu_k)} \frac{|\lambda - \xi_k|}{|\Psi_k|}\,|\mathrm{d}\lambda| \\
&\quad \leq C_3 \cdot |\lambda_k - \varkappa_{k,1}| \\
&\qquad + C_4 \cdot (|\xi_k - \varkappa_{k,*}| + |\varkappa_{k,1} - \varkappa_{k,2}|) \cdot C_{\text{arcosh}} \cdot \left|\frac{\lambda_k - \varkappa_{k,*}}{\frac{1}{2}(\varkappa_{k,1} - \varkappa_{k,2})} - 1\right|
\end{aligned}$$

$$= C_3 \cdot |\lambda_k - \varkappa_{k,1}|$$
$$+ C_4 \cdot (|\xi_k - \varkappa_{k,*}| + |\varkappa_{k,1} - \varkappa_{k,2}|) \cdot C_{\text{arcosh}} \cdot \left| \frac{\lambda_k - \varkappa_{k,1}}{\frac{1}{2}(\varkappa_{k,1} - \varkappa_{k,2})} \right|$$
$$= \left((C_3 + 2\,C_4\,C_{\text{arcosh}}) + 2\,C_4\,C_{\text{arcosh}} \left| \frac{\xi_k - \varkappa_{k,*}}{\varkappa_{k,1} - \varkappa_{k,2}} \right| \right) \cdot |\lambda_k - \varkappa_{k,1}|.$$

Thus the estimate claimed in (3) holds (with $C_1 := 2\,C_4\,C_{\text{arcosh}}$ and $C_2 := C_3 + 2\,C_4\,C_{\text{arcosh}}$).

For (4). If $\varkappa_{k,1} = \varkappa_{k,2}$ holds, the integrand equals 1 (because we had required $\xi_k = \varkappa_{k,*}$ in this case), and thus the integral equals $\text{vol}(U_{k,\delta}) \in \ell^{\infty}_{-2,6}(k)$. Thus we now suppose $\varkappa_{k,1} \neq \varkappa_{k,2}$, consider the case $k > 0$, and choose $\delta' > 0$ (independently of k) so that $U_{k,\delta} \subset B(\lambda_{k,0}, k\delta')$ holds. We then compute the integral $\int_{B(\lambda_{k,0},k\delta')} \frac{|\lambda-\xi_k|^2}{|\lambda-\varkappa_{k,1}|\cdot|\lambda-\varkappa_{k,2}|}\, \mathrm{d}^2\lambda$. We abbreviate $A := \frac{|\xi_k-\varkappa_{k,*}|}{|\varkappa_{k,1}-\varkappa_{k,2}|}$ and split the domain of integration $B(\lambda_{k,0}, k\delta')$ into four parts:

$$M_1 := B(\varkappa_{k,1}, \tfrac{1}{2}|\varkappa_{k,1} - \varkappa_{k,2}|)$$
$$M_2 := B(\varkappa_{k,2}, \tfrac{1}{2}|\varkappa_{k,1} - \varkappa_{k,2}|)$$
$$M_3 := B(\varkappa_{k,*}, \max\{|\xi_k - \varkappa_{k,*}|, |\varkappa_{k,1} - \varkappa_{k,2}|\}) \setminus (M_1 \cup M_2)$$
$$M_4 := B(\lambda_{k,0}, k\delta') \setminus (M_1 \cup M_2 \cup M_3).$$

For $\nu \in \{1, 2\}$ and $\lambda \in M_\nu$ we have

$$|\lambda - \xi_k| \leq |\lambda - \varkappa_{k,\nu}| + |\varkappa_{k,\nu} - \varkappa_{k,*}| + |\varkappa_{k,*} - \xi_k|$$
$$\leq \frac{1}{2}|\varkappa_{k,1} - \varkappa_{k,2}| + \frac{1}{2}|\varkappa_{k,1} - \varkappa_{k,2}| + A\,|\varkappa_{k,1} - \varkappa_{k,2}|$$
$$= (A + 1) \cdot |\varkappa_{k,1} - \varkappa_{k,2}|$$

and

$$|\lambda - \varkappa_{k,3-\nu}| \geq |\varkappa_{k,3-\nu} - \varkappa_{k,\nu}| - |\lambda - \varkappa_{k,\nu}| \geq \frac{1}{2}|\varkappa_{k,1} - \varkappa_{k,2}|,$$

and therefore

$$\frac{|\lambda - \xi_k|^2}{|\lambda - \varkappa_{k,1}| \cdot |\lambda - \varkappa_{k,2}|} \leq 2\,(A + 1)^2 \cdot |\varkappa_{k,1} - \varkappa_{k,2}| \cdot \frac{1}{|\lambda - \varkappa_{k,\nu}|}.$$

Thus we obtain

$$\begin{aligned}
&\int_{M_\nu} \frac{|\lambda - \xi_k|^2}{|\lambda - \varkappa_{k,1}| \cdot |\lambda - \varkappa_{k,2}|} \, \mathrm{d}^2\lambda \\
&\leq 2\,(A+1)^2 \cdot |\varkappa_{k,1} - \varkappa_{k,2}| \cdot \int_{M_\nu} \frac{1}{|\lambda - \varkappa_{k,\nu}|} \, \mathrm{d}^2\lambda \\
&\leq 2\,(A+1)^2 \cdot |\varkappa_{k,1} - \varkappa_{k,2}| \cdot \int_{r=0}^{\frac{1}{2}|\varkappa_{k,1}-\varkappa_{k,2}|} \int_{\varphi=0}^{2\pi} \frac{1}{r}\, r \, \mathrm{d}r \, \mathrm{d}\varphi \\
&= 2\pi\,(A+1)^2\, |\varkappa_{k,1} - \varkappa_{k,2}|^2 \\
&\leq 8\pi\,(|\xi_k - \varkappa_{k,*}|^2 + |\varkappa_{k,1} - \varkappa_{k,2}|^2).
\end{aligned} \tag{16.17}$$

For $\lambda \in M_3$ we have

$$|\lambda - \xi_k| \leq |\lambda - \varkappa_{k,*}| + |\varkappa_{k,*} - \xi_k| \leq 2\,\max\{|\xi_k - \varkappa_{k,*}|, |\varkappa_{k,1} - \varkappa_{k,2}|\}$$

and for $\nu \in \{1, 2\}$

$$|\lambda - \varkappa_{k,\nu}| \geq \frac{1}{2}\,|\varkappa_{k,1} - \varkappa_{k,2}|,$$

and therefore

$$\frac{|\lambda - \xi_k|^2}{|\lambda - \varkappa_{k,1}| \cdot |\lambda - \varkappa_{k,2}|} \leq 16\,\frac{(\max\{|\xi_k - \varkappa_{k,*}|, |\varkappa_{k,1} - \varkappa_{k,2}|\})^2}{|\varkappa_{k,1} - \varkappa_{k,2}|^2}.$$

It follows that we have

$$\begin{aligned}
&\int_{M_3} \tfrac{|\lambda-\xi_k|^2}{|\lambda-\varkappa_{k,1}|\cdot|\lambda-\varkappa_{k,2}|}\, \mathrm{d}^2\lambda \leq 16\,\tfrac{(\max\{|\xi_k-\varkappa_{k,*}|,|\varkappa_{k,1}-\varkappa_{k,2}|\})^2}{|\varkappa_{k,1}-\varkappa_{k,2}|^2} \cdot \mathrm{vol}(M_3) \\
&\leq 16\,\tfrac{(\max\{|\xi_k-\varkappa_{k,*}|,|\varkappa_{k,1}-\varkappa_{k,2}|\})^2}{|\varkappa_{k,1}-\varkappa_{k,2}|^2} \cdot \mathrm{vol}(B(\varkappa_{k,*}, \max\{|\xi_k - \varkappa_{k,*}|, |\varkappa_{k,1} - \varkappa_{k,2}|\})) \\
&= 16\,\tfrac{(\max\{|\xi_k-\varkappa_{k,*}|,|\varkappa_{k,1}-\varkappa_{k,2}|\})^2}{|\varkappa_{k,1}-\varkappa_{k,2}|^2} \cdot \pi\,(\max\{|\xi_k - \varkappa_{k,*}|, |\varkappa_{k,1} - \varkappa_{k,2}|\})^2 \\
&\leq 16\pi\left(|\varkappa_{k,1} - \varkappa_{k,2}|^2 + \tfrac{|\xi_k-\varkappa_{k,*}|^4}{|\varkappa_{k,1}-\varkappa_{k,2}|^2}\right).
\end{aligned} \tag{16.18}$$

Finally, for $\lambda \in M_4$ we have

$$|\lambda - \xi_k| \leq |\lambda - \varkappa_{k,*}| + |\varkappa_{k,*} - \xi_k| \leq |\lambda - \varkappa_{k,*}| + |\lambda - \varkappa_{k,*}| = 2\,|\lambda - \varkappa_{k,*}|$$

and for $\nu \in \{1, 2\}$

$$|\lambda - \varkappa_{k,\nu}| \geq |\lambda - \varkappa_{k,*}| - |\varkappa_{k,*} - \varkappa_{k,\nu}| = |\lambda - \varkappa_{k,*}| - \frac{1}{2}\,|\varkappa_{k,1} - \varkappa_{k,2}| \geq \frac{1}{2}\,|\lambda - \varkappa_{k,*}|$$

and hence

$$\frac{|\lambda - \xi_k|^2}{|\lambda - \varkappa_{k,1}| \cdot |\lambda - \varkappa_{k,2}|} \leq 16.$$

Thus we obtain

$$\int_{M_4} \frac{|\lambda-\xi_k|^2}{|\lambda-\varkappa_{k,1}|\cdot|\lambda-\varkappa_{k,2}|}\, \mathrm{d}^2\lambda \leq 16 \cdot \mathrm{vol}(M_4) \leq 16 \cdot \mathrm{vol}(B(\lambda_{k,0}, k\delta')) = 16 \cdot \pi\, (k\delta')^2 \leq 64\, \pi\, k^2 \tag{16.19}$$

It follows from Eqs. (16.17), (16.18) and (16.19), and the fact that $\varkappa_{k,1} - \varkappa_{k,2} \in \ell^2_{-1,3}(k)$ holds, that there exists $C > 0$ so that

$$\int_{U_{k,\delta}} \frac{|\lambda - \xi_k|^2}{|\lambda - \varkappa_{k,1}| \cdot |\lambda - \varkappa_{k,2}|}\, \mathrm{d}^2\lambda \leq C \cdot \max\left\{1, \frac{|\xi_k - \varkappa_{k,*}|}{|\varkappa_{k,1} - \varkappa_{k,2}|}\right\}^4 \cdot (|\varkappa_{k,1} - \varkappa_{k,2}|^2 + k^2)$$

holds. □

Proposition 16.5 *Let $N \in \mathbb{N}$ be as in Lemma 16.1.*

(1) There exist $C > 0$ and a sequence $b_k \in \ell^2_{-1,3}(k)$ (depending on Σ) such that for any $k \in \mathbb{Z}$ with $|k| > N$ and for any $(\lambda_k^o, \mu_k^o), (\lambda_k, \mu_k) \in \widehat{U}'_{k,\delta}$ (in the case $\varkappa_{k,1} = \varkappa_{k,2}$ we require that (λ_k, μ_k) and (λ_k^o, μ_k^o) lie in the same connected component of $\widehat{U}'_{k,\delta}$) we have

$$\left| \int_{(\lambda_k^o,\mu_k^o)}^{(\lambda_k,\mu_k)} \frac{1}{\mu - \mu^{-1}}\, \mathrm{d}\lambda - 4\mathrm{i}(-1)^k\, \lambda_{k,0}^{\varrho/2}\, \ln\left(\frac{\lambda_k - \varkappa_{k,*} + \Psi_k(\lambda_k, \mu_k)}{\lambda_k^o - \varkappa_{k,*} + \Psi_k(\lambda_k^o, \mu_k^o)}\right)\right| \leq C \cdot \left(|\lambda_k - \lambda_k^o| + b_k\right) \cdot \left|\ln\left(\frac{\lambda_k - \varkappa_{k,*} + \Psi_k(\lambda_k, \mu_k)}{\lambda_k^o - \varkappa_{k,*} + \Psi_k(\lambda_k^o, \mu_k^o)}\right)\right|.$$

Here we integrate along any path that is contained in $\widehat{U}'_{k,\delta}$, set $\varrho = 1$ for $k > 0$ and $\varrho = 3$ for $k < 0$, and $\ln(z)$ is the branch of the complex logarithm function with $\ln(1) = 2\pi \mathrm{i} m$, where $m \in \mathbb{Z}$ is the winding number of the path of integration around the pair of branch points $\varkappa_{k,1}$, $\varkappa_{k,2}$.

(2) *There exists* $C > 0$ *so that for any* $k \in \mathbb{Z}$ *with* $|k| > N$ *and for any points* $(\lambda_k^o, \mu_k^o), (\lambda_k, \mu_k) \in \widehat{U}'_{k,\delta}$ *as in (1) we have*

$$\int_{(\lambda_k^o,\mu_k^o)}^{(\lambda_k,\mu_k)} \frac{1}{|\mu-\mu^{-1}|}\,|\mathrm{d}\lambda| \leq C\,\lambda_{k,0}^{\varrho/2} \cdot \left| \ln\left(\frac{\lambda_k - \varkappa_{k,*} + \Psi_k(\lambda_k,\mu_k)}{\lambda_k^o - \varkappa_{k,*} + \Psi_k(\lambda_k^o,\mu_k^o)} \right) \right|, \tag{16.20}$$

where we integrate along an admissible path (Definition 16.2), and set $\varrho = 1$ *for* $k > 0$ *and* $\varrho = 3$ *for* $k < 0$.
Moreover, for all k *with* $\varkappa_{k,1} \neq \varkappa_{k,2}$ *we have*

$$\int_{\varkappa_{k,1}}^{\varkappa_{k,2}} \frac{1}{\mu-\mu^{-1}}\,\mathrm{d}\lambda = -4(-1)^k\pi\,\lambda_{k,0}^{\varrho/2} + \ell^2_{-1,3}(k), \tag{16.21}$$

$$\int_{\varkappa_{k,1}}^{\varkappa_{k,2}} \frac{1}{|\mu-\mu^{-1}|}\,|\mathrm{d}\lambda| = 4\pi\,\lambda_{k,0}^{\varrho/2} + \ell^2_{-1,3}(k), \tag{16.22}$$

where the path of integration is a straight line from $\varkappa_{k,1}$ *to* $\varkappa_{k,2}$ *in the* λ*-plane.*

(3) *There exist* $C_1, C_2 > 0$ *so that for any* $k \in \mathbb{Z}$, *for any* $(\lambda_k^o, \mu_k^o), (\lambda_k, \mu_k) \in \widehat{U}_{k,\delta}$ *as in (1), and for arbitrary* $\xi_k \in U_{k,\delta}$ *if* $\varkappa_{k,1} \neq \varkappa_{k,2}$, *whereas we set* $\xi_k := \varkappa_{k,*}$ *if* $\varkappa_{k,1} = \varkappa_{k,2}$, *we have*

$$\int_{\varkappa_{k,1}}^{(\lambda_k,\mu_k)} \frac{|\lambda-\xi_k|}{|\mu-\mu^{-1}|}\,|\mathrm{d}\lambda| \leq \left(C_1 \left| \frac{\xi_k - \varkappa_{k,*}}{\varkappa_{k,1}-\varkappa_{k,2}} \right| + C_2 \right) \cdot \lambda_{k,0}^{\varrho/2} \cdot |\lambda_k - \varkappa_{k,1}|,$$

where the path of integration is a straight line from $\varkappa_{k,1}$ *to* λ_k *in the* λ*-plane, we set* $\varrho = 1$ *for* $k > 0$ *and* $\varrho = 3$ *for* $k < 0$, *and the expression* $\left|\frac{\xi_k-\varkappa_{k,*}}{\varkappa_{k,1}-\varkappa_{k,2}}\right|$ *is to be read as* 0 *in the case* $\varkappa_{k,1} = \varkappa_{k,2}$.

(4) *There exists* $C > 0$ *so that for any* $k \in \mathbb{Z}$ *and* ξ_k *as in (3) we have*

$$\int_{U_{k,\delta}} \frac{|\lambda-\xi_k|^2}{|\Delta(\lambda)^2-4|}\,\mathrm{d}^2\lambda \leq C\,\lambda_{k,0}^{\varrho} \cdot \left(|\xi_k - \varkappa_{k,*}|^2 + \frac{|\xi_k-\varkappa_{k,*}|^4}{|\varkappa_{k,1}-\varkappa_{k,2}|^2} + \ell^\infty_{-2,6}(k) \right),$$

where $\mathrm{d}^2\lambda$ *again denotes the 2-dimensional Lebesgue measure on* $\mathbb{C}$. *Here we again put* $\varrho = 1$ *for* $k > 0$ *and* $\varrho = 3$ *for* $k < 0$; *and in the case* $\varkappa_{k,1} = \varkappa_{k,2}$, *the expression* $\frac{|\xi_k-\varkappa_{k,*}|^4}{|\varkappa_{k,1}-\varkappa_{k,2}|^2}$ *is to be read as equal to* 0.

Proof For (1). We may suppose without loss of generality that the path of integration is composed of an admissible path connecting (λ_k^o, μ_k^o) to (λ_k, μ_k) and $|m|$ circles around $\varkappa_{k,1}$, $\varkappa_{k,2}$ (with the orientation given by the sign of m). Note that each of the circles is also an admissible path.

If the point (λ, μ) is on the trace of the path of integration, we then have because of Eq. (16.5) in Definition 16.2

$$|\lambda - \varkappa_{k,\nu}| \leq \max\left\{2\,|\varkappa_{k,1} - \varkappa_{k,2}|,\ |\lambda_k^o - \varkappa_{k,*}|,\ |\lambda_k - \varkappa_{k,*}|\right\}$$
$$\leq |\lambda_k - \lambda_k^o| + \widetilde{b}_k \quad \text{with} \quad \widetilde{b}_k := 2\,|\varkappa_{k,1} - \varkappa_{k,2}| + |\lambda_k^o - \varkappa_{k,*}| \in \ell^2_{-1,3}(k).$$

Therefore it follows from Lemma 16.1 that there exist constants $C_1, C_2 > 0$ and $a_k \in \ell^2_{-1,3}(k)$ so that we have for such points

$$\left|\frac{1}{\mu-\mu^{-1}} - 4\mathrm{i}(-1)^k\,\lambda_{k,0}^{\varrho/2}\,\frac{1}{\Psi_k(\lambda,\mu)}\right| \leq \frac{C_1\cdot(|\lambda-\varkappa_{k,1}|+|\lambda-\varkappa_{k,2}|)+a_k}{|\Psi_k(\lambda,\mu)|} \tag{16.23}$$

$$\leq C_2\cdot\frac{|\lambda_k-\lambda_k^o|+b_k}{|\Psi_k(\lambda,\mu)|} \quad \text{with} \quad b_k := a_k + \widetilde{b}_k. \tag{16.24}$$

We obtain by integration

$$\left|\int_{(\lambda_k^o,\mu_k^o)}^{(\lambda_k,\mu_k)} \frac{1}{\mu-\mu^{-1}}\,\mathrm{d}\lambda - 4\mathrm{i}(-1)^k\,\lambda_{k,0}^{\varrho/2}\int_{(\lambda_k^o,\mu_k^o)}^{(\lambda_k,\mu_k)}\frac{1}{\Psi_k}\,\mathrm{d}\lambda\right|$$
$$\leq C_2\cdot\left(|\lambda_k-\lambda_k^o|+b_k\right)\cdot\int_{(\lambda_k^o,\mu_k^o)}^{(\lambda_k,\mu_k)}\frac{1}{|\Psi_k|}\,\mathrm{d}\lambda.$$

(1) now follows from Lemma 16.4(1),(2).

For (2). It follows from Eq. (16.23) that there exists $C_3 > 0$ so that we have for $(\lambda, \mu) \in \widehat{U}_{k,\delta}$

$$\frac{1}{|\mu-\mu^{-1}|} \leq C_3\,\lambda_{k,0}^{\varrho/2}\,\frac{1}{|\Psi_k(\lambda,\mu)|}$$

and therefore

$$\int_{(\lambda_k^o,\mu_k^o)}^{(\lambda_k,\mu_k)}\frac{1}{|\mu-\mu^{-1}|}\,|\mathrm{d}\lambda| \leq C_3\,\lambda_{k,0}^{\varrho/2}\int_{(\lambda_k^o,\mu_k^o)}^{(\lambda_k,\mu_k)}\frac{1}{|\Psi_k(\lambda,\mu)|}\,|\mathrm{d}\lambda|.$$

By Lemma 16.4(2),(1) there exists $C_4 > 0$ with

$$\int_{(\lambda_k^o,\mu_k^o)}^{(\lambda_k,\mu_k)}\frac{1}{|\Psi_k(\lambda,\mu)|}\,|\mathrm{d}\lambda| \leq C_4\cdot\left|\int_{(\lambda_k^o,\mu_k^o)}^{(\lambda_k,\mu_k)}\frac{1}{\Psi_k(\lambda,\mu)}\,\mathrm{d}\lambda\right| = C_4\cdot\left|\ln\left(\frac{\lambda_k-\varkappa_{k,*}+\Psi_k(\lambda_k,\mu_k)}{\lambda_k^o-\varkappa_{k,*}+\Psi_k(\lambda_k^o,\mu_k^o)}\right)\right|,$$

and Eq. (16.20) follows from these estimates.

Equations (16.21) and (16.22) follow from part (1) resp. from Eq. (16.20) by plugging in $(\lambda_k^o, \mu_k^o) = \varkappa_{k,1}$ and $(\lambda_k, \mu_k) = \varkappa_{k,2}$ and noting that $\varkappa_{k,2} - \varkappa_{k,1} \in \ell^2_{-1,3}(k)$ and $\ln\left(\frac{\varkappa_{k,2}-\varkappa_{k,*}-\Psi_k(\varkappa_{k,2})}{\varkappa_{k,1}-\varkappa_{k,*}-\Psi_k(\varkappa_{k,1})}\right) = \ln(-1) = i\pi$ holds.

For (3). It follows from (16.24) (applied with $\lambda_k^o = \varkappa_{k,1}$) that we have

$$\frac{|\lambda - \xi_k|}{|\mu - \mu^{-1}|} \leq C_5 \cdot \big(|\lambda_k - \varkappa_{k,1}| + b_k\big) \cdot \frac{|\lambda - \xi_k|}{|\Psi_k(\lambda)|} \leq C_6 \cdot \lambda_{k,0}^{\varrho/2} \cdot \frac{|\lambda - \xi_k|}{|\Psi_k(\lambda)|}$$

with constants $C_5, C_6 > 0$ and a sequence $(b_k) \in \ell^2_{-1,3}(k)$. By integration we thus obtain

$$\int_{\varkappa_{k,1}}^{(\lambda_k,\mu_k)} \frac{|\lambda - \xi_k|}{|\mu - \mu^{-1}|}\, |d\lambda| \leq C_6 \cdot \lambda_{k,0}^{\varrho/2} \cdot \int_{\varkappa_{k,1}}^{(\lambda_k,\mu_k)} \frac{|\lambda - \xi_k|}{|\Psi_k|}\, |d\lambda|.$$

The claimed estimate now follows from the application of Lemma 16.4(3).

For (4). From Eq. (16.2) we obtain for $\lambda \in U_{k,\delta}$, because $z \mapsto z^{-1}$ is Lipschitz continuous near $z = -\frac{1}{16}$,

$$\begin{aligned} &\left|\lambda_{k,0}^{-\varrho} \cdot \frac{(\lambda - \varkappa_{k,1}) \cdot (\lambda - \varkappa_{k,2})}{\Delta(\lambda)^2 - 4} - (-16)\right| \\ &\quad \leq C_7 \cdot \begin{Bmatrix} k^{-1} & \text{if } k > 0 \\ k^3 & \text{if } k < 0 \end{Bmatrix} \cdot \big(|\lambda - \varkappa_{k,1}| + |\lambda - \varkappa_{k,2}|\big) + \ell^2_{0,0}(k) \end{aligned}$$

and therefore

$$\begin{aligned} &\left|\frac{1}{\Delta(\lambda)^2 - 4} - \left(-\frac{16\,\lambda_{k,0}^{\varrho}}{(\lambda - \varkappa_{k,1}) \cdot (\lambda - \varkappa_{k,2})}\right)\right| \\ &\leq \frac{\lambda_{k,0}^{\varrho}}{|\lambda - \varkappa_{k,1}| \cdot |\lambda - \varkappa_{k,2}|} \cdot \left(C_7 \cdot \begin{Bmatrix} k^{-1} & \text{if } k > 0 \\ k^3 & \text{if } k < 0 \end{Bmatrix} \cdot \big(|\lambda - \varkappa_{k,1}| + |\lambda - \varkappa_{k,2}|\big) + \ell^2_{0,0}(k)\right), \end{aligned}$$

whence

$$\left|\frac{1}{\Delta(\lambda)^2 - 4}\right| \leq C_8\, \lambda_{k,0}^{\varrho}\, \frac{1}{|\lambda - \varkappa_{k,1}| \cdot |\lambda - \varkappa_{k,2}|}$$

follows. By integration and application of Lemma 16.4(4), we obtain

$$\begin{aligned} \int_{U_{k,\delta}} \frac{|\lambda - \xi_k|^2}{|\Delta(\lambda)^2 - 4|}\, d^2\lambda &\leq C_8\, \lambda_{k,0}^{\varrho} \int_{U_{k,\delta}} \frac{|\lambda - \xi_k|^2}{|\lambda - \varkappa_{k,1}| \cdot |\lambda - \varkappa_{k,2}|}\, d^2\lambda \\ &\leq C_9\, \lambda_{k,0}^{\varrho} \left(|\xi_k - \varkappa_{k,*}|^2 + \frac{|\xi_k - \varkappa_{k,*}|^4}{|\varkappa_{k,1} - \varkappa_{k,2}|^2} + \ell^\infty_{-2,6}(k)\right). \end{aligned}$$

□

Chapter 17
Asymptotic Behavior of 1-Forms on the Spectral Curve

We continue our preparations for the construction of the Jacobi coordinates on the spectral curve Σ. One important step in this construction is to obtain a basis $(\omega_n)_{n\in\mathbb{Z}}$ of the space of square-integrable, holomorphic[1] 1-forms on Σ which is dual to a given homology basis $(A_n, B_n)_{n\in\mathbb{Z}}$ on Σ in the sense that $\int_{A_k} \omega_\ell = \delta_{k,\ell}$ holds. Because we will need to assess the asymptotic behavior of the ω_n, a general statement of existence (like in [Fe-K-T, Theorem 3.8, p. 28]) is not enough. Rather we will explicitly construct ω_n in the form $\omega_n = \frac{\Phi_n(\lambda)}{\mu-\mu^{-1}}\,\mathrm{d}\lambda$, where Φ_n is a holomorphic function on $\mathbb{C}^*$ equal to a suitable linear combination of infinite products. This presentation of Φ_n will give us the asymptotic estimates we need.

In the present chapter, we therefore study holomorphic 1-forms on Σ and in particular the asymptotic behavior of holomorphic 1-forms $\omega = \frac{\Phi(\lambda)}{\mu-\mu^{-1}}\,\mathrm{d}\lambda$, where $\Phi(\lambda)$ is an infinite product of the kind we are interested in for the construction of the ω_n.

More specifically, the following Proposition 17.1 shows that any square-integrable holomorphic 1-form ω is anti-invariant with respect to the hyperelliptic involution and therefore of the form $\omega = \frac{\Phi(\lambda)}{\mu-\mu^{-1}}\,\mathrm{d}\lambda$ with some holomorphic function $\Phi : \mathbb{C}^* \to \mathbb{C}$. Conversely, the same Proposition gives a necessary condition and a (different) sufficient condition for the asymptotic behavior of the function Φ so that the holomorphic 1-form $\frac{\Phi(\lambda)}{\mu-\mu^{-1}}\,\mathrm{d}\lambda$ is square-integrable.

Propositions 17.2–17.4 describe a specific construction method for square-integrable holomorphic 1-forms $\omega = \frac{\Phi(\lambda)}{\mu-\mu^{-1}}\,\mathrm{d}\lambda$, where the function Φ is given as an infinite product, so that Φ has one zero in every excluded domain with a single exception; we in particular describe the asymptotic behavior of such 1-forms. Note

[1]The requirements of square-integrability and holomorphy need to be modified near singular points of Σ, see the precise statements below.

S. Klein, *A Spectral Theory for Simply Periodic Solutions of the Sinh-Gordon Equation*, Lecture Notes in Mathematics 2229,
https://doi.org/10.1007/978-3-030-01276-2_17

that the asymptotic behavior of such 1-forms is "better" than what can be shown for square-integrable 1-forms in the general case. (See Remark 17.5.)

As in the previous chapter, we fix a holomorphic function $\Delta : \mathbb{C}^* \to \mathbb{C}$ with $\Delta - \Delta_0 \in \mathrm{As}(\mathbb{C}^*, \ell^2_{0,0}, 1)$, and thereby the associated spectral curve Σ. We continue to use the notations of Chap. 16. In particular we let $\varkappa_{k,\nu}$ be the zeros of $\Delta^2 - 4$ as before, interpret $\varkappa_{k,\nu}$ also as points on Σ, and put $\varkappa_{k,*} := \frac{1}{2}(\varkappa_{k,1} + \varkappa_{k,2})$. We also continue to use the possibly punctured excluded domains $U'_{k,\delta}$ and $\widehat{U}'_{k,\delta}$ defined by Eq. (16.4).

To avoid difficulties, we will suppose from now on:

$$\boxed{\Delta^2 - 4 \text{ does not have any zeros of order } \geq 3.} \tag{17.1}$$

Because $\Delta^2 - 4$ has asymptotically and totally only two zeros in every excluded domain, this condition excludes only the case where more than two zeros of $\Delta^2 - 4$ combine in a single point in the "compact part" of Σ. It is a consequence of this condition that Σ does not have any singularities other than ordinary double points.

We then enumerate the zeros $\varkappa_{k,\nu}$ of $\Delta^2 - 4$ in such a way that if $\varkappa \in \mathbb{C}^*$ is a zero of $\Delta^2 - 4$ of order 2, there exists $k \in \mathbb{Z}$ with $\varkappa = \varkappa_{k,1} = \varkappa_{k,2}$ (even for $|k|$ small), and we define

$$S := \{ k \in \mathbb{Z} \,|\, \varkappa_{k,1} = \varkappa_{k,2} \}. \tag{17.2}$$

With this definition, $\{ \varkappa_{k,\nu} \,|\, k \in S \}$ is the set of singular points of Σ, and therefore

$$\Sigma' := \Sigma \setminus \{ \varkappa_{k,\nu} \,|\, k \in S \} \tag{17.3}$$

is the regular set of Σ, a Riemann surface that is open and dense in Σ.

We denote the space of holomorphic 1-forms on Σ by $\Omega(\Sigma)$ (in the singular points of Σ, the holomorphy of $\omega \in \Omega(\Sigma)$ is considered in the normalization $\widehat{\Sigma}$ of Σ), and the space of square-integrable 1-forms on Σ by $L^2(\Sigma, T^*\Sigma)$.

The following proposition shows that any square-integrable, holomorphic 1-form ω on Σ is anti-symmetric with respect to the hyperelliptic involution of Σ, and provides a sufficient criterion for the square-integrability of a holomorphic 1-form.

Proposition 17.1

(1) *Let $\omega \in \Omega(\Sigma) \cap L^2(\Sigma, T^*\Sigma)$. Then there exists a holomorphic function $\Phi : \mathbb{C}^* \to \mathbb{C}$ so that*

$$\omega = \frac{\Phi(\lambda)}{\mu - \mu^{-1}}\, \mathrm{d}\lambda \tag{17.4}$$

holds. We have $\Phi \in \mathrm{As}(\mathbb{C}^*, \ell^2_{1,-3}, 1)$.

(2) *Let $\Phi \in \mathrm{As}(\mathbb{C}^*, \ell^2_{3/2,-5/2}, 1)$ be given and put $\omega := \frac{\Phi}{\mu-\mu^{-1}}\,\mathrm{d}\lambda$. Then ω is a meromorphic 1-form on Σ that is holomorphic on $\Sigma \setminus \{ \varkappa_{k,*} \,|\, k \in S \}$, and $\omega|\widehat{V}_\delta \in L^2(\widehat{V}_\delta, T^*\widehat{V}_\delta)$ holds.*

(3) *In the setting of (2), suppose that the following condition additionally holds: There exists a finite set* $T \subset \mathbb{Z}$ *so that* Φ *has a zero* ξ_k *in* $U_{k,\delta}$ *for every* $k \in \mathbb{Z} \setminus T$*, and there exists some* $C_\xi > 0$ *so that* $|\xi_k - \varkappa_{k,*}| \leq C_\xi \cdot |\varkappa_{k,1} - \varkappa_{k,2}|$ *holds for all* $k \in \mathbb{Z} \setminus T$.
Then ω *is holomorphic on* $\Sigma \setminus \{ \varkappa_{k,*} \,|\, k \in T \cap S \}$*, and square-integrable on* $\Sigma \setminus \bigcup_{k \in T \cap S} \widehat{U}_{k,\delta}$.

Proof For (1). Let $\sigma : (\lambda, \mu) \mapsto (\lambda, \mu^{-1})$ be the hyperelliptic involution of Σ. Then we have $\omega = \omega_+ + \omega_-$ with $\omega_\pm := \frac{1}{2}(\omega \pm \sigma^* \omega)$. ω_+ is σ-invariant, and therefore can be regarded as a square-integrable, holomorphic 1-form $f_+(\lambda)\, \mathrm{d}\lambda$ on $\mathbb{C}^*$. Because the holomorphic function f_+ is square-integrable on $\mathbb{C}^*$, it is identically zero, and therefore $\omega = \omega_-$ is σ-anti-invariant.

Hence $(\mu - \mu^{-1}) \cdot \omega$ is a σ-invariant, holomorphic 1-form, and can thus be regarded as a holomorphic 1-form on $\mathbb{C}^*$. Therefore there exists a holomorphic function $\Phi : \mathbb{C}^* \to \mathbb{C}$ with $(\mu - \mu^{-1}) \cdot \omega = \Phi(\lambda)\, \mathrm{d}\lambda$, and with this Φ Eq. (17.4) holds.

It remains to show $\Phi \in \mathrm{As}(\mathbb{C}^*, \ell^2_{1,-3}, 1)$. For this we may suppose without loss of generality that $\delta > 0$ is chosen so small that even the excluded domains $U_{k,3\delta}$ do not overlap. Because of Proposition 9.4(1) it then suffices to show that $\Phi | V_{3\delta} \in \mathrm{As}(V_{3\delta}, \ell^2_{1,-3}, 1)$ holds. For each $k \in \mathbb{Z}$, let $\lambda_k \in S_k \cap V_{3\delta}$ be such that $\frac{|\Phi(\lambda)|}{w(\lambda)}$ attains its maximum on the compact set $S_k \cap V_{3\delta}$ at λ_k (where S_k is the annulus defined by Eq. (9.1)). Let M_k be the topological annulus

$$M_k := \begin{cases} \{ \lambda \in \mathbb{C}^* \,|\, \delta k \leq |\lambda - \lambda_k| \leq 2\delta k \} & \text{for } k > 0 \\ \{ \lambda \in \mathbb{C}^* \,|\, \delta k^{-3} \leq |\lambda - \lambda_k| \leq 2\delta k^{-3} \} & \text{for } k < 0 \end{cases}$$

for $|k|$ sufficiently large. Then only the M_k of consecutive indices can intersect, and M_k is contained in $(S_{k-1} \cup S_k \cup S_{k+1}) \cap V_\delta$.

We now first look at $k > 0$. Because of the mean value property for the holomorphic function $f(\lambda) := \left(\frac{\Phi}{\mu - \mu^{-1}}\right)^2 = \frac{\Phi^2}{\Delta^2 - 4}$, we have for any $r \in [\delta k, 2\delta k]$

$$f(\lambda_k) = \frac{1}{2\pi} \int_0^{2\pi} f(\lambda_k + r e^{it})\, \mathrm{d}t$$

and therefore

$$\begin{aligned} f(\lambda_k) &= \frac{1}{\delta k} \int_{r=\delta k}^{2\delta k} \frac{1}{2\pi} \int_{t=0}^{2\pi} f(\lambda_k + r e^{it})\, \mathrm{d}t\, \mathrm{d}r \\ &= \frac{1}{2\pi \delta k} \int_{r=\delta k}^{2\delta k} \int_{t=0}^{2\pi} \frac{1}{r} f(\lambda_k + r e^{it})\, \mathrm{d}t\, r\, \mathrm{d}r \\ &= \frac{1}{2\pi \delta k} \int_{M_k} \frac{1}{r} f(\lambda)\, \mathrm{d}^2\lambda, \end{aligned}$$

where $d^2\lambda$ here and in the sequel denotes the Lebesgue measure on $\mathbb{C}$, and whence

$$|f(\lambda_k)| \leq \frac{1}{2\pi\delta k} \cdot \frac{1}{\delta k} \cdot \int_{M_k} |f(\lambda)|\, d^2\lambda$$

follows. Because of $\omega \in L^2(\Sigma, T^*\Sigma)$, we have in particular $f|V_\delta \in L^1(V_\delta)$. The M_k are contained in V_δ, and each point of V_δ is covered by at most two of the M_k. Therefore we obtain:

$$|f(\lambda_k)| \in \frac{1}{k^2} \cdot \ell^1(k) = \ell^1_2(k)$$

and hence

$$\frac{\Phi(\lambda_k)}{\sqrt{\Delta(\lambda_k)^2 - 4}} \in \ell^2_1(k)$$

holds. Because of $\lambda_k \in V_{3\delta}$, $\sqrt{\Delta(\lambda_k)^2 - 4}$ is comparable to $w(\lambda_k)$, and thus we obtain

$$\frac{\Phi(\lambda_k)}{w(\lambda_k)} \in \ell^2_1(k).$$

It follows from the definition of λ_k that $\Phi|V_{3\delta} \in \mathrm{As}_\infty(V_{3\delta}, \ell^2_1, 1)$ holds. A similar argument yields $\Phi|V_{3\delta} \in \mathrm{As}_0(V_{3\delta}, \ell^2_{-3}, 1)$, and thus $\Phi|V_{3\delta} \in \mathrm{As}(V_{3\delta}, \ell^2_{1,-3}, 1)$. By Proposition 9.4(1), $\Phi \in \mathrm{As}(\mathbb{C}^*, \ell^2_{1,-3}, 1)$ follows.

For (2). It is clear that ω is a meromorphic 1-form on Σ, and that it is holomorphic except at the zeros of $\mu - \mu^{-1}$, i.e. at the $\varkappa_{k,\nu}$. If $\varkappa_{k,\nu}$ is a regular branch point of Σ, i.e. a simple zero of $\Delta^2 - 4$, then both $\mu - \mu^{-1}$ and $d\lambda$ have a simple zero at this point, and therefore ω is holomorphic also at these points. It follows that ω is holomorphic on $\Sigma \setminus \{\, \varkappa_{k,*} \,|\, k \in S \,\}$.

Because of $\Phi \in \mathrm{As}(\mathbb{C}^*, \ell^2_{3/2,-5/2}, 1)$ and because $\mu - \mu^{-1} = \sqrt{\Delta(\lambda)^2 - 4}$ is on $\widehat{V_\delta}$ comparable to $w(\lambda)$, we have $\frac{\Phi}{\mu-\mu^{-1}}|\widehat{V_\delta} \in \mathrm{As}(\widehat{V_\delta}, \ell^2_{3/2,-5/2}, 0)$ and therefore $\frac{\Phi^2}{\Delta^2-4}|V_\delta \in \mathrm{As}(V_\delta, \ell^1_{3,-5}, 0)$. Because of $\mathrm{vol}(S_k \cap V_\delta) \in \ell^\infty_{-3,5}$, it follows that $\frac{\Phi^2}{\Delta^2-4} \in L^1(V_\delta)$ and therefore $\omega|\widehat{V_\delta} \in L^2(\widehat{V_\delta}, T^*\widehat{V_\delta})$ holds.

For (3). The additional condition ensures that $\xi_k = \varkappa_{k,*}$ holds for every $k \in (\mathbb{Z} \setminus T) \cap S$; this fact implies together with (17.1) that ω is holomorphic on $\Sigma \setminus \{\, \varkappa_{k,*} \,|\, k \in T \cap S \,\}$. Moreover, these hypotheses ensure that $\frac{\Phi^2}{\Delta^2-4}$ has at most simple poles on every $U_{k,\delta}$ with $k \in \mathbb{Z} \setminus (T \cap S)$, and therefore ω is square-integrable on each individual excluded domain $\widehat{U}_{k,\delta}$ with $k \in \mathbb{Z} \setminus (T \cap S)$. Because ω is also square-integrable on $\widehat{V_\delta}$ by (2), it only remains to show that the sum of the integrals of ω^2 on $\widehat{U}_{k,\delta}$, $k \in \mathbb{Z} \setminus (T \cap S)$ is finite.

For $\lambda \in U_{k,\delta}$ we have

$$|\Phi(\lambda)| \leq \max_{U_{k,\delta}} |\Phi'| \cdot |\lambda - \xi_k|.$$

We now note that $\Phi \in \mathrm{As}(\mathbb{C}^*, \ell^2_{3/2,-5/2}, 1)$ implies $\Phi' \in \mathrm{As}(\mathbb{C}^*, \ell^2_{5/2,-11/2}, 1)$ by Proposition 9.4(3), and therefore there exists a sequence $(a_k) \in \ell^2_{5/2,-11/2}(k)$ so that for every $\lambda \in U_{k,\delta}$ we have

$$|\Phi(\lambda)| \leq a_k \cdot |\lambda - \xi_k|.$$

We now have by Proposition 16.5(4)

$$\left| \int_{\widehat{U}_{k,\delta}} \omega^2 \right| \leq 2 \int_{U_{k,\delta}} \frac{|\Phi(\lambda)|^2}{|\Delta(\lambda)^2 - 4|} \, \mathrm{d}^2\lambda \leq 2\, a_k^2 \cdot \int_{U_{k,\delta}} \frac{|\lambda - \xi_k|^2}{|\Delta^2 - 4|} \, \mathrm{d}^2\lambda$$
$$\leq C\, a_k^2\, \lambda_{k,0}^{\varrho} \cdot \left(|\xi_k - \varkappa_{k,*}|^2 + \frac{|\xi_k - \varkappa_{k,*}|^4}{|\varkappa_{k,1} - \varkappa_{k,2}|^2} + \ell^\infty_{-2,6}(k) \right).$$

We have $|\xi_k - \varkappa_{k,*}|^2 \in \ell^\infty_{-2,6}(k)$ and $\frac{|\xi_k - \varkappa_{k,*}|^2}{|\varkappa_{k,1} - \varkappa_{k,2}|^2} \leq C_\xi^2$. It follows that we have

$$\left| \int_{\widehat{U}_{k,\delta}} \omega^2 \right| \leq C\, a_k^2\, \ell^\infty_{-2,6}(k) \cdot \ell^\infty_{-2,6}(k) \in \ell^1_{1,1}(k) \subset \ell^1(k).$$

□

In the following proposition we investigate holomorphic functions $\Phi : \mathbb{C}^* \to \mathbb{C}$ which are infinite products and have one zero in every excluded domain $U_{k,\delta}$ with a single exception $U_{n,\delta}$, $n \in \mathbb{Z}$. To show asymptotic estimates for such holomorphic functions, we will apply Proposition 10.1. (By adding one more linear factor corresponding a zero in the "missing" excluded domain $U_{n,\delta}$, we obtain a function of the type of the holomorphic functions c studied in Proposition 10.1.) One more degree of freedom is provided by multiplying Φ with λ^ϱ for some $\varrho \in \mathbb{Z}$. We will see in Proposition 17.3 that there is exactly one choice of ϱ so that the associated holomorphic 1-form $\frac{\Phi(\lambda)}{\mu - \mu^{-1}}\, \mathrm{d}\lambda$ becomes square-integrable on Σ (or on $\Sigma \setminus \widehat{U}_{n,\delta}$ if $n \in S$). In Chap. 18 we will use square-integrable holomorphic 1-forms of this type to construct the canonical basis of $\Omega(\Sigma)$.

Proposition 17.2 *We fix $R_0 > 0$. Let $\varrho \in \mathbb{Z}$, $n \in \mathbb{Z}$, and $(\xi_k)_{k \in \mathbb{Z}\setminus\{n\}}$ with $\xi_k - \lambda_{k,0} \in \ell^2_{-1,3}(\mathbb{Z}\setminus\{n\})$, $\|\xi_k - \lambda_{k,0}\|_{\ell^2_{-1,3}(\mathbb{Z}\setminus\{n\})} \leq R_0$ and $\xi_k = \varkappa_{k,*}$ for all $k \in S\setminus\{n\}$ be given.*

The function

$$\Phi_{n,\xi,\varrho}(\lambda) := \begin{cases} \lambda^{\varrho} \cdot (\lambda - \xi_0) \cdot \prod_{k\in\mathbb{N}\setminus\{n\}} \frac{\xi_k-\lambda}{16\pi^2 k^2} \cdot \prod_{k\in\mathbb{N}} \frac{\lambda-\xi_{-k}}{\lambda} & \text{if } n > 0 \\ \lambda^{\varrho-1} \cdot (\lambda - \xi_0) \cdot \prod_{k\in\mathbb{N}} \frac{\xi_k-\lambda}{16\pi^2 k^2} \cdot \prod_{k\in\mathbb{N}\setminus\{-n\}} \frac{\lambda-\xi_{-k}}{\lambda} & \text{if } n \leq 0 \end{cases} \tag{17.5}$$

is a holomorphic function on $\mathbb{C}^*$ *with the following asymptotic properties:*

(1) (Asymptotic behavior of $\Phi_{n,\xi,\varrho}$.) *We have*

$$\Phi_{n,\xi,\varrho} - \Phi_{n,0,\varrho} \in \mathrm{As}_{\infty}(\mathbb{C}^*, \ell^2_{-2\varrho+1}, 1)$$
$$\text{and} \quad \Phi_{n,\xi,\varrho} - \tau_{\xi}^{-2}\, \Phi_{n,0,\varrho} \in \mathrm{As}_0(\mathbb{C}^*, \ell^2_{2\varrho+1}, 1)$$

with

$$\Phi_{n,0,\varrho} := \begin{cases} 4 \cdot \lambda^{\varrho} \cdot \frac{16\pi^2 n^2}{\lambda_{n,0}-\lambda} \cdot c_0(\lambda) & \text{if } n > 0 \\ 4 \cdot \lambda^{\varrho} \cdot \frac{1}{\lambda-\lambda_{n,0}} \cdot c_0(\lambda) & \text{if } n \leq 0 \end{cases} \tag{17.6}$$

and

$$\tau_{\xi} := \left(\prod_{k\in\mathbb{Z}\setminus\{n\}} \frac{\lambda_{k,0}}{\xi_k} \right)^{1/2} .$$

In particular $\Phi_{n,\xi,\varrho} \in \mathrm{As}(\mathbb{C}^*, \ell^{\infty}_{-2\varrho+1,2\varrho+1}, 1)$ *holds.*

(2) (Asymptotic comparison of two functions of the form $\frac{\Phi_{n,\xi,\varrho}(\lambda)}{\lambda-\xi_k}$ on $U_{k,\delta}$.) *Let two sequences* $(\xi_k^{[1]})$, $(\xi_k^{[2]})$ *of the kind of* (ξ_k) *above be given. We denote the quantities defined above associated to* $(\xi_k^{[\nu]})$ *by the superscript* $^{[\nu]}$ *(for* $\nu \in \{1, 2\}$*).*
Then there exists a constant $C > 0$ *(dependent only on* R_0*) so that we have for every* $k \in \mathbb{Z} \setminus \{n\}$ *and every* $\lambda \in U_{k,\delta}$ *if* $k > 0$

$$\left| \frac{\Phi^{[1]}_{n,\xi,\varrho}(\lambda)}{\lambda - \xi_k^{[1]}} - \frac{\Phi^{[2]}_{n,\xi,\varrho}(\lambda)}{\lambda - \xi_k^{[2]}} \right| \leq C\, r_k$$

and if $k < 0$

$$\left| \frac{(\tau^{[1]})^2 \cdot \Phi^{[1]}_{n,\xi,\varrho}(\lambda)}{\lambda - \xi_k^{[1]}} - \frac{(\tau^{[2]})^2 \cdot \Phi^{[2]}_{n,\xi,\varrho}(\lambda)}{\lambda - \xi_k^{[2]}} \right| \leq C\, r_k ,$$

where the sequence $(r_k) \in \ell^2_{2-2\varrho,2\varrho-2}(k)$ *is defined for* $n > 0$ *by*

$$r_k := \begin{cases} \frac{n^2\,k^{2\varrho}}{|n^2-k^2|}\,(a_j * \frac{1}{|j|})_k & \text{if } k > 0 \\ k^{2-2\varrho}\,(a_j * \frac{1}{|j|})_k & \text{if } k < 0 \end{cases},$$

and for $n < 0$ *by*

$$r_k := \begin{cases} k^{2\varrho-2}\,(a_j * \frac{1}{|j|})_k & \text{if } k > 0 \\ \frac{n^2\,k^{4-2\varrho}}{|n^2-k^2|}\,(a_j * \frac{1}{|j|})_k & \text{if } k < 0 \end{cases},$$

with $(a_k) \in \ell^2_{0,0}(k)$ *defined by*

$$a_k := \begin{cases} k^{-1}\,|\xi_k^{[1]} - \xi_k^{[2]}| & \text{if } k > 0 \\ k^3\,|\xi_k^{[1]} - \xi_k^{[2]}| & \text{if } k < 0 \end{cases}. \tag{17.7}$$

We have $\|r_k\|_{\ell^2_{2-2\varrho,2\varrho-2}} \leq \|\xi_k^{[1]} - \xi_k^{[2]}\|_{\ell^2_{-1,3}}$.

(3) (Asymptotic estimate of $\frac{\Phi_{n,\xi,\varrho}(\lambda)}{\lambda-\xi_k}$ on $U_{k,\delta}$.) *Let* (ξ_k) *be as before and let* (r_k) *be as in (2) applied to* $\xi_k^{[1]} = \xi_k$ *and* $\xi_k^{[2]} = \lambda_{k,0}$.
Then there exist $C_1, C_2 > 0$ *(dependent only on* R_0*) such that for every* $k \in \mathbb{Z} \setminus \{n\}$ *and* $\lambda \in U_{k,\delta}$ *we have*

$$\text{if } n, k > 0: \left| \frac{\Phi_{n,\xi,\varrho}(\lambda)}{\lambda-\xi_k} - \frac{8(-1)^k\,\pi^2\,n^2\cdot\xi_k^{\varrho}}{\lambda_{n,0}-\xi_k} \right| \leq C_1 \cdot \frac{n^2\cdot k^{2\varrho-1}}{|n^2-k^2|} \cdot |\lambda - \xi_k| + C_2 \cdot r_k$$

$$\text{if } n > 0, k < 0: \left| \frac{\Phi_{n,\xi,\varrho}(\lambda)}{\lambda-\xi_k} - \left(-\frac{8(-1)^k\,\pi^2\,n^2\cdot\xi_k^{\varrho-1}}{\tau_\xi^2\cdot(\lambda_{n,0}-\xi_k)} \right) \right| \leq C_1 \cdot k^{5-2\varrho} \cdot |\lambda - \xi_k| + C_2 \cdot r_k$$

$$\text{if } n < 0, k > 0: \left| \frac{\Phi_{n,\xi,\varrho}(\lambda)}{\lambda-\xi_k} - \left(-\frac{(-1)^k\cdot\xi_k^{\varrho}}{2\,(\lambda_{n,0}-\xi_k)} \right) \right| \leq C_1 \cdot k^{2\varrho-3} \cdot |\lambda - \xi_k| + C_2 \cdot r_k$$

$$\text{if } n, k < 0: \left| \frac{\Phi_{n,\xi,\varrho}(\lambda)}{\lambda-\xi_k} - \frac{(-1)^k\cdot\xi_k^{\varrho-1}}{2\,\tau_\xi^2\cdot(\lambda_{n,0}-\xi_k)} \right| \leq C_1 \cdot \frac{n^2\cdot k^{7-2\varrho}}{|n^2-k^2|} \cdot |\lambda - \xi_k| + C_2 \cdot r_k.$$

Proof For the proof, we abbreviate $\Phi := \Phi_{n,\xi,\varrho}$ and $\Phi_0 := \Phi_{n,0,\varrho}$. Moreover, we put $\xi_n := \lambda_{n,0} \in U_{n,\delta}$, and thereby obtain a sequence $\xi_k \in \ell^2_{-1,3}(\mathbb{Z})$.

For (1). By Proposition 10.1

$$c_\xi(\lambda) := \frac{1}{4}\,\tau_\xi\,(\lambda - \xi_0) \cdot \prod_{k=1}^{\infty} \frac{\xi_k - \lambda}{16\,\pi^2\,k^2} \cdot \prod_{k=1}^{\infty} \frac{\lambda - \xi_{-k}}{\lambda}$$

defines a holomorphic function on $\mathbb{C}^*$, and we have

$$\Phi(\lambda) = \begin{cases} \dfrac{4}{\tau_\xi} \cdot \lambda^\varrho \cdot \dfrac{16\pi^2 n^2}{\xi_n - \lambda} \cdot c_\xi(\lambda) & \text{if } n > 0 \\ \dfrac{4}{\tau_\xi} \cdot \lambda^\varrho \cdot \dfrac{1}{\lambda - \xi_n} \cdot c_\xi(\lambda) & \text{if } n < 0 \end{cases} . \tag{17.8}$$

Because c_ξ has a zero at $\lambda = \xi_n$, it follows from this description that Φ is a holomorphic function on $\mathbb{C}^*$. Moreover, we have $c_0(\lambda) = \lambda^{1/2}\sin(\zeta(\lambda)) \in \mathrm{As}(\mathbb{C}^*, \ell^\infty_{-1,1}, 1)$ and by Proposition 10.1

$$c_\xi - \tau_\xi\, c_0 \in \mathrm{As}_\infty(\mathbb{C}^*, \ell^2_{-1}, 1) \quad \text{and} \quad c_\xi - \tau_\xi^{-1}\, c_0 \in \mathrm{As}_0(\mathbb{C}^*, \ell^2_1, 1),$$

and hence $c_\xi \in \mathrm{As}(\mathbb{C}^*, \ell^\infty_{-1,1}, 1)$. Because of Eq. (17.8), it follows that we have

$$\Phi - \Phi_0 \in \mathrm{As}_\infty(\mathbb{C}^*, \ell^2_{-2\varrho+1}, 1) \quad \text{and} \quad \Phi - \tau_\xi^{-2}\, \Phi_0 \in \mathrm{As}_0(\mathbb{C}^*, \ell^2_{2\varrho+1}, 1).$$

Because of $\Phi_0 \in \mathrm{As}(\mathbb{C}^*, \ell^\infty_{-2\varrho+1,2\varrho+1}, 1)$ it follows that $\Phi \in \mathrm{As}(\mathbb{C}^*, \ell^\infty_{-2\varrho+1,2\varrho+1}, 1)$ holds.

For (2). By Eq. (17.8) we have

$$\frac{\Phi^{[1]}(\lambda)}{\lambda - \xi_k^{[1]}} - \frac{\Phi^{[2]}(\lambda)}{\lambda - \xi_k^{[2]}} = \begin{cases} 4\,\lambda^\varrho \cdot \frac{16\pi^2 n^2}{\xi_n - \lambda} \cdot \left(\frac{c_\xi^{[1]}(\lambda)}{\tau_\xi^{[1]} \cdot (\lambda - \xi_k^{[1]})} - \frac{c_\xi^{[2]}(\lambda)}{\tau_\xi^{[2]} \cdot (\lambda - \xi_k^{[2]})} \right) & \text{if } n > 0 \\ 4\,\lambda^\varrho \cdot \frac{1}{\lambda - \xi_n} \cdot \left(\frac{c_\xi^{[1]}(\lambda)}{\tau_\xi^{[1]} \cdot (\lambda - \xi_k^{[1]})} - \frac{c_\xi^{[2]}(\lambda)}{\tau_\xi^{[2]} \cdot (\lambda - \xi_k^{[2]})} \right) & \text{if } n < 0 \end{cases} . \tag{17.9}$$

In the case $n > 0$ we note that we have for $k \in \mathbb{Z}$ and $\lambda \in U_{k,\delta}$

$$|\lambda|^\varrho \leq \begin{cases} C_3 \cdot k^{2\varrho} & \text{if } k > 0 \\ C_3 \cdot k^{-2\varrho} & \text{if } k < 0 \end{cases} \quad \text{and} \quad \frac{1}{|\xi_n - \lambda|} \leq \begin{cases} C_4 \cdot \dfrac{1}{|n^2 - k^2|} & \text{if } k > 0 \\ C_4 \cdot \dfrac{1}{n^2} & \text{if } k < 0 \end{cases} . \tag{17.10}$$

Moreover, by Corollary 10.3(2) there exists $C_5 > 0$ so that with $(a_k) \in \ell^2_{0,0}(k)$ defined by Eq. (17.7) we have for $k > 0$

$$\left| \frac{c^{[1]}(\lambda)}{\tau^{[1]} \cdot (\lambda - \xi_k^{[1]})} - \frac{c^{[2]}(\lambda)}{\tau^{[2]} \cdot (\lambda - \xi_k^{[2]})} \right| \leq C_5 \cdot \left(a_k * \frac{1}{|k|} \right) \tag{17.11}$$

and for $k < 0$

$$\left| \frac{c^{[1]}(\lambda)}{(\tau^{[1]})^{-1} \cdot (\lambda - \xi_k^{[1]})} - \frac{c^{[2]}(\lambda)}{(\tau^{[2]})^{-1} \cdot (\lambda - \xi_k^{[2]})} \right| \leq C_5\, k^2 \cdot \left(a_k * \frac{1}{|k|} \right). \tag{17.12}$$

In the case of $k > 0$ we obtain for $\lambda \in U_{k,\delta}$ by plugging the estimates (17.10) and (17.11) into Eq. (17.9):

$$\left| \frac{\Phi^{[1]}(\lambda)}{\lambda - \xi_k^{[1]}} - \frac{\Phi^{[2]}(\lambda)}{\lambda - \xi_k^{[2]}} \right| \leq 4 \cdot C_3\, k^{2\varrho} \cdot 16\pi^2 n^2 \cdot \frac{C_4}{|n^2 - k^2|} \cdot C_5 \left(a_k * \frac{1}{|k|} \right) \leq C \cdot r_k$$

with $C := 16\,\pi^2\, C_3\, C_4\, C_5$, whereas for $k < 0$ we similarly obtain

$$\left| (\tau^{[1]})^2 \cdot \tfrac{\Phi^{[1]}(\lambda)}{\lambda - \xi_k^{[1]}} - \tfrac{(\tau^{[2]})^2 \cdot \Phi^{[2]}(\lambda)}{\lambda - \xi_k^{[2]}} \right| \leq 4 \cdot C_3\, k^{-2\varrho} \cdot 16\pi^2 n^2 \cdot C_4\, \tfrac{1}{n^2} \cdot C_5\, k^2 \left(a_k * \frac{1}{|k|} \right) \leq C\, r_k.$$

The case $n < 0$ is handled analogously.

For (3). Suppose that $k \in \mathbb{Z} \setminus \{n\}$ is given. Here we only consider the case $n, k > 0$; the other cases are shown in an analogous way. For given $\lambda \in U_{k,\delta}$ we have

$$\begin{aligned} &\left| \frac{\Phi(\lambda)}{\lambda - \xi_k} - \frac{8(-1)^k\, \pi^2\, n^2 \cdot \xi_k^{\varrho}}{\lambda_{n,0} - \xi_k} \right| \\ &\leq \left| \frac{\Phi(\lambda)}{\lambda - \xi_k} - \Phi'(\xi_k) \right| + \left| \Phi'(\xi_k) - \Phi_0'(\lambda_{k,0}) \right| \\ &\quad + \left| \Phi_0'(\lambda_{k,0}) - \frac{8(-1)^k\, \pi^2\, n^2 \cdot \xi_k^{\varrho}}{\lambda_{n,0} - \xi_k} \right|. \end{aligned} \tag{17.13}$$

Because of Eq. (17.8) we have

$$\left(\frac{\Phi(\lambda)}{\lambda - \xi_k} \right)' = \frac{4 \cdot 16\pi^2 n^2}{\tau_\xi} \cdot \left(\left(\frac{\lambda^{\varrho}}{\xi_n - \lambda} \right)' \cdot \frac{c_\xi(\lambda)}{\lambda - \xi_k} + \frac{\lambda^{\varrho}}{\xi_n - \lambda} \cdot \left(\frac{c_\xi(\lambda)}{\lambda - \xi_k} \right)' \right).$$

Because $\frac{c_\xi(\lambda)}{\lambda - \xi_k} = O(1)$ and $\left(\frac{c_\xi(\lambda)}{\lambda - \xi_k} \right)' = O(k^{-1})$ uniformly on all the excluded domains, it follows that there exists $C_6 > 0$ with

$$\left| \left(\frac{\Phi(\lambda)}{\lambda - \xi_k} \right)' \right| \leq C_6 \cdot \frac{n^2 \cdot k^{2\varrho - 1}}{|n^2 - k^2|}.$$

Because $\frac{\Phi(\lambda)}{\lambda-\xi_k}$ is extended holomorphically at $\lambda = \xi_k$ by the value $\Phi'(\xi_k)$, it follows that

$$\left|\frac{\Phi(\lambda)}{\lambda - \xi_k} - \Phi'(\xi_k)\right| \leq C_6 \cdot \frac{n^2 \cdot k^{2\varrho-1}}{|n^2 - k^2|} \cdot |\lambda - \xi_k| \tag{17.14}$$

holds.

Moreover, again because the holomorphic function $\frac{\Phi(\lambda)}{\lambda-\xi_k}$ resp. $\frac{\Phi_0(\lambda)}{\lambda-\lambda_{k,0}}$ is extended holomorphically at $\lambda = \xi_k$ resp. at $\lambda = \lambda_{k,0}$ by the value $\Phi'(\xi_k)$ resp. by $\Phi_0'(\lambda_{k,0})$, it follows from (2) that

$$\left|\Phi'(\xi_k) - \Phi_0'(\lambda_{k,0})\right| \leq r_k \tag{17.15}$$

holds.

Finally, we have by Eq. (17.6)

$$\Phi'_{n,0,\varrho}(\lambda_{k,0}) = 4\,\lambda_{k,0}^{\varrho} \cdot \frac{16\pi^2 n^2}{\lambda_{n,0} - \lambda_{k,0}} \cdot \underbrace{c_0'(\lambda_{k,0})}_{=(-1)^k/8} = \frac{8(-1)^k\,\pi^2\,n^2\,\lambda_{k,0}^{\varrho}}{\lambda_{n,0} - \lambda_{k,0}}$$

and therefore with constants $C_7, C_8 > 0$

$$\begin{aligned}\left|\Phi'_{n,0,\varrho}(\lambda_{k,0}) - \tfrac{8(-1)^k\,\pi^2\,n^2\cdot\xi_k^{\varrho}}{\lambda_{n,0}-\xi_k}\right| &\leq C_7 \cdot \tfrac{n^2\,k^{2\varrho-2}}{|n^2-k^2|} \cdot |\xi_k - \lambda_{k,0}| \\ &= C_7 \cdot \tfrac{n^2\,k^{2\varrho-1}}{|n^2-k^2|} \cdot a_k \leq C_8 \cdot r_k. \end{aligned} \tag{17.16}$$

By applying the estimates (17.14), (17.15) and (17.16) to (17.13), we obtain the claimed statement. □

As explained at the beginning of the present chapter, we are interested in holomorphic 1-forms $\omega = \frac{\Phi_{n,\xi,\varrho}(\lambda)}{\mu-\mu^{-1}}\,\mathrm{d}\lambda$, where $\Phi_{n,\xi,\varrho}$ is as in Proposition 17.2, and such that ω is square-integrable on Σ (at least away from the singular points of Σ). The following proposition addresses the question when such an ω is in fact square-integrable. It turns out that a necessary condition for this to be the case is that the exponent ϱ occurring in the definition of $\Phi_{n,\xi,\varrho}$ in Eq. (17.5) equals $\varrho = -1$. For ω to be actually square-integrable, it is further necessary that the zeros ξ_k of $\Phi_{n,\xi,\varrho}$ are "not too far removed" from the center $\varkappa_{k,*}$ of the branch points, in comparison to the size $|\varkappa_{k,1} - \varkappa_{k,2}|$ of the "handle" on Σ defined by $\varkappa_{k,1}$ and $\varkappa_{k,2}$. More specifically, the required condition is that there exist a constant $C_\xi > 0$ so that $|\xi_k - \varkappa_{k,*}| \leq C_\xi \cdot |\varkappa_{k,1} - \varkappa_{k,2}|$ holds for all $k \in \mathbb{Z} \setminus \{n\}$, compare Proposition 17.1(3).

Proposition 17.3 *In the setting of Proposition 17.2, the function* $f_{n,\xi,\varrho}$ *defined by*

$$f_{n,\xi,\varrho} := \frac{\Phi_{n,\xi,\varrho}}{\mu - \mu^{-1}}$$

is a meromorphic function on Σ *with the following asymptotic properties:*

(1) With

$$f_{n,0,\varrho} := \begin{cases} -2\mathrm{i}\,\lambda^{\varrho+1/2} \cdot \frac{16\pi^2 n^2}{\lambda_{n,0}-\lambda} & \text{if } n > 0 \\ -2\mathrm{i}\,\lambda^{\varrho+1/2} \cdot \frac{1}{\lambda-\lambda_{n,0}} & \text{if } n < 0 \end{cases}$$

we have

$$(f_{n,\xi,\varrho} - f_{n,0,\varrho})|\widehat{V}_\delta \in \mathrm{As}_\infty(\widehat{V}_\delta, \ell^2_{-2\varrho+1}, 0)$$
$$\text{and} \quad (f_{n,\xi,\varrho} - \tau_\xi^{-2}\, f_{n,0,\varrho})|\widehat{V}_\delta \in \mathrm{As}_0(\widehat{V}_\delta, \ell^2_{2\varrho+1}, 0);$$

in particular $f_{n,\xi,\varrho}|\widehat{V}_\delta \in \mathrm{As}(\widehat{V}_\delta, \ell^\infty_{-2\varrho+1,2\varrho+1}, 0)$.

(2) $f_{n,\xi,\varrho}|\widehat{V}_\delta \in L^2(\widehat{V}_\delta)$ *if and only if* $\varrho = -1$. *If* $\varrho = -1$ *holds and there exists* $C_\xi > 0$ *so that* $|\xi_k - \varkappa_{k,*}| \leq C_\xi \cdot |\varkappa_{k,1} - \varkappa_{k,2}|$ *holds for all* $k \in \mathbb{Z} \setminus \{n\}$, *then we have* $f_{n,\xi,\varrho} \in L^2(\Sigma)$ *if* $n \notin S$, *and* $f_{n,\xi,\varrho} \in L^2(\Sigma \setminus \widehat{U}_{n,\delta})$ *if* $n \in S$.

Proof We continue to use the notations from the proof of Proposition 17.2, and also abbreviate $f := f_{n,\xi,\varrho}$ and $f_0 := f_{n,0,\varrho}$.

For (1). We have

$$\mu_0 - \mu_0^{-1} = \sqrt{\Delta_0^2 - 4} = 2i \cdot \sin(\zeta(\lambda)) = 2i\,\lambda^{-1/2}\, c_0(\lambda)$$

and therefore

$$f_0 = \frac{\Phi_0}{2\mathrm{i}\,\lambda^{-1}\, c_0} = \frac{\Phi_0}{\mu_0 - \mu_0^{-1}},$$

whence

$$\begin{aligned} f - f_0 &= \frac{\Phi}{\mu - \mu^{-1}} - \frac{\Phi_0}{\mu_0 - \mu_0^{-1}} \\ &= \Phi \cdot \left(\frac{1}{\mu - \mu^{-1}} - \frac{1}{\mu_0 - \mu_0^{-1}} \right) + (\Phi - \Phi_0) \cdot \frac{1}{\mu_0 - \mu_0^{-1}}. \end{aligned}$$

follows. Because of

$$\left((\mu - \mu^{-1}) - (\mu_0 - \mu_0^{-1})\right)\Big|\, \widehat{V}_\delta \in \mathrm{As}(\widehat{V}_\delta, \ell^2_{0,0}, 1)$$

and Proposition 17.2(1), we obtain $(f - f_0)|\widehat{V}_\delta \in \mathrm{As}_\infty(\widehat{V}_\delta, \ell^2_{-2\varrho+1}, 0)$. A similar calculation yields $(f - \tau_\xi^{-2} f_0)|\widehat{V}_\delta \in \mathrm{As}_0(\widehat{V}_\delta, \ell^2_{2\varrho+1}, 0)$.

For (2). By Proposition 17.1(1), a necessary condition for f to be square-integrable is that $\Phi \in \mathrm{As}(\mathbb{C}^*, \ell^2_{1,-3}, 1)$ holds. We have $\Phi \in \mathrm{As}(\mathbb{C}^*, \ell^\infty_{-2\varrho+1,2\varrho+1}, 1)$ by Proposition 17.2(1) (and no better asymptotic can hold). Because $\ell^\infty_{-2\varrho+1,2\varrho+1} \subset \ell^2_{1,-3}$ holds if and only if the inequalities $-2\varrho+1 > 1+\frac{1}{2}$, i.e. $\varrho < -\frac{1}{4}$ and $2\varrho + 1 > -3 + \frac{1}{2}$, i.e. $\varrho > -\frac{7}{4}$ holds, we see that f cannot be square-integrable for $\varrho \neq -1$.

Let us now suppose $\varrho = -1$. Then we have $\Phi \in \mathrm{As}(\mathbb{C}^*, \ell^\infty_{3,-1}, 1)$ by Proposition 17.2(1), and therefore in particular $\Phi \in \mathrm{As}(\mathbb{C}^*, \ell^2_{3/2,-5/2}, 1)$. Therefore $f|\widehat{V}_\delta \in L^2(\widehat{V}_\delta)$ then holds.

If moreover there exists $C_\xi > 0$ so that $|\xi_k - \varkappa_{k,*}| \leq C_\xi \cdot |\varkappa_{k,1} - \varkappa_{k,2}|$ holds for all $k \in \mathbb{Z} \setminus \{n\}$, then we have $f \in L^2(\Sigma)$ if $n \notin S$, and $f \in L^2(\Sigma \setminus \widehat{U}_{n,\delta})$ if $n \in S$ by Proposition 17.1(3) (applied with $T = \{n\}$). □

The preceding proposition showed that $\omega = \frac{\Phi_{n,\xi,\varrho}}{\mu-\mu^{-1}}\, d\lambda$ is square-integrable only if $\varrho = -1$ holds and there exists $C_\xi > 0$ so that $|\xi_k - \varkappa_{k,*}| \leq C_\xi \cdot |\varkappa_{k,1} - \varkappa_{k,2}|$ holds for all $k \in \mathbb{Z} \setminus \{n\}$. For this reason we will subsequently consider $\Phi_{n,\xi,\varrho}$ only where these requirements are satisfied. In the following, final proposition on asymptotic estimates we study the asymptotic behavior of path integrals of holomorphic 1-forms ω of this type.

For this purpose we introduce the following notation: Within the space Div of all (classical) asymptotic divisors (regarded as point multi-sets in $\mathbb{C}^* \times \mathbb{C}^*$), we consider the subspace of those divisors whose support is contained in the spectral curve Σ resp. in the regular surface Σ' obtained from Σ by puncturing at the singularities (compare Eq. (17.3)):

$$\mathsf{Div}(\Sigma) := \{ D \in \mathsf{Div} \,|\, \mathrm{supp}(D) \subset \Sigma \} \text{ resp. } \mathsf{Div}(\Sigma') := \{ D \in \mathsf{Div} \,|\, \mathrm{supp}(D) \subset \Sigma' \}. \tag{17.17}$$

Then $\mathsf{Div}(\Sigma)$ is an (infinite-dimensional) complex subvariety of Div, and $\mathsf{Div}(\Sigma')$ is an open and dense subset of $\mathsf{Div}(\Sigma)$. We also consider the corresponding open and dense subsets of tame divisors (see Definition 12.4(1)):

$$\mathsf{Div}_{tame}(\Sigma) := \mathsf{Div}(\Sigma) \cap \mathsf{Div}_{tame} \quad \text{and} \quad \mathsf{Div}_{tame}(\Sigma') := \mathsf{Div}(\Sigma') \cap \mathsf{Div}_{tame}. \tag{17.18}$$

Proposition 17.4 *Let $C_\xi > 0$. In the present proposition we use the notations of Propositions 17.2 and 17.3, and consider 1-forms $\omega = f_{n,\xi,\varrho=-1}\, d\lambda$, where $n \in \mathbb{Z}$, $\varrho = -1$ and $|\xi_k - \varkappa_{k,*}| \leq C_\xi \cdot |\varkappa_{k,1} - \varkappa_{k,2}|$ holds for all $k \in \mathbb{Z} \setminus \{n\}$. By Proposition 17.3(2) any such ω is square-integrable on Σ resp. on $\Sigma \setminus \widehat{U}_{n,\delta}$ for $n \notin S$ resp. for $n \in S$. If $n \notin S$, $\omega \in \Omega(\Sigma)$ holds, whereas for $n \in S$, we have $\omega \in \Omega(\Sigma \setminus \{\varkappa_{n,*}\})$ and $\varkappa_{n,*}$ is a regular point of ω in the sense of Serre [Se, IV.9, p. 68ff.].*

For any path integral occurring in the sequel, we suppose that the path of integration runs entirely in Σ', and that for $|k|$ large, the path runs entirely in $\widehat{U}_{k,\delta}$.

We let $N \in \mathbb{N}$ be as in Lemma 16.1.

(1) *There exists a sequence $(b_n) \in \ell^2_{-1,-1}(n)$ (depending only on Σ, C_ξ and δ) so that for every $\omega = f_{n,\xi,\varrho=-1}\,\mathrm{d}\lambda$ of the type considered in the present proposition with $|n| > N$ and every $(\lambda_n^o, \mu_n^o), (\lambda_n, \mu_n) \in \widehat{U}'_{n,\delta}$ we have*

$$\text{if } n>0: \left| \int_{(\lambda_n^o,\mu_n^o)}^{(\lambda_n,\mu_n)} \omega - (-2\mathrm{i})\,\lambda_{n,0}^{1/2} \cdot \ln\left(\frac{\lambda_n - \varkappa_{n,*} + \Psi_n(\lambda_n,\mu_n)}{\lambda_n^o - \varkappa_{n,*} + \Psi_n(\lambda_n^o,\mu_n^o)} \right) \right|$$

$$\leq b_n \cdot \left| \ln\left(\frac{\lambda_n - \varkappa_{n,*} + \Psi_n(\lambda_n,\mu_n)}{\lambda_n^o - \varkappa_{n,*} + \Psi_n(\lambda_n^o,\mu_n^o)} \right) \right|$$

$$\text{if } n<0: \left| \int_{(\lambda_n^o,\mu_n^o)}^{(\lambda_n,\mu_n)} \omega - \frac{2\mathrm{i}}{\tau_\xi^2} \cdot \lambda_{n,0}^{-1/2} \cdot \ln\left(\frac{\lambda_n - \varkappa_{n,*} + \Psi_n(\lambda_n,\mu_n)}{\lambda_n^o - \varkappa_{n,*} + \Psi_n(\lambda_n^o,\mu_n^o)} \right) \right|$$

$$\leq b_n \cdot \left| \ln\left(\frac{\lambda_n - \varkappa_{n,*} + \Psi_n(\lambda_n,\mu_n)}{\lambda_n^o - \varkappa_{n,*} + \Psi_n(\lambda_n^o,\mu_n^o)} \right) \right| .$$

Here we integrate along any path that is contained in $\widehat{U}_{n,\delta}$, and $\ln(z)$ is the branch of the complex logarithm function with $\ln(1) = 2\pi\mathrm{i}m$, where $m \in \mathbb{Z}$ is the winding number of the path of integration around the pair of branch points $\varkappa_{k,1}$, $\varkappa_{k,2}$.

(2) *Let $R_0 > 0$ be given. There exists a constant $C > 0$, depending only on Σ, C_ξ and R_0, so that we have for every ω of the type considered here, every divisor $D = \{(\lambda_k, \mu_k)\} \in \mathsf{Div}(\Sigma')$ with $\|\lambda_k - \varkappa_{k,1}\|_{\ell^2_{-1,3}} \leq R_0$ and every $k \in \mathbb{Z}\setminus\{n\}$*

$$\text{if } n,k>0: \left| \int_{\varkappa_{k,1}}^{(\lambda_k,\mu_k)} \omega \right| \leq C \cdot \frac{n^2}{|n^2-k^2|} \cdot a_k$$

$$\text{if } n>0,\ k<0: \left| \int_{\varkappa_{k,1}}^{(\lambda_k,\mu_k)} \omega \right| \leq C \cdot \frac{1}{k^2} \cdot a_k$$

$$\text{if } n<0,\ k>0: \left| \int_{\varkappa_{k,1}}^{(\lambda_k,\mu_k)} \omega \right| \leq C \cdot \frac{1}{k^2} \cdot a_k$$

$$\text{if } n,k<0: \left| \int_{\varkappa_{k,1}}^{(\lambda_k,\mu_k)} \omega \right| \leq C \cdot \frac{n^2}{|n^2-k^2|} \cdot a_k ,$$

where the sequence $a_k \in \ell^2_{0,0}(k)$ *is given by*

$$a_k := \begin{cases} \frac{1}{k}|\lambda_k - \varkappa_{k,1}| & \textit{for } k > 0 \\ k^3\,|\lambda_k - \varkappa_{k,1}| & \textit{for } k < 0 \end{cases}, \tag{17.19}$$

and where the paths of integration of the above integrals run entirely in $\widehat{U}_{k,\delta}$ *and do not wind around the pair of branch points* $\varkappa_{k,1}, \varkappa_{k,2}$.

(3) *For every* ω *of the type considered here and every divisor* $D = \{(\lambda_k, \mu_k)\} \in \mathsf{Div}(\Sigma')$, *the sum* $\sum_{k\in\mathbb{Z}\setminus\{n\}} \int_{\varkappa_{k,1}}^{(\lambda_k,\mu_k)} \omega$ *converges absolutely, where the paths of integration of the integrals are as in (2). For every* $R_0 > 0$, *and with the associated sequence* $(a_k) \in \ell^2_{0,0}(k)$ *as in (2) there exist constants* $C, C_1 > 0$ *so that we have if* $\|\lambda_k - \varkappa_{k,1}\|_{\ell^2_{-1,3}} \leq R_0$

$$\sum_{k\in\mathbb{Z}\setminus\{n\}} \left| \int_{\varkappa_{k,1}}^{(\lambda_k,\mu_k)} \omega \right| \leq C\,n \left(a_k * \frac{1}{|k|} \right)_n + C_1 \in \ell^2_{-1,-1}(n).$$

(4) *Let* B *be any cycle or path of integration on* Σ *that avoids the branch points and the singularities of* Σ. *Then there exists a constant* $C_B > 0$ *(dependent on* Σ, B *and* C_ξ*) so that we have for every* ω *of the type considered here*

$$\left| \int_B \omega \right| \leq C_B .$$

Proof For (1). We again abbreviate $\Phi := \Phi_{\xi,n,\varrho=-1}$, $f := f_{\xi,n,\varrho=-1} = \frac{\Phi}{\mu-\mu^{-1}}$ and $\omega = f\,\mathrm{d}\lambda$, and use the quantities defined in Propositions 17.2 and 17.3.

By Corollary 10.3(1) we have for $\lambda \in U_{n,\delta}$:

$$\begin{aligned} &\text{if } n>0: \left| \frac{\Phi(\lambda)}{\lambda-\lambda_{n,0}} - \frac{(-1)^n\,\tau_\xi}{8} \right| \leq C_1\,\frac{|\lambda-\lambda_{n,0}|}{n} + r_n \\ &\text{if } n<0: \left| \frac{\Phi(\lambda)}{\lambda-\lambda_{n,0}} - \frac{(-1)^n\,\tau_\xi^{-1}\,\lambda_{n,0}^{-1}}{8} \right| \leq C_1\,|\lambda-\lambda_{n,0}|\,n^5 + r_n \end{aligned} \tag{17.20}$$

with a sequence $(r_n) \in \ell^2_{0,-2}(n)$. One possible choice of (r_n) for given (ξ_k) is

$$r_n = \frac{1}{8}\,\lambda_{n,0}^{-1} + \begin{cases} C_2 \cdot \left(a_k * \frac{1}{|k|}\right)_n & \text{for } n>0 \\ C_2\,n^2 \cdot \left(a_k * \frac{1}{|k|}\right)_n & \text{for } n<0 \end{cases}$$

with

$$a_k := \begin{cases} k^{-1}\cdot|\xi_k - \lambda_{k,0}| & \text{for } k>0 \\ k^3\cdot|\xi_k-\lambda_{k,0}| & \text{for } k<0 \end{cases}.$$

Because of

$$|\xi_k - \lambda_{k,0}| \leq |\xi_k - \varkappa_{k,*}| + |\varkappa_{k,*} - \lambda_{k,0}| \leq C_\xi \cdot |\varkappa_{k,1} - \varkappa_{k,2}| + |\varkappa_{k,*} - \lambda_{k,0}|,$$

it follows that the sequence (a_k) and therefore also the sequence (r_n) can be chosen such as to depend only on C_ξ and the $\varkappa_{k,\nu}$.

Moreover, there exists constants $C_3, C_4 > 0$ (dependent only on δ) so that we have for $\lambda \in U_{n,\delta}$

$$\begin{aligned}
\text{if } n > 0: \left|\frac{16\pi^2 n^2}{\lambda} - 1\right| &= \left|\frac{\lambda - 16\pi^2 n^2 - \lambda}{\lambda}\right| \leq C_3 \cdot \frac{|\lambda - \lambda_{n,0}|}{\lambda_{n,0}} \\
&\leq C_4 \cdot \frac{|\lambda - \lambda_{n,0}|}{n} \\
\text{if } n < 0: \left|\frac{1}{\lambda} - \frac{1}{\lambda_{n,0}}\right| &= \left|\frac{\lambda - \lambda_{n,0}}{\lambda \cdot \lambda_{n,0}}\right| \leq C_3 \cdot \frac{|\lambda - \lambda_{n,0}|}{\lambda_{n,0}^2} \leq C_4 \cdot |\lambda - \lambda_{n,0}| \cdot n^5.
\end{aligned} \tag{17.21}$$

By applying the asymptotic estimates (17.20) and (17.21) to Eq. (17.8), we obtain with another constant $C_5 > 0$

$$\begin{aligned}
&\text{if } n > 0: \left|\Phi(\lambda) - \left(-\frac{(-1)^n}{2}\right)\right| \leq C_5 \left(\frac{|\lambda - \lambda_{n,0}|}{n} + r_n\right) \\
&\text{if } n < 0: \left|\Phi(\lambda) - \frac{(-1)^n}{2\,\tau_\xi^2} \cdot \lambda_{n,0}^{-2}\right| \leq C_5 \cdot \left(|\lambda - \lambda_{n,0}| \cdot n^7 + r_n \cdot n^2\right),
\end{aligned}$$

and therefore by multiplication with $\frac{1}{|\mu - \mu^{-1}|}$

$$\begin{aligned}
&\text{if } n > 0: \left|\frac{\Phi(\lambda)}{\mu - \mu^{-1}} - \left(-\frac{(-1)^n}{2}\right)\frac{1}{\mu - \mu^{-1}}\right| \leq C_5 \left(\frac{|\lambda - \lambda_{n,0}|}{n} + r_n\right)\frac{1}{|\mu - \mu^{-1}|} \\
&\text{if } n < 0: \left|\frac{\Phi(\lambda)}{\mu - \mu^{-1}} - \frac{(-1)^n}{2\,\tau_\xi^2} \cdot \lambda_{n,0}^{-2} \cdot \frac{1}{\mu - \mu^{-1}}\right| \leq C_5 \cdot \left(|\lambda - \lambda_{n,0}| \cdot n^7 + r_n \cdot n^2\right)\frac{1}{|\mu - \mu^{-1}|}.
\end{aligned}$$

By integration along a given path from (λ_n^o, μ_n^o) to (λ_n, μ_n) we obtain

$$\begin{aligned}
&\text{if } n > 0: \left|\int_{(\lambda_n^o,\mu_n^o)}^{(\lambda_n,\mu_n)} \frac{\Phi(\lambda)}{\mu-\mu^{-1}}\,d\lambda - \left(-\tfrac{(-1)^n}{2}\right)\int_{(\lambda_n^o,\mu_n^o)}^{(\lambda_n,\mu_n)} \frac{1}{\mu-\mu^{-1}}\,d\lambda\right| \leq s_n \cdot \int_{(\lambda_n^o,\mu_n^o)}^{(\lambda_n,\mu_n)} \frac{1}{|\mu-\mu^{-1}|}\,d\lambda \\
&\text{if } n < 0: \left|\int_{(\lambda_n^o,\mu_n^o)}^{(\lambda_n,\mu_n)} \frac{\Phi(\lambda)}{\mu-\mu^{-1}}\,d\lambda - \tfrac{(-1)^n}{2\,\tau_\xi^2} \cdot \lambda_{n,0}^{-2} \cdot \int_{(\lambda_n^o,\mu_n^o)}^{(\lambda_n,\mu_n)} \frac{1}{\mu-\mu^{-1}}\,d\lambda\right| \leq s_n \cdot \int_{(\lambda_n^o,\mu_n^o)}^{(\lambda_n,\mu_n)} \frac{1}{|\mu-\mu^{-1}|}\,d\lambda,
\end{aligned}$$

where we define the sequence $(s_n) \in \ell^2_{0,-4}(n)$ by

$$s_n := \begin{cases} C_5 \cdot \left(\frac{1}{n} \max_{\lambda \in U_{n,\delta}} |\lambda - \lambda_{n,0}| + r_n\right) & \text{if } n > 0 \\ C_5 \cdot \left(n^7 \cdot \max_{\lambda \in U_{n,\delta}} |\lambda - \lambda_{n,0}| + r_n \cdot n^2\right) & \text{if } n < 0 \end{cases}.$$

Let $m \in \mathbb{Z}$ be the winding number of the given path of integration around the pair of branch points $\varkappa_{n,1}$, $\varkappa_{n,2}$. Then we may suppose without loss of generality that the path of integration is composed of an admissible path (Definition 16.2) and $|m|$ circles (with the orientation given by the sign of m) around $\varkappa_{n,1}$, $\varkappa_{n,2}$. These circles are also admissible paths, and therefore it follows from Proposition 16.5(1),(2) that there exists another sequence $(t_k) \in \ell^2_{-1,3}(k)$ that depends only on Σ so that we have for $n > 0$

$$\begin{aligned} &\left| \int_{(\lambda_n^o,\mu_n^o)}^{(\lambda_n,\mu_n)} \frac{\Phi(\lambda)}{\mu - \mu^{-1}}\, d\lambda - (-2i)\, \lambda_{n,0}^{1/2} \ln\left(\frac{\lambda_n - \varkappa_{n,*} + \Psi_n(\lambda_n, \mu_n)}{\lambda_n^o - \varkappa_{n,*} + \Psi_n(\lambda_n^o, \mu_n^o)} \right) \right| \\ &\quad \le C_6 \cdot (\lambda_{n,0}^{1/2} s_n + |\lambda_n - \varkappa_{n,1}| + t_n) \cdot \left| \ln\left(\frac{\lambda_n - \varkappa_{n,*} + \Psi_n(\lambda_n, \mu_n)}{\lambda_n^o - \varkappa_{n,*} + \Psi_n(\lambda_n^o, \mu_n^o)} \right) \right| \\ &\quad \le b_n \cdot \left| \ln\left(\frac{\lambda_n - \varkappa_{n,*} + \Psi_n(\lambda_n, \mu_n)}{\lambda_n^o - \varkappa_{n,*} + \Psi_n(\lambda_n^o, \mu_n^o)} \right) \right| \end{aligned}$$

and for $n < 0$

$$\begin{aligned} &\left| \int_{(\lambda_n^o,\mu_n^o)}^{(\lambda_n,\mu_n)} \frac{\Phi(\lambda)}{\mu - \mu^{-1}}\, d\lambda - \frac{2i}{\tau_\xi^2} \cdot \lambda_{n,0}^{-1/2} \cdot \ln\left(\frac{\lambda_n - \varkappa_{n,*} + \Psi_n(\lambda_n, \mu_n)}{\lambda_n^o - \varkappa_{n,*} + \Psi_n(\lambda_n^o, \mu_n^o)} \right) \right| \\ &\quad \le C_6 \cdot (s_n\, \lambda_{n,0}^{3/2} + (|\lambda_n - \varkappa_{n,1}| + t_n)\, \lambda_{n,0}^{-2}) \cdot \left| \ln\left(\frac{\lambda_n - \varkappa_{n,*} + \Psi_n(\lambda_n, \mu_n)}{\lambda_n^o - \varkappa_{n,*} + \Psi_n(\lambda_n^o, \mu_n^o)} \right) \right| \\ &\quad \le b_n \cdot \left| \ln\left(\frac{\lambda_n - \varkappa_{n,*} + \Psi_n(\lambda_n, \mu_n)}{\lambda_n^o - \varkappa_{n,*} + \Psi_n(\lambda_n^o, \mu_n^o)} \right) \right|, \end{aligned}$$

where $(b_n) \in \ell^2_{-1,-1}(n)$ is defined by

$$b_n := \begin{cases} C_6 \cdot (\lambda_{n,0}^{1/2} s_n + |\lambda_n - \varkappa_{n,1}| + t_n) & \text{for } n > 0 \\ C_6 \cdot (s_n\, \lambda_{n,0}^{3/2} + (|\lambda_n - \varkappa_{n,1}| + t_n)\, \lambda_{n,0}^{-2}) & \text{for } n < 0 \end{cases}.$$

Thus the claimed statement follows.

For (2). To simplify notation, we consider only the case $n, k > 0$ in the sequel. The other three cases with respect to the signs of n and k are handled analogously. Because of the hypothesis on the path of integration, we may suppose without loss of generality that the path of integration from $\varkappa_{k,1}$ to (λ_k, μ_k) is the lift of the straight line $[\varkappa_{k,1}, \lambda_k]$ in $\mathbb{C}^*$; this is in particular an admissible path in the sense of Definition 16.2.

By Proposition 17.2(3) there exist constants $C_1, C_2 > 0$ (depending only on C_ξ and R_0) so that we have for $\lambda \in [\varkappa_{k,1}, \lambda_k]$

$$\left| \frac{\Phi(\lambda)}{\lambda - \xi_k} - \frac{8(-1)^k \pi^2 n^2}{\xi_k \cdot (\lambda_{n,0} - \xi_k)} \right| \leq \frac{n^2}{k^2 \cdot |n^2 - k^2|} \cdot \left(C_1 \cdot \frac{|\lambda - \xi_k|}{k} + C_2 \cdot \left(b_j * \frac{1}{|j|} \right)_k \right)$$

with

$$b_j := \begin{cases} j^{-1}\, |\xi_j - \lambda_{j,0}| & \text{if } j > 0 \\ j^3\, |\xi_j - \lambda_{j,0}| & \text{if } j < 0 \end{cases}.$$

We have

$$\begin{aligned} |\lambda - \xi_k| &\leq |\lambda - \varkappa_{k,1}| + |\varkappa_{k,1} - \varkappa_{k,*}| + |\varkappa_{k,*} - \xi_k| \\ &\leq R_0 \cdot k + \tfrac{1}{2} |\varkappa_{k,1} - \varkappa_{k,2}| + C_\xi\, |\varkappa_{k,1} - \varkappa_{k,2}| \\ &\leq R_0 \cdot k + (C_\xi + \tfrac{1}{2}) \cdot |\varkappa_{k,1} - \varkappa_{k,2}| \end{aligned}$$

and

$$|\xi_j - \lambda_{k,0}| \leq |\xi_j - \varkappa_{k,*}| + |\varkappa_{k,*} - \lambda_{k,0}| \leq C_\xi \cdot |\varkappa_{k,1} - \varkappa_{k,2}| + |\varkappa_{k,*} - \lambda_{k,0}|.$$

Therefore it follows that there exists a constant $C_3 > 0$ (dependent only on R_0, C_ξ and the $\varkappa_{k,\nu}$) so that we have

$$\left| \frac{\Phi(\lambda)}{\lambda - \xi_k} \right| \leq C_3 \cdot \frac{n^2}{k^2 \cdot |n^2 - k^2|}.$$

By multiplying this inequality with $\frac{\lambda - \xi_k}{\mu - \mu^{-1}}$, we obtain:

$$\left| \frac{\Phi(\lambda)}{\mu - \mu^{-1}} \right| \leq C_3 \cdot \frac{n^2}{k^2 \cdot |n^2 - k^2|} \cdot \frac{|\lambda - \xi_k|}{|\mu - \mu^{-1}|}.$$

By integration along the straight line from $\varkappa_{k,1}$ to (λ_k, μ_k) this yields

$$\left| \int_{\varkappa_{k,1}}^{(\lambda_k, \mu_k)} \frac{\Phi(\lambda)}{\mu - \mu^{-1}}\, d\lambda \right| \leq C_3 \cdot \frac{n^2}{k^2 \cdot |n^2 - k^2|} \cdot \int_{\varkappa_{k,1}}^{(\lambda_k, \mu_k)} \frac{|\lambda - \xi_k|}{|\mu - \mu^{-1}|}\, |d\lambda|.$$

Because $\left| \frac{\xi_k - \varkappa_{k,*}}{\varkappa_{k,1} - \varkappa_{k,2}} \right| \leq C_\xi$ holds, Proposition 16.5(3) implies that there exists a constant $C_4 > 0$ (depending only on C_ξ) so that we have

$$\int_{\varkappa_{k,1}}^{(\lambda_k, \mu_k)} \frac{|\lambda - \xi_k|}{|\mu - \mu^{-1}|}\, |d\lambda| \leq C_4 \cdot k \cdot |\lambda_k - \varkappa_{k,1}|,$$

and from this estimate we obtain with $C_5 := C_3 \cdot C_4$

$$\left| \int_{\varkappa_{k,1}}^{(\lambda_k,\mu_k)} \frac{\Phi(\lambda)}{\mu - \mu^{-1}} \, \mathrm{d}\lambda \right| \leq C_5 \cdot \frac{n^2}{k \cdot |n^2 - k^2|} \cdot |\lambda_k - \varkappa_{k,1}| .$$

In consideration of the definition of (a_k), this gives the claimed estimate.

For (3). Let us consider the case $n > 0$. Then we have by (2) with the quantities C and (a_k) defined there:

$$\begin{aligned}
\sum_{k \in \mathbb{Z} \setminus \{n\}} \left| \int_{\varkappa_{k,1}}^{(\lambda_k,\mu_k)} \omega \right| &= \sum_{\substack{k \geq 0 \\ k \neq n}} \left| \int_{\varkappa_{k,1}}^{(\lambda_k,\mu_k)} \omega \right| + \sum_{k<0} \left| \int_{\varkappa_{k,1}}^{(\lambda_k,\mu_k)} \omega \right| \\
&\leq \sum_{\substack{k \geq 0 \\ k \neq n}} C \frac{n^2}{|n^2 - k^2|} a_k + \sum_{k<0} \frac{C}{k^2} a_k \\
&= C\, n^2 \sum_{\substack{k \geq 0 \\ k \neq n}} \underbrace{\frac{1}{n+k}}_{\leq 1/n} a_k \frac{1}{|n-k|} + C \sum_{k<0} \frac{1}{k^2} a_k \\
&\leq C\, n \left(a_k * \frac{1}{|k|} \right)_n + C \left\| \frac{1}{k^2} \right\|_{\ell^2} \cdot \|a_k\|_{\ell^2} \\
&\leq C\, n \left(a_k * \frac{1}{|k|} \right)_n + C_6
\end{aligned}$$

with a constant $C_6 > 0$. The case $n < 0$ is treated analogously.

For (4). We let c_ξ be as in the proof of Proposition 17.2(1). Then there exists a constant $C_7 > 0$ (dependent only on B and C_ξ) so that

$$|c_\xi(\lambda)| \leq C_7 \cdot |c_0(\lambda)|$$

holds for all (λ, μ) on the trace of B. By Eq. (17.8) it follows that there exists a constant $C_8 > 0$ so that

$$|\Phi(\lambda)| \leq C_8 \cdot |c_\xi(\lambda)| \leq C_8 \cdot C_7 \cdot |c_0(\lambda)|$$

holds for all (λ, μ) on the trace of B. We thus have

$$\left| \int_B \omega \right| \leq \int_B \frac{|\Phi(\lambda)|}{|\mu - \mu^{-1}|} \, |\mathrm{d}\lambda| \leq C_8 \cdot C_7 \cdot C_9 \cdot C_{10}$$

with

$$C_9 := \max_{\lambda \in \mathrm{tr}(B)} |c_0(\lambda)| \quad \text{and} \quad C_{10} := \int_B \frac{1}{|\mu - \mu^{-1}|} \, d\lambda .$$

Both C_9 and C_{10} are finite (positive) numbers that depend only on B, because the trace of B is compact and avoids the zeros of $\mu - \mu^{-1}$. This yields the claimed estimate with $C_B := C_7 \cdot C_8 \cdot C_9 \cdot C_{10}$. □

Remark 17.5 In the setting of the preceding propositions, suppose that $\frac{\Phi_{n,\xi,\varrho}}{\mu - \mu^{-1}} \, d\lambda$ is square-integrable, and therefore $\varrho = -1$ holds by Proposition 17.3(2). By Proposition 17.2(1) we then obtain $\Phi_{n,\xi,\varrho} \in \mathrm{As}(\mathbb{C}^*, \ell^\infty_{3,-1}, 1)$. Note that this is an improved statement for 1-forms constructed in the way of Propositions 17.2–17.4 over the general statement $\Phi \in \mathrm{As}(\mathbb{C}^*, \ell^2_{1,-3}, 1)$ in Proposition 17.1(1), because of $\ell^\infty_{3,-1} \subsetneq \ell^2_{1,-3}$.

Chapter 18
Construction of the Jacobi Variety for the Spectral Curve

In the present chapter, we will construct Jacobi coordinates for the spectral curve Σ, which can have singularities, is non-compact, and has always infinite arithmetic genus and generally infinite geometric genus. For the construction of the Jacobi variety, the estimates proven in the preceding two chapters will play an important role.

As an introduction, we review the construction of the Jacobi variety for *compact* Riemann surfaces (see for example [Fa-K, Section III.6]): Let X be a compact Riemann surface, say of genus $g \geq 1$, and let $(A_k, B_k)_{k=1,\dots,g}$ be a canonical homology basis of X, i.e. (A_k, B_k) is a basis of the homology group $H_1(X, \mathbb{Z})$ with the intersection properties $A_k \times B_\ell = \delta_{k\ell}$ (Kronecker delta), $A_k \times A_\ell = 0 = B_k \times B_\ell$ for $k, \ell \in \{1, \dots, g\}$. Then there exists a canonical basis $(\omega_k)_{k=1,\dots,g}$ of the vector space $\Omega(X)$ of holomorphic 1-forms on X that is dual to (A_k) in the sense that $\int_{A_k} \omega_\ell = \delta_{k\ell}$ holds. To any given positive divisor $D = \{P_1, \dots, P_g\}$ of degree g on X, we then associate the quantity

$$\widetilde{\varphi}(D) := \left(\sum_{k=1}^{g} \int_{P_0}^{P_k} \omega_\ell \right)_{\ell=1,\dots,g} \in \mathbb{C}^g,$$

where $P_0 \in X$ is the "origin point", which we hold fixed. Because these integrals depend on the homology class of the paths of integration from P_0 to P_k we choose, the quantity $\widetilde{\varphi}(D)$ is well-defined only modulo the *period lattice*

$$\Gamma := \left\langle \left(\int_{A_k} \omega_\ell \right)_{\ell=1,\dots,g}, \left(\int_{B_k} \omega_\ell \right)_{\ell=1,\dots,g} \right\rangle_{\mathbb{Z}} \subset \mathbb{C}^g.$$

S. Klein, *A Spectral Theory for Simply Periodic Solutions of the Sinh-Gordon Equation*, Lecture Notes in Mathematics 2229,
https://doi.org/10.1007/978-3-030-01276-2_18

Thus we obtain the *Jacobi variety* $\mathrm{Jac}(X) := \mathbb{C}^g/\Gamma$ of X, and by projecting the values of $\widetilde{\varphi}$ onto $\mathrm{Jac}(X)$ the *Abel map*

$$\varphi : \mathrm{Div}_g(X) \to \mathrm{Jac}(X),$$

where $\mathrm{Div}_g(X)$ denotes the space of positive divisors of degree g on X.

To obtain a Jacobi variety and an Abel map for the spectral curve Σ by this strategy, we need to generalize the construction described above in two directions: First we need to deal with the fact that Σ is not compact and its homology group is generally infinite dimensional, and second we need to handle the fact that Σ can have singularities.

As consequence of the fact that the homology group of Σ is infinite dimensional, the space of square-integrable, holomorphic[1] 1-forms on Σ is also infinite dimensional. Therefore the space $\mathbb{C}^g$ occurring in the treatment of the compact case as the universal cover (or the tangent space) of the Jacobi variety needs to be replaced by a suitable Banach space adapted to Σ, the period lattice Γ will also be infinite dimensional, the positive divisors of genus g will likewise be replaced by positive divisors of infinitely many points, and the sum of integrals defining the Abel map will be a sum of infinitely many terms. To make sure that the latter sum converges absolutely, we need to impose an asymptotic condition on the divisors we consider, and that condition of course is the asymptotic condition for the space Div, see Definition 8.3. For such divisors, we will use the estimates of the preceding two chapters to show that the Abel map, which is now defined for $D = \{(\lambda_k, \mu_k)\} \in \mathsf{Div}$ via

$$\widetilde{\varphi}(D) := \left(\sum_{k\in\mathbb{Z}} \int_{(\lambda_k^o, \mu_k^o)}^{(\lambda_k, \mu_k)} \omega_\ell \right)_{\ell\in\mathbb{Z}}$$

with fixed points $(\lambda_k^o, \mu_k^o) \in \widehat{U}_{k,\delta}$, is indeed convergent and maps into the right Banach space.

Regarding singularities of Σ there is the problem that if (λ_*, μ_*) is a singular point of Σ and A_k is a member of the homology basis of Σ that encircles this point, then the corresponding member ω_k of the canonical basis of the 1-forms on Σ will not be holomorphic at (λ_*, μ_*), but only regular in the sense of Serre [Se, IV.9, p. 68ff.]. Moreover, ω_k will not be square-integrable near (λ_*, μ_*). To handle this phenomenon, we follow the approach of Rosenlicht [Ro] who constructed generalized Jacobi varieties by admitting only divisors whose support is contained in the regular set Σ' of Σ, see also [Se, Chapter V]. (Rosenlicht in fact considered a more general setting where he constructed a family of generalized Jacobi maps depending on a modulus $\mathfrak{m}$ on the singular set of Σ, but for our purposes the

[1] Again the concepts of square-integrability and holomorphy need to be modified near singularities of Σ.

case $\mathfrak{m} = 0$ will suffice.) In this sense we will in fact construct the Jacobi variety of Σ'. Considering only divisors whose support avoids the singular points of Σ is actually sufficient for our purposes: The main reason why we are interested in Jacobi coordinates for the spectral curve Σ is to describe the motion of spectral divisors under translation of the potential (this motion turns out to be linear in the Jacobi coordinates), and because divisor points that are singular points of Σ remain stationary under translation, it suffices to consider the regular points of the divisor in this context.

We continue to use the notations of the preceding two chapters. In particular Σ is a spectral curve defined via Eq. (3.8) by a holomorphic function $\Delta : \mathbb{C}^* \to \mathbb{C}$ with $\Delta - \Delta_0 \in \mathrm{As}(\mathbb{C}^*, \ell^2_{0,0}, 1)$, we denote the zeros of $\Delta^2 - 4$ by $\varkappa_{k,\nu}$ as before, interpret $\varkappa_{k,\nu}$ also as points on Σ and put $\varkappa_{k,*} := \frac{1}{2}(\varkappa_{k,1} + \varkappa_{k,2})$. We continue to exclude singularities of higher order on Σ as we did in Chap. 17 by (17.1), i.e. we require that

$$\Delta^2 - 4 \text{ does not have any zeros of order } \geq 3.$$

As in Chap. 17 (see before Eq. (17.2)) we then suppose that the zeros $\varkappa_{k,\nu}$ of $\Delta^2 - 4$ (corresponding to the branch points and singularities of Σ) are numbered in such a way that if $\varkappa$ is a zero of order 2 of $\Delta^2 - 4$, then we have $\varkappa_{k,1} = \varkappa_{k,2} = \varkappa$ for some $k \in \mathbb{Z}$.

Moreover we again consider the possibly punctured excluded domains $U'_{k,\delta}$ and $\widehat{U}'_{k,\delta}$ defined by Eq. (16.4), the set $S \subset \mathbb{Z}$ of indices for which $\varkappa_{k,1} = \varkappa_{k,2}$ is a double point of Σ (Eq. (17.2)) and the regular set $\Sigma' = \Sigma \setminus \{ \varkappa_{k,\nu} \mid k \in S \}$ of Σ. We also consider the spaces $\mathsf{Div}(\Sigma)$ and $\mathsf{Div}(\Sigma')$ of (classical) asymptotic divisors with support in Σ resp. in Σ'.

We now fix a basis of the homology of Σ'. For every $k \in \mathbb{Z}$ there is a non-trivial cycle A_k in Σ' encircling the pair of points $(\varkappa_{k,1}, \varkappa_{k,2})$. Note that in Σ' (unlike in Σ) this is true regardless of whether $k \notin S$ (then $\varkappa_{k,1}, \varkappa_{k,2}$ is a pair of branch points) or $k \in S$ (then $\varkappa_{k,1} = \varkappa_{k,2}$ is a puncture of Σ') holds. For $k \neq 0$, there exists only for $k \notin S$ another cycle B_k that encircles the pair of points $(\varkappa_{k,1}, \varkappa_{k=0,1})$. A final cycle B_0 comes from the observation that $\sqrt{\lambda}$ is a global parameter on $\widehat{V}_\delta$. Because the Riemann surface associated to $\sqrt{\lambda}$ has branch points in $\lambda = 0$ and $\lambda = \infty$, we see that there is another non-trivial cycle B_0 in $\widehat{V}_\delta \subset \Sigma'$ encircling these two branch points. Of these cycles we require that their intersection numbers satisfy $A_k \times A_\ell = 0 = B_k \times B_\ell$ and $A_k \times B_\ell = \delta_{k\ell}$ (Kronecker delta) for all $k, \ell \neq 0$, then we call (A_k, B_k) a *canonical basis of the homology* of Σ'. We can choose the cycles such that for $|k|$ large, A_k runs entirely in $\widehat{U}'_{k,\delta}$, and B_k runs in $\widehat{U}'_{k,\delta} \cup \widehat{U}'_{k=0,\delta} \cup \widehat{V}_\delta$. See Fig. 18.1.

We note that with the canonical basis (A_k, B_k) defined in this way, we have for any 1-form $\omega \in \Omega(\Sigma)$ and any $k \in \mathbb{Z} \setminus S$

$$\int_{A_k} \omega = 2 \cdot \int_{\varkappa_{k,1}}^{\varkappa_{k,2}} \omega, \tag{18.1}$$

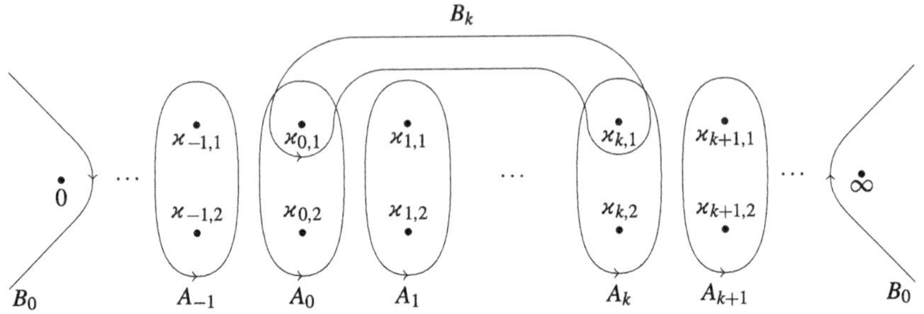

Fig. 18.1 The canonical basis (A_k, B_k) of the homology of Σ'

where on the right hand side the path of integration is the lift of the straight line $[\varkappa_{k,1}, \varkappa_{k,2}]$ to Σ.

In the case $k \in S$, we have an analogue to the preceding equation: Consider a meromorphic 1-form ω on Σ that is regular (in the sense of Serre) at $\varkappa_{k,*}$; the latter means that regarded as a meromorphic 1-form on the normalization $\widehat{\Sigma}$ of Σ, the pole order of ω at each of the two points above $\varkappa_{k,*}$ in $\widehat{\Sigma}$ is at most 1 (the order of the zero of $\mu - \mu^{-1}$ there), and that the sum of the residues of ω in these points is zero, see [Se, IV.9, p. 68ff.]. Note that for $n \in S$ the 1-form ω constructed in Proposition 17.4 is regular at $\varkappa_{n,*}$ and holomorphic on $\Sigma \setminus \{\varkappa_{n,*}\}$. If any meromorphic 1-form ω is regular at $\varkappa_{k,*}$, then we have

$$\int_{A_k} \omega = 0 \iff \omega \text{ is holomorphic at } \varkappa_{k,*}, \tag{18.2}$$

where for the question of holomorphy at $\varkappa_{k,*}$, ω is again regarded as a 1-form on the normalization $\widehat{\Sigma}$.

Our first step on the way towards Jacobi coordinates for Σ' is to obtain a canonical basis $(\omega_n)_{n\in\mathbb{Z}}$ of holomorphic 1-forms $\omega_n \in \Omega(\Sigma')$ that is dual to the basis of the homology (A_k, B_k) in the sense that for all $k, n \in \mathbb{Z}$

$$\int_{A_k} \omega_n = \delta_{kn}$$

holds. It will turn out that for $n \in \mathbb{Z} \setminus S$, ω_n is holomorphic on Σ, and for $n \in S$, ω_n is holomorphic on $\Sigma \setminus \{\varkappa_{n,*}\}$ and regular in $\varkappa_{n,*}$. For $n \notin S$ we will have $\omega_n \in L^2(\Sigma, T^*\Sigma)$, whereas for $n \in S$, we will have $\omega_n \in L^2(\Sigma \setminus \widehat{U}_{n,\delta}, T^*(\Sigma \setminus \widehat{U}_{n,\delta}))$ for every $\delta > 0$.

In the case $S = \varnothing$, the existence of such a basis follows from general results on open Riemann surfaces (of infinite genus) which are parabolic in the sense of Ahlfors and Nevanlinna, as they are described in [Fe-K-T, Chapter 1]. Indeed Σ is then parabolic (see the proof of Proposition 18.1 below), so the existence of the basis (ω_n) follows from [Fe-K-T, Theorem 3.8, p. 28]. However, this argument does not apply to the case $S \neq \varnothing$, where Σ is no longer smooth. Moreover it turns out that even for $S = \varnothing$ the information on the asymptotic behavior of ω_k we obtain from this general result (by Proposition 17.1(1)) is not sufficient to prove that the infinite sum defining the Jacobi coordinates converges. Therefore we need to carry out an explicit construction of the ω_n.

As preparation we need one consequence of the general results on parabolic Riemann surfaces:

Proposition 18.1 *Suppose that $\omega_1, \omega_2 \in \Omega(\Sigma) \cap L^2(\Sigma, T^*\Sigma)$ are given so that*

$$\int_{A_k} \omega_1 = \int_{A_k} \omega_2$$

holds for all $k \in \mathbb{Z}$. Then we have $\omega_1 = \omega_2$.

Remark 18.2 This proposition generalizes to 1-forms that are holomorphic on Σ with the exception of finitely many poles in singular points of Σ, that are regular in these poles, and that are square-integrable on $\Sigma \setminus \bigcup_{k \in I} \widehat{U}_{k,\delta}$ (where $I \subset S$ is the finite set of indices $k \in S$ for which $\widehat{U}_{k,\delta}$ contains a pole of the 1-form). The reason is that if ω_1, ω_2 are 1-forms of this kind with $\int_{A_k}(\omega_1 - \omega_2) = 0$ for all $k \in \mathbb{Z}$, then this hypothesis implies that $\omega_1 - \omega_2$ is holomorphic on all of Σ by (18.2); because $\bigcup_{k \in I} \widehat{U}_{k,\delta}$ is relatively compact in Σ, this also means that $\omega_1 - \omega_2$ is square-integrable on Σ. Proposition 18.1 then shows that $\omega_1 - \omega_2 = 0$ holds.

Note that this generalization is applicable to the members ω_n of the canonical basis constructed in Theorem 18.3 for $n \in S$, because they have a regular pole in $\varkappa_{n,*}$ and are otherwise holomorphic on Σ, and they are square-integrable on $\Sigma \setminus \widehat{U}_{n,\delta}$.

Proof (of Proposition 18.1) We first note that the normalization $\widehat{\Sigma}$ of Σ is parabolic in the sense of Ahlfors and Nevanlinna; this means that $\widehat{\Sigma}$ has a harmonic exhaustion function h, i.e. h is a continuous, proper, non-negative function on $\widehat{\Sigma}$ that is harmonic on the complement of a compact subset (see [Fe-K-T, Definition 3.1, p. 25]). Indeed, $h := \left|\log(|\lambda|)\right|$ is such a function on $\widehat{\Sigma}$, so $\widehat{\Sigma}$ is parabolic.

It follows by [Fe-K-T, Proposition 3.6, p. 27] that $\widehat{\Sigma}$ then also has an exhaustion function with finite charge, whence it follows by [Fe-K-T, Proposition 2.10, p. 22] that for any square-integrable, holomorphic 1-form $\widehat{\omega}$ on $\widehat{\Sigma}$, the condition $\int_{A_k} \widehat{\omega} = 0$ for all $k \in \mathbb{Z}$ implies $\widehat{\omega} = 0$. Because $\widehat{\Sigma} \to \Sigma$ is a one-sheeted covering, we have the analogous statement for Σ: If ω is a square-integrable, holomorphic 1-form on Σ (where for the question of holomorphy we again consider ω on $\widehat{\Sigma}$ in the singular points of Σ) and $\int_{A_k} \omega = 0$ holds for all $k \in \mathbb{Z}$, then we have $\omega = 0$.

By applying this statement to $\omega := \omega_1 - \omega_2$, the claimed result follows. □

We next proceed to the explicit construction of the canonical basis (ω_n) of the holomorphic 1-forms. As explained above, we need this explicit representation to be able to apply the results of Chap. 16 to obtain a finer description of the asymptotic behavior of the ω_n; it is also needed for the treatment of the case $S \neq \varnothing$.

The strategy for the explicit construction is to use the estimates of Proposition 16.5 and Propositions 17.2–17.4 to construct holomorphic 1-forms that have zeros in all excluded domains $\widehat{U}_{k,\delta}$ except for $k = n$, and to use an argument based on the Banach Fixed Point Theorem to adjust the position of these zeros such that $\int_{A_k} \omega_n = 0$ holds for $k \neq n$ with $|k|$ large. Multiplication with an appropriate constant factor ensures $\int_{A_n} \omega_n = 1$, whereas an additional (finite) linear combination of 1-forms of the described kind is needed to achieve $\int_{A_k} \omega_n = 0$ also for small values of $|k|$.

Theorem 18.3

(1) There exists $R_0 > 0$ so that for every $C_\xi \geq R_0$ there exists $N \in \mathbb{N}$ with the following property:
For every $n \in \mathbb{Z}$ and every given finite sequence $(\xi_{n,k})_{|k|\leq N, k\neq n}$ with $|\xi_{n,k} - \varkappa_{k,*}| \leq C_\xi \cdot |\varkappa_{k,1} - \varkappa_{k,2}|$ there exists an extension of this sequence to an infinite sequence $(\xi_{n,k})_{k\in\mathbb{Z}\setminus\{n\}}$ still with $|\xi_{n,k} - \varkappa_{k,*}| \leq C_\xi \cdot |\varkappa_{k,1} - \varkappa_{k,2}|$, so that:

(a) With $\widetilde{\Phi}_n := \Phi_{n,\xi_{n,k},\varrho=-1}$ (see Proposition 17.2) and $\widetilde{\omega}_n := \frac{\widetilde{\Phi}_n(\lambda)}{\mu-\mu^{-1}}\,\mathrm{d}\lambda$, we have:

(i) If $n \notin S$, then $\widetilde{\omega}_n \in \Omega(\Sigma) \cap L^2(\Sigma, T^*\Sigma)$ holds.
(ii) If $n \in S$, then $\widetilde{\omega}_n \in \Omega(\Sigma \setminus \{\varkappa_{n,*}\}) \cap L^2(\Sigma \setminus \widehat{U}_{n,\delta}, T^*(\Sigma \setminus \widehat{U}_{n,\delta}))$ for every $\delta > 0$, and $\widetilde{\omega}_n$ is regular in $\varkappa_{n,*}$.

(b) We have

$$\int_{A_k} \widetilde{\omega}_n = 0 \quad \textit{for every } k \in \mathbb{Z}\setminus\{n\} \textit{ with } (|k| > N \textit{ or } k \in S).$$

(2) Let $R_0 > 0$ be as in (1), $C_\xi \geq R_0$, and $N \in \mathbb{N}$ corresponding to C_ξ as in (1). We fix pairwise unequal $(\xi_k)_{|k|\leq N}$ so that $|\xi_k - \varkappa_{k,*}| \leq C_\xi \cdot |\varkappa_{k,1} - \varkappa_{k,2}|$ holds for all $|k| \leq N$. We apply (1) for every $n \in \mathbb{Z}$, yielding a sequence $(\xi_{n,k})_{k\in\mathbb{Z}\setminus\{n\}}$ with $\xi_{n,k} = \xi_k$ for $|k| \leq N$, $k \neq n$ so that the associated function $\widetilde{\Phi}_n$ and the 1-form $\widetilde{\omega}_n$ have the properties (a) and (b) from (1).
Then for every $n \in \mathbb{Z}$ there exist numbers $s_{n,\ell} \in \mathbb{C}$ for $|\ell| \leq N$ (where $s_{n,\ell} = 0$ for every $\ell \in S$ with $|\ell| \leq N$ and $\ell \neq n$) and for $|n| > N$ also $s_{n,n} \in \mathbb{C}$, such that the 1-form

$$\omega_n := \left\{ \begin{array}{ll} \sum_{|\ell|\leq N} s_{n,\ell}\,\widetilde{\omega}_\ell & \text{if } |n| \leq N \\ s_{n,n}\,\widetilde{\omega}_n + \sum_{|\ell|\leq N} s_{n,\ell}\,\widetilde{\omega}_\ell & \text{if } |n| > N \end{array} \right\} = \frac{\Phi_n(\lambda)}{\mu - \mu^{-1}}\,\mathrm{d}\lambda$$

with

$$\Phi_n := \begin{cases} \sum_{|\ell|\leq N} s_{n,\ell} \cdot \widetilde{\Phi}_\ell & \text{if } |n| \leq N \\ s_{n,n} \cdot \widetilde{\Phi}_n + \sum_{|\ell|\leq N} s_{n,\ell} \cdot \widetilde{\Phi}_\ell & \text{if } |n| > N \end{cases}$$

has the properties (1)(a)(i)–(ii) and satisfies

$$\int_{A_k} \omega_n = \delta_{kn} \quad \text{for all } k, n \in \mathbb{Z}.$$

Here we have $\Phi_n \in \mathrm{As}(\mathbb{C}^*, \ell^\infty_{3,-1}, 1)$.
The constants $s_{n,n}$ *have the following asymptotic behavior for* $|n| \to \infty$*:*

$$s_{n,n} = \begin{cases} \frac{1}{4\pi} \cdot \lambda_{n,0}^{-1/2} + \ell^2_1(n) & \text{for } n > N \\ \frac{1}{4\pi} \cdot \tau^2_{\xi_{n,k}} \cdot \lambda_{n,0}^{1/2} + \ell^2_1(n) & \text{for } n < -N \end{cases}, \tag{18.3}$$

and the constants $s_{n,\ell}$ *with* $|\ell| \leq N$ *are also of the order of* $|n|^{-1}$ *for* $n \to \pm\infty$*. We have* $s_{n,\ell} = 0$ *for* $\ell \in S$, $\ell \neq n$.

Remark 18.4 The 1-forms ω_n with $n \in \mathbb{Z}$ from Theorem 18.3(2) are for $S = \varnothing$ the canonical basis of $\Omega(\Sigma) \cap L^2(\Sigma, T^*\Sigma)$ associated to the basis of the homology (A_n, B_n) of Σ. If $n \in S$ holds, then ω_n is not holomorphic, but only regular in $\varkappa_{n,*}$, but $(\omega_n)_{n\in\mathbb{Z}}$ is still the appropriate replacement for the basis of $\Omega(\Sigma) \cap L^2(\Sigma, T^*\Sigma)$ in the case where the spectral curve Σ has singularities.

In either case, Theorem 18.3 shows that any ω_n is a finite linear combination of 1-forms of the form $\frac{\Phi_k(\lambda)}{\mu-\mu^{-1}}\, d\lambda$, where the holomorphic functions $\Phi_k : \mathbb{C}^* \to \mathbb{C}$ are infinite products of the type considered in Proposition 17.2, and therefore have asymptotically and totally exactly one zero in every excluded domain $\widehat{U}_{j,\delta}$, with the exception of $\widehat{U}_{k,\delta}$.

A natural question is if ω_n itself can be represented in the form $\frac{\Phi_n(\lambda)}{\mu-\mu^{-1}}\, d\lambda$ with the function Φ_n being an infinite product as in Proposition 17.2, i.e. without a linear combination. If this were the case, then ω_n would have asymptotically and totally exactly one zero in every excluded domain, except for $\widehat{U}_{n,\delta}$. Unfortunately however, the answer to this question is negative in general, as the following argument shows:

It follows from Proposition 17.2(1) that the holomorphic function Φ_n from Theorem 18.3(2) has the following asymptotic behavior: There exist constants $z_{n,\infty}, z_{n,0} \in \mathbb{C}$ such that the asymptotic relations $\Phi_n - z_{n,\infty}\,\Phi_0 \in \mathrm{As}_\infty(\mathbb{C}^*, \ell^2_3, 1)$ and $\Phi_n - z_{n,0}\Phi_0 \in \mathrm{As}_0(\mathbb{C}^*, \ell^2_{-1}, 1)$ hold with the comparison function $\Phi_0(\lambda) := \frac{c_0(\lambda)}{\lambda\cdot(\lambda-\lambda_{0,0})}$. If the constants $z_{n,\infty}$ and $z_{n,0}$ are both non-zero, then it can indeed be shown that $\Phi_n(\lambda)$ itself can be represented in the form $s_n \cdot \Phi_{n,\xi_{n,k},-1}$ with a constant $s_n \in \mathbb{C}^*$ and a sequence $(\xi_{n,k})_{k\in\mathbb{Z}\setminus\{n\}}$ with $|\xi_{n,k} - \varkappa_{k,*}| \leq C_\xi \cdot |\varkappa_{k,1} - \varkappa_{k,2}|$.

However the case that either of the constants $z_{n,\infty}$ or $z_{n,0}$ vanishes can occur, depending on the specific shape of the spectral curve Σ. (For $n \to \infty$, $z_{n,\infty}$ grows of the order n^2, and therefore is non-zero for sufficiently large n, but $z_{n,0}$ remains bounded, and there is no general reason preventing it from becoming zero. For $n \to -\infty$, the same is true with the roles of $z_{n,\infty}$ and $z_{n,0}$ reversed.) This corresponds to the case that one or more zeros $\xi_{n,k}$ from the compact part of Σ (i.e. with $|k|$ small) have moved out to ∞ resp. to 0. It is for this reason that a product representation of the form $s_n \cdot \Phi_{n,\xi_{n,k},-1}$ is not possible for Φ_n in general.

Proof (of Theorem 18.3) For (1). For an arbitrary sequence $(\xi_{n,k})$ with $\xi_{n,k}-\lambda_{k,0} \in \ell^2_{-1,3}(k)$, $\Phi_{n,\xi_{n,k}} := \Phi_{n,\xi_{n,k},\varrho=-1}$ is a holomorphic function by Proposition 17.2, and $\omega_{n,\xi_{n,k}} := \frac{\Phi_{n,\xi_{n,k}}(\lambda)}{\mu-\mu^{-1}}\,d\lambda$ is a holomorphic 1-form on Σ' by Proposition 17.4. If moreover $|\xi_{n,k} - \varkappa_{k,*}| \leq C_\xi \cdot |\varkappa_{k,1} - \varkappa_{k,2}|$ holds for some $C_\xi > 0$, then Proposition 17.3(2) yields that $\omega_{n,\xi_{n,k}} \in L^2(\Sigma, T^*\Sigma)$ holds if $n \notin S$, whereas $\omega_{n,\xi_{n,k}} \in L^2(\Sigma \setminus \widehat{U}_{n,\delta}, T^*(\Sigma \setminus \widehat{U}_{n,\delta}))$ holds for all $\delta > 0$ if $n \in S$. We also note that for $k \in S$ with $k \neq n$, the condition $|\xi_{n,k} - \varkappa_{k,*}| \leq C_\xi \cdot |\varkappa_{k,1} - \varkappa_{k,2}|$ implies $\xi_{n,k} = \varkappa_{k,*}$; therefore both $\Phi_{n,\xi_{n,k}}(\lambda)\,d\lambda$ and $\mu - \mu^{-1}$ have a zero of order 2 at $\varkappa_{k,*}$, whence it follows that $\omega_{n,\xi_{n,k}}$ is holomorphic in $\varkappa_{k,*}$. Thus $\int_{A_k} \omega_{n,\xi_{n,k}} = 0$ holds automatically for $k \in S \setminus \{n\}$ by (18.2).

In the sequel we fix $R_0 > 0$ and let $C_\xi \geq R_0$ be given. We also fix $N \in \mathbb{N}$. It will turn out in the course of the proof how to choose R_0 and N (the latter depending on C_ξ) so that the claimed statement holds. We will always choose N at least as large as the N in Lemma 16.1, so that Proposition 16.5(1),(2) is applicable to all $k \in \mathbb{Z}$ with $|k| > N$.

Moreover, we let $n \in \mathbb{Z}$ be given. We then consider the Banach space

$$\mathfrak{B} := \left\{ (a_k)_{k\in\mathbb{Z}\setminus\{n\}} \;\middle|\; \exists C \geq 0\, \forall k \in \mathbb{Z} \setminus \{n\} \,:\, |a_k| \leq C \cdot |\varkappa_{k,1} - \varkappa_{k,2}| \right\}$$

with the norm

$$\|a_k\|_{\mathfrak{B}} := \sup_{k\in\mathbb{Z}\setminus(\{n\}\cup S)} \left| \frac{a_k}{\varkappa_{k,1} - \varkappa_{k,2}} \right| \quad \text{for } (a_k) \in \mathfrak{B}.$$

Note that for any $(a_k) \in \mathfrak{B}$, we have $a_k = 0$ for all $k \in S$. We also suppose that a finite sequence $(\xi^{[0]}_{n,k})_{|k|\leq N, k\neq n}$ with $|\xi^{[0]}_{n,k} - \varkappa_{k,*}| \leq C_\xi \cdot |\varkappa_{k,1} - \varkappa_{k,2}|$ is given, and consider the translated subspace disc in $\mathfrak{B}$

$$\mathfrak{B}_{C_\xi} := \{ (\xi_{n,k})_{k\in\mathbb{Z}\setminus\{n\}} \mid (\xi_{n,k} - \varkappa_{k,*}) \in \mathfrak{B},\ \|\xi_{n,k} - \varkappa_{k,*}\|_{\mathfrak{B}} \leq C_\xi,\ \xi_{n,k} = \xi^{[0]}_{n,k} \text{ for } |k| \leq N \}.$$

We now associate to each sequence $(\xi_{n,k}) \in \mathfrak{B}_{C_\xi}$ a new sequence $(\widetilde{\xi}_{n,k})_{k\in\mathbb{Z}\setminus\{n\}}$ in the following way: For $k \in \mathbb{Z}\setminus\{n\}$, we define the holomorphic function $\Phi_{n,\xi_{n,k};k} : \mathbb{C}^* \to \mathbb{C}$ by

$$\Phi_{n,\xi_{n,k};k}(\lambda) := \frac{\Phi_{n,\xi_{n,k}}(\lambda)}{\lambda - \xi_{n,k}}.$$

We then define $\widetilde{\xi}_{n,k}$ in the following way: For $|k| \leq N$ we put $\widetilde{\xi}_{n,k} := \xi_{n,k} = \xi^{[0]}_{n,k}$, for $|k| > N$ with $k \in S$ we put $\widetilde{\xi}_{n,k} := \varkappa_{k,*}$, and for $|k| > N$ with $k \notin S$ we put

$$\widetilde{\xi}_{n,k} := \frac{\displaystyle\int_{A_k} \lambda \cdot \frac{\Phi_{n,\xi_{n,k};k}(\lambda)}{\mu - \mu^{-1}}\, d\lambda}{\displaystyle\int_{A_k} \frac{\Phi_{n,\xi_{n,k};k}(\lambda)}{\mu - \mu^{-1}}\, d\lambda} = \varkappa_{k,*} + \frac{\displaystyle\int_{A_k} (\lambda - \varkappa_{k,*}) \cdot \frac{\Phi_{n,\xi_{n,k};k}(\lambda)}{\mu - \mu^{-1}}\, d\lambda}{\displaystyle\int_{A_k} \frac{\Phi_{n,\xi_{n,k};k}(\lambda)}{\mu - \mu^{-1}}\, d\lambda}. \tag{18.4}$$

We will show below that (for suitable conditions on R_0 and N) the map $(\xi_{n,k}) \mapsto (\widetilde{\xi}_{n,k})$ we just defined has exactly one fixed point $(\xi^*_{n,k})$. The sequence $(\xi^*_{n,k})$ has the following property: For any $k \in \mathbb{Z}$ with $|k| > N$ and $k \notin S$ we have

$$\begin{aligned}
\int_{A_k} \frac{\Phi_{n,\xi^*_{n,k}}(\lambda)}{\mu-\mu^{-1}}\, d\lambda &= \int_{A_k} (\lambda - \xi^*_{n,k}) \cdot \frac{\Phi_{n,\xi^*_{n,k};k}(\lambda)}{\mu-\mu^{-1}}\, d\lambda \\
&= \int_{A_k} \lambda \cdot \frac{\Phi_{n,\xi^*_{n,k};k}(\lambda)}{\mu-\mu^{-1}}\, d\lambda - \underbrace{\xi^*_{n,k}}_{=\widetilde{\xi}^*_{n,k}} \cdot \int_{A_k} \frac{\Phi_{n,\xi^*_{n,k};k}(\lambda)}{\mu-\mu^{-1}}\, d\lambda \\
&= \int_{A_k} \lambda \cdot \frac{\Phi_{n,\xi^*_{n,k};k}(\lambda)}{\mu-\mu^{-1}}\, d\lambda - \frac{\displaystyle\int_{A_k} \lambda \cdot \frac{\Phi_{n,\xi^*_{n,k};k}(\lambda)}{\mu-\mu^{-1}}\, d\lambda}{\displaystyle\int_{A_k} \frac{\Phi_{n,\xi^*_{n,k};k}(\lambda)}{\mu-\mu^{-1}}\, d\lambda} \cdot \int_{A_k} \frac{\Phi_{n,\xi^*_{n,k};k}(\lambda)}{\mu-\mu^{-1}}\, d\lambda \\
&= 0.
\end{aligned} \tag{18.5}$$

Moreover, for any $k \in \mathbb{Z}$ with $|k| > N$ and $k \in S$, the 1-form $\frac{\Phi_{n,\xi^*_{n,k}}(\lambda)}{\mu-\mu^{-1}}\, d\lambda$ extends holomorphically in $\varkappa_{k,*} = \xi^*_{n,k}$, and therefore we have

$$\int_{A_k} \frac{\Phi_{n,\xi^*_{n,k}}(\lambda)}{\mu - \mu^{-1}}\, d\lambda = 0$$

also in this case. This shows the form $\widetilde{\omega}_n := \frac{\Phi_{n,\xi^*_{n,k}}(\lambda)}{\mu-\mu^{-1}}\, d\lambda$ then has the properties (a) and (b) from the theorem (1).

We will now show that if N is chosen large enough (depending on C_ξ), then $F : (\xi_{n,k}) \mapsto (\widetilde{\xi}_{n,k})$ maps $\mathfrak{B}_{C_\xi}$ into itself and is a contraction. Therefore, by the Banach Fixed Point Theorem, there then exists one and only one fixed point of this map; this fixed point yields the solution to the problem of the part (1) of the theorem, as was just explained.

We begin by showing that the map F maps $\mathfrak{B}_{C_\xi}$ into $\mathfrak{B}_{C_\xi}$. We let $(\xi_{n,k}) \in \mathfrak{B}_{C_\xi}$ be given and put $(\widetilde{\xi}_{n,k}) := F((\xi_{n,k}))$. Then we have to show that

$$|\widetilde{\xi}_{n,k} - \varkappa_{k,*}| \leq C_\xi \cdot |\varkappa_{k,1} - \varkappa_{k,2}| \tag{18.6}$$

holds for all $k \in \mathbb{Z}$ with $|k| > N$ and $k \notin S$.

In the sequel, we will consider the case where n and k are positive; the other cases are handled analogously. By Proposition 16.5(2) and Eq. (18.1) we have

$$\int_{A_k} \frac{1}{\mu - \mu^{-1}}\, \mathrm{d}\lambda = -8\pi\, (-1)^k\, \lambda_{k,0}^{1/2} + \ell^2_{-1}(k)$$

$$\int_{A_k} \frac{1}{|\mu - \mu^{-1}|}\, |\mathrm{d}\lambda| = 8\pi\, \lambda_{k,0}^{1/2} + \ell^2_{-1}(k),$$

therefore we may suppose that N is chosen large enough so that for $k > N$ we have

$$\frac{1}{2} \int_{A_k} \frac{1}{|\mu - \mu^{-1}|}\, |\mathrm{d}\lambda| \leq \left| \int_{A_k} \frac{1}{\mu - \mu^{-1}}\, \mathrm{d}\lambda \right| \leq \int_{A_k} \frac{1}{|\mu - \mu^{-1}|}\, |\mathrm{d}\lambda|. \tag{18.7}$$

Moreover, Proposition 17.2(3) shows that there exist constants $C_1, C_2 > 0$ (depending only on C_ξ) so that we have for $\lambda \in U_{k,\delta}$

$$\left| \Phi_{n,\xi_{n,k};k}(\lambda) - \frac{8(-1)^k \pi^2 n^2}{\xi_{n,k} \cdot (\lambda_{n,0} - \xi_{n,k})} \right| \leq C_1 \cdot \frac{n^2}{k^2 \cdot |n^2 - k^2|} \cdot \frac{|\lambda - \xi_{n,k}|}{k} + C_2 \cdot r_k \tag{18.8}$$

with the sequence $(r_k) \in \ell^2_0(k)$ given by

$$r_k = \left(a_j * \frac{1}{|j|} \right)_k \quad \text{with} \quad a_k := \begin{cases} k^{-1} \cdot |\xi_{n,k} - \lambda_{k,0}| & \text{for } k > 0 \\ k^3 \cdot |\xi_{n,k} - \lambda_{k,0}| & \text{for } k < 0 \end{cases}.$$

We may suppose without loss of generality that the projection of the cycle A_k onto the λ-plane is the straight line $[\varkappa_{k,1}, \varkappa_{k,2}] \subset U_{k,\delta}$, traversed twice in opposite directions. Then we have for any λ in the projection of A_k, i.e. for $\lambda \in [\varkappa_{k,1}, \varkappa_{k,2}]$

$$|\lambda - \xi_{n,k}| \leq |\lambda - \varkappa_{k,*}| + |\varkappa_{k,*} - \xi_{n,k}| \leq \left(\tfrac{1}{2} + C_\xi \right) \cdot |\varkappa_{k,1} - \varkappa_{k,2}|,$$

and we also have

$$|\xi_{n,k} - \lambda_{k,0}| \leq |\xi_{n,k} - \varkappa_{k,*}| + |\varkappa_{k,*} - \lambda_{k,0}| \leq C_\xi \cdot |\varkappa_{k,1} - \varkappa_{k,2}| + |\varkappa_{k,*} - \lambda_{k,0}|.$$

We now define sequences $(a'_k), (r'_k) \in \ell^2_0(k)$ by

$$a'_k := \begin{cases} k^{-1} \cdot (C_\xi \cdot |\varkappa_{k,1} - \varkappa_{k,2}| + |\varkappa_{k,*} - \lambda_{k,0}|) & \text{for } k > 0 \\ k^3 \cdot (C_\xi \cdot |\varkappa_{k,1} - \varkappa_{k,2}| + |\varkappa_{k,*} - \lambda_{k,0}|) & \text{for } k < 0 \end{cases}$$

and

$$r'_k := C_1 \left(C_\xi + \tfrac{1}{2} \right) \frac{|\varkappa_{k,1} - \varkappa_{k,2}|}{k} + C_2 \left(a'_j * \frac{1}{|j|} \right)_k .$$

It then follows from (18.8) that we have for $\lambda \in [\varkappa_{k,1}, \varkappa_{k,2}]$

$$\left| \Phi_{n,\xi_{n,k};k}(\lambda) - \frac{8(-1)^k \pi^2 n^2}{\xi_{n,k} \cdot (\lambda_{n,0} - \xi_{n,k})} \right| \leq \frac{n^2}{k^2 \cdot |n^2 - k^2|} \cdot r'_k. \tag{18.9}$$

Note that the sequence $(r'_k) \in \ell^2_0(k)$ depends only on C_ξ (and the spectral curve Σ), but neither on n nor on the specific sequence $(\xi_{n,k}) \in \mathfrak{B}_{C_\xi}$ under consideration. Therefore we may suppose that N is chosen large enough so that $|r'_k| \leq \frac{1}{2}$ holds for $k > N$.

From Eq. (18.9) we obtain by multiplication with $\frac{1}{\mu - \mu^{-1}}$, integration, and application of Eq. (18.7) for $k > N$

$$\begin{aligned} &\frac{2\pi^2 n^2}{|\xi_{n,k}| \cdot |\lambda_{n,0} - \xi_{n,k}|} \cdot \int_{A_k} \frac{1}{|\mu - \mu^{-1}|} \, |\mathrm{d}\lambda| \leq \left| \int_{A_k} \frac{\Phi_{n,\xi_{n,k};k}(\lambda)}{\mu - \mu^{-1}} \, \mathrm{d}\lambda \right| \\ &\leq \int_{A_k} \frac{|\Phi_{n,\xi_{n,k};k}(\lambda)|}{|\mu - \mu^{-1}|} \, |\mathrm{d}\lambda| \leq \frac{16\pi^2 n^2}{|\xi_{n,k}| \cdot |\lambda_{n,0} - \xi_{n,k}|} \cdot \int_{A_k} \frac{1}{|\mu - \mu^{-1}|} \, |\mathrm{d}\lambda|. \end{aligned} \tag{18.10}$$

For any λ in the projection of A_k to the λ-plane, i.e. for $\lambda \in [\varkappa_{k,1}, \varkappa_{k,2}]$ we have $|\lambda - \varkappa_{k,*}| \leq \frac{1}{2} |\varkappa_{k,1} - \varkappa_{k,2}|$, and therefore it follows from (18.10) that we have

$$\begin{aligned} &\left| \int_{A_k} (\lambda - \varkappa_{k,*}) \cdot \frac{\Phi_{n,\xi_{n,k};k}(\lambda)}{\mu - \mu^{-1}} \, \mathrm{d}\lambda \right| \\ &\qquad \leq \tfrac{1}{2} |\varkappa_{k,1} - \varkappa_{k,2}| \cdot \frac{16\pi^2 n^2}{|\xi_{n,k}| \cdot |\lambda_{n,0} - \xi_{n,k}|} \cdot \int_{A_k} \frac{1}{|\mu - \mu^{-1}|} \, |\mathrm{d}\lambda|. \end{aligned} \tag{18.11}$$

From the estimates Eqs. (18.11) and (18.10) we now obtain

$$|\widetilde{\xi}_{n,k} - \varkappa_{k,*}| \overset{(18.4)}{=} \left| \frac{\displaystyle\int_{A_k} (\lambda - \varkappa_{k,*}) \cdot \frac{\Phi_{n,\xi_{n,k};k}(\lambda)}{\mu - \mu^{-1}} \, \mathrm{d}\lambda}{\displaystyle\int_{A_k} \frac{\Phi_{n,\xi_{n,k};k}(\lambda)}{\mu - \mu^{-1}} \, \mathrm{d}\lambda} \right| \leq 4 \cdot |\varkappa_{k,1} - \varkappa_{k,2}|.$$

By a similar argument for the case where k is negative, we also obtain for $k < -N$

$$|\widetilde{\xi}_{n,k} - \varkappa_{k,*}| \leq 4 \cdot |\varkappa_{k,1} - \varkappa_{k,2}|,$$

and therefore Eq. (18.6) holds if $C_\xi \geq 4$. This shows that (at least) for $C_\xi \geq 4$, there exists a sufficiently large $N \in \mathbb{N}$ such that $(\xi_{n,k}) \in \mathfrak{B}_{C_\xi}$ implies $(\widetilde{\xi}_{n,k}) \in \mathfrak{B}_{C_\xi}$. The case $n < 0$ is handled in an analogous manner. Therefore the map $F : (\xi_{n,k}) \mapsto (\widetilde{\xi}_{n,k})$ indeed maps $\mathfrak{B}_{C_\xi}$ into $\mathfrak{B}_{C_\xi}$.

We now show that $F : \mathfrak{B}_{C_\xi} \to \mathfrak{B}_{C_\xi}$ is a contraction, if R_0 and N are suitably chosen. We let $(\xi^{[1]}_{n,k}), (\xi^{[2]}_{n,k}) \in \mathfrak{B}_{C_\xi}$ be given, and put $\widetilde{\xi}^{[\nu]}_{n,k} := F(\xi^{[\nu]}_{n,k})$ for $\nu \in \{1, 2\}$. We again consider the case of positive n and k. Then we have for $k > N$, $k \notin S$ by Eq. (18.4)

$$\begin{aligned}
\widetilde{\xi}^{[1]}_{n,k} - \widetilde{\xi}^{[2]}_{n,k} &= \frac{\displaystyle\int_{A_k} (\lambda - \varkappa_{k,*}) \cdot \frac{\Phi_{n,\xi^{[1]}_{n,k};k}(\lambda)}{\mu-\mu^{-1}}\, \mathrm{d}\lambda}{\displaystyle\int_{A_k} \frac{\Phi_{n,\xi^{[2]}_{n,k};k}(\lambda)}{\mu-\mu^{-1}}\, \mathrm{d}\lambda} - \frac{\displaystyle\int_{A_k} (\lambda - \varkappa_{k,*}) \cdot \frac{\Phi_{n,\xi^{[2]}_{n,k};k}(\lambda)}{\mu-\mu^{-1}}\, \mathrm{d}\lambda}{\displaystyle\int_{A_k} \frac{\Phi_{n,\xi^{[2]}_{n,k};k}(\lambda)}{\mu-\mu^{-1}}\, \mathrm{d}\lambda} \\
&= \frac{\displaystyle\int_{A_k} (\lambda - \varkappa_{k,*}) \cdot \frac{\Phi_{n,\xi^{[1]}_{n,k};k}(\lambda)-\Phi_{n,\xi^{[2]}_{n,k};k}(\lambda)}{\mu-\mu^{-1}}\, \mathrm{d}\lambda}{\displaystyle\int_{A_k} \frac{\Phi_{n,\xi^{[1]}_{n,k};k}(\lambda)}{\mu-\mu^{-1}}\, \mathrm{d}\lambda} \\
&\quad + \frac{\displaystyle\int_{A_k} (\lambda - \varkappa_{k,*}) \cdot \frac{\Phi_{n,\xi^{[2]}_{n,k};k}\lambda)}{\mu-\mu^{-1}}\, \mathrm{d}\lambda}{\displaystyle\int_{A_k} \frac{\Phi_{n,\xi^{[2]}_{n,k};k}(\lambda)}{\mu-\mu^{-1}}\, \mathrm{d}\lambda} \cdot \frac{\displaystyle\int_{A_k} \frac{\Phi_{n,\xi^{[2]}_{n,k};k}(\lambda)-\Phi_{n,\xi^{[1]}_{n,k};k}(\lambda)}{\mu-\mu^{-1}}\, \mathrm{d}\lambda}{\displaystyle\int_{A_k} \frac{\Phi_{n,\xi^{[1]}_{n,k};k}(\lambda)}{\mu-\mu^{-1}}\, \mathrm{d}\lambda}.
\end{aligned} \tag{18.12}$$

We now note that by Proposition 17.2(2) we have for any $\lambda \in U_{k,\delta}$

$$|\Phi_{n,\xi^{[1]}_{n,k};k}(\lambda) - \Phi_{n,\xi^{[2]}_{n,k};k}(\lambda)| \leq C_3 \cdot r_k,$$

where $C_3 > 0$ is a constant depending only on C_ξ and $(r_k) \in \ell^2_4(k)$ is the sequence given by

$$r_k := \frac{n^2}{k^2 \cdot |n^2 - k^2|} \left(a_j * \frac{1}{|j|} \right)_k \quad \text{with} \quad a_k := \begin{cases} k^{-1}\, |\xi^{[1]}_{n,k} - \xi^{[2]}_{n,k}| & \text{if } k > 0 \\ k^3\, |\xi^{[1]}_{n,k} - \xi^{[2]}_{n,k}| & \text{if } k < 0 \end{cases}.$$

By the definition of $\|\cdot\|_{\mathfrak{B}}$, we have for every k

$$|\xi^{[1]}_{n,k} - \xi^{[2]}_{n,k}| \leq \|\xi^{[1]}_{n,k} - \xi^{[2]}_{n,k}\|_{\mathfrak{B}} \cdot |\varkappa_{k,1} - \varkappa_{k,2}|,$$

and thus we see that we have

$$a_k \leq \|\xi_{n,k}^{[1]} - \xi_{n,k}^{[2]}\|_{\mathfrak{B}} \cdot a_k' \quad \text{with} \quad a_k' := \begin{cases} k^{-1}\,|\varkappa_{k,1} - \varkappa_{k,2}| & \text{if } k > 0 \\ k^3\,|\varkappa_{k,1} - \varkappa_{k,2}| & \text{if } k < 0 \end{cases}$$

and therefore

$$r_k \leq \|\xi_{n,k}^{[1]} - \xi_{n,k}^{[2]}\|_{\mathfrak{B}} \cdot r_k' \quad \text{with} \quad r_k' := \frac{n^2}{k^2 \cdot |n^2 - k^2|} \left(a_j' * \frac{1}{|j|}\right)_k .$$

Note that again, the sequences (a_k') and therefore also (r_k') depend only on C_ξ, but not on n or on the specific sequences $(\xi_{n,k}^{[\nu]}) \in \mathfrak{B}_{C_\xi}$ under consideration. We now have

$$|\Phi_{n,\xi_{n,k}^{[1]};k}(\lambda) - \Phi_{n,\xi_{n,k}^{[2]};k}(\lambda)| \leq C_3\, r_k \leq C_3\, \|\xi_{n,k}^{[1]} - \xi_{n,k}^{[2]}\|_{\mathfrak{B}} \cdot r_k'$$

and therefore by integration

$$\left| \int_{A_k} \frac{\Phi_{n,\xi_{n,k}^{[2]};k}(\lambda) - \Phi_{n,\xi_{n,k}^{[1]};k}(\lambda)}{\mu - \mu^{-1}}\, d\lambda \right| \leq C_3\, \|\xi_{n,k}^{[1]} - \xi_{n,k}^{[2]}\|_{\mathfrak{B}} \cdot r_k' \cdot \int_{A_k} \frac{1}{|\mu - \mu^{-1}|}\, |d\lambda|, \tag{18.13}$$

whence also

$$\left| \int_{A_k} (\lambda - \varkappa_{k,*}) \cdot \frac{\Phi_{n,\xi_{n,k}^{[1]};k}(\lambda) - \Phi_{n,\xi_{n,k}^{[2]};k}(\lambda)}{\mu - \mu^{-1}}\, d\lambda \right|$$
$$\leq \frac{1}{2}\, C_3\, |\varkappa_{k,1} - \varkappa_{k,2}| \cdot \|\xi_{n,k}^{[1]} - \xi_{n,k}^{[2]}\|_{\mathfrak{B}} \cdot r_k' \cdot \int_{A_k} \frac{1}{|\mu - \mu^{-1}|}\, |d\lambda| \tag{18.14}$$

follows.

By applying the estimates (18.10), (18.11), (18.13) and (18.14) to Eq. (18.12), and cancelling, we now obtain that there exists $C_4 > 0$ (dependent only on C_ξ) with

$$|\widetilde{\xi}_{n,k}^{[1]} - \widetilde{\xi}_{n,k}^{[2]}| \leq C_4 \cdot \left(a_j' * \frac{1}{|j|}\right)_k \cdot |\varkappa_{k,1} - \varkappa_{k,2}| \cdot \|\xi_{n,k}^{[1]} - \xi_{n,k}^{[2]}\|_{\mathfrak{B}} . \tag{18.15}$$

The analogous calculation applies for $k < -N$, yielding literally the same estimate as (18.15) for this case, and therefore we obtain

$$\|\widetilde{\xi}_{n,k}^{[1]} - \widetilde{\xi}_{n,k}^{[2]}\|_{\mathfrak{B}} \leq L \cdot \|\xi_{n,k}^{[1]} - \xi_{n,k}^{[2]}\|_{\mathfrak{B}} \quad \text{with} \quad L := C_4 \cdot \max_{|k|>N} \left| \left(a_j' * \frac{1}{|j|}\right)_k \right| ;$$

note that $a'_j * \frac{1}{|j|} \in \ell^2_{0,0} \subset \ell^\infty$ holds and therefore L is finite. This shows that the map F is Lipschitz continuous with the Lipschitz constant L. Because the sequence $a'_j * \frac{1}{|j|}$ is an ℓ^2-sequence which depends on C_ξ and nothing else, and also C_4 depends only on C_ξ, we can choose N so large that $\max_{|k|>N} \left|\left(a'_j * \frac{1}{|j|}\right)_k\right| \leq \frac{1}{2C_4}$ holds. Then we have $L \leq \frac{1}{2}$, hence with this choice of N, F is a contraction, completing the proof of (1).

For (2). By (1), the 1-forms $\widetilde{\omega}_n$ have the property that

$$\int_{A_k} \widetilde{\omega}_n = 0 \quad \text{holds for all } k, n \in \mathbb{Z} \text{ with } (|k| > N \text{ or } k \in S) \text{ and } k \neq n. \tag{18.16}$$

We consider the linear space

$$V := \left\{ \omega \in \Omega(\Sigma) \cap L^2(\Sigma, T^*\Sigma) \,\middle|\, \int_{A_k} \omega = 0 \text{ for } |k| > N \text{ or } k \in S \right\}.$$

Let $I := \{ k \in \mathbb{Z} \setminus S \,|\, |k| \leq N \}$ and $m := \#I \leq 2N+1$. Then it follows from Proposition 18.1 that the linear map $f : V \to \mathbb{C}^m,\ \omega \mapsto \left(\int_{A_\ell} \omega\right)_{\ell \in I}$ is injective and that therefore $\dim(V) \leq m$ holds. On the other hand, the holomorphic 1-forms $\widetilde{\omega}_\ell$ with $\ell \in I$ are in $\Omega(\Sigma) \cap L^2(\Sigma, T^*\Sigma)$ and therefore in V by (1), and they are linear independent (because $\widetilde{\omega}_\ell$ does not vanish at ξ_ℓ, but all the $\widetilde{\omega}_{\ell'}$ with $\ell' \neq \ell$ do). Therefore we have $\dim(V) = m$ and $(\widetilde{\omega}_\ell)_{\ell \in I}$ is a basis of V. The linear map $f : V \to \mathbb{C}^m$ is therefore bijective, which means that for every finite sequence $(z_\ell)_{\ell \in I}$ there exists one and only one $\omega \in V$ with $\int_{A_\ell} \omega = z_\ell$ for all $\ell \in I$.

Now let $n \in \mathbb{Z}$ be given. If $|n| \leq N$ and $n \notin S$ holds, then the preceding statement shows immediately that there exists $\omega_n \in V$ such that

$$\int_{A_k} \omega_n = \delta_{kn} \quad \text{holds for all } k \in \mathbb{Z};$$

because $(\widetilde{\omega}_\ell)_{\ell \in I}$ is a basis of V, it follows that there exist numbers $s_{n,\ell} \in \mathbb{C}$ for $\ell \in I$ so that $\omega_n = \sum_{\ell \in I} s_{n,\ell}\, \widetilde{\omega}_\ell$ holds.

On the other hand, if either $n \in S$ or $|n| > N$ holds, we put $s_{n,n} := \left(\int_{A_n} \widetilde{\omega}_n\right)^{-1}$. There then exists $\widehat{\omega} \in V$ so that

$$\int_{A_\ell} \widehat{\omega} = -s_{n,n} \cdot \int_{A_\ell} \widetilde{\omega}_n \quad \text{holds for all } \ell \in I;$$

because $(\widetilde{\omega}_\ell)_{\ell \in I}$ is a basis of V, it follows that there exist numbers $s_{n,\ell} \in \mathbb{C}$ for $\ell \in I$ so that $\widehat{\omega} = \sum_{\ell \in I} s_{n,\ell}\, \widetilde{\omega}_\ell$ holds; we then have

$$\int_{A_k} \omega_n = \delta_{kn} \quad \text{for all } k \in \mathbb{Z}$$

with

$$\omega_n := s_{n,n} \cdot \widetilde{\omega}_n + \widehat{\omega} = s_{n,n} \cdot \widetilde{\omega}_n + \sum_{\ell \in I} s_{n,\ell}\, \widetilde{\omega}_\ell .$$

In either case this shows the claimed representation of ω_n. It follows from Proposition 17.2(1) that $\Phi_n \in \mathrm{As}(\mathbb{C}^*, \ell^\infty_{3,-1}, 1)$ holds.

It remains to show the statements on the asymptotic behavior of $s_{n,n}$ and $s_{n,\ell}$, and for this it suffices to consider the case $|n| > N$. We have

$$\Phi_{n,\xi_{n,k}}(\lambda) = \begin{cases} -\frac{4}{\tau_{\xi_{n,k}}} \cdot 16\pi^2 n^2 \cdot \lambda^{-1} \cdot \frac{c_{\xi_{n,k}}(\lambda)}{\lambda - \lambda_{n,0}} & \text{if } n > N \\ \frac{4}{\tau_{\xi_{n,k}}} \cdot \lambda^{-1} \cdot \frac{c_{\xi_{n,k}}(\lambda)}{\lambda - \lambda_{n,0}} & \text{if } n < -N \end{cases} \tag{18.17}$$

where $c_{\xi_{n,k}}$ and $\tau_{\xi_{n,k}}$ are as in the proof of Proposition 17.2(1). By Corollary 10.3(1), there exists a constant $C_5 > 0$ and a sequence $(r_k) \in \ell^2_{0,-2}(k)$ so that we have for $n, k \in \mathbb{Z}$ with $|n| > N$ and $\lambda \in U_{k,\delta}$

$$\begin{cases} \left| \frac{c_{\xi_{n,k}}(\lambda)}{\tau_{\xi_{n,k}} \cdot (\lambda - \xi_{n,k})} - \frac{(-1)^k}{8} \right| \le C_5 \frac{|\lambda - \xi_{n,k}|}{k} + r_k & \text{if } k > 0 \\ \left| \frac{c_{\xi_{n,k}}(\lambda)}{\lambda - \xi_{n,k}} - \left(-\tau_{\xi_{n,k}}^{-1} \frac{(-1)^k}{8} \lambda_{n,0}^{-1} \right) \right| \le C_5 \, |\lambda - \xi_{n,k}| \, k^5 + r_k & \text{if } k < 0 \end{cases},$$

here we put $\xi_{n,n} := \lambda_{n,0}$ and the sequence (r_k) can be chosen independently of n because $\left| \frac{\xi_{n,k} - \varkappa_{k,*}}{\varkappa_{k,1} - \varkappa_{k,2}} \right|$ is bounded independently of n. By setting $k = n$ in the preceding estimate we see that for $\lambda \in [\varkappa_{k,1}, \varkappa_{k,2}]$ we have

$$\begin{cases} \left| \frac{c_{\xi_{n,k}}(\lambda)}{\tau_{\xi_{n,k}} \cdot (\lambda - \lambda_{n,0})} - \frac{(-1)^k}{8} \right| \le r'_n & \text{if } n > N \\ \left| \frac{c_{\xi_{n,k}}(\lambda)}{\lambda - \lambda_{n,0}} - \left(-\tau_{\xi_{n,k}}^{-1} \frac{(-1)^k}{8} \lambda_{n,0}^{-1} \right) \right| \le r'_n & \text{if } n < -N \end{cases}$$

with the sequence $(r'_k) \in \ell^2_{0,-2}(k)$ defined by

$$r'_k := \begin{cases} C_5 \left(C_\xi + \frac{1}{2} \right) \frac{|\varkappa_{k,1} - \varkappa_{k,2}|}{k} + r_k & \text{for } k > 0 \\ C_5 \left(C_\xi + \frac{1}{2} \right) |\varkappa_{k,1} - \varkappa_{k,2}| \, k^5 + r_k & \text{for } k < 0 \end{cases}.$$

By applying this estimate, and also the fact that we have $\frac{16\pi^2 n^2}{\lambda} - 1 \in \ell^2_1(n)$ (for the case $n > 0$) resp. $\frac{1}{\lambda} - \frac{1}{\lambda_{n,0}} \in \ell^2_{-1}(n)$ (for the case $n < 0$) for $\lambda \in [\varkappa_{k,1}, \varkappa_{k,2}]$, to Eq. (18.17), we obtain

$$\begin{cases} \left|\Phi_{n,\xi_{n,k}}(\lambda) - \left(-\frac{(-1)^k}{2}\right)\right| \in \ell^2(n) & \text{for } n > N \\ \left|\Phi_{n,\xi_{n,k}}(\lambda) - \left(-\frac{(-1)^k}{2}\,\tau_{\xi_{n,k}}^{-2}\,\lambda_{n,0}^{-2}\right)\right| \in \ell^2_{-4}(n) & \text{for } n < -N. \end{cases} \tag{18.18}$$

By Proposition 16.5(2) we have

$$\int_{A_n} \frac{1}{\mu-\mu^{-1}}\,\mathrm{d}\lambda \overset{(18.1)}{=} 2\cdot\int_{\varkappa_{n,1}}^{\varkappa_{n,2}} \frac{1}{\mu-\mu^{-1}}\,\mathrm{d}\lambda = \begin{cases} -8\pi(-1)^n\,\lambda_{n,0}^{1/2} + \ell^2_{-1}(n) & \text{for } n > N \\ -8\pi(-1)^n\,\lambda_{n,0}^{3/2} + \ell^2_3(n) & \text{for } n < -N \end{cases}$$

and therefore we derive from Eq. (18.18)

$$\int_{A_n} \frac{\Phi_{n,\xi_{n,k}}(\lambda)}{\mu-\mu^{-1}}\,\mathrm{d}\lambda = \begin{cases} 4\pi\cdot\lambda_{n,0}^{1/2} + \ell^2_{-1}(n) & \text{for } n > N \\ 4\pi\cdot\tau_{\xi_{n,k}}^{-2}\cdot\lambda_{n,0}^{-1/2} + \ell^2_{-1}(n) & \text{for } n < -N \end{cases}$$

Thus we obtain

$$s_{n,n} = \left(\int_{A_n} \frac{\Phi_{n,\xi_{n,k}}(\lambda)}{\mu-\mu^{-1}}\,\mathrm{d}\lambda\right)^{-1} = \begin{cases} \frac{1}{4\pi}\cdot\lambda_{n,0}^{-1/2} + \ell^2_1(n) & \text{for } n > N \\ \frac{1}{4\pi}\cdot\tau_{\xi_{n,k}}^{2}\cdot\lambda_{n,0}^{1/2} + \ell^2_1(n) & \text{for } n < -N. \end{cases}$$

We now calculate the $s_{n,\ell}$. For this purpose, let P be the "restricted period matrix" of the $\widetilde{\omega}_\ell$, i.e. let $P = (p_{k,\ell})_{k,\ell\in I}$ be the $(m\times m)$-matrix with

$$p_{k,\ell} := \int_{A_k} \widetilde{\omega}_\ell\,.$$

Note that the matrix P is invertible because of Proposition 18.1 and the fact that $(\widetilde{\omega}_\ell)_{\ell\in I}$ is a basis of V. We have

$$\omega_n = s_{n,n}\cdot\widetilde{\omega}_n + \sum_{\ell\in I} s_{n,\ell}\cdot\widetilde{\omega}_\ell$$

and therefore for $k \in I$

$$0 = \int_{A_k} \omega_n = s_{n,n}\cdot\int_{A_k}\widetilde{\omega}_n + \sum_{\ell\in I} p_{k,\ell}\cdot s_{n,\ell}\,.$$

Hence we have as an equality of $\mathbb{C}^m$-vectors

$$(s_{n,\ell})_{\ell\in I} = -s_{n,n}\,P^{-1}\cdot\left(\int_{A_k}\widetilde{\omega}_n\right)_{k\in I}.$$

Because the $\Phi_{n,\xi_{n,k}}$ are uniformly bounded on the "compact part of Σ" $\bigcup_{|k|\leq N} \widehat{S}_k$, and $s_{n,n}$ is of the order of n^{-1}, it follows from this equation that also $s_{n,\ell}$ is with respect to n of the order n^{-1}. □

After having constructed the canonical basis (ω_n) of the 1-forms in the preceding theorem, the next step for the construction of the Jacobi variety and Abel map for Σ is to define the Banach space $\widetilde{\mathrm{Jac}}(\Sigma)$ which will serve as the covering space of the Jacobi variety (and also plays the role of a tangent space for $\mathrm{Jac}(\Sigma)$), and the map $\widetilde{\varphi}$ which will turn out to be the lift of the Abel map $\varphi : \mathsf{Div}(\Sigma') \to \mathrm{Jac}(\Sigma)$; we will also study the asymptotic behavior of these objects.

In the construction of the Abel map we only permit divisors whose support is contained in the regular set Σ' of Σ. One way to handle divisor points in singularities would be to desingularize Σ at these points (as Hitchin does when constructing his spectral curve in [Hi]). By this process, one would remove the ω_n and the coordinates of the Jacobi variety corresponding to singular divisor points, and obtain a Jacobi variety associated to a Banach space for the remaining coordinates.

Proposition 18.5 *We fix a divisor $D^o \in \mathsf{Div}(\Sigma')$ (which will be used as the origin point for the definition of the Abel map) and denote by $\mathfrak{C}_{D^o}$ the set of sequences $(\gamma_k)_{k\in\mathbb{Z}}$ where each γ_k is a curve in Σ' running from a point $(\lambda_k^o, \mu_k^o) \in \Sigma'$ to another point $(\lambda_k, \mu_k) \in \Sigma'$, such that $(\lambda_k^o, \mu_k^o)_{k\in\mathbb{Z}}$ equals the support of D^o and $D := \{(\lambda_k, \mu_k) \,|\, k \in \mathbb{Z}\} \in \mathsf{Div}(\Sigma')$ holds; moreover for large $|k|$ the curve γ_k runs entirely in $\widehat{U}_{k,\delta}$, and there is a number $m_\gamma \in \mathbb{N}$ (depending on γ but not on k) so that the winding number of any γ_k around any branch point or puncture of Σ' is at most m_γ. In this situation we call D the divisor induced by the sequence $(\gamma_k)_{k\in\mathbb{Z}}$. We also let $(\omega_n)_{n\in\mathbb{Z}}$ be the canonical basis of 1-forms from Theorem 18.3(2).*

(1) For $n \in \mathbb{Z}$ and $(\gamma_k)_{k\in\mathbb{Z}} \in \mathfrak{C}_{D^o}$ the infinite sum

$$\sum_{k\in\mathbb{Z}} \int_{\gamma_k} \omega_n \tag{18.19}$$

converges absolutely in $\mathbb{C}$, and we define the map

$$\widetilde{\varphi}_n : \mathfrak{C}_{D^o} \to \mathbb{C},\ (\gamma_k)_{k\in\mathbb{Z}} \mapsto \sum_{k\in\mathbb{Z}} \int_{\gamma_k} \omega_n . \tag{18.20}$$

For every $(\gamma_k) \in \mathfrak{C}_{D^o}$ there exist sequences $(b_n), (c_n) \in \ell^2(n)$ so that for $n \in \mathbb{Z}$ with $|n| > N$ (where $N \in \mathbb{N}$ is as in Theorem 18.3(2)) we have

$$\widetilde{\varphi}_n((\gamma_k)) = \frac{\mathrm{sign}(n)}{2\pi \mathrm{i}} \cdot \ln\left(\frac{\lambda_n - \varkappa_{n,*} + \Psi_n(\lambda_n, \mu_n)}{\lambda_n^o - \varkappa_{n,*} + \Psi_n(\lambda_n^o, \mu_n^o)}\right) \cdot (1+b_n) + c_n , \tag{18.21}$$

where Ψ_n *is as in Lemma 16.1, and* $\ln(z)$ *is the branch of the complex logarithm function with* $\ln(1) = 2\pi i m_n$ *with* $m_n \in \mathbb{Z}$ *being the winding number of* γ_n *around the pair of branch points* $\varkappa_{n,1}$, $\varkappa_{n,2}$.

(2) *We consider the topological vector space*

$$\widetilde{\mathrm{Jac}}(\Sigma) := \{ (a_n)_{n\in\mathbb{Z}} \,\big|\, a_n \cdot (\varkappa_{n,1} - \varkappa_{n,2}) \in \ell^2_{-1,3}(n) \},$$

whose topology is induced by the semi-norm

$$\|a_n\|_{\widetilde{\mathrm{Jac}}(\Sigma)} := \big\|a_n \cdot (\varkappa_{n,1} - \varkappa_{n,2})\big\|_{\ell^2_{-1,3}} \quad \text{for } (a_n) \in \widetilde{\mathrm{Jac}}(\Sigma).$$

If $S \neq \varnothing$ *holds, then* $\widetilde{\mathrm{Jac}}(\Sigma)$ *is non-Hausdorff. If* $S = \varnothing$ *holds, then* $\|\cdot\|_{\widetilde{\mathrm{Jac}}(\Sigma)}$ *is a norm and* $\widetilde{\mathrm{Jac}}(\Sigma)$ *is a Banach space. We have* $\ell^\infty(\mathbb{Z}) \subset \widetilde{\mathrm{Jac}}(\Sigma)$, *and for any* $(\gamma_k) \in \mathfrak{C}_{D^o}$

$$(\widetilde{\varphi}_n((\gamma_k)))_{n\in\mathbb{Z}} \in \widetilde{\mathrm{Jac}}(\Sigma).$$

Thereby we obtain the map

$$\widetilde{\varphi} : \mathfrak{C}_{D^o} \to \widetilde{\mathrm{Jac}}(\Sigma),\ (\gamma_k) \mapsto (\widetilde{\varphi}_n(\gamma_k))_{n\in\mathbb{Z}}.$$

(3) *For any* $k \in \mathbb{Z} \setminus S$ *we have*

$$\left(\int_{B_k} \omega_n\right)_{n\in\mathbb{Z}} \in \widetilde{\mathrm{Jac}}(\Sigma).$$

Remark 18.6 The semi-norm and the topology of $\widetilde{\mathrm{Jac}}(\Sigma)$ do not contain any information on the coordinates a_k in $\widetilde{\mathrm{Jac}}(\Sigma)$ with $k \in S$. This means that the present construction of the Jacobi variety is not very well suited for studying spectral curves with infinitely many singularities, when the corresponding divisor points are outside these singularities (or at least, when infinitely many of them are). Such divisors occur, for example, as spectral divisors for minimal surfaces in S^3 consisting of infinitely many bubbletons. (On the spectral curve Σ_0 of the vacuum, a bubbleton is induced by moving one divisor point that was in a singularity of Σ_0 away from this singularity.)

Proof (of Proposition 18.5) For (1). By Theorem 18.3, we have

$$\omega_n = \frac{\Phi_n(\lambda)}{\mu - \mu^{-1}}\, d\lambda,$$

where the holomorphic function Φ_n is a finite linear combination of functions $\widetilde{\Phi}_\ell$, $\ell \in \{-N, \dots, N\} \cup \{n\}$ of the type studied in Proposition 17.2. In other words, ω_n itself is a linear combination of 1-forms $\widetilde{\omega}_\ell$ (with $\ell \in \{-N, \dots, N\} \cup \{n\}$) of the type studied in Proposition 17.4. The corresponding sequences $(\xi_{\ell,k})_{k\in\mathbb{Z}\setminus\{\ell\}}$

of zeros of $\widetilde{\Phi}_\ell$ satisfy the estimate $|\xi_{\ell,k} - \varkappa_{k,*}| \leq C_\xi \cdot |\varkappa_{k,1} - \varkappa_{k,2}|$ with a constant $C_\xi > 0$ that is independent of ℓ and k, and therefore the estimates in Proposition 17.2–17.4 apply uniformly to all $\widetilde{\Phi}_\ell$ resp. $\widetilde{\omega}_\ell$.

For the proof of the convergence of the infinite sum (18.19) for some given $n \in \mathbb{Z}$, we note that we have

$$\int_{(\lambda_k^o,\mu_k^o)}^{(\lambda_k,\mu_k)} \omega_n = \int_{\varkappa_{k,1}}^{(\lambda_k,\mu_k)} \omega_n - \int_{\varkappa_{k,1}}^{(\lambda_k^o,\mu_k^o)} \omega_n ,$$

where the paths of integration are chosen suitably on the right hand side. Therefore we may suppose without loss of generality that $(\lambda_k^o, \mu_k^o) = \varkappa_{k,1}$ holds for all $k \neq n$. Let $(\gamma_k) \in \mathfrak{C}_{D^o}$ be given. Because we have $\int_{A_k} \widetilde{\omega}_\ell = 0$ for $|k|$ large, $k \neq n$, the question of the convergence of the sum $\sum_{k\in\mathbb{Z}\setminus\{n\}} \left|\int_{\gamma_k} \widetilde{\omega}_\ell\right|$ does not depend on how many times the curve γ_k winds around the pair of branch points $\varkappa_{k,1}$, $\varkappa_{k,2}$, and therefore Proposition 17.4(3) shows that the sum $\sum_{k\in\mathbb{Z}\setminus\{n\}} \left|\int_{\gamma_k} \widetilde{\omega}_\ell\right|$ converges. Because ω_n is a finite linear combination of the $\widetilde{\omega}_\ell$, it follows that the infinite sum (18.19) converges absolutely.

For proving the asymptotic estimate (18.21) we let $(\gamma_k) \in \mathfrak{C}_{D^o}$ be given and consider $n \in \mathbb{Z}$ with $|n| > N$. The constants $C_k > 0$ occurring in the sequel depend on (γ_k) but not on n. We have

$$\widetilde{\varphi}_n((\gamma_k)) = \int_{\gamma_n} \omega_n + \sum_{|k|\leq N} \int_{\gamma_k} \omega_n + \sum_{\substack{|k|>N \\ k\neq n}} \int_{\gamma_k} \omega_n . \tag{18.22}$$

In the sequel we will estimate the three summands on the right hand side of this equation separately. For this purpose we will use the fact that by Theorem 18.3(2) there exist constants $s_{n,n} \in \mathbb{C}$ and $s_{n,\ell} \in \mathbb{C}$ for $|\ell| \leq N$ so that

$$\omega_n = s_{n,n} \cdot \widetilde{\omega}_n + \sum_{|\ell|\leq N} s_{n,\ell} \cdot \widetilde{\omega}_\ell \tag{18.23}$$

holds; here the constants $s_{n,n}$ and $s_{n,\ell}$ are of order n^{-1}, hence there exists $C_1 > 0$ so that we have

$$|s_{n,n}|,\ |s_{n,\ell}| \leq C_1 \cdot \frac{1}{|n|} . \tag{18.24}$$

For the first summand in (18.22) we note that by Theorem 18.3(2) there exists a sequence $(b_n^{[1]}) \in \ell^2_{1,1}(n)$ so that

$$s_{n,n} = \begin{cases} \frac{1}{4\pi} \cdot \lambda_{n,0}^{-1/2} + b_n^{[1]} & \text{for } n > N \\ \frac{1}{4\pi} \cdot \tau_{\xi n,k}^2 \cdot \lambda_{n,0}^{1/2} + b_n^{[1]} & \text{for } n < -N \end{cases}$$

holds, and by Proposition 17.4(1) there exists a sequence $(b_n^{[2]}) \in \ell^2_{-1,-1}(n)$ so that we have

$$\int_{\gamma_n} \widetilde{\omega}_n = \begin{cases} (-2\mathrm{i}\,\lambda_{n,0}^{1/2} + b_n^{[2]}) \cdot \ln\left(\frac{\lambda_n - \varkappa_{n,*} + \Psi_n(\lambda_n,\mu_n)}{\lambda_n^o - \varkappa_{n,*} + \Psi_n(\lambda_n^o,\mu_n^o)}\right) & \text{for } n > N \\ \left(\frac{2\mathrm{i}}{\tau_{\xi_{n,k}}^2} \cdot \lambda_{n,0}^{-1/2} + b_n^{[2]}\right) \cdot \ln\left(\frac{\lambda_n - \varkappa_{n,*} + \Psi_n(\lambda_n,\mu_n)}{\lambda_n^o - \varkappa_{n,*} + \Psi_n(\lambda_n^o,\mu_n^o)}\right) & \text{for } n < N \end{cases},$$

where the choice of the branch of the complex logarithm corresponds to the winding number m_n of γ_n around the pair of branch points $\varkappa_{n,1}$, $\varkappa_{n,2}$ by $\ln(1) = 2\pi\mathrm{i}m_n$. By multiplying the preceding two equations, we obtain that there exist sequences $(b_n^{[3]}), (b_n) \in \ell^2_{0,0}(n)$ with

$$\begin{aligned} s_{n,n} \cdot \int_{\gamma_n} \widetilde{\omega}_n &= \begin{cases} \left(\frac{1}{2\pi\mathrm{i}} + b_n^{[3]}\right) \cdot \ln\left(\frac{\lambda_n - \varkappa_{n,*} + \Psi_n(\lambda_n,\mu_n)}{\lambda_n^o - \varkappa_{n,*} + \Psi_n(\lambda_n^o,\mu_n^o)}\right) & \text{for } n > N \\ \left(-\frac{1}{2\pi\mathrm{i}} + b_n^{[3]}\right) \cdot \ln\left(\frac{\lambda_n - \varkappa_{n,*} + \Psi_n(\lambda_n,\mu_n)}{\lambda_n^o - \varkappa_{n,*} + \Psi_n(\lambda_n^o,\mu_n^o)}\right) & \text{for } n < -N \end{cases} \\ &= \frac{\operatorname{sign}(n)}{2\pi\mathrm{i}} \cdot \ln\left(\frac{\lambda_n - \varkappa_{n,*} + \Psi_n(\lambda_n,\mu_n)}{\lambda_n^o - \varkappa_{n,*} + \Psi_n(\lambda_n^o,\mu_n^o)}\right) \cdot (1 + b_n). \end{aligned} \tag{18.25}$$

Moreover because of $\int_{A_n} \widetilde{\omega}_\ell = 0$ for $|\ell| \leq N$, the value of $\left|\int_{\gamma_n} \widetilde{\omega}_\ell\right|$ does not depend on the winding number of γ_n around the pair of branch points $\varkappa_{n,1}$, $\varkappa_{n,2}$. Therefore it follows from Proposition 17.4(2) that there exists a constant $C_2 > 0$ with $\left|\int_{\gamma_n} \widetilde{\omega}_\ell\right| \leq C_2 \cdot \frac{1}{n^2}$ for $|\ell| \leq N$. Because of (18.24) it follows that there exists a constant $C_3 > 0$ with

$$\sum_{|\ell| \leq N} |s_{n,\ell}| \cdot \left|\int_{\gamma_n} \widetilde{\omega}_\ell\right| \leq C_3 \cdot \frac{1}{n^3}.$$

We thus obtain by combining the preceding estimates with Eq. (18.23) that there exists a sequence $(c_n^{[1]}) \in \ell^2_{0,0}(n)$ with

$$\int_{\gamma_n} \omega_n = \frac{\operatorname{sign}(n)}{2\pi\mathrm{i}} \cdot \ln\left(\frac{\lambda_n - \varkappa_{n,*} + \Psi_n(\lambda_n,\mu_n)}{\lambda_n^o - \varkappa_{n,*} + \Psi_n(\lambda_n^o,\mu_n^o)}\right) \cdot (1 + b_n) + c_n^{[1]}. \tag{18.26}$$

For the second summand in (18.22) we note that by applying Proposition 17.4(4) to all the finitely many paths of integration γ_k with $|k| \leq N$ it follows that there exists $C_4 > 0$ so that we have for all $|k| \leq N$ and $\ell \in \mathbb{Z}$

$$\left|\int_{\gamma_k} \widetilde{\omega}_\ell\right| \leq C_4;$$

via Eq. (18.23) we obtain

$$\begin{aligned}\left|\sum_{|k|\leq N}\int_{\gamma_k}\omega_n\right| &= |s_{n,n}|\cdot\sum_{|k|\leq N}\left|\int_{\gamma_k}\widetilde{\omega}_n\right| + \sum_{|\ell|\leq N}|s_{n,\ell}|\cdot\left(\sum_{|k|\leq N}\left|\int_{\gamma_k}\widetilde{\omega}_\ell\right|\right)\\ &\leq \left(|s_{n,n}| + \sum_{|\ell|\leq N}|s_{n,\ell}|\right)\cdot(2N+1)\cdot C_4\\ &=: c_n^{[2]},\end{aligned} \tag{18.27}$$

where the sequence $(c_n^{[2]})$ is of order n^{-1} by (18.24), in particular $(c_n^{[2]}) \in \ell^2_{0,0}(n)$ holds.

For the third summand in (18.22) we note that for $|k| > N,\ k \neq n$ we have $\int_{A_k}\widetilde{\omega}_\ell = 0$ for all $\ell \in \{-N,\dots,N\}\cup\{n\}$, and therefore the value of $\left|\int_{\gamma_k}\widetilde{\omega}_\ell\right|$ does not depend on the winding number of γ_k around the pair of branch points $\varkappa_{k,1},\ \varkappa_{k,2}$ for such k,ℓ. It therefore follows from Proposition 17.4(3) that for $\ell \in \{-N,\dots,N\}\cup\{n\}$ we have

$$\sum_{\substack{|k|>N\\ k\neq n}}\left|\int_{\gamma_k}\widetilde{\omega}_\ell\right| \in \ell^2_{-1,-1}(n),$$

via Eqs. (18.23) and (18.24) we conclude

$$c_n^{[3]} := \left|\sum_{\substack{|k|>N\\ k\neq n}}\int_{\gamma_k}\omega_n\right| \in \ell^2_{0,0}(n). \tag{18.28}$$

By plugging the estimates (18.26), (18.27) and (18.28) into (18.22), we see that (18.21) holds with $b_n \in \ell^2(n)$ as above and $c_n := c_n^{[1]} + c_n^{[2]} + c_n^{[3]} \in \ell^2(n)$.

For (2). It is clear that $\widetilde{\mathrm{Jac}}(\Sigma)$ is a linear space, that $\|\cdot\|_{\widetilde{\mathrm{Jac}}(\Sigma)}$ is a semi-norm on $\widetilde{\mathrm{Jac}}(\Sigma)$ and that in the case $S = \varnothing$ (i.e. $\varkappa_{k,1} - \varkappa_{k,2} \neq 0$ for all k), this semi-norm becomes a norm and $\widetilde{\mathrm{Jac}}(\Sigma)$ a Banach space. We have $\varkappa_{k,1} - \varkappa_{k,2} \in \ell^2_{-1,3}(k)$ and therefore $\ell^\infty(\mathbb{Z}) \subset \widetilde{\mathrm{Jac}}(\Sigma)$. To prove $(\widetilde{\varphi}_n((\gamma_k))) \in \widetilde{\mathrm{Jac}}(\Sigma)$, we need to show that

$$\widetilde{\varphi}_n((\gamma_k))\cdot(\varkappa_{n,1} - \varkappa_{n,2}) \in \ell^2_{-1,3}(n)$$

holds. If we have $n \in S$ and therefore $\varkappa_{n,1} = \varkappa_{n,2}$, there is nothing to show, so we consider $n \in \mathbb{Z}\setminus S$ now. We may again suppose without loss of generality that

$(\lambda_n^o, \mu_n^o) = \varkappa_{n,1}$ holds for $n \in \mathbb{Z} \setminus S$. Then we have

$$\ln\left(\frac{\lambda_n - \varkappa_{n,*} + \Psi_n(\lambda_n, \mu_n)}{\lambda_n^o - \varkappa_{n,*} + \Psi_n(\lambda_n^o, \mu_n^o)}\right) = \ln\left(\frac{\lambda_n - \varkappa_{n,*} + \Psi_n(\lambda_n, \mu_n)}{\frac{1}{2}(\varkappa_{k,1} - \varkappa_{k,2})}\right)$$
$$= \operatorname{arcosh}\left(\frac{\lambda_n - \varkappa_{n,*}}{\frac{1}{2}(\varkappa_{k,1} - \varkappa_{k,2})}\right) + 2\pi \mathrm{i}\, m_n, \qquad (18.29)$$

where $m_n \in \mathbb{Z}$ again is the winding number of γ_n around the pair of branch points $\varkappa_{n,1}, \varkappa_{n,2}$. Here we denote by $\ln(z)$ the branch of the complex logarithm function with $\ln(1) = 2\pi \mathrm{i} m_n$, and by $\operatorname{arcosh}(z)$ the branch of the complex area cosine hyperbolicus function with $\operatorname{arcosh}(1) = 0$.

Because of $(\gamma_k) \in \mathfrak{C}_{D^o}$, the sequence $(m_n)_{n\in\mathbb{Z}}$ is bounded, and therefore it follows from (1) and the estimate for arcosh of Eq. (16.16) that there exist constants $C_5, C_6 > 0$ with

$$|\widetilde{\varphi}_n((\gamma_k))| \leq C_5 \cdot \left|\operatorname{arcosh}\left(\frac{\lambda_n - \varkappa_{n,*}}{\frac{1}{2}(\varkappa_{n,1} - \varkappa_{n,2})}\right)\right|$$
$$+ C_6 \leq 2\, C_5 \cdot C_{\mathrm{arcosh}} \cdot \left|\frac{\lambda_n - \varkappa_{n,1}}{\varkappa_{n,1} - \varkappa_{n,2}}\right| + C_6. \qquad (18.30)$$

Thus we have

$$|\widetilde{\varphi}_n((\gamma_k))| \cdot |\varkappa_{n,1} - \varkappa_{n,2}| \leq 2\, C_5\, C_{\mathrm{arcosh}} \cdot |\lambda_n - \varkappa_{n,1}| + C_6 \cdot |\varkappa_{n,1} - \varkappa_{n,2}|$$
$$\in \ell^2_{-1,3}(n \in \mathbb{Z} \setminus S)$$

and therefore $(\widetilde{\varphi}_n((\gamma_k))) \in \widetilde{\mathrm{Jac}}(\Sigma)$.

For (3). For every fixed $k \in \mathbb{Z}$, $\int_{B_k} \widetilde{\omega}_n$ is bounded with respect to n by Proposition 17.4(4). Because both $s_{n,n}$ and the $s_{n,\ell}$ with $|\ell| \leq N$ are of order n^{-1} by Theorem 18.3(2), it follows that

$$\int_{B_k} \omega_n = s_{n,n} \cdot \int_{B_k} \widetilde{\omega}_n + \sum_{|\ell|\leq N} s_{n,\ell} \int_{B_k} \widetilde{\omega}_\ell$$

is of order n^{-1}. Thus we have $\left(\int_{B_k} \omega_n\right)_{n\in\mathbb{Z}} \in \ell^\infty_{1,1}(n) \subset \ell^\infty(n) \subset \widetilde{\mathrm{Jac}}(\Sigma)$ by (2).

□

With the following theorem, the construction of the Jacobi variety and the Abel map of Σ is completed:

Theorem 18.7 *We again fix an "origin divisor" $D^o \in \mathsf{Div}(\Sigma')$ and use the notations of Proposition 18.5.*

(1) For $k, n \in \mathbb{Z}$ *we let*

$$\alpha_n^{[k]} := \int_{A_k} \omega_n = \delta_{k,n} \quad \text{and if } k \notin S \quad \beta_n^{[k]} := \int_{B_k} \omega_n .$$

Then we have $(\alpha_n^{[k]})_n \in \widetilde{\mathrm{Jac}}(\Sigma)$ *for every* $k \in \mathbb{Z}$ *and* $(\beta_n^{[k]})_n \in \widetilde{\mathrm{Jac}}(\Sigma)$ *for every* $k \in \mathbb{Z} \setminus S$.
We let Γ *be the abelian group corresponding to the periods of all closed loops in* $\mathfrak{C}_{D^o}$, *i.e.*

$$\Gamma := \left\{ \sum_{k \in \mathbb{Z}} (a_k\, \alpha^{[k]} + b_k\, \beta^{[k]}) \;\middle|\; \begin{array}{c} a_k, b_k \in \mathbb{Z} \\ \exists N, m \in \mathbb{N}\ \forall k \in \mathbb{Z}, |k| > N \,:\, |a_k| \leq m,\ b_k = 0 \end{array} \right\} .$$

Then Γ *is an abelian subgroup of* $\widetilde{\mathrm{Jac}}(\Sigma)$. *We will call* Γ *the* period lattice *of* Σ *(although it is not a discrete subset of* $\widetilde{\mathrm{Jac}}(\Sigma)$*).*
We call the topological quotient space $\mathrm{Jac}(\Sigma) := \widetilde{\mathrm{Jac}}(\Sigma)/\Gamma$ *the* Jacobi variety *of* Σ, *and we denote the canonical projection by* $\pi : \widetilde{\mathrm{Jac}}(\Sigma) \to \mathrm{Jac}(\Sigma)$.

(2) Let $\tau : \mathfrak{C}_{D^o} \to \mathsf{Div}(\Sigma')$ *be the map that associates to each* $(\gamma_k) \in \mathfrak{C}_{D^o}$ *the divisor* $D \in \mathsf{Div}$ *induced by* (γ_k). *Then there exists one and only one map* $\varphi : \mathsf{Div}(\Sigma') \to \mathrm{Jac}(\Sigma)$ *with* $\varphi \circ \tau = \pi \circ \widetilde{\varphi}$, *i.e. so that the following diagram commutes:*

$$\begin{array}{ccc} \mathfrak{C}_{D^o} & \xrightarrow{\ \widetilde{\varphi}\ } & \widetilde{\mathrm{Jac}}(\Sigma) \\ {\scriptstyle \tau}\downarrow & & \downarrow{\scriptstyle \pi} \\ \mathsf{Div}(\Sigma') & \xrightarrow[\ \varphi\ ]{} & \mathrm{Jac}(\Sigma) \end{array}$$

We call φ *the* Abel map *of* Σ.

(3) Change of the origin divisor D^o *corresponds to a linear translation of* φ. *More specifically, if we let* $D^{oo} \in \mathsf{Div}(\Sigma')$ *be another divisor, and denote the Abel maps with the origin divisor* D^o *resp.* D^{oo} *by* φ_{D^o} *resp. by* $\varphi_{D^{oo}}$, *then we have for any* $D \in \mathsf{Div}(\Sigma')$

$$\varphi_{D^{oo}}(D) = \varphi_{D^o}(D) - \varphi_{D^o}(D^{oo}).$$

Proof For (1). Because each sequence $(\alpha_n^{[k]})_{n \in \mathbb{Z}}$ has exactly one non-zero element, we have $(\alpha_n^{[k]})_{n \in \mathbb{Z}} \in \ell^\infty(n) \subset \widetilde{\mathrm{Jac}}(\Sigma)$ by Proposition 18.5(2). Moreover, we have $(\beta_n^{[k]})_{n \in \mathbb{Z}} \in \widetilde{\mathrm{Jac}}(\Sigma)$ by Proposition 18.5(3).

Because of the definition of $\mathfrak{C}_{D^o}$ (in Proposition 18.5) and the fact that the homology group of $\widehat{U}_{k,\delta}$ is generated by A_k for $|k|$ large, it is clear that Γ is the group of periods in $\mathfrak{C}_{D^o}$. Because $(\alpha_n^{[k]})_{n \in \mathbb{Z}}$ is the k-th standard unit vector of the space of sequences $\mathbb{C}^{\mathbb{Z}}$, it also follows from the definition of Γ that $\Gamma \subset \ell^\infty(n) \subset \widetilde{\mathrm{Jac}}(\Sigma)$; for the inclusion $\ell^\infty \subset \widetilde{\mathrm{Jac}}(\Sigma)$ see Proposition 18.5(2).

For (2). Let $(\gamma_k), (\widetilde{\gamma}_k) \in \mathfrak{C}_{D^o}$ be given such that $\tau((\gamma_k)) = \tau((\widetilde{\gamma}_k)) =: D \in \mathsf{Div}(\Sigma')$ holds. We need to show $\widetilde{\varphi}((\gamma_k)) - \widetilde{\varphi}((\widetilde{\gamma}_k)) \in \Gamma$. Without loss of generality we may suppose that the curves γ_k and $\widetilde{\gamma}_k$ are parameterized on the interval $[0, 1]$. Because (γ_k) and $(\widetilde{\gamma}_k)$ induce the same divisor D, there exist permutations $h_0, h_1 : \mathbb{Z} \to \mathbb{Z}$ such that for every $k \in \mathbb{Z}$ we have $\gamma_k(0) = \gamma_{h_0(k)}(0) =: (\lambda_k^o, \mu_k^o)$ and $\gamma_k(1) = \gamma_{h_1(k)}(1) =: (\lambda_k, \mu_k)$; here we have $D^o = \{(\lambda_k^o, \mu_k^o)\}$ and $D = \{(\lambda_k, \mu_k)\}$. By the definition of $\mathfrak{C}_{D^o}$ (in Proposition 18.5) there exists $N \in \mathbb{N}$ so that γ_k and $\widetilde{\gamma}_k$ run entirely in $\widehat{U}_{k,\delta}$ for $|k| > N$. Then $h_0(k) = h_1(k) = k$ holds for all $k \in \mathbb{Z}$ with $|k| > N$, and therefore $h_0|\{-N, \dots, N\}$ and $h_1|\{-N, \dots, N\}$ are permutations of $\{-N, \dots, N\}$.

It follows that the curves $\gamma_k(t)$ and the curves $\widetilde{\gamma}_k(1-t)$, where k runs through all of $\{-N, \dots, N\}$, together define a cycle Z_* on Σ, hence Z_* is a $\mathbb{Z}$-linear combination of some A_k and B_k, say

$$Z_* = \sum_{|j| \leq N'} (a_j\, A_j + b_j\, B_j) \quad \text{with} \quad N' \in \mathbb{N},\ a_j, b_j \in \mathbb{Z}.$$

Moreover for each individual $k \in \mathbb{Z}$ with $|k| > N$, the curves $\gamma_k(t)$ and $\widetilde{\gamma}_k(1-t)$ form a cycle Z_k on Σ that runs entirely in $\widehat{U}_{k,\delta}$, and is therefore a multiple of A_k, say $Z_k = a_k' \cdot A_k$ with $a_k' \in \mathbb{Z}$; here there exists a constant $m > 0$ with $|a_k'| \leq m$ for all k because $(\gamma_k), (\widetilde{\gamma}_k) \in \mathfrak{C}_{D^o}$ holds. Therefore

$$\begin{aligned} \widetilde{\varphi}((\gamma_k)) - \widetilde{\varphi}((\widetilde{\gamma}_k)) &= \left(\int_{Z_*} \omega_n + \sum_{|k|>N} \int_{Z_k} \omega_n \right)_{n \in \mathbb{Z}} \\ &= \sum_{|j| \leq N'} (a_j\, \alpha^{[j]} + b_j\, \beta^{[j]}) + \sum_{|k|>N} a_k'\, \alpha^{[k]} \in \Gamma \end{aligned}$$

holds.

For (3). We write $D = \{(\lambda_k, \mu_k)\}$, $D^o = \{(\lambda_k^o, \mu_k^o)\}$ and $D^{oo} = \{(\lambda_k^{oo}, \mu_k^{oo})\}$ with the usual asymptotic enumeration of asymptotic divisors. For every $k \in \mathbb{Z}$ we let $\gamma_k^{[1]}$ be a path in Σ connecting (λ_k^o, μ_k^o) to $(\lambda_k^{oo}, \mu_k^{oo})$, and let $\gamma_k^{[2]}$ be a path connecting $(\lambda_k^{oo}, \mu_k^{oo})$ to (λ_k, μ_k). We choose these paths so that $(\gamma_k^{[1]}) \in \mathfrak{C}_{D^o}$ and $(\gamma_k^{[2]}) \in \mathfrak{C}_{D^{oo}}$ holds. For $k \in \mathbb{Z}$, let γ_k be the concatenation of the paths $\gamma_k^{[1]}$ and $\gamma_k^{[2]}$, then we also have $(\gamma_k) \in \mathfrak{C}_{D^o}$.

For every $k \in \mathbb{Z}$ and $n \in \mathbb{Z}$ we have

$$\int_{\gamma_k} \omega_n = \int_{\gamma_k^{[1]}} \omega_n + \int_{\gamma_k^{[2]}} \omega_n$$

and therefore

$$\widetilde{\varphi}_{D^o}((\gamma_k)) = \widetilde{\varphi}_{D^o}((\gamma_k^{[1]})) + \widetilde{\varphi}_{D^{oo}}((\gamma_k^{[2]})).$$

Because the divisor induced by (γ_k), $(\gamma_k^{[1]})$ resp. $(\gamma_k^{[2]})$ is D, D^{oo} resp. D, we conclude

$$\varphi_{D^o}(D) = \varphi_{D^o}(D^{oo}) + \varphi_{D^{oo}}(D).$$

□

Remark 18.8 Note that the "period lattice" Γ from Theorem 18.7 is not discrete in $\widetilde{\mathrm{Jac}}(\Sigma)$, not even for $S = \varnothing$, and is therefore in fact not a lattice. (We have $\|(\alpha_n^{[k]})_n\|_{\widetilde{\mathrm{Jac}}(\Sigma)} = \frac{1}{k} \cdot |\varkappa_{k,1} - \varkappa_{k,2}| \in \ell^2(k)$ for $k > 0$, and therefore the $(\alpha_n^{[k]})_n \in \Gamma$ accumulate near $0 \in \widetilde{\mathrm{Jac}}(\Sigma)$.) For this reason we view the Jacobi variety $\mathrm{Jac}(\Sigma) = \widetilde{\mathrm{Jac}}(\Sigma)/\Gamma$ only as a quotient topological space, not as an (infinite-dimensional) manifold. The situation is similar to the one encountered by McKean and Trubowitz in [McK-T] concerning the Jacobi variety for the integrable system associated to Hill's operator: There the period lattice is also not discrete in the respective Banach space, and the Jacobi variety is compact (topologically, it is a product of infinitely many circles) and therefore does not carry the structure of an infinite dimensional manifold, see the discussion in [McK-T, p. 154].

Another peculiarity concerning our period lattice Γ is that in the case $S \neq \varnothing$ it no longer spans $\widetilde{\mathrm{Jac}}(\Sigma)$ over $\mathbb{R}$. (So even if Γ were discrete, it would not be a *maximal* discrete abelian subgroup of $\widetilde{\mathrm{Jac}}(\Sigma)$.) The reason is that for $k \in S$, there is no cycle B_k. Geometrically, one can think of a family of spectral curves with $k \notin S$ where the two branch points $\varkappa_{k,1}$ and $\varkappa_{k,2}$ "close up" to each other; in the limit we then have $\varkappa_{k,1} = \varkappa_{k,2}$ and therefore $k \in S$. This process of "closing up" a pair of branch points causes the period corresponding to the cycle B_k to go to infinity, and this is the reason why there is no period in Γ corresponding to B_k in the limit spectral curve, where $k \in S$ holds. As a consequence, $\mathrm{Jac}(\Sigma)$ is not compact for $S \neq \varnothing$.

Remark 18.9 The Abel map $\varphi : \mathsf{Div}(\Sigma') \to \mathrm{Jac}(\Sigma)$ described in Theorem 18.7 is not a local diffeomorphism in general (not even near non-special divisors), because at least for a divisor $D = \{(\lambda_k, \mu_k)\} \in \mathsf{Div}(\Sigma')$ that does *not* satisfy the condition

$$|\lambda_k - \varkappa_{k,*}| \leq C_\xi \cdot |\varkappa_{k,1} - \varkappa_{k,2}| \tag{18.31}$$

for any $C_\xi > 0$, the Banach space structures of $\mathsf{Div}(\Sigma')$ near D (locally diffeomorphic to $\ell^2_{-1,3}$) and of $\mathrm{Jac}(\Sigma)$ near $\varphi(D)$ (locally diffeomorphic to $\widetilde{\mathrm{Jac}}(\Sigma)$) do not match.

For some applications it would probably be preferable to have a Jacobi variety and associated Abel map for Σ that is a local diffeomorphism near every non-special, asymptotic divisor. This is true in particular where one is interested in divisors with points close to (but not in) singularities of the spectral curve, as they occur for example as spectral divisors of bubbletons (note that the condition (18.31) forces $\lambda_k = \varkappa_{k,*}$ for all $k \in S$). To obtain a Jacobi variety of this kind, one would need to expand the space of 1-forms considered; more specifically, one might work with 1-forms that are square-integrable only on $\widehat{V}_\delta$ for a fixed $\delta > 0$ and are regular on Σ. This would permit one to do away with the condition (18.31), also for the sequence $(\xi_{n,k})$ of the 1-forms $\widetilde{\omega}_n$ of Theorem 18.3(1). However the proof of the analogue of Theorem 18.3 would become more difficult in this setting.

The space $\widetilde{\mathrm{Jac}}(\Sigma)$ plays the role of a tangent space for the Jacobi variety $\mathrm{Jac}(\Sigma)$. In our setting where the period lattice Γ is not discrete, the tangent space of $\mathrm{Jac}(\Sigma)$ is not unique however, and similarly as it is the case for Hill's equation as studied by McKean and Trubowitz in [McK-T], we need to pass to a larger tangent space so that the flow of translations of the potential (which we will study via the Jacobi variety in Chap. 19) is tangential to $\mathrm{Jac}(\Sigma)$. This corresponds to a larger space of curve tuples $\mathfrak{C}_{D^o}$ and a larger Banach space $\widetilde{\mathrm{Jac}}(\Sigma)$, as we construct in the following proposition.

In fact McKean and Trubowitz construct in [McK-T] an entire ascending family of tangent spaces for their Jacobi variety, which correspond to the higher flows of the integrable system associated with Hill's equation. In our setting we cannot define more than the first extension described in the following proposition, because our potentials are only once weakly differentiable, in contrast to the infinitely differentiable potentials in [McK-T].

Proposition 18.10 *We again fix an origin divisor $D^o \in \mathsf{Div}(\Sigma')$. We denote by $\mathfrak{C}^{(1)}_{D^o}$ the set of sequences $(\gamma_k)_{k\in\mathbb{Z}}$ where each γ_k is a curve in Σ' running from a point $(\lambda_k^o, \mu_k^o) \in \Sigma'$ to another point $(\lambda_k, \mu_k) \in \Sigma'$, such that $(\lambda_k^o, \mu_k^o)_{k\in\mathbb{Z}}$ equals the support of D^o and $D := \{(\lambda_k, \mu_k) \,|\, k \in \mathbb{Z}\} \in \mathsf{Div}(\Sigma')$ holds; moreover for large $|k|$ the curve γ_k runs entirely in $\widehat{U}_{k,\delta}$, and there is a number $m_\gamma \in \mathbb{N}$ (depending on γ but not on k) so that the winding number of any γ_k around any branch point or puncture of Σ' is at most $m_\gamma \cdot |k|$.*

For $n \in \mathbb{Z}$ and $(\gamma_k)_{k\in\mathbb{Z}} \in \mathfrak{C}^{(1)}_{D^o}$ the infinite sum

$$\sum_{k\in\mathbb{Z}} \int_{\gamma_k} \omega_n \tag{18.32}$$

converges absolutely in $\mathbb{C}$, and we define the map

$$\widetilde{\varphi}^{(1)}_n : \mathfrak{C}^{(1)}_{D^o} \to \mathbb{C}, \ (\gamma_k)_{k\in\mathbb{Z}} \mapsto \sum_{k\in\mathbb{Z}} \int_{\gamma_k} \omega_n . \tag{18.33}$$

For every $(\gamma_k) \in \mathfrak{C}^{(1)}_{D^o}$ there exist sequences $(b_n), (c_n) \in \ell^2(n)$ so that for $n \in \mathbb{Z}$ with $|n| > N$ (where $N \in \mathbb{N}$ is as in Theorem 18.3(2)) we have

$$\widetilde{\varphi}_n((\gamma_k)) = \frac{\mathrm{sign}(n)}{2\pi \mathrm{i}} \cdot \ln\left(\frac{\lambda_n - \varkappa_{n,*} + \Psi_n(\lambda_n, \mu_n)}{\lambda_n^o - \varkappa_{n,*} + \Psi_n(\lambda_n^o, \mu_n^o)}\right) \cdot (1 + b_n) + c_n , \tag{18.34}$$

where Ψ_n is as in Lemma 16.1, and $\ln(z)$ is the branch of the complex logarithm function with $\ln(1) = 2\pi \mathrm{i} m_n$ with $m_n \in \mathbb{Z}$ being the winding number of γ_n around the pair of branch points $\varkappa_{n,1}$, $\varkappa_{n,2}$.

Moreover with

$$\widetilde{\mathrm{Jac}}^{(1)}(\Sigma) := \{\, (a_n)_{n\in\mathbb{Z}} \,\big|\, a_n \cdot (\varkappa_{n,1} - \varkappa_{n,2}) \in \ell^2_{-2,2}(n) \,\},$$

the map $\widetilde{\varphi}^{(1)} := (\widetilde{\varphi}_n^{(1)})_{n\in\mathbb{Z}}$ *maps into* $\widetilde{\mathrm{Jac}}^{(1)}(\Sigma)$*. We have* $\ell^\infty_{-1,-1}(n) \subset \widetilde{\mathrm{Jac}}^{(1)}(\Sigma)$ *and* $\widetilde{\mathrm{Jac}}(\Sigma) \subset \widetilde{\mathrm{Jac}}^{(1)}(\Sigma)$.

Proof The proof is mostly analogous to that of Proposition 18.5(1),(2). The only substantial difference is in the treatment of $\widetilde{\varphi}_n^{(1)}((\gamma_k))$ in the proof that $\widetilde{\varphi}^{(1)}$ maps into $\widetilde{\mathrm{Jac}}^{(1)}(\Sigma)$ in Proposition 18.5(2). If we suppose that $(\gamma_k) \in \mathfrak{C}^{(1)}_{D^o}$ is given, we still have as in Eq. (18.29)

$$\ln\left(\frac{\lambda_n - \varkappa_{n,*} + \Psi_n(\lambda_n,\mu_n)}{\lambda_n^o - \varkappa_{n,*} + \Psi_n(\lambda_n^o,\mu_n^o)}\right) = \operatorname{arcosh}\left(\frac{\lambda_n - \varkappa_{n,*}}{\frac{1}{2}(\varkappa_{k,1} - \varkappa_{k,2})}\right) + 2\pi\mathrm{i}\, m_n,$$

where m_n is the winding number of γ_n around the pair of branch points $\varkappa_{n,1}$, $\varkappa_{n,2}$. But in contrast to the situation of Proposition 18.5, the sequence $(m_n)_{n\in\mathbb{Z}}$ is no longer bounded here, rather there exists a constant $m_\gamma \in \mathbb{N}$ so that $|m_n| \leq m_\gamma \cdot |n|$ holds for all n. Thus we obtain instead of (18.30) that there exist $C_5, C_6 > 0$ so that

$$|\widetilde{\varphi}_n^{(1)}((\gamma_k))| \leq C_5 \left|\operatorname{arcosh}\left(\frac{\lambda_n - \varkappa_{n,*}}{\frac{1}{2}(\varkappa_{n,1} - \varkappa_{n,2})}\right)\right| + C_6\,|n| \leq 2\,C_5\,C_{\mathrm{arcosh}} \left|\frac{\lambda_n - \varkappa_{n,1}}{\varkappa_{n,1} - \varkappa_{n,2}}\right| + C_6\,|n|$$

and therefore

$$\begin{aligned} |\widetilde{\varphi}_n^{(1)}((\gamma_k))| \cdot |\varkappa_{n,1} - \varkappa_{n,2}| &\leq 2\,C_5\,C_{\mathrm{arcosh}}\,|\lambda_n - \varkappa_{n,1}| \\ &\quad + C_6\,|\varkappa_{n,1} - \varkappa_{n,2}| \cdot |n| l \in \ell^2_{-2,2}(n \in \mathbb{Z} \setminus S), \end{aligned}$$

whence $(\widetilde{\varphi}_n^{(1)}((\gamma_k))) \in \widetilde{\mathrm{Jac}}^{(1)}(\Sigma)$ follows. □

Chapter 19
The Jacobi Variety and Translations of the Potential

Let (Σ, D) be spectral data corresponding to a simply periodic solution $u : X \to \mathbb{C}$ of the sinh-Gordon equation, where $X \subset \mathbb{C}$ is a horizontal strip with $0 \in X$.

Throughout this entire work, we have constructed the spectral data (Σ, D) corresponding to u with respect to the monodromy at the base point $z = 0$. In the present chapter, we would now like to describe the spectral data (Σ_{z_0}, D_{z_0}) corresponding to the monodromy based at some other point $z_0 \in X$, or equivalently, the spectral data (in the previous sense, via the monodromy at the base point $z = 0$) of the translated potential $z \mapsto u(z + z_0)$.

First we note that the spectral curve does not change at all under such a translation, i.e. $\Sigma_{z_0} = \Sigma$ holds. Indeed by Eq. (3.3), the monodromy M_{z_0} satisfies with respect to z_0 the differential equation

$$\mathrm{d}_{z_0} M_{z_0}(\lambda) = [\alpha_\lambda(z_0), M_{z_0}(\lambda)],$$

where α_λ is the connection form associated to the potential u as in Eq. (3.2). If we denote by $F_\lambda(z)$ the extended frame associated to u, then $\widetilde{M}_{z_0}(\lambda) := F_\lambda(z_0) \cdot M(\lambda) \cdot F_\lambda(z_0)^{-1}$ satisfies the same differential equation with the same initial value $\widetilde{M}_{z_0=0}(\lambda) = M(\lambda)$ at $z_0 = 0$. Hence we have $M_{z_0}(\lambda) = F_\lambda(z_0) \cdot M(\lambda) \cdot F_\lambda(z_0)^{-1}$. Therefore the eigenvalues of the monodromy, and thus the spectral curve, do not depend on z_0.

But the spectral divisor $D = \{ (\lambda_k, \mu_k) \mid k \in \mathbb{Z} \}$ does change under translation. We will describe the motion of the divisor points on Σ in terms of the image of the divisor under the Abel map, i.e. in the Jacobi variety, via ordinary differential equations for the translation in x-direction and in y-direction. Similarly as is well-known for finite-type potentials, it will turn out that both translations correspond to linear motions, that is to constant vector fields, in the Jacobi variety. In the case of the translation in x-direction, the direction of that linear motion is precisely a

S. Klein, *A Spectral Theory for Simply Periodic Solutions of the Sinh-Gordon Equation*, Lecture Notes in Mathematics 2229,
https://doi.org/10.1007/978-3-030-01276-2_19

lattice direction of the Jacobi variety, corresponding to the fact that the potential u is periodic in the x-direction. These results are explicated in Theorem 19.1 below.

In the sequel, we consider either a potential $(u, u_y) \in \mathsf{Pot}$, or a simply periodic solution of the sinh-Gordon equation $u : X \to \mathbb{C}$ with period 1 defined on a horizontal strip $X \subset \mathbb{C}$ with 0 in the interior of X. In the former case we consider only the translation in x-direction (all references to translations in the direction of y and associated objects are then to be disregarded), and in the latter case, we consider both translations in x-direction and in y-direction. We let $(\Sigma, D(x))$ resp. $(\Sigma, D(y))$ be the spectral data of the potential translated in x-direction resp. y-direction, i.e. of $u(z+x)$ resp. of $u(z+\mathrm{i}y)$. Like in Chap. 18 we continue to require that the spectral curve Σ does not have any singularities besides ordinary double points, i.e. that $\Delta^2 - 4$ does not have any zeros of order ≥ 3, where $\Delta = \mu + \mu^{-1}$ is the trace function of Σ. We then let $(\omega_n)_{n\in\mathbb{Z}}$ be the canonical basis of $\Omega(\Sigma)$ (Theorem 18.3(2)), $\mathrm{Jac}(\Sigma)$ be the Jacobi variety and $\varphi : \mathsf{Div}(\Sigma') \to \mathrm{Jac}(\Sigma)$ the Abel map of Σ (Theorem 18.7), where we choose $D^o := D(0)$ as the origin divisor for the construction of the Abel map.

In the sequel we will look at the derivatives $\frac{\partial \varphi_n}{\partial x}$ and $\frac{\partial \varphi_n}{\partial y}$ of the n-th Jacobi coordinate φ_n. For these derivatives to make sense, we need to define Jacobi coordinates φ_n of $D(x)$ resp. $D(y)$ at least for small $|x|$ resp. $|y|$ for all $n \in \mathbb{Z}$. For this purpose we write $D(x) = \{\lambda_k(x), \mu_k(x)\}$ for small $|x|$ and then consider for fixed x and all $k \in \mathbb{Z}$ the curve $\gamma_{x,k} : [0, x] \to \Sigma,\ t \mapsto (\lambda_k(t), \mu_k(t))$. Because the map $\mathsf{Pot} \to \mathsf{Div}$ is asymptotically close to the Fourier transform of the potential, $\gamma_{x=1,k}$ winds $|k|$ times around the pair of branch points $\varkappa_{k,1}$, $\varkappa_{k,2}$ for $|k|$ large; it follows that we do not have $(\gamma_{x,k}) \in \mathfrak{C}_{D^o}$ (Proposition 18.5), but we do have $(\gamma_{x,k}) \in \mathfrak{C}^{(1)}_{D^o}$ (Proposition 18.10). Therefore we can define Jacobi coordinates for the translation in x-direction in the vicinity of $D(0)$ by

$$\varphi_n(x) := \widetilde{\varphi}_n^{(1)}(\gamma_{x,k}) \quad \text{for } n \in \mathbb{Z},$$

where $\widetilde{\varphi}_n^{(1)}$ is as in Proposition 18.10. A similar construction applies for the translation in y-direction; here it is relevant that for sufficiently small $|y|$ and large $|k|$, the divisor point $(\lambda_k(y), \mu_k(y))$ remains in the excluded domain $\widehat{U}_{k,\delta}$ because the asymptotic estimates then apply to the translated potentials uniformly in y.

We denote by $\frac{\partial \varphi_n}{\partial x}$ resp. $\frac{\partial \varphi_n}{\partial y}$ the derivative of the Jacobi coordinate $\varphi_n(x)$ resp. $\varphi_n(y)$ with respect to x resp. y.

Theorem 19.1 *There exist sequences $a_n^x, a_n^y \in \ell^2_{-1,-1}(n)$ (dependent only on the spectral curve Σ) so that under translation of the potential u in the direction of x resp. y, the Jacobi coordinates φ_n ($n \in \mathbb{Z}$) follow the differential equations*

$$\frac{\partial \varphi_n}{\partial x} = n + a_n^x,$$

$$\frac{\partial \varphi_n}{\partial y} = -\mathrm{i}|n| + a_n^y.$$

Moreover we have $a_n^x = 0$ *for* $|n|$ *large, and for every* $n \in \mathbb{Z}$, $\frac{\partial \varphi_n}{\partial x}$ *corresponds to a member of the period lattice, i.e. there exists a cycle* Z_n *of* Σ' *so that* $\frac{\partial \varphi_n}{\partial x} = \int_{Z_n} \omega_n$ *holds. (No similar statements apply to the sequence* (a_n^y) *in general.)*

The statement of the preceding theorem that $\frac{\partial \varphi_n}{\partial x}$ corresponds to a member of the period lattice is an expression of the fact that the corresponding potential u is periodic in x-direction. This is the reason why there is no analogous statement for $\frac{\partial \varphi_n}{\partial y}$.

The proof of Theorem 19.1 is the objective of the remainder of the present chapter. At the heart of the proof of Theorem 19.1 is a general construction of linear flows in the Picard variety of a Riemann surface X (the space of isomorphy classes of line bundles on X) known as the *Krichever construction*. In the sequel we will carry out the proof in our specific situation; we will then discuss the relationship of the proof to the Krichever construction in Remark 19.10.

Proposition 19.2 *Suppose that the spectral divisor* D *is tame. Under translation of the potential* u *in the direction of* x *resp.* y, *the Jacobi coordinates* φ_n $(n \in \mathbb{Z})$ *of the spectral divisor follow the differential equations*

$$\frac{\partial \varphi_n}{\partial x} = -\sum_{k \in \mathbb{Z}} \alpha_{21}^x(\lambda_k) \cdot \Phi_n(\lambda_k) \cdot \frac{1}{c'(\lambda_k)}$$

$$\frac{\partial \varphi_n}{\partial y} = -\sum_{k \in \mathbb{Z}} \alpha_{21}^y(\lambda_k) \cdot \Phi_n(\lambda_k) \cdot \frac{1}{c'(\lambda_k)}$$

with

$$\alpha_{21}^x(\lambda) = \frac{1}{4}(\lambda\tau + \tau^{-1}) \quad \textit{and} \quad \alpha_{21}^y(\lambda) = \frac{\mathrm{i}}{4}(-\lambda\tau + \tau^{-1}), \quad \textit{where} \quad \tau := \mathrm{e}^{-u/2}.$$

Proof We have

$$\frac{\partial \varphi_n}{\partial x} = \sum_{k \in \mathbb{Z}} \frac{\partial \varphi_n}{\partial \lambda_k} \cdot \frac{\partial \lambda_k}{\partial x}. \tag{19.1}$$

From the construction of the lift of the Abel map $\widetilde{\varphi}^{(1)}$ in Proposition 18.10 it follows that the "canonical map", i.e. the derivative of the Jacobi coordinate φ_n with respect to the divisor coordinate λ_k is given by

$$\frac{\partial \varphi_n}{\partial \lambda_k} = \frac{\omega_n}{\mathrm{d}\lambda}(\lambda_k, \mu_k) = \frac{\Phi_n(\lambda_k)}{\mu_k - \mu_k^{-1}}, \tag{19.2}$$

where the holomorphic function Φ_n is as in Theorem 18.3(2). To calculate $\frac{\partial \lambda_k}{\partial x}$, we regard $c(\lambda) = c(\lambda, x)$ and $\lambda_k = \lambda_k(x)$ also as functions of x. By definition of the spectral divisor, we have $c(\lambda_n(x), x) = 0$, and by differentiating this equation we obtain

$$\frac{\partial \lambda_k}{\partial x} = -\left(\frac{\partial c}{\partial \lambda}\bigg|_{\lambda=\lambda_k} \right)^{-1} \cdot \frac{\partial c}{\partial x}\bigg|_{\lambda=\lambda_k}, \tag{19.3}$$

where $\frac{\partial c}{\partial \lambda}\big|_{\lambda=\lambda_k} \neq 0$ holds because D is tame. By Eq. (3.3), the monodromy M_{z_0} satisfies with respect to z_0 the differential equation

$$\mathrm{d}_{z_0} M_{z_0}(\lambda) = [\alpha_\lambda(z_0), M_{z_0}(\lambda)],$$

where α_λ is the connection form associated to the potential u as in Eq. (3.2), and therefore we have

$$\frac{\partial c}{\partial x} = ([\alpha, M])_{21} = \alpha_{21}^x\, a + \alpha_{22}^x\, c - \alpha_{11}^x\, c - \alpha_{21}^x\, d.$$

Because of $c(\lambda_n(x), x) = 0$ it follows that

$$\frac{\partial c}{\partial x}\bigg|_{\lambda=\lambda_n} = \alpha_{21}^x(\lambda_k)\,(a(\lambda_k) - d(\lambda_k)) = \alpha_{21}^x(\lambda_k)\,(\mu_k - \mu_k^{-1}).$$

By plugging this equation into Eq. (19.3), we obtain

$$\frac{\partial \lambda_k}{\partial x} = -\frac{1}{c'(\lambda_k)} \cdot \alpha_{21}^x(\lambda_k) \cdot (\mu_k - \mu_k^{-1}), \tag{19.4}$$

and by plugging Eqs. (19.2) and (19.4) into Eq. (19.1), we obtain

$$\begin{aligned} \frac{\partial \varphi_n}{\partial x} &= \sum_{k\in\mathbb{Z}} \frac{\Phi_n(\lambda_k)}{\mu_k - \mu_k^{-1}} \cdot \frac{-1}{c'(\lambda_k)} \cdot \alpha_{21}^x(\lambda_k) \cdot (\mu_k - \mu_k^{-1}) \\ &= -\sum_{k\in\mathbb{Z}} \alpha_{21}^x(\lambda_k) \cdot \Phi_n(\lambda_k) \cdot \frac{1}{c'(\lambda_k)}. \end{aligned}$$

The differential equation for $\frac{\partial \varphi_n}{\partial y}$ is proven in literally the same way, by just replacing every x with y. □

We also note the following:

Corollary 19.3 *If $D = \{(\lambda_k, \mu_k)\} \in \mathsf{Div}$ is given, and $\widetilde{D} = \{(\widetilde{\lambda}_k, \widetilde{\mu}_k)\}$ is the image of D under the hyperelliptic involution $\sigma : \Sigma \to \Sigma$, then we have $\frac{\partial \widetilde{\lambda}_n}{\partial x} = -\frac{\partial \lambda_n}{\partial x}$ and $\frac{\partial \widetilde{\lambda}_n}{\partial y} = -\frac{\partial \lambda_n}{\partial y}$.*

Proof The statement for $\frac{\partial \lambda_n}{\partial x}$ follows from Eq. (19.4):

We first note that τ is determined by the λ_k up to sign by the trace formula in Theorem 11.2(3), and is therefore up to sign invariant under the hyperelliptic involution σ. Both $\alpha_{21}^x(\lambda_n)$ and $c'(\lambda_n)$ depend only on the λ_k and on τ, and they both change their sign when τ changes sign. It follows that $-\frac{1}{c'(\lambda_k)} \cdot \alpha_{21}^x(\lambda_n)$ is invariant under σ. Because $\mu_k - \mu_k^{-1}$ changes sign under σ, it follows from Eq. (19.4) that $\frac{\partial \lambda_n}{\partial x}$ changes sign under σ.

The statement for $\frac{\partial \lambda_n}{\partial y}$ is again shown in the same way. □

The presentation of the differential equations for translating the potential in Proposition 19.2 is not yet satisfactory, because their right hand side appears to vary with λ_k, whereas we are expecting that the Jacobi coordinates change linearly under translation, as explained at the beginning of the chapter, that is, $\frac{\partial \varphi_n}{\partial x}$ and $\frac{\partial \varphi_n}{\partial y}$ should be constant. Moreover, the description in Proposition 19.2 is applicable only if the spectral divisor D is tame. In the following Proposition 19.6 we will show that the translational flows of the Jacobi coordinates are indeed linear, and also dispose of the restriction to tame spectral divisors D.

In preparation for Proposition 19.6 we define residues in 0 and in ∞ for meromorphic 1-forms on $\widehat{V}_\delta$.

For this purpose we let $\widehat{K}_r(0)$ be a cycle in Σ that is obtained by lifting a counter-clockwise parameterization of two traversals of a circle of radius $r > 0$ around 0 in $\mathbb{C}^*$ to Σ. We also put $\widehat{K}_r(\infty) := -\widehat{K}_{1/r}(0)$. Then we choose a sequence $(r_n)_{n\geq 1}$ with $r_n > 0$ and $\lim_{n\to\infty} r_n = 0$, such that $\widehat{K}_{r_n}(0)$ and $\widehat{K}_{r_n}(\infty)$ are contained in $\widehat{V}_\delta$ for all $n \geq 1$. For any meromorphic 1-form η defined on $\widehat{V}_\delta$ we then define

$$\mathrm{Res}_0(\eta) := \frac{1}{2\pi \mathrm{i}} \lim_{n\to\infty} \int_{\widehat{K}_{r_n}(0)} \eta$$

$$\text{and} \quad \mathrm{Res}_\infty(\eta) := \frac{1}{2\pi \mathrm{i}} \lim_{n\to\infty} \int_{\widehat{K}_{r_n}(\infty)} \eta = -\frac{1}{2\pi \mathrm{i}} \lim_{n\to\infty} \int_{\widehat{K}_{1/r_n}(0)} \eta,$$

provided that these limits exist for all possible choices of the sequence (r_n). If this is the case, then the values of the limits do not depend on the choice of the sequence (r_n) used to calculate them. It should be noted that because $\widehat{K}_r(0)$ resp. $\widehat{K}_r(\infty)$ winds twice around $\lambda = 0$ resp. $\lambda = \infty$, we have for the meromorphic 1-form $\frac{\mathrm{d}\lambda}{\lambda}$ on Σ

$$\mathrm{Res}_0\left(\frac{\mathrm{d}\lambda}{\lambda}\right) = 2 \quad \text{and} \quad \mathrm{Res}_\infty\left(\frac{\mathrm{d}\lambda}{\lambda}\right) = -2. \tag{19.5}$$

The orientation of the paths of integration was chosen such that if η is in fact a meromorphic 1-form on Σ, then

$$\sum_P \mathrm{Res}_P(\eta) + \mathrm{Res}_0(\eta) + \mathrm{Res}_\infty(\eta) = 0 \tag{19.6}$$

holds; here the sum runs over all $P \in \Sigma$ where η has a pole.

Lemma 19.4

(1) Let h be a holomorphic function on $\widehat{V}_\delta$.

(a) If $h \in \mathrm{As}_0(\widehat{V}_\delta, \ell^2_{-1}, 0)$ holds, then we have $\mathrm{Res}_0(h \cdot \omega_n) = 0$.

(b) If $h \in \mathrm{As}_\infty(\widehat{V}_\delta, \ell^2_{-1}, 0)$ holds, then we have $\mathrm{Res}_\infty(h \cdot \omega_n) = 0$.

(2) (a) There exists a sequence $(a_n)_{n\in\mathbb{Z}}$ (dependent only on the spectral curve Σ) with $(a_n)_{n>0} \in O(n^{-1})$ and $(a_n)_{n<0} \in \ell^2_{-1}(n)$ such that for every $h \in \mathrm{As}_0(\widehat{V}_\delta, \ell^\infty_{-1}, 0)$ for which $(h \cdot \sqrt{\lambda})(0) := \lim_{\lambda\to 0, \lambda\in V_\delta}\left(h(\lambda)\cdot\sqrt{\lambda}\right)$ exists in $\mathbb{C}$ we have

$$\mathrm{Res}_0(h \cdot \omega_n) = \begin{cases} \left(h\cdot\sqrt{\lambda}\right)(0)\cdot a_n & \text{for } n>0 \\ \left(h\cdot\sqrt{\lambda}\right)(0)\cdot(-4in + a_n) & \text{for } n<0 \end{cases}. \tag{19.7}$$

(b) There exists a sequence $(a_n)_{n\in\mathbb{Z}}$ (dependent only on the spectral curve Σ) with $(a_n)_{n>0} \in \ell^2_{-1}(n)$ and $(a_n)_{n<0} \in O(n^{-1})$ such that for every $h \in \mathrm{As}_\infty(\widehat{V}_\delta, \ell^\infty_{-1}, 0)$ for which $(\frac{h}{\sqrt{\lambda}})(\infty) := \lim_{\lambda\to\infty, \lambda\in V_\delta}\left(\frac{h(\lambda)}{\sqrt{\lambda}}\right)$ exists in $\mathbb{C}$ we have

$$\mathrm{Res}_\infty(h \cdot \omega_n) = \begin{cases} \left(\frac{h}{\sqrt{\lambda}}\right)(\infty)\cdot(-4in + a_n) & \text{for } n>0 \\ \left(\frac{h}{\sqrt{\lambda}}\right)(\infty)\cdot a_n & \text{for } n<0 \end{cases}. \tag{19.8}$$

Remark 19.5 In Eqs. (19.7) and (19.8), the factors $-4in + a_n$ resp. a_n are essentially the value of $2\,\omega_n$ at $\lambda = 0$ resp. $\lambda = \infty$ in a sense that is made precise by the proof of the lemma below. We write this factor differently for $n > 0$ resp. $n < 0$ to obtain a sequence a_n that becomes small for $|n| \to \infty$. The different behavior of a_n for $n \to \infty$ and for $n \to -\infty$ (for example in Lemma 19.4(2)(a) we have $a_n = O(n^{-1})$ for $n > 0$ but $a_n \in \ell^2_{-1}(n)$ for $n < 0$) occurs because in one case the λ_n approach the point where the residue is taken, whereas in the other case the λ_n are far away from this point.

Proof (of Lemma 19.4) At first, we consider in both (1) and (2) the case where the residue of $h \cdot \omega_n$ is calculated at $\lambda = 0$. For this purpose, we suppose that $h \in \mathrm{As}_0(\widehat{V}_\delta, \ell^\infty_{-1}, 0)$ holds and that for $\widetilde{h} := h \cdot \sqrt{\lambda} \in \mathrm{As}_0(\widehat{V}_\delta, \ell^\infty_0, 0)$, the limit $\widetilde{h}(0) := \lim_{\lambda\to 0, \lambda\in V_\delta} \widetilde{h}(\lambda)$ exists in $\mathbb{C}$.

By Theorem 18.3, we have $\omega_n = \frac{\Phi_n(\lambda)}{\mu-\mu^{-1}}\,d\lambda$, where Φ_n is a linear combination of product functions of the kind investigated in Proposition 17.2. By Proposition 17.3(1) it therefore follows that we have $f_n := \frac{\Phi_n(\lambda)}{\mu-\mu^{-1}}|\widehat{V}_\delta \in \mathrm{As}_0(\widehat{V}_\delta, \ell^\infty_{-1}, 0)$. Thus $\widetilde{f}_n := \sqrt{\lambda}\cdot f_n$ is holomorphic on $\widehat{V}_\delta$ with $\widetilde{f}_n \in \mathrm{As}(\widehat{V}_\delta, \ell^\infty_0, 0)$. The asymptotic estimate of Proposition 17.3(1) (with $\varrho = -1$) moreover shows that the limit $\widetilde{f}_n(0) := \lim_{\lambda\to 0, \lambda\in V_\delta} \widetilde{f}_n(\lambda)$ exists in $\mathbb{C}$.

We now obtain

$$\begin{aligned}\mathrm{Res}_0(h\cdot\omega_n) &= \mathrm{Res}_0\left(h\cdot\frac{\Phi_n(\lambda)}{\mu-\mu^{-1}}\,\mathrm{d}\lambda\right) = \mathrm{Res}_0\left(\frac{1}{\sqrt{\lambda}}\,\widetilde{h}(\lambda)\cdot\frac{1}{\sqrt{\lambda}}\,\widetilde{f}_n(\lambda)\,\mathrm{d}\lambda\right)\\ &= \widetilde{h}(0)\cdot\widetilde{f}_n(0)\cdot\mathrm{Res}_0\left(\frac{\mathrm{d}\lambda}{\lambda}\right)\overset{(19.5)}{=}2\cdot\widetilde{h}(0)\cdot\widetilde{f}_n(0).\end{aligned} \tag{19.9}$$

From this calculation, (1)(a) follows: If we have $h\in\mathrm{As}_0(\widehat{V}_\delta,\ell^2_{-1},0)$, then we have $\widetilde{h}\in\mathrm{As}_0(\widehat{V}_\delta,\ell^2_0,0)$ and therefore $\widetilde{h}(0)=0$. Thus we obtain $\mathrm{Res}_0(h\cdot\omega_n)=0$ from Eq. (19.9).

For the proof of (2)(a) we put

$$a_n := \begin{cases} 2\,\widetilde{f}_n(0) & \text{for } n>0\\ 2\widetilde{f}_n(0)+4\mathrm{i}n & \text{for } n<0\end{cases};$$

it follows from Eq. (19.9) that with this choice of (a_n), Eq. (19.7) holds. Because $\widetilde{f}_n(0)$ depends only on the spectral curve Σ, not on h, also the sequence (a_n) depends only on the spectral curve. It remains to show that (a_n) has the claimed asymptotic behavior, i.e. that $(a_n)_{n>0}\in O(n^{-1})$ and $(a_n)_{n<0}\in\ell^2_{-1}(n)$ holds.

For this purpose we need to determine $\widetilde{f}_n(0)$ more precisely; it suffices to consider $|n|>N$, where $N\in\mathbb{N}$ is the constant from Theorem 18.3(2). From Theorem 18.3(2) and Proposition 17.3(1) it follows that for $n>N$ we have

$$\widetilde{f}_n-(-2\mathrm{i})\cdot\left(s_{n,n}\,\tau_{\xi_n}^{-2}\,\frac{16\pi^2n^2}{\lambda_{n,0}-\lambda}+\sum_{|\ell|\leq N}s_{n,\ell}\,\tau_{\xi_\ell}^{-2}\,\frac{16\pi^2\ell^2}{\lambda_{\ell,0}-\lambda}\right)\in\mathrm{As}_0(\widehat{V}_\delta,\ell^2_0,0),$$

where $s_{n,n}$ and $s_{n,\ell}$ are the constants from Theorem 18.3(2), and therefore

$$\widetilde{f}_n(0)=(-2\mathrm{i})\cdot\left(s_{n,n}\,\tau_{\xi_n}^{-2}\,\frac{16\pi^2n^2}{\lambda_{n,0}}+\sum_{|\ell|\leq N}s_{n,\ell}\,\tau_{\xi_\ell}^{-2}\,\frac{16\pi^2\ell^2}{\lambda_{\ell,0}}\right).$$

Because the $s_{n,n}$ and the $s_{n,\ell}$ are of order n^{-1} by Theorem 18.3(2), it follows that we have $\widetilde{f}_n(0)\in O(n^{-1})$ and hence $a_n=2\,\widetilde{f}_n(0)\in O(n^{-1})$ for $n>0$.

For $n<-N$ we have again by Theorem 18.3(2) and Proposition 17.3(1)

$$\widetilde{f}_n(0)=(-2\mathrm{i})\cdot\left(s_{n,n}\,\tau_{\xi_n}^{-2}\,\frac{-1}{\lambda_{n,0}}+\sum_{|\ell|\leq N}s_{n,\ell}\,\tau_{\xi_\ell}^{-2}\,\frac{-1}{\lambda_{\ell,0}}\right).$$

By the asymptotic description of $s_{n,n}$ in Eq. (18.3) and again the fact that $s_{n,\ell}$ is of order $|n|^{-1}$ for $|\ell| \leq N$ we conclude

$$\widetilde{f}_n(0) = -\frac{1}{2\pi \mathrm{i}} \lambda_{n,0}^{-1/2} + \ell^2_{-1}(n) = -2\mathrm{i}n + \ell^2_{-1}(n)$$

and therefore $a_n = 2\,\widetilde{f}_n(0) + 4\mathrm{i}n \in \ell^2_{-1}(n)$ for $n < 0$.

To also obtain the residue in $\lambda = \infty$ in (1)(b) and (2)(b), we could apply (1)(a) resp. (2)(a) to the function $\check{h} := h \circ (\lambda \mapsto \lambda^{-1})$ on the surface $\check{\Sigma} := ((\lambda, \mu) \mapsto (\lambda^{-1}, \mu))(\Sigma)$. But to avoid technical difficulties in phrasing this transformation, we rather carry out a calculation analogous to the one for $\lambda = 0$. We suppose that $h \in \mathrm{As}_\infty(\widehat{V}_\delta, \ell^\infty_{-1}, 0)$ holds, and that for $\widetilde{h} := \frac{h}{\sqrt{\lambda}} \in \mathrm{As}_\infty(\widehat{V}_\delta, \ell^\infty_0, 0)$ the limit $\widetilde{h}(\infty) := \lim_{\lambda \to \infty, \lambda \in V_\delta} \widetilde{h}(\lambda)$ exists in $\mathbb{C}$. By Proposition 17.3(1), $\frac{\Phi_n(\lambda)}{\mu - \mu^{-1}} | \widehat{V}_\delta \in \mathrm{As}_\infty(\widehat{V}_\delta, \ell^\infty_3, 0)$ holds, and thus $\widetilde{f}_n := \lambda^{3/2} \cdot \frac{\Phi_n(\lambda)}{\mu - \mu^{-1}} | \widehat{V}_\delta \in \mathrm{As}_\infty(\widehat{V}_\delta, \ell^\infty_0, 0)$; it also follows from Proposition 17.3(1) that the limit $\widetilde{f}_n(\infty) := \lim_{\lambda \to \infty, \lambda \in V_\delta} \widetilde{f}_n(\lambda)$ exists in $\mathbb{C}$. We thus obtain

$$\begin{aligned} \mathrm{Res}_\infty(h \cdot \omega_n) &= \mathrm{Res}_\infty\left(h \cdot \frac{\Phi_n(\lambda)}{\mu - \mu^{-1}}\, \mathrm{d}\lambda\right) = \mathrm{Res}_\infty\left(\sqrt{\lambda}\, \widetilde{h}(\lambda) \cdot \frac{1}{\lambda^{3/2}}\, \widetilde{f}_n(\lambda)\, \mathrm{d}\lambda\right) \\ &= \widetilde{h}(\infty) \cdot \widetilde{f}_n(\infty) \cdot \mathrm{Res}_\infty\left(\frac{\mathrm{d}\lambda}{\lambda}\right) \overset{(19.5)}{=} -2 \cdot \widetilde{h}(\infty) \cdot \widetilde{f}_n(\infty). \end{aligned} \tag{19.10}$$

Similarly as before, (1)(b) immediately follows from Eq. (19.10), and we calculate

$$\widetilde{f}_n(\infty) = \begin{cases} 2\mathrm{i}n + \ell^2_{-1}(n) & \text{for } n > 0 \\ O(n^{-1}) & \text{for } n < 0 \end{cases},$$

whence (2)(b) also follows. □

The following proposition is crucial. It shows in particular that the translations of the potential correspond to linear motions in the Jacobi variety.

Proposition 19.6 *Let* $M(\lambda) = \begin{pmatrix} a(\lambda) & b(\lambda) \\ c(\lambda) & d(\lambda) \end{pmatrix}$ *be the monodromy associated to the spectral divisor* D*, and let* α^x_{21} *and* α^y_{21} *be as in Proposition 19.2. For* $n \in \mathbb{Z}$*, we then consider the meromorphic* 1*-forms on* Σ

$$\chi^x_n := \alpha^x_{21} \cdot \frac{\mu - d}{c} \cdot \omega_n \quad \text{and} \quad \chi^y_n := \alpha^y_{21} \cdot \frac{\mu - d}{c} \cdot \omega_n.$$

Then the residues $\mathrm{Res}_0(\chi_n)$ *and* $\mathrm{Res}_\infty(\chi_n)$ *depend only on the spectral curve* Σ*, not on the spectral divisor* D *under consideration. Moreover, there exist sequences* $(a^x_n)_{n \in \mathbb{Z}}, (a^y_n)_{n \in \mathbb{Z}} \in \ell^2_{-1,-1}(n)$ *(depending only on the spectral curve* Σ*), so that* *the Jacobi coordinates* φ_n $(n \in \mathbb{Z})$ *satisfy the following differential equations under translation of the potential* u *in the direction of* x *resp.* y*:*

$$\frac{\partial \varphi_n}{\partial x} = \mathrm{Res}_0(\chi_n^x) + \mathrm{Res}_\infty(\chi_n^x) = n + a_n^x$$

$$\frac{\partial \varphi_n}{\partial y} = \mathrm{Res}_0(\chi_n^y) + \mathrm{Res}_\infty(\chi_n^y) = -\mathrm{i}|n| + a_n^y .$$

Proof Because the set of tame divisors on Σ' is dense in $\mathsf{Div}(\Sigma')$ and the right hand side of the differential equations claimed does not depend on the spectral divisor D, it suffices to consider the case where the spectral divisor D is tame. For the same reason, we may further suppose without loss of generality that no point in the support of D is a branch point of Σ'.

We let α_{21} be one of the functions α_{21}^x, α_{21}^y, and correspondingly let χ_n be one of χ_n^x, χ_n^y. χ_n is a meromorphic 1-form on Σ; because we supposed that D is tame, χ_n has simple poles in the points (λ_k, μ_k), $k \in \mathbb{Z}$ comprising the support of D, and no other poles. We compute the residue of χ_n in (λ_k, μ_k): We have

$$\chi_n = \alpha_{21} \cdot \frac{\mu - d}{c} \cdot \omega_n = \frac{\alpha_{21}(\lambda) \cdot (\mu - d(\lambda)) \cdot \Phi_n(\lambda)}{\mu - \mu^{-1}} \cdot \frac{\mathrm{d}\lambda}{c(\lambda)} .$$

Because (λ_k, μ_k) is not a branch point of Σ', the function $\frac{\alpha_{21}(\lambda)\cdot(\mu - d(\lambda))\cdot\Phi_n(\lambda)}{\mu-\mu^{-1}}$ is holomorphic at (λ_k, μ_k). Moreover, $c(\lambda)$ has a zero of order 1 and $d(\lambda_k) = \mu_k^{-1}$ holds, and thus we obtain

$$\mathrm{Res}_{(\lambda_k,\mu_k)}(\chi_n) = \frac{\alpha_{21}(\lambda_k)\cdot(\mu_k - d(\lambda_k))\cdot\Phi_n(\lambda_k)}{\mu_k - \mu_k^{-1}} \cdot \frac{1}{c'(\lambda_k)} = \alpha_{21}(\lambda_k) \cdot \Phi_n(\lambda_k) \cdot \frac{1}{c'(\lambda_k)} .$$

By Proposition 19.2 and Eq. (19.6) we therefore obtain

$$\frac{\partial \varphi_n}{\partial x} = -\sum_{k\in\mathbb{Z}} \mathrm{Res}_{(\lambda_k,\mu_k)}(\chi_n) = \mathrm{Res}_0(\chi_n) + \mathrm{Res}_\infty(\chi_n).$$

We now proceed to calculate the residues of χ_n in $\lambda = 0$ and in $\lambda = \infty$. We will see that these residues do not depend on the divisor D. For this, the trace formula of Theorem 11.2(3) turns out to be crucial: This formula shows that the constant $\tau = \mathrm{e}^{-u(0)/2}$ occurring in the component α_{21} of the connection form α associated to the potential u equals the constant $\tau = \left(\prod_{k\in\mathbb{Z}} \frac{\lambda_{k,0}}{\lambda_k}\right)^{1/2}$ occurring in the description of the asymptotic behavior of c given in Proposition 10.1.

By Corollary 11.4(6) we have

$$\frac{\mu - d(\lambda)}{c(\lambda)} - \frac{\mathrm{i}}{\tau\sqrt{\lambda}} \in \mathrm{As}_\infty(\widehat{V}_\delta, \ell_1^2, 0) \quad \text{and} \quad \frac{\mu - d(\lambda)}{c(\lambda)} - \frac{\mathrm{i}}{\tau^{-1}\sqrt{\lambda}} \in \mathrm{As}_0(\widehat{V}_\delta, \ell_{-1}^2, 0),$$

and therefore

$$\alpha_{21}(\lambda) \cdot \left(\frac{\mu - d(\lambda)}{c(\lambda)} - \frac{\mathrm{i}}{\tau\sqrt{\lambda}} \right) \in \mathrm{As}_\infty(\widehat{V}_\delta, \ell^2_{-1}, 0)$$
$$\text{and} \quad \alpha_{21}(\lambda) \cdot \left(\frac{\mu - d(\lambda)}{c(\lambda)} - \frac{\mathrm{i}}{\tau^{-1}\sqrt{\lambda}} \right) \in \mathrm{As}_0(\widehat{V}_\delta, \ell^2_{-1}, 0).$$

We therefore obtain by Lemma 19.4(1),(2)

$$\begin{aligned} \mathrm{Res}_0(\chi_n) &= \mathrm{Res}_0\left(\alpha_{21} \cdot \frac{\mu - d}{c} \cdot \omega_n\right) = \mathrm{Res}_0\left(\alpha_{21} \cdot \frac{\mathrm{i}}{\tau^{-1}\sqrt{\lambda}} \cdot \omega_n\right) \\ &= \left(\frac{\alpha_{21} \cdot \mathrm{i}}{\tau^{-1}}\right)(0) \cdot \begin{cases} O(n^{-1}) & \text{for } n > 0 \\ -4\mathrm{i}n + \ell^2_{-1}(n) & \text{for } n < 0 \end{cases} \end{aligned}$$

and

$$\begin{aligned} \mathrm{Res}_\infty(\chi_n) &= \mathrm{Res}_\infty\left(\alpha_{21} \cdot \frac{\mu - d}{c} \cdot \omega_n\right) = \mathrm{Res}_\infty\left(\alpha_{21} \cdot \frac{\mathrm{i}}{\tau\sqrt{\lambda}} \cdot \omega_n\right) \\ &= \left(\frac{\alpha_{21} \cdot \mathrm{i}}{\tau \cdot \lambda}\right)(\infty) \cdot \begin{cases} -4\mathrm{i}n + \ell^2_{-1}(n) & \text{for } n > 0 \\ O(n^{-1}) & \text{for } n < 0 \end{cases}, \end{aligned}$$

where the sequences in $O(n^{-1})$ and $\ell^2_{-1}(n)$ occurring here depend only on the spectral curve Σ, not on the divisor D.

When the translation in x direction is concerned, we have $\alpha^x_{21} = \frac{1}{4}(\lambda\tau + \tau^{-1})$ and therefore

$$\left(\frac{\alpha^x_{21} \cdot \mathrm{i}}{\tau^{-1}}\right)(0) = \left(\frac{\alpha^x_{21} \cdot \mathrm{i}}{\tau \cdot \lambda}\right)(\infty) = \frac{\mathrm{i}}{4},$$

whence

$$\frac{\partial \varphi_n}{\partial x} = \mathrm{Res}_0(\chi^x_n) + \mathrm{Res}_\infty(\chi^x_n) = \frac{\mathrm{i}}{4} \cdot (-4\mathrm{i}n + \ell^2_{-1,-1}(n)) = n + \ell^2_{-1,-1}(n)$$

follows, again with an $\ell^2_{-1,-1}(n)$-sequence that depends only on Σ. On the other hand, for the translation in y direction, we have $\alpha^y_{21} = \frac{\mathrm{i}}{4}(-\lambda\tau + \tau^{-1})$ and therefore

$$\left(\frac{\alpha^y_{21} \cdot \mathrm{i}}{\tau^{-1}}\right)(0) = -\frac{1}{4} \quad \text{and} \quad \left(\frac{\alpha^y_{21} \cdot \mathrm{i}}{\tau \cdot \lambda}\right)(\infty) = \frac{1}{4},$$

whence

$$\frac{\partial \varphi_n}{\partial y} = \mathrm{Res}_0(\chi_n^y) + \mathrm{Res}_\infty(\chi_n^y) = \begin{cases} \frac{1}{4} \cdot (-4in + \ell^2_{-1}(n)) & \text{for } n > 0 \\ -\frac{1}{4} \cdot (-4in + \ell^2_{-1}(n)) & \text{for } n < 0 \end{cases}$$
$$= -\mathrm{i}|n| + \ell^2_{-1,-1}(n)$$

follows, where the $\ell^2_{-1,-1}(n)$-sequence depends only on Σ. □

For the translation in x-direction, we expect a better result still. Our potentials are periodic in that direction, so we expect that $\frac{\partial \varphi_n}{\partial x}$ is an *exact*, not only an asymptotically approximate multiple of a lattice vector of the Jacobi variety of Σ. (For the translation in y-direction we cannot expect a similar result, because u is not periodic in the direction of y.) Because the coordinates (λ_n, μ_n) of the spectral divisor corresponding to some $(u, u_y) \in \mathsf{Pot}$ are asymptotically close to the n-th Fourier coefficient of u_z resp. $u_{\bar{z}}$, we more specifically expect that $\frac{\partial \varphi_n}{\partial x} = n$ holds at least for $|n|$ large.

The periodicity of the potential u corresponds to the existence of the global function μ on the spectral curve Σ, and we base our proof of the expected statement on the existence of a global logarithm $\ln(\mu)$ of that function on a surface $\widetilde{\Sigma}$ with boundary obtained from Σ by cutting along certain curves. The following lemma serves to establish the topological prerequisites for the cutting process.

Lemma 19.7 *The branch points resp. singularities $\varkappa_{k,\nu}$ of the spectral curve Σ can be numbered in such a way that besides the previous requirements ($\varkappa_{k,\nu} \in \widehat{U}_{k,\delta}$ for $|k|$ large and $\varkappa$ is a double point of Σ if and only if there exists $k \in S$ with $\varkappa_{k,1} = \varkappa_{k,2} = \varkappa$) the following holds:*

For every $k \in \mathbb{Z}$ there exists a curve ϑ_k in the λ-plane, which for $k \notin S$ connects $\varkappa_{k,1}$ to $\varkappa_{k,2}$ whereas for $k \in S$ is a small circle around $\varkappa_{k,*}$ that does not include any other branch points of Σ, such that none of the ϑ_k intersect, and such that for every $k \in \mathbb{Z}$ we have

$$\int_{\vartheta_k} \frac{\mathrm{d}\mu}{\mu} = 0.$$

Proof We consider the potential $(u, u_y) \in \mathsf{Pot}$ corresponding to the spectral data (Σ, D^o), and the family $u_t := (t \cdot u, t \cdot u_y) \in \mathsf{Pot}$ with $t \in [0, 1]$ describing the deformation of the vacuum ($u = 0$) into that potential. For $t \in [0, 1]$ we let $\Sigma(t)$ be the spectral curve corresponding to u_t. The set $\{ u_t \,|\, t \in [0, 1] \}$ is relatively compact in Pot, therefore the asymptotic estimates apply uniformly to this set of potentials. Hence there exist continuous functions $\varkappa_{k,\nu}(t) : [0, 1] \to \mathbb{C}^*$ so that $\varkappa_{k,\nu}(0) = \lambda_{k,0}$ holds (these points are the double points of the spectral curve $\Sigma(0) = \Sigma_0$ of the vacuum) and so that for every $t \in [0, 1]$, the $\varkappa_{k,\nu}(t)$ are the branch points resp. singularities of $\Sigma(t)$; moreover there exists $N \in \mathbb{N}$ so that $\varkappa_{k,\nu}(t) \in U_{k,\delta}$ holds for all $k \in \mathbb{Z}$ with $|k| > N$ and for all $t \in [0, 1]$. Note that it is possible to choose these functions as continuous even if some of the $\varkappa_{k,\nu}(t)$ coincide for some $t \in [0,1]$.

Further there exist for every $t \in [0,1]$ continuous curves $\vartheta_{k,t} : [0,1] \to \mathbb{C}^*$ which connect $\varkappa_{k,1}(t)$ to $\varkappa_{k,2}(t)$. They can be chosen such that they depend continuously also on $t \in [0,1]$, that they run entirely within $U_{k,\delta}$ for $|k| > N$ and that they are constant if $\varkappa_{k,1}(t) = \varkappa_{k,2}(t)$ holds. Moreover after possibly modifying the $\varkappa_{k,\nu}(t)$ past the times t where two $\varkappa_{k,\nu}(t)$ intersect, it is possible to choose the $\vartheta_{k,t}$ so that for every $t \in [0,1]$ no two $\vartheta_{k,t}$ intersect.

Note that for every $t \in [0,1]$ the lift onto $\Sigma(t)$ of $\vartheta_{k,t}$ traversed once in the usual direction and then once in the opposite direction is a closed curve through $\varkappa_{k,1}(t)$ and $\varkappa_{k,2}(t)$ that meets no other branch points of $\Sigma(t)$, and therefore $\int_{\vartheta_{k,t}} \frac{\mathrm{d}\mu}{\mu}$ is in any event an integer multiple of $\mathrm{i}\pi$ (here μ denotes the corresponding parameter of the hyperelliptic curve $\Sigma(t)$). Because this integral depends continuously on t, and we have $\int_{\vartheta_{k,0}} \frac{\mathrm{d}\mu}{\mu} = 0$ because $\vartheta_{k,0}$ is constant, it follows that

$$\int_{\vartheta_{k,t}} \frac{\mathrm{d}\mu}{\mu} = 0 \quad \text{holds for all } t \in [0,1] \text{ and } k \in \mathbb{Z}.$$

Because of the hypothesis that $\Sigma = \Sigma(1)$ does not have any singularities other than ordinary double points, no more than two of the $\varkappa_{k,\nu}(t=1)$ can coincide. However it is possible for $\varkappa_{k_1,\nu_1}(t=1) = \varkappa_{k_2,\nu_2}(t=1)$ to hold, where $k_1 \neq k_2$ and $\nu_1, \nu_2 \in \{1,2\}$. In this case, we put

$$\varkappa_{k_1,1} := \varkappa_{k_1,\nu_1}(t=1) \qquad \varkappa_{k_2,1} := \varkappa_{k_1,3-\nu_1}(t=1)$$

$$\varkappa_{k_1,2} := \varkappa_{k_2,\nu_2}(t=1) \qquad \varkappa_{k_2,2} := \varkappa_{k_2,3-\nu_2}(t=1),$$

and we let ϑ_{k_1} be a small circle around $\varkappa_{k_1,1} = \varkappa_{k_1,2}$ that does not encircle any other branch points of Σ. Moreover if $\varkappa_{k_2,1} \neq \varkappa_{k_2,2}$ we let ϑ_{k_2} be a curve close and homologically equivalent to the concatenation of $\vartheta_{k_1,t=1}$ and $\vartheta_{k_2,t=1}$ in the appropriate direction to connect $\varkappa_{k_2,1}$ with $\varkappa_{k_2,2}$, but which avoids the circle ϑ_{k_1} around $\varkappa_{k_1,1} = \varkappa_{k_1,2}$; if $\varkappa_{k_2,1} = \varkappa_{k_2,2}$ we let ϑ_{k_2} again be a small circle around $\varkappa_{k_2,1} = \varkappa_{k_2,2}$ that does not encircle any other branch points of Σ.

In all other cases we let $\varkappa_{k,\nu} := \varkappa_{k,\nu}(t=1)$, and if $\varkappa_{k,1} \neq \varkappa_{k,2}$ holds we let $\vartheta_k := \vartheta_{k,t=1}$, whereas for $\varkappa_{k,1} = \varkappa_{k,2}$ we let ϑ_k be a small circle around $\varkappa_{k,1} = \varkappa_{k,2}$ that does not encircle any other branch points of Σ. □

Proposition 19.8 *Suppose that the branch points $\varkappa_{k,\nu}$ of Σ are numbered as in Lemma 19.7, and that the cycles A_k from the canonical basis (A_k, B_k) of the homology of Σ are chosen as the lift onto Σ of two traversals of the curves ϑ_k from Lemma 19.7.*

Then we let $\widetilde{\Sigma}$ be the surface with boundary that is obtained from Σ by cutting along all the cycles A_n. $\widetilde{\Sigma}$ contains as boundaries two copies of each A_n, which *we denote by* A_n^+ *and* A_n^- *(as cycles with the same orientation as* A_n*), where* B_n *runs from* A_n^+ *to* A_n^-*. We view* $\widehat{V}_\delta$ *as a subset of* $\widetilde{\Sigma}$*.*

On $\widetilde{\Sigma}$, $\ln(\mu)$ *exists as a global, holomorphic function. For* $|n|$ *large, and corresponding points* μ^+ *and* μ^- *on* A_n^+ *resp. on* A_n^-*, we have* $\ln(\mu^+) - \ln(\mu^-) = -2\pi \mathrm{i} n$*. Moreover* $\big(\ln(\mu) - \mathrm{i}\,\zeta(\lambda)\big)|\widehat{V}_\delta \in \mathrm{As}(\widehat{V}_\delta, \ell^2_{0,0}, 0)$ *holds.*

Proof By Lemma 19.7 we have for any $k \in \mathbb{Z}$

$$\int_{A_k} \mathrm{d}(\ln(\mu)) = 2\int_{\vartheta_k} \frac{\mathrm{d}\mu}{\mu} = 0. \tag{19.11}$$

For any cycle Z on Σ,

$$\int_Z \mathrm{d}(\ln(\mu)) = \int_Z \frac{\mathrm{d}\mu}{\mu}$$

is $2\pi \mathrm{i}$ times the winding number of $\mu \circ Z$ around 0. If the cycle Z passes from an excluded domain $\widehat{U}_{k,\delta}$ to $\widehat{U}_{k\pm 1,\delta}$, μ changes from values near ± 1 to values near ∓ 1, and therefore we have for the cycle $Z = B_k$ with $|k|$ large (this cycle connects $\widehat{U}_{k,\delta}$ to $\widehat{U}_{0,\delta}$)

$$\int_{B_k} \mathrm{d}(\ln(\mu)) = 2\pi \mathrm{i} k. \tag{19.12}$$

Because the homology group of $\widetilde{\Sigma}$ is generated by the $A_k^\pm$, it follows from Eq. (19.11) that $\ln(\mu)$ is a global holomorphic function on $\widetilde{\Sigma}$. Moreover, it follows from Eq. (19.12) that the values of $\ln(\mu)$ at corresponding points of A_n^+ and A_n^- differ by $-2\pi \mathrm{i} n$.

Finally, concerning the asymptotic behavior of $\ln(\mu)$ near $\lambda = 0$, $\lambda = \infty$, we note that by Corollary 11.4(4), we have $(\mu - \mu_0)|\widehat{V}_\delta \in \mathrm{As}(\widehat{V}_\delta, \ell^2_{0,0}, 1)$ and therefore $\left(\frac{\mu-\mu_0}{\mu_0}\right)\Big|\,\widehat{V}_\delta \in \mathrm{As}(\widehat{V}_\delta, \ell^2_{0,0}, 0)$ because on $\widehat{V}_\delta$, μ_0 is comparable to $w(\lambda)$. Here from we obtain by the equality

$$\ln(\mu) - \ln(\mu_0) = \ln\left(\frac{\mu}{\mu_0}\right) = \ln\left(1 + \frac{\mu - \mu_0}{\mu_0}\right)$$

and the Taylor expansion $\ln(1+z) = z + O(z^2)$ of the logarithm near $z = 1$ that $\ln(\mu) - \ln(\mu_0) \in \mathrm{As}(\widehat{V}_\delta, \ell^2_{0,0}, 0)$ holds. Because of $\ln(\mu_0) = \mathrm{i}\zeta(\lambda)$ (see Eq. (4.7)), the claimed asymptotic behavior follows. □

Proposition 19.9 *Under translation of the potential* u *in the direction of* x*, the Jacobi coordinates* φ_n *with* $|n|$ *large satisfy the differential equation*

$$\frac{\partial \varphi_n}{\partial x} = n.$$

Proof For $n \in \mathbb{Z}$ we consider the holomorphic 1-form $\eta_n := \ln(\mu) \cdot \omega_n$ on the surface with boundary $\widetilde{\Sigma} \supset \widehat{V}_\delta$ from Proposition 19.8. We claim that

$$\mathrm{Res}_0(\eta_n) = \mathrm{Res}_0(\chi_n^x) \quad \text{and} \quad \mathrm{Res}_\infty(\eta_n) = \mathrm{Res}_\infty(\chi_n^x) \tag{19.13}$$

holds. Indeed, by the calculations in the proof of Proposition 19.6 we have

$$\left(\alpha_{21}^x \frac{\mu - d}{c} - \frac{\mathrm{i}\sqrt{\lambda}}{4}\right)\Bigg| \widehat{V}_\delta \in \mathrm{As}_\infty(\widehat{V}_\delta, \ell^2_{-1}, 0)$$

$$\text{and} \quad \left(\alpha_{21}^x \frac{\mu - d}{c} - \frac{\mathrm{i}}{4\sqrt{\lambda}}\right)\Bigg| \widehat{V}_\delta \in \mathrm{As}_0(\widehat{V}_\delta, \ell^2_{-1}, 0),$$

and therefore

$$\left(\alpha_{21}^x \frac{\mu - d}{c} - \mathrm{i}\,\zeta(\lambda)\right)\Bigg| \widehat{V}_\delta \in \mathrm{As}(\widehat{V}_\delta, \ell^2_{-1,-1}, 0)$$

holds. By Proposition 19.8 we also have

$$(\ln(\mu) - \mathrm{i}\,\zeta(\lambda))|\,\widehat{V}_\delta \in \mathrm{As}(\widehat{V}_\delta, \ell^2_{0,0}, 0),$$

and thus we obtain

$$\left(\alpha_{21}^x \frac{\mu - d}{c} - \ln(\mu)\right)\Bigg| \widehat{V}_\delta \in \mathrm{As}(\widehat{V}_\delta, \ell^2_{-1,-1}, 0).$$

It therefore follows from Lemma 19.4(1) that Eqs. (19.13) hold.

Due to the topology of $\widetilde{\Sigma}$, we have

$$\sum_{k \in \mathbb{Z}} \frac{1}{2\pi\mathrm{i}} \left(\int_{A_k^+} \eta_n - \int_{A_k^-} \eta_n \right) + \mathrm{Res}_0(\eta_n) + \mathrm{Res}_\infty(\eta_n) = 0. \tag{19.14}$$

For $|k|$ large, the function values of $\ln(\mu)$ at corresponding points of A_k^+ and A_k^- differ by $-2\pi\mathrm{i}k$ (see Proposition 19.8), and thus

$$\int_{A_k^+} \eta_n - \int_{A_k^-} \eta_n = -2\pi\mathrm{i}k \cdot \int_{A_k} \omega_n = -2\pi\mathrm{i}k \cdot \delta_{k,n}$$

holds. Therefore it follows from Eq. (19.14) for $|n|$ large that

$$\mathrm{Res}_0(\eta_n) + \mathrm{Res}_\infty(\eta_n) = n \tag{19.15}$$

holds.

From Proposition 19.6 we now obtain

$$\frac{\partial \varphi_n}{\partial x} = \mathrm{Res}_0(\chi_n^x) + \mathrm{Res}_\infty(\chi_n^x) \overset{(19.13)}{=} \mathrm{Res}_0(\eta_n) + \mathrm{Res}_\infty(\eta_n) \overset{(19.15)}{=} n.$$

□

Proof (of Theorem 19.1) In view of Propositions 19.6 and 19.9 it only remains to show that for $n \in \mathbb{Z}$ there exists a cycle Z_n on Σ' so that $\frac{\partial \varphi_n}{\partial x} = \int_{Z_n} \omega_n$ holds. To show this we choose a tame divisor D on Σ; by Theorem 15.1 there exists a periodic potential $(u, u_y) \in \mathsf{Pot}$ with spectral data (Σ, D). Because this potential is periodic in x-direction, the flow corresponding to x-translation in the Jacobi variety is also periodic, and therefore the Jacobi coordinate φ_n changes under the flow of x-translation along a period of (u, u_y) by a lattice vector of the Jacobi variety. Thus there exists a cycle Z_n on Σ' so that $\frac{\partial \varphi_n}{\partial x} = \int_{Z_n} \omega_n$ holds. □

Remark 19.10 The preceding proofs of this chapter bears a relationship to a construction principle for linear flows on the Picard variety of a compact Riemann surface X that is due to Krichever [Kr].

We now describe this construction principle in general terms for a compact Riemann surface X. We mark points $x_1, \dots, x_n \in X$ on X and fix local coordinates $z_1, \dots, z_n$ of X around these points with $z_k(x_k) = 0$. Let H be the linear space of meromorphic function germs near $0 \in \mathbb{C}$ (i.e. H is the space of Laurent series in one complex variable with positive convergence radius). For every $(h_1, \dots, h_n) \in H^n$ there exist neighborhoods U_k of x_k in X such that $z_k^* h_k$ is a holomorphic function on $U_k \setminus \{x_k\}$. With the additional neighborhood $U_0 := X \setminus \{x_1, \dots, x_n\}$, $\mathcal{U} := \{U_0, U_1, \dots, U_n\}$ is an open covering of X, and $(z_k^* h_k)_{k=1,\dots,n}$ defines a cocycle with respect to the covering $\mathcal{U}$ and thus induces an element in the cohomology group $H^1(X, \mathcal{O})$ (where $\mathcal{O}$ denotes the sheaf of holomorphic functions on X). Thereby we obtain a map $\psi : H^n \to H^1(X, \mathcal{O})$.

The short exact sequence of sheaves

$$0 \longrightarrow \mathbb{Z} \longrightarrow \mathcal{O} \longrightarrow \mathcal{O}^* \longrightarrow 0$$

(where $\mathcal{O}^*$ denotes the subsheaf of $\mathcal{O}$ of invertible function germs, and the map $\mathcal{O} \longrightarrow \mathcal{O}^*$ is $f \mapsto \exp(f)$) induces a long exact sequence of cohomology groups, see [For, Theorem 15.12, p. 123]:

$$0 \longrightarrow H^0(X, \mathbb{Z}) \longrightarrow H^0(X, \mathcal{O}) \longrightarrow H^0(X, \mathcal{O}^*) \longrightarrow H^1(X, \mathbb{Z}) \longrightarrow$$
$$\longrightarrow H^1(X, \mathcal{O}) \longrightarrow H^1(X, \mathcal{O}^*) \longrightarrow H^2(X, \mathbb{Z}) \longrightarrow H^2(X, \mathcal{O}) \longrightarrow \dots.$$

The sequence $0 \longrightarrow H^0(X, \mathbb{Z}) \longrightarrow H^0(X, \mathcal{O}) \longrightarrow H^0(X, \mathcal{O}^*) \longrightarrow 0$ splits off, and moreover we have $H^2(X, \mathcal{O}) = 0$. Thus we obtain the exactness of the sequence

$$0 \longrightarrow H^1(X, \mathbb{Z}) \longrightarrow H^1(X, \mathcal{O}) \longrightarrow H^1(X, \mathcal{O}^*) \longrightarrow H^2(X, \mathbb{Z}) \longrightarrow 0.$$

Here we note that the group $H^1(X, \mathcal{O}^*)$ is the Picard variety of X, which is isomorphic to the space of isomorphy classes of line bundles on X, and the map $H^1(X, \mathcal{O}^*) \longrightarrow H^2(X, \mathbb{Z})$ in the above sequence is the degree map of line bundles. Therefore the Jacobi variety $H^1(X, \mathcal{O})$ of X is the Lie algebra of the Picard variety $H^1(X, \mathcal{O}^*)$.

For this reason, any $h = (h_1, \ldots, h_n) \in H^n$ defines a one-parameter subgroup $(L_h(t))_{t\in\mathbb{R}}$ in $H^1(X, \mathcal{O}^*)$ so that $\psi(h) \in H^1(X, \mathcal{O})$ is the tangent vector to this family at $t = 0$. Explicitly, the cocycle defining the line bundle $L_h(t)$ on X is given by $(z_k^* \exp(t\, h_k))_{k=1,\ldots,n}$. The property that $L_h(t)$ is a one-parameter subgroup means that we have $L_h(t + t') = L_h(t) \otimes L_h(t')$ for $t, t' \in \mathbb{R}$. Moreover it can be shown that the flow $L_h(t)$ thus induced by some $h \in H^n$ is periodic with period 1 (i.e. $L_h(t + 1) = L_h(t)$ for all t) if and only if the Mittag-Leffler distribution given by h is solvable by means of a multi-valued, meromorphic function on X whose function values at a point differ by elements of $2\pi\mathrm{i}\mathbb{Z}$.

The Krichever construction applies to the situation of the present chapter in the following way: We carry out the analogous construction on the spectral curve Σ or on the surface $\widetilde{\Sigma}$ constructed in Proposition 19.8, which unlike the previous surface X are non-compact and can have singularities. We mark the two points 0 and ∞, and consider the local coordinates $\sqrt{\lambda}$ and $\frac{1}{\sqrt{\lambda}}$ around these points. This choice of marked points and coordinates singles out the spectral curve of the integrable system for the sinh-Gordon equation among spectral curves of other integrable systems. Let $h_x = (h_{x,0}, h_{x,\infty})$ resp. $h_y = (h_{y,0}, h_{y,\infty})$ be the pair of Laurent series near $0 \in \mathbb{C}$ which each have a pole of order 1 and whose principal part is characterized by

$$\operatorname{Res}_0(h_{x,0}) = \operatorname{Res}_0\left(\alpha_{21}^x \cdot \frac{\mu - d}{c}\right) \quad \text{and} \quad \operatorname{Res}_0(h_{x,\infty}) = \operatorname{Res}_\infty\left(\alpha_{21}^x \cdot \frac{\mu - d}{c}\right)$$

resp. the analogous equations for h_y; here Proposition 19.6 shows that these residues do not depend on the monodromy $M(\lambda) = \left(\begin{smallmatrix} a & b \\ c & d \end{smallmatrix}\right)$ with spectral curve Σ used in the equations. Then the spectral divisor D of this monodromy corresponds to the line bundle Λ with the holomorphic section $(\frac{\mu-d}{c}, 1)$; when we translate the divisor in x-direction resp. in y-direction, the translated divisor corresponds to the line bundle $\Lambda \otimes L_{h_x}(t)$ resp. $\Lambda \otimes L_{h_y}(t)$. This is the interpretation of Proposition 19.6 in the context of the Krichever construction. Concerning the periodicity of the x-translation, Proposition 19.9 shows that the function $\ln(\mu)$ constructed on $\widetilde{\Sigma}$ in Proposition 19.8 solves the Mittag-Leffler distribution given by h_x as a multi-valued, meromorphic function, whose function values at a point differ by an integer multiple of $2\pi\mathrm{i}$. Thus it follows that the 1-parameter group $L_{h_x}(t)$ induced by the x-translation is periodic, as we expected.

Chapter 20
Asymptotics of Spectral Data for Potentials on a Horizontal Strip

As a final result, we study the asymptotic behavior of the spectral data (Σ, D) corresponding to a simply periodic solution $u : X \to \mathbb{C}$ of the sinh-Gordon equation defined on an entire horizontal strip $X \subset \mathbb{C}$ with positive height. Because such a solution is real analytic on the interior of X, we expect a far better asymptotic for such spectral data than for the spectral data of Cauchy data potentials (u, u_y) with only the weak requirements $u \in W^{1,2}([0,1])$, $u_y \in L^2([0,1])$ we have been using throughout most of the paper. More specifically, we expect both the distance of branch points $\varkappa_{k,1} - \varkappa_{k,2}$ of the spectral curve Σ and the distance of the corresponding spectral divisor points to the branch points to fall off exponentially for $k \to \pm\infty$.

The following theorem shows that this expectation is correct:

Theorem 20.1 *Let $X \subset \mathbb{C}$ be a closed, horizontal strip in $\mathbb{C}$ of height $2y_0$ with $y_0 > 0$, i.e. $X = \{z \in \mathbb{C} \,\big|\, |\operatorname{Im}(z)| \leq y_0\}$, and let $u : X \to \mathbb{C}$ be a simply periodic solution of the sinh-Gordon equation $\Delta u + \sinh(u) = 0$, such that for every $y \in [-y_0, y_0]$ (even on the boundary!) we have $u(\cdot + \mathrm{i}y) \in \mathsf{Pot}$ and $\|u(\cdot + \mathrm{i}y)\|_{\mathsf{Pot}} \leq C_1$ with $C_1 > 0$. We let Σ be the spectral curve corresponding to u (with branch points $\varkappa_{n,\nu}$, and $\varkappa_{n,*} := \frac{1}{2}(\varkappa_{n,1} + \varkappa_{n,2})$) and let $D := \{(\lambda_n, \mu_n)\}_{n\in\mathbb{Z}}$ be the spectral divisor of u along the real axis.*

Then there exists a constant $C > 0$ and a sequence $(s_n)_{n\in\mathbb{Z}} \in \ell^2_{0,0}(n)$ of real numbers so that

$$|\varkappa_{n,1} - \varkappa_{n,2}| \leq C\,\mathrm{e}^{-2\pi\,(1-s_n)\,|n|\,y_0}, \tag{20.1}$$

$$|\lambda_n - \varkappa_{n,*}| \leq C\,\mathrm{e}^{-2\pi\,(1-s_n)\,|n|\,y_0}, \tag{20.2}$$

$$\text{and} \quad |\mu_n - (-1)^n| \leq C\,\mathrm{e}^{-\pi\,(1-s_n)\,|n|\,y_0}. \tag{20.3}$$

S. Klein, *A Spectral Theory for Simply Periodic Solutions of the Sinh-Gordon Equation*, Lecture Notes in Mathematics 2229,
https://doi.org/10.1007/978-3-030-01276-2_20

Proof For $y \in [-y_0, y_0]$ we let $D(y) := \{(\lambda_n(y), \mu_n(y))\}_{n\in\mathbb{Z}}$ be the spectral divisor of $u(\cdot + \mathrm{i}y) \in \mathsf{Pot}$, then $D(0) = D$ holds. Note that all the divisors $D(y)$ lie on the same spectral curve Σ as we noted at the beginning of Chap. 19.

If Σ has a singularity at some $\varkappa_{n,*}$ with $n \in \mathbb{Z}$, and $\lambda_n(y) = \varkappa_{n,*}$ holds for some $y \in [-y_0, y_0]$, then we will also have $\lambda_n(y) = \varkappa_{n,*}$ for all $y \in [-y_0, y_0]$; conversely if there exists $y \in [-y_0, y_0]$ with $\lambda_n(y) \neq \varkappa_{n,*}$, then $\lambda_n(y)$ will avoid the singularity $\varkappa_{n,*}$ for all times $y \in [-y_0, y_0]$. For this reason we may suppose without loss of generality that $D(y) \in \mathsf{Div}(\Sigma')$ holds for all $y \in [-y_0, y_0]$.

For given $n \in \mathbb{Z}$, we let Ψ_n be as in Lemma 16.1 and define the abbreviation

$$\Theta_n^{\pm}(\lambda, \mu) := \lambda - \varkappa_{n,*} \pm \Psi_n(\lambda, \mu).$$

We clearly have

$$\Theta_n^{+}(\lambda, \mu) + \Theta_n^{-}(\lambda, \mu) = 2 \cdot (\lambda - \varkappa_{n,*}) \tag{20.4}$$

and because we have $\Psi_n \circ \sigma = -\Psi_n$ (where $\sigma : \Sigma \to \Sigma$ denotes the hyperelliptic involution of Σ), we have

$$\Theta_n^{\pm} \circ \sigma = \Theta_n^{\mp}. \tag{20.5}$$

Moreover a straight-forward calculation using Eq. (16.1) shows

$$\Theta_n^{+}(\lambda, \mu) \cdot \Theta_n^{-}(\lambda, \mu) = (\lambda - \varkappa_{n,*})^2 - \Psi_n(\lambda)^2 = \frac{1}{4}\,(\varkappa_{n,1} - \varkappa_{n,2})^2. \tag{20.6}$$

We now let $y_1 := \operatorname{sign}(n) \cdot y_0$, and choose the divisor $D^o := D(y_1) \in \mathsf{Div}(\Sigma')$ as origin divisor; with this divisor we consider Jacobi coordinates $\varphi_n(D(y))$ for $D(y)$ like in Chap. 19. We then have on one hand by Proposition 18.10

$$\varphi_n(D(0)) = \frac{\operatorname{sign}(n)}{2\pi \mathrm{i}} \cdot \ln\left(\frac{\Theta_n^{+}(\lambda_n(0), \mu_n(0))}{\Theta_n^{+}(\lambda_n(y_1), \mu_n(y_1))}\right) \cdot (1 + a_n) + b_n$$

with sequences $(a_n), (b_n) \in \ell^2_{0,0}(n)$, on the other hand by Theorem 19.1

$$\varphi_n(D(0)) = \varphi_n(D(0)) - \varphi_n(D(y_1)) = (-\mathrm{i}|n| + c_n) \cdot (-y_1)$$

with a sequence $c_n \in \ell^2_{-1,-1}(n)$. By comparing these two equations, it follows that we have

$$\begin{aligned}\ln\left(\frac{\Theta_n^{+}(\lambda_n(0), \mu_n(0))}{\Theta_n^{+}(\lambda_n(y_1), \mu_n(y_1))}\right) &= \frac{2\pi \mathrm{i}\,\operatorname{sign}(n) \cdot \big((\mathrm{i}\,|n| - c_n)\, y_1 - b_n\big)}{1 + a_n} \\ &= -2\pi n y_1 + r_n = -2\pi |n| y_0 + r_n\end{aligned}$$

with

$$r_n := -2\pi n y_1 \cdot \left(\frac{1}{1+a_n} - 1 \right) - \frac{2\pi \mathrm{i}\, \mathrm{sign}(n)\, (c_n\, y_0 + b_n)}{1+a_n} \in \ell^2_{-1,-1}(n),$$

and therefore

$$\Theta_n^+(\lambda_n(0), \mu_n(0)) = \mathrm{e}^{-2\pi |n| y_0 + r_n} \cdot \Theta_n^+(\lambda_n(y_1), \mu_n(y_1)). \tag{20.7}$$

We now also consider the divisor $\widetilde{D}(0) := \sigma(D(0))$. Because of Corollary 19.3, the y-translational flow $\widetilde{D}(y)$ of $\widetilde{D}(0)$ is defined for all times $y \in [-y_0, y_0]$, and we have

$$\widetilde{D}(y) = \sigma(D(-y)). \tag{20.8}$$

Repeating the calculation leading to Eq. (20.7) with $\widetilde{D}(y)$ in the place of $D(y)$, we obtain

$$\Theta_n^+(\widetilde{\lambda}_n(0), \widetilde{\mu}_n(0)) = \mathrm{e}^{-2\pi |n| y_0 + \widetilde{r}_n} \cdot \Theta_n^+(\widetilde{\lambda}_n(y_1), \widetilde{\mu}_n(y_1))$$

with another sequence $\widetilde{r}_n \in \ell^2_{-1,-1}(n)$. Because of Eqs. (20.8) and (20.5), it follows that we have

$$\Theta_n^-(\lambda_n(0), \mu_n(0)) = \mathrm{e}^{-2\pi |n| y_0 + \widetilde{r}_n} \cdot \Theta_n^-(\lambda_n(-y_1), \mu_n(-y_1)). \tag{20.9}$$

By taking the sum of Eqs. (20.7) and (20.9) and applying Eq. (20.4), we obtain

$$2 \cdot (\lambda_n(0) - \varkappa_{n,*}) = \mathrm{e}^{-2\pi |n| y_0} \cdot \left(\mathrm{e}^{r_n} \cdot \Theta_n^+(\lambda_n(y_1), \mu_n(y_1)) + \mathrm{e}^{\widetilde{r}_n} \cdot \Theta_n^-(\lambda_n(-y_1), \mu_n(-y_1)) \right). \tag{20.10}$$

We have $\Theta_n^\pm(\lambda_n(\pm y_1), \mu_n(\pm y_1)) \in \ell^2_{-1,3}(n)$ and therefore in any case

$$|\Theta_n^\pm(\lambda_n(\pm y_1), \mu_n(\pm y_1))| \leq C_1 \cdot |n| = C_1 \cdot \mathrm{e}^{\widehat{r}_n} \tag{20.11}$$

with $\widehat{r}_n := \ln |n| \in \ell^2_{-1,-1}(n)$ and a constant $C_1 > 0$. By plugging this into Eq. (20.10) we see that

$$|\lambda_n(0) - \varkappa_{n,*}| \leq C_2 \cdot \mathrm{e}^{-2\pi |n| (1-s_n) y_0}$$

holds with another constant $C_2 > 0$ and with

$$s_n := \frac{1}{2\pi |n|\, y_0} \cdot (\max\{\mathrm{Re}(r_n), \mathrm{Re}(\widetilde{r}_n)\} + \widehat{r}_n) \in \ell^2_{0,0}(n),$$

showing Eq. (20.2).

Moreover, by taking the product of Eqs. (20.7) and (20.9), we obtain

$$\begin{aligned}&\Theta_n^+(\lambda_n(0),\mu_n(0))\cdot\Theta_n^-(\lambda_n(0),\mu_n(0))\\&\quad=\mathrm{e}^{-4\pi|n|y_0+r_n+\widetilde{r}_n}\cdot\Theta_n^+(\lambda_n(y_1),\mu_n(y_1))\cdot\Theta_n^-(\lambda_n(-y_1),\mu_n(-y_1)),\end{aligned}$$

whence we obtain by applying Eq. (20.6) and taking the square root

$$\tfrac{1}{2}(\varkappa_{n,1}-\varkappa_{n,2})=\mathrm{e}^{-2\pi|n|y_0+(r_n+\widetilde{r}_n)/2}\cdot\left(\Theta_n^+(\lambda_n(y_1),\mu_n(y_1))\cdot\Theta_n^-(\lambda_n(-y_1),\mu_n(-y_1))\right)^{1/2}.$$

Taking absolute values and again applying the estimate (20.11), we obtain

$$|\varkappa_{n,1}-\varkappa_{n,2}|\leq C_3\cdot\mathrm{e}^{-2\pi\,(1-s_n')\,|n|\,y_0}$$

with a constant $C_3>0$ and

$$s_n':=\frac{1}{2\pi\,|n|\,y_0}\cdot\left(\frac{\mathrm{Re}(r_n+\widetilde{r}_n)}{2}+\widehat{r}_n\right)\in\ell^2_{0,0}(n).$$

This shows Eq. (20.1).

Finally, because of $\Delta'\in\mathrm{As}(\mathbb{C}^*,\ell^\infty_{1,-3},1)$ we have

$$\begin{aligned}\left(\mu_n-(-1)^n\right)^2&=\mu_n\cdot\left(\mu_n+\mu_n^{-1}-2(-1)^n\right)=\mu_n\cdot(\Delta(\lambda_n)-\Delta(\varkappa_{n,1}))\\&=\mu_n\cdot\int_{\varkappa_{n,1}}^{\lambda_n}\Delta'(\lambda)\,\mathrm{d}\lambda=\ell^\infty_{1,-3}(n)\cdot(\lambda_n-\varkappa_{n,1})\\&=\ell^\infty_{1,-3}(n)\cdot\left((\lambda_n-\varkappa_{n,*})-\tfrac{1}{2}(\varkappa_{n,1}-\varkappa_{n,2})\right)\end{aligned}$$

and therefore

$$|\mu_n-(-1)^n|\leq C_4\cdot\mathrm{e}^{-\pi\,(1-s_n'')\,|n|\,y_0}$$

with another constant $C_4>0$ and a sequence $(s_n'')\in\ell^2_{0,0}(n)$, concluding the proof of Eq. (20.3). □

Chapter 21
Perspectives

In this book we studied (singularity-free) simply periodic solutions $u : X \to \mathbb{C}$ of the sinh-Gordon equation via their spectral data (Σ, D) in the style of Bobenko. In particular we gained insight into the asymptotic behavior of the spectral data. As we saw in Chap. 2, real-valued such solutions give rise to minimal immersions $f : X \to S^3$, and similarly to constant mean curvature (CMC) immersions into $\mathbb{R}^3$ and H^3. In this situation, umbilical points of the immersion correspond to coordinate singularities of the solution u. In the following, I sketch how the research described in this book might be extended to solutions u with such coordinate singularities, and why the corresponding minimal immersions with umbilical points are of interest.

We thus now suppose that $u : N \setminus S \to \mathbb{C}$ is a real-valued, simply periodic solution of the sinh-Gordon equation on a horizontal strip $N \subset \mathbb{C}$ which has coordinate singularities of the type that arises from umbilical points of the corresponding minimal surface at the points of the discrete set $S \subset N$. We again suppose without loss of generality that the period of u is 1. By choosing an base point $z_0 \in N$ for the monodromy so that the line through z_0 does not pass through a singularity of u, we can still define spectral data for u relative to this base point, which have the properties described in this book. When one then varies z_0 so as to approach a singularity of u, the spectral curve will not change, but the spectral divisor will change, and in the limit have a different behavior, representing the fact that a singularity has been encountered. One would need to characterise that limit behavior. In the limit, both the asymptotic behavior and other properties of the spectral divisor (in particular non-specialness) could change. It would also be desirable at this point to solve the associated inverse problem at least for the monodromy, i.e. to reconstruct the monodromy from the spectral data in the presence of singularities of u.

S. Klein, *A Spectral Theory for Simply Periodic Solutions of the Sinh-Gordon Equation*, Lecture Notes in Mathematics 2229,
https://doi.org/10.1007/978-3-030-01276-2_21

The next question arises when one considers base points resp. periods of u that lie on opposite sides of a singularity. Here one would need to study whether and how the spectral data on the two sides of the singularity are related. This would hopefully permit to describe the translational flow of the spectral data across singularities of u, and thereby enabling solving the inverse problem of reconstructing the solution u from its spectral data even where u has isolated singularities.

One reason why minimal resp. CMC immersions with umbilical points are of interest is because of their relationship to *compact* (immersed) minimal resp. CMC surfaces in the 3-dimensional space forms, for example minimal immersions $f : M \to S^3$ of a *compact Riemann surface* M into S^3. Indeed any such immersion of a surface M of genus g has $3g - 3$ umbilical points (counted with multiplicity), so if one wishes to study compact minimal surfaces by means of spectral theory, one has to address the behavior of spectral data near the corresponding singularities of the solution u.

For compact minimal surfaces of genus g, the case of $g = 0$ is of no interest, as only the round sphere can occur by a result due to Hopf [Ho]. The case $g = 1$ has been classified completely by the work of Pinkall and Sterling [Pi-S] resp. Hitchin [Hi]. Therefore one would be interested mostly in the case $g \geq 2$. For $g \geq 2$ there has been progress in the study of spectral data for surfaces with additional symmetries, in particular for Lawson surfaces by Heller et al. (see the Introduction for references), but there are no results yet on spectral data for general minimal surfaces of genus $g \geq 2$, much less a classification or parameterization of the moduli space for any such g.

When one thus tries to understand a compact minimal surface of genus $g \geq 2$ by means of spectral theory, one could consider cylindrical pieces between two umbilical points. The corresponding solution of the sinh-Gordon equation has no singularities, and thus the theory described in this book is applicable for it. Next one might consider pieces that do contain an umbilical point, and here the investigation proposed above would be applicable. The final, and perhaps largest hurdle, would be to understand the way in which these pieces are "glued together", to form the closed compact surface. It is possible that here the theory of quadratic differentials developed by Strebel [St], applied to the Hopf differential $E dz^2$ might be useful.

In this humble author's opinion, setting out on the path sketched here might well be a worthwhile endeavor if one seeks to understand minimal surfaces with umbilical points, or even (as a very distant goal) compact minimal resp. CMC surfaces.

Appendix A
Some Infinite Sums and Products

In the present appendix we prove convergence results and estimates for some infinite sums and products. These results are required for the proofs in Chap. 10. We shall work in a setting that is modelled after the one "half" of the setting of Chaps. 4 and 6, in that we will consider sequences $(\lambda_k)_{k\geq 1}$ only for positive indices k, corresponding to the "half" of the sequence that goes to ∞. We will work in a somewhat more general setting than in the main text in that we fix a sequence $(\lambda_{k,0})_{k\in\mathbb{Z}}$ in $\mathbb{C}^*$ with

$$\lambda_{k,0} = ck^2 + O(1) \tag{A.1}$$

for $k \geq 1$ and a fixed constant $c \in \mathbb{C}^*$, and then define "excluded domains" for $\delta > 0$ and $k \geq 1$ for the purposes of this appendix by

$$U_{k,\delta} := \{\, \lambda \in \mathbb{C} \,\big|\, |\lambda - \lambda_{k,0}| < \delta\, k \,\};$$

we also put $U_\delta := \bigcup_{k\geq 1} U_{k,\delta}$ and $V_\delta := \mathbb{C} \setminus U_\delta$. Proposition 6.2(2) shows that this definition of excluded domains is essentially equivalent to the one in Definition 6.1.

In an analogous vein, we define the annuli S_k for $k \geq 1$ for the purposes of this appendix by

$$S_k := \{\, \lambda \in \mathbb{C} \,\big|\, |c|\,(k - \tfrac{1}{2})^2 \leq |\lambda| \leq |c|\,(k + \tfrac{1}{2})^2 \,\}.$$

Again, this definition is equivalent to the one in Eq. (9.1). The S_k intersect only on their boundary, and we have $\bigcup_{k\geq 1} S_k = \{\, \lambda \in \mathbb{C} \,\big|\, |\lambda| \geq |c|/4 \,\}$.

Proposition A.1 (Holomorphic Functions Defined by Infinite Sums) *Let $R_0 > 0$ and sequences $(\lambda_k)_{k\geq 1}$, $(\lambda_k^{[1]})_{k\geq 1}$, $(\lambda_k^{[2]})_{k\geq 1}$ with $\|\lambda_k - \lambda_{k,0}\|_{\ell^2_{-1}}$, $\|\lambda_k^{[\nu]} - \lambda_{k,0}\|_{\ell^2_{-1}} \leq R_0$ be given, so that there exists $0 < \delta_0 \leq \frac{|c|}{4}$ such that*

S. Klein, *A Spectral Theory for Simply Periodic Solutions of the Sinh-Gordon Equation*, Lecture Notes in Mathematics 2229,
https://doi.org/10.1007/978-3-030-01276-2

$\lambda_k, \lambda_k^{[1]}, \lambda_k^{[2]} \in U_{k,\delta_0}$ holds for every $k \geq 1$ with at most finitely many exceptions. Also let a further sequence $(a_k)_{k\geq 1}$ be given.

(1) *$\sum_{k=1}^{\infty} \frac{1}{\lambda-\lambda_k}$ defines a holomorphic function on $V_0 := \mathbb{C} \setminus \{\lambda_k \mid k \in \mathbb{N}\}$. For every $\delta > \delta_0$ there exist $R, C > 0$ (depending only on δ, δ_0 and R_0) such that for every $\lambda \in V_\delta$ with $|\lambda| \geq R$ we have*

$$\left| \sum_{k=1}^{\infty} \frac{1}{\lambda - \lambda_k} \right| \leq C \cdot |\lambda|^{-1/2}.$$

(2) *If $a_k \in \ell^2_{-1}(k)$ and $\|a_k\|_{\ell^2_{-1}} \leq R_0$ holds, then $\sum_{k=1}^{\infty} \frac{a_k}{\lambda-\lambda_k}$ defines a holomorphic function on V_0. For every $\delta > \delta_0$ there exists a constant $C > 0$ (depending only on δ, δ_0 and R_0), such that with the sequence $(r_n)_{n\geq 1} \in \ell^2(n)$ given by*

$$r_n = C \cdot \left(\frac{|a_k|}{k} * \frac{1}{|k|} \right)_n$$

(where we put $\frac{1}{0} := 1$, and extend $\frac{|a_k|}{k} = 0$ for $k \leq 0$), we have for every $n \geq 1$ and $\lambda \in S_n \cap V_\delta$

$$\sum_{k=1}^{\infty} \frac{|a_k|}{|\lambda - \lambda_k|} \leq r_n. \tag{A.2}$$

(3) *If in fact $a_k \in \ell^2(k)$ and $\|a_k\|_{\ell^2} \leq R_0$ holds in the situation of (2), then there exists a constant $C > 0$ (depending only on δ, δ_0 and R_0), such that* (A.2) *holds with the sequence $(r_n)_{n\geq 1} \in \ell^2_1(n)$ given by*

$$r_n = \frac{C}{n} \cdot \left(|a_k| * \frac{1}{|k|} \right)_n . \tag{A.3}$$

(4) *If $a_k \in \ell^2_{-1}(k)$ and $\|a_k\|_{\ell^2_{-1}} \leq R_0$ holds, then $\sum_{k=1}^{\infty} \frac{a_k}{(\lambda-\lambda_k^{[1]})\cdot(\lambda-\lambda_k^{[2]})}$ defines a holomorphic function on $\mathbb{C} \setminus \{\lambda_k^{[1]}, \lambda_k^{[2]} \mid k \in \mathbb{N}\}$. For every $\delta > \delta_0$ there exists a constant $C > 0$ (depending only on δ, δ_0 and R_0) such that with the sequence $(r_n)_{n\geq 1} \in \ell^2_1(n)$ given by Eq.* (A.3) *we have for every $n \geq 1$ and $\lambda \in S_n \cap V_\delta$*

$$\sum_{k=1}^{\infty} \frac{|a_k|}{|\lambda - \lambda_k^{[1]}| \cdot |\lambda - \lambda_k^{[2]}|} \leq r_n.$$

Proof We may suppose without loss of generality that $\lambda_k, \lambda_k^{[1]}, \lambda_k^{[2]} \in U_{k,\delta_0}$ holds for all $k \geq 1$. We then have

$$|\lambda_k - ck^2| \leq |\lambda_k - \lambda_{k,0}| + |\lambda_{k,0} - ck^2| \leq \delta_0 \cdot k + C_1 \leq \frac{|c|}{4} \cdot k + C_1 \tag{A.4}$$

with a constant $C_1 > 0$.

To prepare the proof of the proposition, we show the following estimate: There exists $N \in \mathbb{N}$, such that for every $n \geq N$, $\lambda \in S_n$ and $k \in \mathbb{N}$, we have

$$|\lambda - \lambda_k| \geq \frac{|c|}{4} \cdot |n^2 - k^2|. \tag{A.5}$$

Indeed, let $n, k \in \mathbb{N}$ be given. In the case $k < n$, the $\lambda \in S_n$ for which $|\lambda - ck^2|$ is minimal is $\lambda = c(n - \frac{1}{2})^2$, and therefore we have

$$\begin{aligned}
|\lambda - \lambda_k| &\geq |\lambda - ck^2| - |\lambda_k - ck^2| \geq \left|c(n - \tfrac{1}{2})^2 - ck^2\right| - |ck^2 - \lambda_k| \\
&\overset{(A.4)}{\geq} \left(|c|\,(n - \tfrac{1}{2})^2 - |c|\,k^2\right) - \frac{|c|}{4} k - C_1 = |c| \cdot (n^2 - n + \tfrac{1}{4} - k^2 - \tfrac{1}{4}k) - C_1 \\
&\geq \frac{|c|}{4} \cdot (n^2 - k^2) + \frac{|c|}{4} \cdot (3n^2 - 4n + 1 - 3k^2 - k) - C_1 \\
&\overset{(*)}{\geq} \frac{|c|}{4} \cdot (n^2 - k^2) + \frac{|c|}{4} \cdot (3n^2 - 4n + 1 - 3(n-1)^2 - (n-1)) - C_1 \\
&= \frac{|c|}{4} \cdot (n^2 - k^2) + \frac{|c|}{4} \cdot (n - 1) - C_1 \overset{(\dagger)}{\geq} \frac{|c|}{4} \cdot (n^2 - k^2),
\end{aligned}$$

where the $\geq$ sign marked $(*)$ is true because of $k \leq n - 1$, and the $\geq$ sign marked $(\dagger)$ holds for all $n \geq \frac{4}{|c|} C_1 + 1$. Therefore (A.5) holds for such n and all $k < n$.

In the case $k = n$ there is nothing to show. And in the case $k > n$, the $\lambda \in S_n$ for which $|\lambda - ck^2|$ is minimal is $\lambda = c(n + \frac{1}{2})^2$, and therefore we have

$$\begin{aligned}
|\lambda - \lambda_k| &\geq |\lambda - ck^2| - |\lambda_k - ck^2| \geq \left|c(n + \tfrac{1}{2})^2 - ck^2\right| - |ck^2 - \lambda_k| \\
&\overset{(A.4)}{\geq} \left(|c|\,k^2 - |c|\,(n + \tfrac{1}{2})^2\right) - \frac{|c|}{4} \cdot k - C_1 \\
&= |c|\,(k^2 - n^2) - |c|\,(n + \tfrac{1}{4}\,k) - (C_1 + \tfrac{1}{4}\,|c|) \\
&= \tfrac{|c|}{4}\,(k^2 - n^2) + |c| \left(\tfrac{1}{2}\,(k^2 - n^2) - n\right) + \tfrac{|c|}{4} \left(\tfrac{1}{2}\,(k^2 - n^2) - k + \tfrac{1}{2}\right) \\
&\quad + \left(\tfrac{1}{8}\,(k^2 - n^2) - (C_1 + \tfrac{3}{8}\,|c|)\right).
\end{aligned}$$

Because of $k \geq n+1$ we have

$$k^2 - n^2 \geq (n+1)^2 - n^2 = 2n + 1 \geq 2n \quad \text{and thus} \quad \tfrac{1}{2}(k^2 - n^2) - n \geq 0$$

and because the function $f(x) = \frac{1}{2}x^2 - x$ is monotonously increasing for $x \geq 1$,

$$\tfrac{1}{2}(k^2 - n^2) - k + \tfrac{1}{2} = \tfrac{1}{2}k^2 - k - \tfrac{1}{2}n^2 + \tfrac{1}{2} \geq \tfrac{1}{2}(n+1)^2 - (n+1) - \tfrac{1}{2}n^2 + \tfrac{1}{2} = 0.$$

Thus we obtain

$$|\lambda - \lambda_k| \geq \tfrac{|c|}{4}(k^2 - n^2) + (\tfrac{1}{4}n - (C_1 + \tfrac{3}{8}|c|)) \geq \tfrac{|c|}{4}(k^2 - n^2),$$

where the last $\geq$ sign holds whenever $n \geq 4\,C_1 + \frac{3}{2}|c|$, and therefore (A.5) holds for these n and $k > n$.

The preceding calculations show that (A.5) holds for all $k \in \mathbb{N}$ and $n \geq N$, if we choose N as the smallest integer larger than $\max\{\frac{4}{|c|}C_1 + 1, 4\,C_1 + \frac{3}{2}|c|\}$.

We now turn to the proof of (1)–(4). For proving the estimates claimed in the respective parts of the proposition, it suffices to consider $\lambda \in S_n$ with $n \geq N$; the estimates can then be extended to $\lambda \in \mathbb{C}$ by enlarging the constant C via an argument of compactness.

For (1). The series $\sum_{k=1}^{\infty} \frac{1}{\lambda - \lambda_k}$ converges absolutely and locally uniformly for $\lambda \in V_0$, and thus defines a holomorphic function. Now let $\delta > \delta_0$ and $\lambda \in V_\delta$ be given. We write

$$\left|\sum_{k=1}^{\infty} \frac{1}{\lambda - \lambda_k}\right| \leq \left|\sum_{k=1}^{\infty} \frac{1}{\lambda - ck^2}\right| + \sum_{k=1}^{\infty}\left|\frac{1}{\lambda - \lambda_k} - \frac{1}{\lambda - ck^2}\right|. \tag{A.6}$$

To estimate $\sum_{k=1}^{\infty} \frac{1}{\lambda - ck^2}$, we use the partial fraction decomposition of the cotangent function

$$\pi \cot(\pi z) = z^{-1} + \sum_{k=1}^{\infty} \frac{2z}{z^2 - k^2} \quad \text{for } z \in \mathbb{C} \setminus \{k\pi \,|\, k \in \mathbb{Z}\},$$

which implies

$$\sum_{k=1}^{\infty} \frac{1}{\lambda - ck^2} = \frac{1}{2\sqrt{c\,\lambda}}\left(\pi \cot(\pi\sqrt{c^{-1}\lambda}) - \sqrt{c\,\lambda^{-1}}\right) = O\left(|\lambda|^{-1/2}\right) \tag{A.7}$$

for $\lambda \in V_\delta$; note that $\cot(\pi\sqrt{c^{-1}\lambda})$ is bounded on V_δ.

To estimate the second summand in (A.6), we note that we have for $\lambda \in S_n \cap V_\delta$ with $n \geq N$ if $k \neq n$ by (A.4) and (A.5)

$$\left|\frac{1}{\lambda - \lambda_k} - \frac{1}{\lambda - ck^2}\right| = \left|\frac{\lambda_k - ck^2}{(\lambda - \lambda_k)(\lambda - ck^2)}\right| \leq \frac{16}{|c|^2} \frac{\frac{|c|}{4}k + C_1}{|n^2 - k^2|^2} \leq C_2 \cdot \frac{k}{|n^2 - k^2|^2}$$

with a constant $C_2 > 0$, and if $k = n$

$$\left|\frac{1}{\lambda - \lambda_n} - \frac{1}{\lambda - cn^2}\right| = \left|\frac{\lambda_n - cn^2}{(\lambda - \lambda_n)(\lambda - cn^2)}\right| \leq \frac{n\,\delta_0}{(n\,(\delta - \delta_0)) \cdot (n\,\delta)} \leq \frac{1}{n\,(\delta - \delta_0)}.$$

Thus we obtain

$$\begin{aligned} \left|\sum_{k=1}^{\infty}\left(\frac{1}{\lambda - \lambda_k} - \frac{1}{c\lambda - k^2}\right)\right| &\leq \frac{1}{n\,(\delta - \delta_0)} + C_2 \sum_{k \neq n} \frac{k}{|n^2 - k^2|^2} \\ &\leq \frac{1}{n\,(\delta - \delta_0)} + \frac{C_2}{n} \sum_{k \neq n} \frac{1}{|n - k|^2} = O(n^{-1}). \end{aligned} \tag{A.8}$$

Because n^{-1} is comparable to $|\lambda|^{-1/2}$ for $\lambda \in S_n$, it follows by plugging (A.7) and (A.8) into (A.6) that

$$\left|\sum_{k=1}^{\infty} \frac{1}{\lambda - \lambda_k}\right| = O(|\lambda|^{-1/2})$$

holds for $\lambda \in V_\delta$, and this is the claimed statement.

For (2). The series $\sum_{k=1}^{\infty} \frac{a_k}{\lambda - \lambda_k}$ converges absolutely and locally uniformly for $\lambda \in V_0$, and thus defines a holomorphic function. For $n \geq N$ and $\lambda \in S_n \cap V_\delta$ we have

$$\begin{aligned} \sum_{k=1}^{\infty} \frac{|a_k|}{|\lambda - \lambda_k|} &\overset{(A.5)}{\leq} \frac{4}{|c|} \sum_{k \neq n} \frac{|a_k|}{|n^2 - k^2|} + \frac{|a_n|}{|\lambda - \lambda_n|} \\ &\leq \frac{4}{|c|} \sum_{k \neq n} \frac{|a_k|}{k} \cdot \frac{1}{|n - k|} + \frac{|a_n|}{n \cdot (\delta - \delta_0)} \\ &= \frac{4}{|c|}\left(\frac{|a_k|}{k} * \frac{1}{|k|}\right)_n + \frac{1}{\delta - \delta_0} \cdot \frac{|a_n|}{n} \leq C \cdot \left(\frac{|a_k|}{k} * \frac{1}{|k|}\right)_n =: r_n, \end{aligned} \tag{A.9}$$

where we put $C := \max\{\frac{4}{|c|}, \frac{1}{\delta - \delta_0}\}$, and we we extend $\frac{|a_k|}{k}$ to $k \in \mathbb{Z}$ by setting $\frac{|a_k|}{k} = 0$ for $k \leq 0$, and again put $\frac{1}{0} := 1$. The convolution of the ℓ^2-sequence $\frac{|a_k|}{k}$

with $\frac{1}{|k|}$ is again an ℓ^2-sequence by the version (7.4) of Young's inequality for the convolution with $\frac{1}{k}$. This shows that $r_n \in \ell^2(n)$ holds.

For (3). We proceed similarly as in (2): For $n \geq N$ and $\lambda \in S_n \cap V_\delta$ we now have

$$\begin{aligned}
\sum_{k=1}^{\infty} \frac{|a_k|}{|\lambda - \lambda_k|} &\overset{(A.5)}{\leq} \frac{4}{|c|} \sum_{k \neq n} \frac{|a_k|}{|n^2 - k^2|} + \frac{|a_n|}{|\lambda - \lambda_n|} \\
&\leq \frac{4}{|c|} \sum_{k \neq n} \frac{|a_k|}{n} \cdot \frac{1}{|n-k|} + \frac{|a_n|}{n \cdot (\delta - \delta_0)} \\
&= \frac{1}{n} \cdot \left(\frac{4}{|c|} \left(|a_k| * \frac{1}{|k|} \right)_n + \frac{1}{\delta - \delta_0} \cdot |a_n| \right) \\
&\leq \frac{C}{n} \cdot \left(|a_k| * \frac{1}{|k|} \right)_n =: r_n,
\end{aligned} \tag{A.10}$$

where we again extend $|a_k|$ to $k \in \mathbb{Z}$ by zero, and where $C > 0$ is a constant. The convolution of $|a_k|$ with $\frac{1}{|k|}$ is an ℓ^2-sequence by (7.4), hence $r_n \in \ell^2_1(n)$ holds.

For (4). For the proof of (4), we choose N so that (A.5) applies for both $(\lambda_n^{[1]})$ and $(\lambda_n^{[2]})$ in the place of (λ_n). Then we have

$$\begin{aligned}
\sum_{k=1}^{\infty} \frac{a_k}{(\lambda - \lambda_k^{[1]}) \cdot (\lambda - \lambda_k^{[2]})} &\overset{(A.5)}{\leq} \frac{16}{|c|^2} \sum_{k \neq n} \frac{|a_k|}{|n^2 - k^2|^2} + \frac{|a_n|}{|\lambda - \lambda_n^{[1]}| \cdot |\lambda - \lambda_n^{[2]}|} \\
&\leq \frac{16}{|c|^2} \cdot \frac{1}{n} \sum_{k \neq n} \frac{|a_k|}{k} \cdot \frac{1}{|n-k|^2} + \frac{|a_n|}{(n\,(\delta - \delta_0))^2} \\
&= \frac{1}{n} \cdot \left(\frac{16}{|c|^2} \left(\frac{|a_k|}{k} * \frac{1}{k^2} \right)_n + \frac{1}{(\delta - \delta_0)^2} \cdot \frac{|a_n|}{n} \right) \\
&\leq \frac{C}{n} \cdot \left(\frac{|a_k|}{k} * \frac{1}{k} \right)_n =: r_n .
\end{aligned}$$

Here, $C > 0$ is again a constant, and similarly as before, we see that $r_n \in \ell^2_1(n)$ holds. □

The relationship between convergent infinite products $\prod_{k=1}^{\infty}(1 + a_k)$ and absolutely convergent series $\sum_{k=1}^{\infty} a_k$ described in the following proposition is very well-known (see, for example, [Co, Corollary VII.5.6 and Lemma VII.5.8, p. 166]), as is the estimate of the product by the ℓ^1-norm of (a_k). We state and prove this result here for the sake of completeness.

Proposition A.2 (Estimating Products by Sums) *Let a sequence $(a_k) \in \ell^1(k)$ be given. The infinite product $\prod_{k=1}^{\infty}(1+a_k)$ converges in $\mathbb{C}^*$, and we have*

$$\left|\prod_{k=1}^{\infty}(1+a_k)-1\right| \leq \exp\left(\|a_k\|_{\ell^1}\right)-1.$$

Moreover, if the (a_k) depend on a parameter so that $\sum_{k=1}^{\infty}|a_k|$ converges uniformly with respect to that parameter, then the convergence of $\prod_{k=1}^{\infty}(1+a_k)$ is also uniform.

Proof For $N \in \mathbb{N}$ we have

$$\prod_{k=1}^{N}(1+a_k)-1=\sum_{k=1}^{N}\left(\sum_{1\leq j_1<\ldots<j_k\leq N} a_{j_1}\cdot\ldots\cdot a_{j_k}\right).$$

Moreover, we have for $1 \leq k \leq N$

$$\begin{aligned}\left|\sum_{1\leq j_1<\ldots<j_k\leq N} a_{j_1}\cdot\ldots\cdot a_{j_k}\right| &\leq \frac{1}{k!}\sum_{1\leq j_1,\ldots,j_k\leq N}|a_{j_1}|\cdot\ldots\cdot|a_{j_k}| \\ &= \frac{1}{k!}\left(\sum_{j=1}^{N}|a_j|\right)^k \leq \frac{1}{k!}\|a_j\|_{\ell^1(j)}^k\end{aligned}$$

and therefore

$$\left|\prod_{k=1}^{N}(1+a_k)-1\right| \leq \sum_{k=1}^{N}\frac{1}{k!}\|a_j\|_{\ell^1(j)}^k \leq \sum_{k=1}^{\infty}\frac{1}{k!}\|a_j\|_{\ell^1(j)}^k = \exp(\|a_j\|_{\ell^1(j)})-1.$$

By taking the limit $N \to \infty$, we obtain the claimed result. □

Proposition A.3 (An Infinite Product) *Let $R_0 > 0$ and sequences $(\lambda_k^{[1]})_{k\geq 1}$, $(\lambda_k^{[2]})_{k\geq 1}$ with $\lambda_k^{[1]}-\lambda_k^{[2]} \in \ell^2_{-1}(k)$ and $\|\lambda_k^{[1]}-\lambda_k^{[2]}\|_{\ell^2_{-1}} \leq R_0$ be given. Then the infinite product $\prod_{k=1}^{\infty}\frac{\lambda_k^{[1]}}{\lambda_k^{[2]}}$ converges in $\mathbb{C}^*$, and there exists a constant $C > 0$, dependent only on R_0, such that*

$$\left|\prod_{k=1}^{\infty}\frac{\lambda_k^{[1]}}{\lambda_k^{[2]}}-1\right| \leq C\cdot\|\lambda_k^{[1]}-\lambda_k^{[2]}\|_{\ell^2_{-1,3}}$$

holds.

Proof We have $\prod_{k=1}^{\infty} \frac{\lambda_k^{[1]}}{\lambda_k^{[2]}} = \prod_{k=1}^{\infty}(1 + a_k)$ with

$$a_k := \frac{\lambda_k^{[1]} - \lambda_k^{[2]}}{\lambda_k^{[2]}}.$$

There exists $C_1 > 0$ with $|\lambda_k^{[\nu]}| \geq C_1 \cdot k^2$ and therefore we have by Cauchy-Schwarz's inequality:

$$\begin{aligned}\|a_k\|_{\ell^1} &= \left\| \frac{\lambda_k^{[1]} - \lambda_k^{[2]}}{\lambda_k^{[2]}} \right\|_{\ell^1} \leq \frac{1}{C_1} \cdot \left\| \frac{\lambda_k^{[1]} - \lambda_k^{[2]}}{k^2} \right\|_{\ell^1} \\ &\leq \frac{1}{C_1} \cdot \left\| \frac{1}{k} \right\|_{\ell^2} \cdot \left\| \frac{\lambda_k^{[1]} - \lambda_k^{[2]}}{k} \right\|_{\ell^2} \leq C_2 \cdot \left\| \lambda_k^{[1]} - \lambda_k^{[2]} \right\|_{\ell^2_{-1}}.\end{aligned}$$

It follows by Proposition A.2(1) that $\prod_{k=1}^{\infty} \frac{\lambda_k^{[1]}}{\lambda_k^{[2]}}$ converges, and that

$$\left| \prod_{k=1}^{\infty} \frac{\lambda_k^{[1]}}{\lambda_k^{[2]}} - 1 \right| \leq \exp\left(\|a_k\|_{\ell^1}\right) - 1 \leq \exp\left(C_2 \cdot \left\| \lambda_k^{[1]} - \lambda_k^{[2]} \right\|_{\ell^2_{-1}} \right) - 1$$

holds. Because exp is Lipschitz continuous on $[0, C_2 \cdot R_0]$, it follows that there exists a constant $C_3 > 0$ so that

$$\left| \prod_{k=1}^{\infty} \frac{\lambda_k^{[1]}}{\lambda_k^{[2]}} - 1 \right| \leq C_3 \cdot \left\| \lambda_k^{[1]} - \lambda_k^{[2]} \right\|_{\ell^2_{-1}}$$

holds. □

Proposition A.4 (Holomorphic Functions Defined by Infinite Products)

(1) Let $(\lambda_k)_{k\geq 1}$ be given with $\lambda_k - \lambda_{k,0} \in \ell^2_{-1}(k)$. Then $f(\lambda) := \prod_{k=1}^{\infty} \left(1 - \frac{\lambda}{\lambda_k}\right)$ defines a holomorphic function on $\mathbb{C}$ with $f(0) = 1$.

(2) Let $R_0 > 0$ and sequences $(\lambda_k^{[1]})_{k\geq 1}, (\lambda_k^{[2]})_{k\geq 1}$ with $\lambda_k^{[1]} - \lambda_k^{[2]} \in \ell^2_{-1}(k)$ and $\|\lambda_k^{[1]} - \lambda_k^{[2]}\|_{\ell^2_{-1}} \leq R_0$ be given, so that there exists $0 < \delta_0 \leq \frac{|c|}{4}$ such that $\lambda_k^{[\nu]} \in U_{k,\delta_0}$ holds for every $k \geq 1$ with at most finitely many exceptions. Then $g(\lambda) := \prod_{k=1}^{\infty} \frac{\lambda_k^{[1]} - \lambda}{\lambda_k^{[2]} - \lambda}$ defines a holomorphic function on $\mathbb{C} \setminus \{ \lambda_k^{[2]} \mid k \geq 1 \}$ with $g(0) = \prod_{k=1}^{\infty} \frac{\lambda_k^{[1]}}{\lambda_k^{[2]}}$, and $h(\lambda) := \prod_{k=1}^{\infty} \frac{(\lambda_k^{[1]})^{-1} - \lambda}{(\lambda_k^{[2]})^{-1} - \lambda}$ defines a holomorphic function on $(\mathbb{C}^* \cup \{\infty\}) \setminus \{ (\lambda_k^{[2]})^{-1} \mid k \geq 1 \}$ with $h(\infty) = 1$.

For $\delta > \delta_0$, there exists a constant $C > 0$, depending only on R_0 and δ, such that with the sequence $(r_n)_{n\in\mathbb{Z}} \in \ell^2(n)$ given by

$$r_n = C \cdot \left(\frac{|\lambda_k^{[1]} - \lambda_k^{[2]}|}{k} * \frac{1}{|k|} \right)_n$$

(where we put $\frac{1}{0} := 1$, and extend $\frac{|\lambda_k^{[1]}-\lambda_k^{[2]}|}{k} = 0$ for $k \leq 0$), we have for every $n \geq 1$ and $\lambda \in S_n \cap V_\delta$

$$|g(\lambda) - 1| \leq r_n \tag{A.11}$$

and

$$|h(\lambda) - 1| \leq r_{-n}. \tag{A.12}$$

Remark A.5 In Proposition A.4(2), in fact a far better estimate than (A.12) is possible regarding the function h: in fact $|h(\lambda) - 1| \leq \widehat{r}_{-n}$ holds with

$$\widehat{r}_n = C \cdot \left(\frac{|\lambda_k^{[1]} - \lambda_k^{[2]}|}{k^3} * \frac{1}{|k|} \right)_n .$$

Proof (of Proposition A.4) For (1). We have $\prod_{k=1}^{\infty}\left(1 - \frac{\lambda}{\lambda_k}\right) = \prod_{k=1}^{\infty}(1 + a_k(\lambda))$ with $a_k(\lambda) := -\frac{\lambda}{\lambda_k}$, and

$$\|a_k(\lambda)\|_{\ell^1(k)} = |\lambda| \cdot \sum_{k=1}^{\infty} \underbrace{\frac{1}{|\lambda_k|}}_{=O(k^{-2})} \leq C_1 \cdot |\lambda|$$

holds with some constant $C_1 > 0$. Therefore it follows from Proposition A.2(1) that the infinite product $\prod_{k=1}^{\infty}(1 + a_k(\lambda))$ converges locally uniformly in λ, hence this product defines a holomorphic function f in $\lambda \in \mathbb{C}$. Moreover, we have $a_k(0) = 0$ for all k, and therefore $f(0) = 1$.

For (2). We have $g(\lambda) = \prod_{k=1}^{\infty}(1 + a_k(\lambda))$ with $a_k(\lambda) := \frac{\lambda_k^{[1]}-\lambda_k^{[2]}}{\lambda_k^{[2]}-\lambda}$. Because we have $|\lambda_k^{[1]} - \lambda_k^{[2]}| \in \ell^2_{-1}(k)$, Proposition A.1(2) is applicable to $\sum_{k=1}^{\infty} |a_k(\lambda)|$. By that Proposition, $\sum_{k=1}^{\infty} a_k(\lambda)$ defines a holomorphic function on $\lambda \in \mathbb{C}\setminus\{\lambda_k^{[2]} \mid k \geq 1\}$. Moreover, if we suppose that $\delta > \delta_0$ is given, there exists a constant $C_2 > 0$ such that with the sequence $(\widetilde{r}_n)_{n\geq 1} \in \ell^2(n)$ given by

$$\widetilde{r}_n = C_2 \cdot \left(\frac{|\lambda_k^{[1]} - \lambda_k^{[2]}|}{k} * \frac{1}{|k|} \right)_n$$

we have for any $n \geq 1$ and $\lambda \in S_n \cap V_\delta$

$$\|a_k(\lambda)\|_{\ell^1} = \sum_{k=1}^{\infty} \frac{|\lambda_k^{[1]} - \lambda_k^{[2]}|}{|\lambda_k^{[2]} - \lambda|} \leq \widetilde{r}_n.$$

By Proposition A.2(1) it follows that

$$|g(\lambda) - 1| \leq \exp(\widetilde{r}_n) - 1 \tag{A.13}$$

holds.

By the variant of Young's inequality for weak ℓ^1-sequences (see (7.4)) there exists a constant $C_3 > 0$ so that we have for every $n \in \mathbb{N}$

$$|\widetilde{r}_n| \leq \|\widetilde{r}_k\|_{\ell^2(k)} \leq C_2 \cdot C_3 \cdot \|\lambda_k^{[1]} - \lambda_k^{[2]}\|_{\ell^2_{-1}(k)} \leq C_2 \cdot C_3 \cdot R_0,$$

and therefore there exists a constant $C_4 > 0$ so that

$$\exp(\widetilde{r}_n) - 1 \leq C_4 \cdot \widetilde{r}_n$$

holds. From the estimate (A.13) we thus obtain that for $n \geq 1$ and $\lambda \in S_n \cap V_\delta$,

$$|g(\lambda) - 1| \leq C_4 \cdot \widetilde{r}_n$$

holds; therefore the claimed estimate (A.11) holds with $C := C_4 \cdot C_2 > 0$.

Similarly, we have $h(\lambda) = \prod_{k=1}^{\infty}(1 + b_k(\lambda))$ with

$$b_k(\lambda) := \frac{(\lambda_k^{[1]})^{-1} - (\lambda_k^{[2]})^{-1}}{(\lambda_k^{[2]})^{-1} - \lambda} = \frac{\lambda_k^{[2]} - \lambda_k^{[1]}}{\lambda_k^{[1]} \cdot (1 - \lambda_k^{[2]} \cdot \lambda)}.$$

We have $|\lambda_k^{[\nu]}| \geq C_5 \cdot k^2$ and for $\lambda \in S_n$, $n \in \mathbb{N}$,

$$|1 - \lambda_k^{[2]} \cdot \lambda| \geq C_6 \cdot |1 - k^2 \cdot n^2| = C_6 \cdot |1 - k\,n| \cdot |1 + k\,n| \geq C_6 \cdot k \cdot |1 - k\,n| \geq C_7 \cdot (k + n),$$

and therefore

$$|b_k(\lambda)| \leq C_8 \cdot \frac{|\lambda_k^{[1]} - \lambda_k^{[2]}|}{k} \cdot \frac{1}{k + n},$$

whence

$$\|b_k(\lambda)\|_{\ell^1} \leq C_8 \cdot \sum_{k=1}^{\infty} \frac{|\lambda_k^{[1]} - \lambda_k^{[2]}|}{k} \cdot \frac{1}{k + n} = C_8 \cdot \left(\frac{|\lambda_k^{[1]} - \lambda_k^{[2]}|}{k} * \frac{1}{|k|} \right)_{-n} = C_9 \cdot r_{-n}$$

follows. By an analogous application of Proposition A.2(1) and Young's inequality for weak ℓ^1-sequences as before, the estimate (A.12) follows (with an appropriately chosen C). □

Remark A.6 Mutatis mutandis, the results of the present appendix remain true if the starting index $k = 1$ of the infinite sums or products is replaced by $k = 0$.

Appendix B
Index of Notations

Latin Letters

A_n	"Small" cycle in the homology basis of Σ, p. 259
$A_n^+,\ A_n^-$	Boundary cycles of the surface $\widetilde{\Sigma}$ from Proposition 19.8 on p. 298
$\mathrm{As}(G, \ell^p_{n,m}, s)$	Space of asymptotic functions on $G \subset \mathbb{C}^*$ or $G \subset \Sigma$, Definition 9.1 on p. 114
$\mathrm{As}_\infty(G_\infty, \ell^p_{n,m}, s)$	Space of asymptotic functions on $G_\infty \subset \mathbb{C}^*$ or $G_\infty \subset \Sigma$ near $\lambda = \infty$, Definition 9.2 on p. 115
$\mathrm{As}_0(G_0, \ell^p_{n,m}, s)$	Space of asymptotic functions on $G_0 \subset \mathbb{C}^*$ or $G_0 \subset \Sigma$ near $\lambda = 0$, Definition 9.2 on p. 115
$a(\lambda)$	Upper-left entry of the monodromy $M(\lambda)$, Eq. (3.5) on p. 26
$a_0(\lambda)$	Upper-left entry of the monodromy $M_0(\lambda)$ of the vacuum, Eq. (4.5) on p. 43
B_n	"Large" cycle in the homology basis of Σ, p. 259
$b(\lambda)$	Upper-right entry of the monodromy $M(\lambda)$, Eq. (3.5) on p. 26
$b_0(\lambda)$	Upper-right entry of the monodromy $M_0(\lambda)$ of the vacuum, Eq. (4.5) on p. 43
$\mathfrak{C}_{D^o}$	The space of sequences of curves emanating from the support of the divisor $D^o \in \mathsf{Div}(\Sigma')$ defined in Proposition 18.5 on p. 275
$\mathfrak{C}^{(1)}_{D^o}$	The larger space of sequences of curves emanating from the support of the divisor $D^o \in \mathsf{Div}(\Sigma')$ defined in Proposition 18.10 on p. 284

S. Klein, *A Spectral Theory for Simply Periodic Solutions of the Sinh-Gordon Equation*, Lecture Notes in Mathematics 2229,
https://doi.org/10.1007/978-3-030-01276-2

$c(\lambda)$	Lower-left entry of the monodromy $M(\lambda)$, Eq. (3.5) on p. 26
$c_0(\lambda)$	Lower-left entry of the monodromy $M_0(\lambda)$ of the vacuum, Eq. (4.5) on p. 43
D	Classical spectral divisor of a potential (u, u_y) or of a monodromy $M(\lambda)$, p. 31
D_0	Classical spectral divisor of the vacuum, Eq. (4.11) on p. 45
$\mathscr{D}$	Generalized spectral divisor of a potential (u, u_y) or of a monodromy $M(\lambda)$, Definition 3.2 on p. 30
$\mathscr{D}_0$	Generalized spectral divisor of the vacuum, p. 45
Div	Space of asymptotic divisors, Definition 8.3 on p. 109 and the following remarks
$\mathsf{Div}(\Sigma)$	Space of asymptotic divisors with support on Σ, Eq. (17.17) on p. 250
$\mathsf{Div}(\Sigma')$	Space of asymptotic divisors with support on Σ', Eq. (17.17) on p. 250
Div_{fin}	Set of asymptotic divisors of finite type, p. 218
Div_{tame}	Space of tame asymptotic divisors, Definition 12.4(1) on p. 170
$\mathsf{Div}_{tame}(\Sigma)$	Space of tame asymptotic divisors with support on Σ, Eq. (17.18) on p. 250
$\mathsf{Div}_{tame}(\Sigma')$	Space of tame asymptotic divisors with support on Σ', Eq. (17.18) on p. 250
$d(\lambda)$	Lower-right entry of the monodromy $M(\lambda)$, Eq. (3.5) on p. 26
$d_0(\lambda)$	Lower-right entry of the monodromy $M_0(\lambda)$ of the vacuum, Eq. (4.5) on p. 43
F, F_λ	Extended frame of a solution u or of a potential (u, u_y), p. 25, p. 48
F_0	Extended frame of the vacuum, Eq. (4.4) on p. 43
$f_{n,\xi,\varrho}$	A meromorphic function on $\mathbb{C}^*$ defined in Proposition 17.3 on p. 249
$\widehat{f}(k)$	Fourier transform of a function f, Eq. (7.5) on p. 87
$\mathrm{Jac}(\Sigma)$	Jacobi variety of Σ, Theorem 18.7(1) on p. 280
$\widetilde{\mathrm{Jac}}(\Sigma)$	Pseudo tangent space for the Jacobi variety of Σ, Proposition 18.5(2) on p. 275
$\widetilde{\mathrm{Jac}}^{(1)}(\Sigma)$	Larger pseudo tangent space for the Jacobi variety of Σ, Proposition 18.10 on p. 284
j_0	A parameter for a generalized divisor $\mathscr{D}$, Proposition 3.3 on p. 31
$\widehat{K}_r(0)$, $\widehat{K}_r(\infty)$	Two circles in Σ around 0 resp. ∞, p. 291

$L^p(M)$	Banach space of p-integrable functions on some domain M
$L^2(\Sigma, T^*\Sigma)$	Space of square-integrable 1-forms on Σ
$\ell^p(M)$	Space of p-summable sequences on an index set $M \subset \mathbb{Z}$
$\ell^p_n(M)$	Space of p-summable sequences shifted by k^n on an index set $M \subset \mathbb{Z}$, p. 87
$\ell^p_{n,m}(M)$	Space of p-summable sequences shifted by k^n resp. k^m on an index set $M \subset \mathbb{Z}$, p. 107
$M(\lambda)$	Monodromy of a solution u or of a potential (u, u_y), p. 25, p. 48
$M_0(\lambda)$	Monodromy of the vacuum, Eq. (4.5) on p. 43
$\mathscr{M}$	Sheaf of meromorphic functions on the spectral curve Σ, p. 27
Mon	Space of asymptotic monodromies, p. 148
Mon_{np}	Space of non-periodic asymptotic monodromies, p. 148
Mon_{τ}	Space of asymptotic monodromies with asymptotic parameter τ, p. 148
$\mathsf{Mon}_{\tau,\upsilon}$	Space of non-periodic asymptotic monodromies with asymptotic parameters τ and υ, p. 148
$\mathscr{O}$	Sheaf of holomorphic functions on the spectral curve Σ, p. 27
$\mathscr{O}_M$	Sheaf of holomorphic (2×2)-matrices which commute with the monodromy $M(\lambda)$, p. 33
$\widehat{\mathscr{O}}$	Direct image of the sheaf of holomorphic functions on the normalization $\widehat{\Sigma}$ of the spectral curve Σ onto Σ, p. 28
Pot	Space of potentials (Cauchy data for the sinh-Gordon equation), Eq. (5.1) on p. 48
Pot_{np}	Space of non-periodic potentials, p. 47
Pot^1_{np}	Space of once more differentiable, non-periodic potentials, Eq. (5.35) on p. 67
Pot_{tame}	Space of tame potentials, Definition 12.4(3) on p. 170
$\mathscr{R}$	The subsheaf of $\widehat{\mathscr{O}}$ over which a generalized divisor $\mathscr{D}$ on Σ is locally free, Proposition 3.3 on p. 31
$\mathscr{R}_M$	Sheaf of eigenvalues of the monodromy $M(\lambda)$, p. 33
$\mathrm{Res}_0(\eta)$, $\mathrm{Res}_\infty(\eta)$	Residue of a 1-form η on Σ at 0 resp. ∞, p. 291
S	$= \{ k \in \mathbb{Z} \mid \varkappa_{k,1} = \varkappa_{k,2} \}$, Eq. (17.2) on p. 240
S^3, $S^3(\varkappa)$	The 3-sphere (of curvature $\varkappa > 0$), p. 21
S_k	Annulus through $\lambda_{k,0}$ in $\mathbb{C}^*$, Eq. (9.1) on p. 114
$\widehat{S}_k$	Annulus through $\lambda_{k,0}$ in Σ, Eq. (9.2) on p. 114
$s_{n,\ell}$, $s_{n,n}$	Numbers describing the member ω_n of the canonical basis of 1-forms on Σ, Theorem 18.3(2) on p. 264
$U_{k,\delta}$	k-th excluded domain in $\mathbb{C}^*$, Definition 6.1 on p. 71

$U'_{k,\delta}$	Possibly punctured k-th excluded domain in $\mathbb{C}^*$, Eq. (16.4) on p. 223
$\widehat{U}_{k,\delta}$	k-th excluded domain in Σ, p. 74
$\widehat{U}'_{k,\delta}$	Possibly punctured k-th excluded domain in Σ, Eq. (16.4) on p. 223
U_δ	Union of all excluded domains in $\mathbb{C}^*$, Definition 6.1 on p. 71
$\widehat{U}_\delta$	Union of all excluded domains in Σ, p. 74
u	Solution of the sinh-Gordon equation $\Delta u + \sinh(u) = 0$, p. 21
(u, u_y)	A potential in Pot or Pot_{np}, i.e. Cauchy data for the sinh-Gordon equation
V_δ	Area outside of the excluded domains in $\mathbb{C}^*$, Definition 6.1 on p. 71
$\widehat{V}_\delta$	Area outside of the excluded domains in Σ, p. 74
v_k	Symplectic basis vector of $T_{(u,u_y)}\mathsf{Pot}$, Eq. (14.3) on p. 191 and Theorem 14.3(1) on p. 192
$W^{n,p}(M)$	Sobolev space of weakly n times differentiable functions with p-integrable derivative on a domain M
$w(\lambda)$	$= \lvert\cos(\zeta(\lambda))\rvert + \lvert\sin(\zeta(\lambda))\rvert$, Eq. (5.4) on p. 50
w_k	Symplectic basis vector of $T_{(u,u_y)}\mathsf{Pot}$, Eq. (14.3) on p. 191 and Theorem 14.3(1) on p. 192

Greek Letters

α, α_λ	The $\mathfrak{sl}(2,\mathbb{C})$-valued flat connection form of a solution u or of a potential (u, u_y), p. 25, p. 48
α_0	The $\mathfrak{sl}(2,\mathbb{C})$-valued flat connection form associated to the vacuum, Eq. (4.1) on p. 41
Γ	Period lattice of Σ, Theorem 18.7(1) on p. 280
$\Delta(\lambda)$	Trace of the monodromy $M(\lambda)$, Eq. (3.6) on p. 26.
$\Delta_0(\lambda)$	Trace of the monodromy $M_0(\lambda)$ of the vacuum, Eq. (4.6) on p. 43.
δf	Variation of the function f defined on Pot, p. 189
$(\delta u, \delta u_y)$	A tangent vector in $T_{(u,u_y)}\mathsf{Pot}$, p. 189
$\delta_{k\ell}$	Kronecker delta, i.e. $\delta_{k\ell} = 1$ for $k = \ell$, $\delta_{k\ell} = 0$ for $k \neq \ell$
$\zeta(\lambda)$	$= \frac{1}{4}(\lambda^{1/2} + \lambda^{-1/2})$, Eq. (4.2) on p. 42
$\widetilde{\zeta}(\lambda)$	$= \frac{1}{4}(\lambda^{1/2} - \lambda^{-1/2})$, Eq. (4.3) on p. 42
η_k	Zero of Δ', Lemma 13.3(1) on p. 177
ϑ_k	Factor for the symplectic basis (v_k, w_k) of $T_{(u,u_y)}\mathsf{Pot}$, Eq. (14.4) on p. 191 and Theorem 14.3(1) on p. 192

$\varkappa_{k,\nu}$	Zero of $\Delta^2 - 4$ (branch point or singularity of the spectral curve Σ), Proposition 6.5(1) on p. 75
$\varkappa_{k,*}$	$= \frac{1}{2}(\varkappa_{k,1} + \varkappa_{k,2})$, Eq. (16.3) on p. 223
Λ	Eigenline bundle of the monodromy $M(\lambda)$, p. 29
λ	Spectral parameter, p. 22
λ_k	λ-parameter of a spectral divisor point, Proposition 6.5(2) on p. 75
$\lambda_{k,0}$	λ-parameter of a divisor point of the vacuum, Eq. (4.8) on p. 44
μ	Eigenvalue of the monodromy $M(\lambda)$, Eq. (3.10) on p. 27
μ_k	μ-parameter of a spectral divisor point, Proposition 6.5(2) on p. 75
$\mu_{k,0}$	μ-parameter of a divisor point of the vacuum, Eq. (4.10) on p. 45
Σ	Spectral curve, Eq. (3.4) on p. 26
Σ'	The spectral curve punctured at its singularities, Eq. (17.3) on p. 240
$\widehat{\Sigma}$	Normalization of the spectral curve, p. 28
$\widetilde{\Sigma}$	Surface obtained from Σ by cutting along the cycles A_n, Proposition 19.8 on p. 298
Σ_0	Spectral curve of the vacuum, Eq. (4.7) on p. 43.
σ	Hyperelliptic involution of the spectral curve, Eq. (3.9) on p. 27
τ	Asymptotic parameter for the component functions $b(\lambda)$, $c(\lambda)$ of the monodromy $M(\lambda)$, Theorem 5.4 on p. 51
υ	Asymptotic parameter for the component function $a(\lambda)$, $d(\lambda)$ of the monodromy $M(\lambda)$, Theorem 5.4 on p. 51
$\Phi_{n,\xi,\varrho}$	Holomorphic function on $\mathbb{C}^*$ defined by an infinite product in Proposition 17.2 on p. 243
φ	Abel map of Σ, Theorem 18.7(2) on p. 280
φ_n	n-th Jacobi coordinate for Σ, Theorem 18.7(4) on p. 280
$\widetilde{\varphi}$	Lift of the Abel map of Σ, Proposition 18.5(2) on p. 275
$\widetilde{\varphi}_n$	Lift of the n-th Jacobi coordinate for Σ, Proposition 18.5(1) on p. 275
$\widetilde{\varphi}^{(1)}$	Lift of the Abel map of Σ for the larger tangent space, Proposition 18.10 on p. 284
$\widetilde{\varphi}_n^{(1)}$	Lift of the n-th Jacobi coordinate for Σ for the larger tangent space, Proposition 18.10 on p. 284
Ψ_k	$= \sqrt{(\lambda - \varkappa_{k,1}) \cdot (\lambda - \varkappa_{k,2})}$, Lemma 16.1 on p. 222
Ω	Symplectic form on Pot, Eq. (14.1) on p. 189

$\Omega(\Sigma)$	Space of holomorphic 1-forms on Σ, p. 240
ω_n	Canonical basis of holomorphic 1-forms on Σ resp. Σ', Theorem 18.3(2) on p. 264
$\widetilde{\omega}_n$	Certain holomorphic 1-forms on Σ that are not quite as good as the member ω_n of the canonical basis, Theorem 18.3(1) on p. 264

Other

$*$	Convolution of two sequences, Eq. (7.3) on p. 86

References

[Ah] L. Ahlfors, Normalintegrale auf offenen Riemannschen Flächen. Ann. Acad. Sci. Fennicae. Ser. A. I. Math.-Phys. **1947**(35), 24 (1947)

[Al] A. Alexandrov, Uniqueness theorems for surfaces in the large, V. Am. Math. Soc. Transl. (2) **21**, 412–416 (1962)

[Ba-1] M. Babich, Real finite-gap solutions of equations connected with the sine-Gordon equation. Leningrad Math. J. **2**, 507–521 (1991)

[Ba-2] M. Babich, Smoothness of real finite-gap solutions of equations connected with the sine-Gordon equation. Leningrad Math. J. **3**, 45–52 (1992)

[Bel] E.D. Belokolos, A.I. Bobenko, V.Z. Enol'skii, A.R. Its, V.B. Matveev, *Algebro-Geometric Approach to Nonlinear Integrable Equations* (Springer, Berlin, 1994)

[Ben-S] C. Bennett, R. Sharpley, *Interpolation of Operators* (Academic Press, Boston, 1988)

[Bo-1] A. Bobenko, All constant mean curvature tori in R^3, S^3, H^3 in terms of theta-functions. Math. Ann. **290**, 209–245 (1991)

[Bo-2] A. Bobenko, Constant mean curvature surfaces and integrable equations. Usp. Mat. Nauk **46**(4), 3–42 (1991)

[Bo-3] A. Bobenko, Surfaces in terms of 2 by 2 matrices. Old and new integrable cases, in *Harmonic Maps and Integrable Systems*, vol. 23, ed. by A.P. Fordy, J.C. Wood (Vieweg Aspects of Mathematics, Braunschweig/Wiesbaden, 1994)

[Bon] J.M. Bony, Principe du maximum, inégalite de Harnack et unicité du problème de Cauchy pour les opérateurs elliptiques dégénérés. Ann. Inst. Fourier (Grenoble) **19**, 277–304 (1969)

[Br] S. Brendle, Embedded minimal tori in S^3 and the Lawson conjecture. Acta Math. **211**, 177–190 (2013)

S. Klein, *A Spectral Theory for Simply Periodic Solutions of the Sinh-Gordon Equation*, Lecture Notes in Mathematics 2229,
https://doi.org/10.1007/978-3-030-01276-2

[Bu-F-P-P] F. Burstall, D. Ferus, F. Pedit, U. Pinkall, Harmonic tori in symmetric spaces commuting Hamiltonian systems and loop algebras. Ann. Math. **138**, 173–212 (1993)

[Cal-I-1] A. Calini, T. Ivey, Finite-gap solutions of the vortex filament equation: genus one solutions and symmetric solutions. J. Nonlinear Sci. **15**, 321–361 (2005)

[Cal-I-2] A. Calini, T. Ivey, Finite-gap solutions of the vortex filament equation: isoperiodic deformations. J. Nonlinear Sci. **17**, 527–567 (2007)

[Car-S-1] E. Carberry, M. Schmidt, The closure of spectral data for constant mean curvature tori in $\mathbb{S}^3$. J. Reine Angew. Math. **721**, 149–166 (2016)

[Car-S-2] E. Carberry, M. Schmidt, The prevalence of tori amongst constant mean curvature planes in $\mathbb{R}^3$. J. Geom. Phys. **106**, 352–366 (2016)

[Co] J. Conway, *Functions of One Complex Variable*, 2nd edn. (Springer, New York 1978)

[Da] G. Darboux, Sur le problème de Pfaff. Bull. Sci. Math. **6**, 14–36, 49–68 (1882)

[deJ-P] T. de Jong, G. Pfister, *Local Analytic Geometry* (Vieweg, Braunschweig 2000)

[Del] C. Delaunay, Sur la surface de révolution dont la courbure moyenne est constante. J. Math. Pures Appl. **6**, 309–320 (1841)

[Do-P-W] J. Dorfmeister, F. Pedit, H. Wu, Weierstrass type representation of harmonic maps into symmetric spaces. Commun. Anal. Geom. **6**, 633–668 (1998)

[Du-K-N] B.A. Dubrovin, I.M. Krichever, S.P. Novikov, *Integrable Systems. I, Dynamical Systems IV*, ed. by V.I. Arnold, S.P. Novikov. Encyclopaedia of Mathematical Sciences, vol. 4, 2nd edn. (Springer, Berlin, 2001), pp. 177–332

[Fa-K] H. Farkas, I. Kra, *Riemann Surfaces*, 2nd edn. (Springer, New York, 1992)

[Fe-K-T] J. Feldman, H. Knörrer, E. Trubowitz, *Riemann Surfaces of Infinite Genus* (American Mathematical Society, Providence, 2003)

[For] O. Forster, *Lectures on Riemann Surfaces* (Springer, New York, 1981)

[Fox-W] D. Fox, J. Wang, Conservation laws for surfaces of constant mean curvature in 3-dimensional space forms, arXiv:1309.6606 (2013)

[Gra] L. Grafakos, *Classical Fourier Analysis*, 2nd edn, (Springer, New York, 2008)

[Gri] P.G. Grinevich, Approximation theorem for the self-focusing nonlinear Schrödinger equation and for the periodic curves in $\mathbb{R}^3$. Physica D **152/153**, 20–27 (2001)

[Gri-S-1] P. Grinevich, M. Schmidt, Period-preserving nonisospectral flows and the moduli spaces of periodic solutions of soliton equations. Physica D **87**, 73–98 (1995)

[Gri-S-2] P. Grinevich, M. Schmidt, Closed curves in $\mathbb{R}^3$: a characterization in terms of curvature and torsion, the Hasimoto map and periodic solutions of the filament equation. SFB 288 preprint no. 254, arXiv:dg-ga/9703020 (1997)

[Har-1] R. Hartshorne, *Algebraic Geometry* (Springer, New York, 1977)

[Har-2] R. Hartshorne, Generalized divisors on Gorenstein curves and a theorem of Noether. J. Math. Kyoto Univ. **26**, 375–386 (1986)

[Has] R. Hasimoto, A soliton on a vortex filament. J. Fluid Mech. **51**, 477–485 (1972)

[Hau-K-S-1] L. Hauswirth, M. Kilian, M.U. Schmidt, Finite type minimal annuli in $\mathbb{S}^2 \times \mathbb{R}$. Ill. J. Math. **57**, 697–741 (2013)

[Hau-K-S-3] L. Hauswirth, M. Kilian, M.U. Schmidt, Mean convex Alexandrov embedded constant mean curvature tori in the 3-sphere. Proc. Lond. Math. Soc. (3) **112**(3), 588–622 (2016)

[Hau-K-S-4] L. Hauswirth, M. Kilian, M.U. Schmidt, On mean-convex Alexandrov embedded surfaces in the 3-sphere. Math. Z. **281**, 483–499 (2015)

[Hele] F. Helein, *Constant Mean Curvature Surfaces, Harmonic Maps and Integrable Systems*. Lectures in Mathematics (Birkhäuser, Basel, 2001)

[Hell-1] S. Heller, Lawson's genus two surface and meromorphic connections. Math. Z. **274**, 745–760 (2013)

[Hell-2] S. Heller, A spectral curve approach to Lawson symmetric CMC surfaces of genus 2. Math. Ann. **360**, 607–652 (2014)

[Hell-H-S] L. Heller, S. Heller, N. Schmitt, The spectral curve theory for (k, l)-symmetric CMC surfaces. J. Geom. Phys. **98**, 201–213 (2015)

[Hell-S] S. Heller, N. Schmitt, Deformations of symmetric CMC surfaces in the 3-sphere, Exp. Math. **24**, 65–75 (2015)

[Hi] N.J. Hitchin, Harmonic maps from a 2-torus to the 3-sphere. J. Differ. Geom. **31**, 627–710 (1990)

[Hi-S-W] N.J. Hitchin, G.B. Segal, R.S. Ward, *Integrable Systems: Twistors, Loop Groups, and Riemann Surfaces*. Oxford Graduate Texts in Mathematics, vol. 4 (Oxford University Press, Oxford, 1999)

[Ho] H. Hopf, *Differential Geometry in the Large*. Lecture Notes in Mathematics, vol. 1000 (Springer, Berlin, 1983)

[I-M-1] A.R. Its, V.B. Matveev, Hill's operator with finitely many gaps. Funct. Anal. Appl. **9**, 65–66 (1975)

[I-M-2] A.R. Its, V.B. Matveev, A class of solutions of the Korteweg-de Vries equation (in Russian). Probl. Mat. Fiz. **8**, 70–92 (1976)

[Kapo-1] N. Kapouleas, Constant mean curvature surfaces in Euclidean three-space. Bull. Am. Math. Soc. (N.S.) **17**, 318–320 (1987)

[Kapo-2] N. Kapouleas, *Complete constant mean curvature surfaces in Euclidean three-space*. Ann. Math. (2) **131**, 239–330 (1990)

[Kapo-3] N. Kapouleas, Compact constant mean curvature surfaces in Euclidean three-space. J. Differ. Geom. **33**, 683–715 (1991)

[Kapp] T. Kappeler, Fibration of the phase space for the Korteweg-de Vries equation. Ann. Inst. Fourier (Grenoble) **41**, 539–575 (1991)

[Kapp-P] T. Kappeler, J. Pöschel, *KdV & KAM*, Ergebnisse der Mathematik und ihrer Grenzgebiete. 3. Folge. A Series of Modern Surveys in Mathematics, vol. 45 (Springer, Berlin, 2003)

[Kapp-T] T. Kappeler, P. Topalov, Global wellposedness of KdV in $H^{-1}(\mathbb{T}, \mathbb{R})$. Duke Math. J. **135**(2), 327–360 (2006)

[Kar-Ko-1] P. Kargaev, E. Korotyaev, The inverse problem for the Hill operator, a direct approach. Invent. Math. **129**(3), 567–593 (1997)

[Kar-Ko-2] P. Kargaev, E. Korotyaev, Erratum: "The inverse problem for the Hill operator, a direct approach". Invent. Math. **138**(1), 227 (1999)

[Ki-S-S] M. Kilian, M. Schmidt, N. Schmitt, Flows of constant mean curvature tori in the 3-sphere: the equivariant case. J. Reine Angew. Math. **707**, 45–86 (2015)

[Kl-K] S. Klein, M. Kilian, On closed finite gap curves in spaceforms I, submitted for publication, arXiv:1801.07032 (2018)

[Kl-L-S-S] S. Klein, E. Lübcke, M. Schmidt, T. Simon, Singular curves and Baker-Akhiezer functions, submitted for publication, arXiv:1609.07011 (2016)

[Kn-1] M. Knopf, Periodic solutions of the sinh-Gordon equation and integrable systems, Dissertation, Mannheim, arXiv:1606.01590 (2013)

[Kn-2] M. Knopf, Darboux coordinates for periodic solutions of the sinh-Gordon equation. J. Geom. Phys. **110**, 60–68 (2016)

[Kn-3] M. Knopf, Periodic solutions of the sinh-Gordon equation and integrable systems. Differ. Geom. Appl. B **54**, 402–436 (2017)

[Ko-1] E. Korotyaev, Estimates of periodic potentials in terms of gap lengths. Commun. Math. Phys. **197**(3), 521–526 (1998)

[Ko-2] E. Korotyaev, Inverse problem and the trace formula for the Hill operator, II. Math. Z. **231**(2), 345–368 (1999)

[Kr] I.M. Krichever, Methods of algebraic geometry in the theory of non-linear equations. Russ. Math. Surv. **32**(6), 185–213 (1977)

[Law-1] H.B. Lawson, *Compact Minimal Surfaces in S^3, Global Analysis (Proceeding of Symposia in Pure Mathematics)*, Berkeley, vol. XV, pp. 275–282 (1968)

[Law-2] H.B. Lawson, The unknottedness of minimal embeddings. Invent. Math. **11**, 183–187 (1970)

[Law-3] H.B. Lawson, Complete minimal surfaces in S^3. Ann. Math. **92**, 335–374 (1970)

[Lax] P. Lax, Periodic solutions of the KdV equation. Commun. Pure Appl. Math. **28**, 141–188 (1975)

[Ma-1] V.A. Marchenko, The periodic Korteweg-de Vries problem. Mat. Sb. (N.S.) **95**(137), 331–356 (1974)

[Ma-2] V.A. Marchenko, *Sturm-Liouville Operators and Applications*. Operator Theory: Advances and Applications, vol. 22 (Birkhäuser, Basel, 1986)

[Ma-O-1] V.A. Marchenko, I.V. Ostrovskii, A characterisation of the spectrum of Hill's operator. Math. USSR Sbornik **26**, 493–554 (1975)

[Ma-O-2] V.A. Marchenko, I.V. Ostrovskii, Approximation of periodic potentials by finite zone potentials (Russian). Vestnik Khar'kov. Gos. Univ. **205**, 4–40 (1980)

[Ma-O-3] V.A. Marchenko, I.V. Ostrovskii, Approximation of periodic by finite-zone potentials. Sel. Math. Sov. **6**, 101–136 (1987)

[Ma-O-4] V.A. Marchenko, I.V. Ostrovskii, Corrections to the article: "Approximation of periodic by finite-zone potentials". Sel. Math. Sov. **7**, 99–100 (1988)

[McI] I. McIntosh, Harmonic tori and their spectral data, in *Surveys on Geometry and Integrable Systems*. Advanced Studies in Pure Mathematics, vol. 51 (Mathematical Society of Japan, Tokyo, 2008), pp. 285–314

[McK] H. McKean, The sine-Gordon and sinh-Gordon equations on the circle. Commun. Pure Appl. Math. **34**, 197–257 (1981)

[McK-M] H. McKean, P. van Moerbeke, The spectrum of Hill's equation. Invent. Math. **30**, 217–274 (1975)

[McK-T] H. McKean, E. Trubowitz, Hill's operator and hyperelliptic function theory in the presence of infinitely many branch points. Commun. Pure Appl. Math. **29**, 143–226 (1976)

[Mü-S-S] W. Müller, M. Schmidt, R. Schrader, Hyperelliptic Riemann surfaces of infinite genus and solutions of the KdV equation. Duke Math. J. **91**, 315–352 (1998)

[Ne] R. Nevanlinna, Quadratisch integrierbare Differentiale auf einer Riemannschen Mannigfaltigkeit. Ann. Acad. Sci. Fennicae. Ser. A. I. Math.-Phys. **1941**(1), 1–34 (1941)

[No] S. Novikov, The periodic problem for the Korteweg-de Vries equation. Funct. Anal. Appl. **8**, 236–246 (1974)

[No-M-P-Z] S. Novikov, S.V. Manakov, L.P. Pitaevskii, V.E. Zakharov, *Theory of Solitons: The Inverse Scattering Method* (Consultants Bureau, New York, 1984)

[Pi-S] U. Pinkall, I. Sterling, On the classification of constant mean curvature tori. Ann. Math. **130**, 407–451 (1989)

[Pö-T] J. Pöschel, E. Trubowitz, *Inverse Spectral Theory* (Academic Press, London, 1987)

[Poh] K. Pohlmeyer, Integrable Hamiltonian systems and interactions through quadratic constraints. Commun. Math. Phys. **46**, 207–221 (1976)

[Ro] M. Rosenlicht, Generalized Jacobian varieties. Ann. Math. **59**, 505–530 (1954)

[Ru-V] A. Ruh, J. Vilms, The tension field of the Gauss map. Trans. Am. Math. Soc. **149**, 569–573 (1970)

[Sch] M. Schmidt, *Integrable Systems and Riemann Surfaces of Infinite Genus* (American Mathematical Society, Providence, 1996)

[Se] J.-P. Serre, *Algebraic Groups and Class Fields* (Springer, New York, 1988)

[St] K. Strebel, *Quadratic Differentials* (Springer, Berlin, 1984)

[T] V. Tkachenko, Spectra of non-selfadjoint Hill's operators and a class of Riemann surfaces. Ann. Math. (2) **143**, 181–231 (1996)

[U] K. Uhlenbeck, Harmonic maps into Lie groups: classical solutions of the chiral model. J. Differ. Geom. **30**, 1–50 (1989)

[W] H. Wente, Counterexample to a conjecture of H. Hopf. Pac. J. Math. **121**, 193–243 (1986)

LECTURE NOTES IN MATHEMATICS Springer

Editorial Policy

1. Lecture Notes aim to report new developments in all areas of mathematics and their applications – quickly, informally and at a high level. Mathematical texts analysing new developments in modelling and numerical simulation are welcome.

   Manuscripts should be reasonably self-contained and rounded off. Thus they may, and often will, present not only results of the author but also related work by other people. They may be based on specialised lecture courses. Furthermore, the manuscripts should provide sufficient motivation, examples and applications. This clearly distinguishes Lecture Notes from journal articles or technical reports which normally are very concise. Articles intended for a journal but too long to be accepted by most journals, usually do not have this "lecture notes" character. For similar reasons it is unusual for doctoral theses to be accepted for the Lecture Notes series, though habilitation theses may be appropriate.

2. Besides monographs, multi-author manuscripts resulting from SUMMER SCHOOLS or similar INTENSIVE COURSES are welcome, provided their objective was held to present an active mathematical topic to an audience at the beginning or intermediate graduate level (a list of participants should be provided).

   The resulting manuscript should not be just a collection of course notes, but should require advance planning and coordination among the main lecturers. The subject matter should dictate the structure of the book. This structure should be motivated and explained in a scientific introduction, and the notation, references, index and formulation of results should be, if possible, unified by the editors. Each contribution should have an abstract and an introduction referring to the other contributions. In other words, more preparatory work must go into a multi-authored volume than simply assembling a disparate collection of papers, communicated at the event.

3. Manuscripts should be submitted either online at www.editorialmanager.com/lnm to Springer's mathematics editorial in Heidelberg, or electronically to one of the series editors. Authors should be aware that incomplete or insufficiently close-to-final manuscripts almost always result in longer refereeing times and nevertheless unclear referees' recommendations, making further refereeing of a final draft necessary. The strict minimum amount of material that will be considered should include a detailed outline describing the planned contents of each chapter, a bibliography and several sample chapters. Parallel submission of a manuscript to another publisher while under consideration for LNM is not acceptable and can lead to rejection.

4. In general, **monographs** will be sent out to at least 2 external referees for evaluation.

   A final decision to publish can be made only on the basis of the complete manuscript, however a refereeing process leading to a preliminary decision can be based on a pre-final or incomplete manuscript.

   Volume Editors of **multi-author works** are expected to arrange for the refereeing, to the usual scientific standards, of the individual contributions. If the resulting reports can be forwarded to the LNM Editorial Board, this is very helpful. If no reports are forwarded or if other questions remain unclear in respect of homogeneity etc, the series editors may wish to consult external referees for an overall evaluation of the volume.

5. Manuscripts should in general be submitted in English. Final manuscripts should contain at least 100 pages of mathematical text and should always include

   - a table of contents;
   - an informative introduction, with adequate motivation and perhaps some historical remarks: it should be accessible to a reader not intimately familiar with the topic treated;
   - a subject index: as a rule this is genuinely helpful for the reader.
   - For evaluation purposes, manuscripts should be submitted as pdf files.

6. Careful preparation of the manuscripts will help keep production time short besides ensuring satisfactory appearance of the finished book in print and online. After acceptance of the manuscript authors will be asked to prepare the final LaTeX source files (see LaTeX templates online: https://www.springer.com/gb/authors-editors/book-authors-editors/manuscriptpreparation/5636) plus the corresponding pdf- or zipped ps-file. The LaTeX source files are essential for producing the full-text online version of the book, see http://link.springer.com/bookseries/304 for the existing online volumes of LNM). The technical production of a Lecture Notes volume takes approximately 12 weeks. Additional instructions, if necessary, are available on request from lnm@springer.com.

7. Authors receive a total of 30 free copies of their volume and free access to their book on SpringerLink, but no royalties. They are entitled to a discount of 33.3 % on the price of Springer books purchased for their personal use, if ordering directly from Springer.

8. Commitment to publish is made by a *Publishing Agreement*; contributing authors of multiauthor books are requested to sign a *Consent to Publish form*. Springer-Verlag registers the copyright for each volume. Authors are free to reuse material contained in their LNM volumes in later publications: a brief written (or e-mail) request for formal permission is sufficient.

Addresses:
Professor Jean-Michel Morel, CMLA, École Normale Supérieure de Cachan, France
E-mail: moreljeanmichel@gmail.com

Professor Bernard Teissier, Equipe Géométrie et Dynamique,
Institut de Mathématiques de Jussieu – Paris Rive Gauche, Paris, France
E-mail: bernard.teissier@imj-prg.fr

Springer: Ute McCrory, Mathematics, Heidelberg, Germany,
E-mail: lnm@springer.com

Printed in the United States
By Bookmasters